Springer Complexity

Springer Complexity is an interdisciplinary program publishing the best research and academic-level teaching on both fundamental and applied aspects of complex systems—cutting across all traditional disciplines of the natural and life sciences, engineering, economics, medicine, neuroscience, social and computer science.

Complex Systems are systems that comprise many interacting parts with the ability to generate a new quality of macroscopic collective behavior the manifestations of which are the spontaneous formation of distinctive temporal, spatial or functional structures. Models of such systems can be successfully mapped onto quite diverse "real-life" situations like the climate, the coherent emission of light from lasers, chemical reaction–diffusion systems, biological cellular networks, the dynamics of stock markets and of the internet, earthquake statistics and prediction, freeway traffic, the human brain, or the formation of opinions in social systems, to name just some of the popular applications.

Although their scope and methodologies overlap somewhat, one can distinguish the following main concepts and tools: self-organization, nonlinear dynamics, synergetics, turbulence, dynamical systems, catastrophes, instabilities, stochastic processes, chaos, graphs and networks, cellular automata, adaptive systems, genetic algorithms and computational intelligence.

The three major book publication platforms of the Springer Complexity program are the monograph series "Understanding Complex Systems" focusing on the various applications of complexity, the "Springer Series in Synergetics", which is devoted to the quantitative theoretical and methodological foundations, and the "SpringerBriefs in Complexity" which are concise and topical working reports, case-studies, surveys, essays and lecture notes of relevance to the field. In addition to the books in these two core series, the program also incorporates individual titles ranging from textbooks to major reference works.

Editorial and Programme Advisory Board

Henry Abarbanel, Institute for Nonlinear Science, University of California, San Diego, USA

Dan Braha, New England Complex Systems Institute and University of Massachusetts, Dartmouth, USA

Péter Érdi, Center for Complex Systems Studies, Kalamazoo College, USA and Hungarian Academy of Sciences, Budapest, Hungary

Karl Friston, Institute of Cognitive Neuroscience, University College London, London, UK

Hermann Haken, Center of Synergetics, University of Stuttgart, Stuttgart, Germany

Viktor Jirsa, Centre National de la Recherche Scientifique (CNRS), Université de la Méditerranée, Marseille, France

Janusz Kacprzyk, System Research, Polish Academy of Sciences, Warsaw, Poland

Kunihiko Kaneko, Research Center for Complex Systems Biology, The University of Tokyo, Tokyo, Japan

Scott Kelso, Center for Complex Systems and Brain Sciences, Florida Atlantic University, Boca Raton, USA

Markus Kirkilionis, Mathematics Institute and Centre for Complex Systems, University of Warwick, Coventry, UK

Jürgen Kurths, Nonlinear Dynamics Group, University of Potsdam, Potsdam, Germany

Andrzej Nowak, Department of Psychology, Warsaw University, Poland

Linda Reichl, Center for Complex Quantum Systems, University of Texas, Austin, USA

Peter Schuster, Theoretical Chemistry and Structural Biology, University of Vienna, Vienna, Austria

Frank Schweitzer, System Design, ETH Zurich, Zürich, Switzerland

Didier Sornette, Entrepreneurial Risk, ETH Zurich, Zürich, Switzerland

Stefan Thurner, Section for Science of Complex Systems, Medical University of Vienna, Vienna, Austria

Understanding Complex Systems

Founding Editor: Scott Kelso

Future scientific and technological developments in many fields will necessarily depend upon coming to grips with complex systems. Such systems are complex in both their composition—typically many different kinds of components interacting simultaneously and nonlinearly with each other and their environments on multiple levels—and in the rich diversity of behavior of which they are capable.

The Springer Series in Understanding Complex Systems series (UCS) promotes new strategies and paradigms for understanding and realizing applications of complex systems research in a wide variety of fields and endeavors. UCS is explicitly transdisciplinary. It has three main goals: First, to elaborate the concepts, methods and tools of complex systems at all levels of description and in all scientific fields, especially newly emerging areas within the life, social, behavioral, economic, neuro- and cognitive sciences (and derivatives thereof); second, to encourage novel applications of these ideas in various fields of engineering and computation such as robotics, nano-technology and informatics; third, to provide a single forum within which commonalities and differences in the workings of complex systems may be discerned, hence leading to deeper insight and understanding.

UCS will publish monographs, lecture notes and selected edited contributions aimed at communicating new findings to a large multidisciplinary audience.

For further volumes:
http://www.springer.com/series/5394

Roderick C. Dewar · Charles H. Lineweaver
Robert K. Niven · Klaus Regenauer-Lieb
Editors

Beyond the Second Law

Entropy Production and Non-equilibrium Systems

Springer

Editors
Roderick C. Dewar
Research School of Biology
Australian National University
Canberra
Australia

Robert K. Niven
School of Engineering and Information Technology
University of New South Wales at ADFA
Canberra
Australia

Charles H. Lineweaver
Research School of Astronomy and Astrophysics and the Research School of Earth Sciences
Australian National University
Canberra
Australia

Klaus Regenauer-Lieb
CSIRO Earth Science and Resource Engineering, School of Earth and Environment
University of Western Australia
Crawley
Australia

ISSN 1860-0832 ISSN 1860-0840 (electronic)
ISBN 978-3-642-40153-4 ISBN 978-3-642-40154-1 (eBook)
DOI 10.1007/978-3-642-40154-1
Springer Heidelberg New York Dordrecht London

Library of Congress Control Number: 2013951124

Printed on acid-free paper

Springer is part of Springer Science+Business Media (www.springer.com)

Preface

The idea for this book grew out of a series of workshops on the Maximum Entropy Production (MaxEP) principle, held annually from 2003 to 2011.[1] These workshops brought together scientists and students interested in the theory and application of MaxEP to non-equilibrium systems across a wide range of disciplines in physics, chemistry and biology.

A first 'state of the art' account of MaxEP research up to 2004 was presented in a previous book[2] within the same Springer series *Understanding Complex Systems*. The present volume provides a timely update on the significant progress, both theoretical and applied, that has been made in this exciting field over the last 9 years. More than that, however, we saw the opportunity to broaden the horizons of MaxEP research—to make connections between MaxEP and other areas of non-equilibrium science, such as the Fluctuation Theorem and the Maximum Entropy (MaxEnt) principle. These areas have largely been developed in isolation from each other, and yet the concepts of entropy and entropy production play a central role in all of them.

History in general, and the history of science in particular, has demonstrated that there is much to be gained when folks with different viewpoints get together and talk to each other. Therefore, a particular aim of the MaxEP 2011 workshop—held at the Australian National University, Canberra and co-organised by the four Editors—was to bring together scientists from traditionally isolated sectors of non-equilibrium science in order to present their work and ideas on entropy and entropy production, with a view to exploring potential connections between them.

This also became the key aim of the present volume. This book contains contributions from participants of MaxEP 2011 as well as others around the globe who are actively engaged in non-equilibrium science—all of them internationally-recognised experts in their respective fields. It is organised into three parts. Part I provides an overview of the landscape of existing non-equilibrium principles beyond the restrictive scope of the second law of thermodynamics; it also offers a

---

[1] National Institute for Agronomy Research, Bordeaux, France (2003–2005), University of Split, Croatia (2006), Max-Planck Institute for Biogeochemistry, Jena (2007–2010), The Australian National University, Canberra, Australia (2011).

[2] Kleidon and Lorenz [1].

tentative road map of potential connections and future research directions within that landscape, based on the material presented in Parts II and III which deal, respectively, with theoretical perspectives on entropy production and applications.

Topics covered include the theoretical basis of MaxEP, non-equilibrium principles associated with Ziegler and Prigogine, the Fluctuation Theorem and related theorems, and MaxEnt, as well as the many applications of these principles to such diverse fields as biogeochemistry, cosmology, crystal growth morphology, Earth system science, evolution of enzyme kinetics, fluid mechanics, land–atmosphere interactions, landscape topography, macroscale technology, planetary climatology, plasma physics and radiative transfer. This volume also brings together a wide variety of analytical and experimental techniques: stability analysis, climate models of varying complexity, fluid mechanics experiments, microbial growth experiments, molecular dynamics and lattice gas simulations, and variational approaches.

Running through it all is the recurring *leitmotiv* of entropy production. We hope that this book will provide readers with an understanding of entropy production as a key unifying concept in non-equilibrium science—one that provides a link between different theoretical approaches as well as between theory and applications.

We thank Thomas Ditzinger at Springer for his kind encouragement and help with this project. We wish to express our warm thanks to the contributing Authors for their hard work and patience in bringing this volume to completion. We are also very grateful to the following reviewers for their advice and expertise: Bjarne Andresen, Debra Bernhardt, Jason Bertram, Thomas Christen, James Dyke, Itai Einav, Chris Essex, Klaus Fraedrich, Kingshuk Ghosh, Guy Houlsby, Ali Karrech, Axel Kleidon, Kevin Knuth, Bernd Noack, Dider Paillard, Salvatore Pascale, Joachim Pelkowski, Carsten Herrmann-Pillath, Angelo Plastino, Ralph Sutherland, Joe Vallino, Ashwin Vaidya, Manolis Veveakis, Xiaolin Wang, Paško Županović and one anonymous reviewer.

Finally, we thank the many participants and support staff of the MaxEP workshops (2003–2011) for their contributions over the years. Their enthusiasm, encouragement and camaraderie have sustained and enriched our own scientific journeys beyond the Second Law.

Canberra and Perth, Australia, October 2013

Roderick C. Dewar
Charles H. Lineweaver
Robert K. Niven
Klaus Regenauer-Lieb

Reference

1. Kleidon, A., Lorenz, R. (eds.) Non-equilibrium Thermodynamics and Entropy Production: Life, Earth and Beyond. Springer, Heidelberg (2005)

Contents

Contributors

Christopher K. Algar Marine Biological Laboratory, Woods Hole, MA, USA

Robert Boschi Meteorologisches Institut, Klima Campus, University of Hamburg, Hamburg, Germany

Rafael L. Bras Georgia Institute of Technology, Atlanta, GA, USA

Sarah J. Brookes Queensland Micro- and Nanotechnology Centre, School of Biomolecular and Physical Sciences, Griffith University, Brisbane, Australia

Milan Brumen Natural Sciences and Mathematics, Medicine, and Health Sciences, University of Maribor, Maribor, Slovenia; Jožef Stefan Institute, Ljubljana, Slovenia

Thomas Christen ABB Corporate Research, Dättwil, Switzerland

Hui Tong Chua School of Mechanical and Chemical Engineering, The University of Western Australia, Crawley, WA, Australia

Roderick C. Dewar Research School of Biology, The Australian National University, Canberra, Australia

Andrej Dobovišek Natural Sciences and Mathematics, Medicine, and Health Sciences, University of Maribor, Maribor, Slovenia

James G. Dyke School of Electronics and Computer Science, University of Southampton, Southampton, UK

Uwe Ehret Institute of Water Resources and River Basin Management Karlsruhe Institute of Technology KIT, Karlsruhe, Germany

Denis J. Evans Research School of Chemistry, Australian National University, Canberra, Australia

Yosuke Fukumura Graduate School of Integrated Arts and Sciences, Hiroshima University, Higashi-Hiroshima, Japan

Oliver Gaede Earth, Environmental and Biological Sciences School, Queensland University of Technology, Brisbane, QLD, Australia

Nuria Fernández González Marine Biological Laboratory, Woods Hole, MA, USA

Peter K. Haff Nicholas School of the Environment, Duke University, Durham, NC, USA

Corentin Herbert National Center for Atmospheric Research, Boulder, CO, USA

Bruce Hobbs School of Earth and Environment, The University of Western Australia, Crawley, WA, Australia

Guy T. Houlsby Department of Engineering Science, University of Oxford, Oxford, UK

Julie A. Huber Marine Biological Laboratory, Woods Hole, MA, USA

Davor Juretić Department of Physics, Faculty of Science, University of Split, Split, Croatia

Ali Karrech School of Earth and Environment, The University of Western Australia, Crawley, Australia

Frank Kassubek ABB Corporate Research, Dättwil, Switzerland

Yohei Kawazura Graduate School of Frontier Sciences, The University of Tokyo, Kashiwa, Chiba, Japan

Axel Kleidon Max-Planck-Institute for Biogeochemistry, Jena, Germany

Daniel Lester CSIRO Mathematics, Informatics and Statistics, Applied Fluid Chaos Group, Highett, VIC, Australia

Charles H. Lineweaver Planetary Science Institute, Research School of Astronomy and Astrophysics and the Research School of Earth Sciences, Australian National University, Canberra, ACT, Australia

Jie Liu School of Earth and Environment, The University of Western Australia, Crawley, WA, Australia

Valerio Lucarini Meteorologisches Institut, Klima Campus, University of Hamburg, Hamburg, Germany; Department of Mathematics and Statistics, University of Reading, Reading, UK

Amos Maritan Department of Physics G. Galilei, University of Padua, Padua, Italy

Leonid M. Martyushev Ural Federal University, Ekaterinburg, Russia; Institute of Industrial Ecology, Russian Academy of Sciences, Ekaterinburg, Russia

Guy Metcalfe CSIRO Mathematics, Informatics and Statistics, Applied Fluid Chaos Group, Highett, VIC, Australia

Veronica Nieves Jet Propulsion Laboratory, California Institute of Technology, Pasadena, CA, USA

Robert K. Niven School of Engineering and Information Technology, The University of New South Wales at ADFA, Canberra, ACT, Australia

Bernd R. Noack Institut PPRIME, CNRS, Université de Poitiers, ENSMA, CEAT, POITIERS Cedex, France

Kevin Oliver National Oceanography Centre, Southampton University of Southampton, Southampton, UK

Alison Ord School of Earth and Environment, The University of Western Australia, Crawley, WA, Australia

Hisashi Ozawa Graduate School of Integrated Arts and Sciences, Hiroshima University, Higashi-Hiroshima, Japan

Didier Paillard Laboratoire des Sciences du Climat et de l'Environnement, IPSL, CEA-CNRS-UVSQ, UMR 8212, Gif-sur-Yvette, France

Salvatore Pascale Meteorologisches Institut, Klima Campus, University of Hamburg, Hamburg, Germany

Thomas Poulet CSIRO Earth Science and Resource Engineering, Bentley, WA, Australia

Klaus Regenauer-Lieb School of Earth and Environment, The University of Western Sydney and CSIRO Earth Science and Resource Engineering, Crawley, Australia

James C. Reid Australian Institute of Bioengineering and Nanotechnology and School of Chemistry and Molecular Biosciences, The University of Queensland, Brisbane, Australia

Ulrike Scherer Institute of Water Resources and River Basin Management Karlsruhe Institute of Technology KIT, Karlsruhe, Germany

Christoph Schrank Earth, Environmental and Biological Sciences School, Queensland University of Technology, Brisbane, QLD, Australia

Debra J. Searles Australian Institute of Bioengineering and Nanotechnology and School of Chemistry and Molecular Biosciences, The University of Queensland, Brisbane, Australia; Queensland Micro- and Nanotechnology Centre, School of Biomolecular and Physical Sciences, Griffith University, Brisbane, Australia

Vladimir D. Seleznev Ural Federal University, Ekaterinburg, Russia

Shinya Shimokawa National Research Institute for Earth Science and Disaster Prevention, Tsukuba, Japan

Mike G. Trefry School of Earth and Environment, The University of Western Australia, Crawley, WA 6009, Australia; CSIRO Land and Water, Floreat Park, WA, Australia

Ashwin Vaidya Department of Mathematical Sciences, Montclair State University, Montclair, NJ, USA

Joseph J. Vallino Marine Biological Laboratory, Woods Hole, MA, USA

Manolis Veveakis CSIRO Earth Science and Resource Engineering, Bentley, WA, Australia

Jingfeng Wang Georgia Institute of Technology, Atlanta, GA, USA

Iain Weaver School of Electronics and Computer Science, University of Southampton, Southampton, UK

Florian Wellmann CSIRO Earth Science and Resource Engineering, Bentley, WA, Australia

Zensho Yoshida Graduate School of Frontier Sciences, The University of Tokyo, Kashiwa, Chiba, Japan

Erwin Zehe Institute of Water Resources and River Basin Management Karlsruhe Institute of Technology KIT, Karlsruhe, Germany

Paško Županović Department of Physics, Faculty of Science, University of Split, Split, Croatia

Part I
Introduction

Chapter 1
Beyond the Second Law: An Overview

Roderick C. Dewar, Charles H. Lineweaver, Robert K. Niven and Klaus Regenauer-Lieb

Abstract The Second Law of Thermodynamics governs the average direction of all non-equilibrium dissipative processes. However it tells us nothing about their actual rates, or the probability of fluctuations about the average behaviour. The last few decades have seen significant advances, both theoretical and applied, in understanding and predicting the behaviour of non-equilibrium systems beyond what the Second Law tells us. Novel theoretical perspectives include various extremal principles concerning entropy production or dissipation, the Fluctuation Theorem, and the Maximum Entropy formulation of non-equilibrium statistical mechanics. However, these new perspectives have largely been developed and applied independently, in isolation from each other. The key purpose of the present book is to bring together these different approaches and identify potential connections between them: specifically, to explore links between hitherto separate theoretical concepts, with entropy production playing a unifying role; and to close the gap between theory and applications. The aim of this overview chapter is to orient and guide the reader towards this end. We begin with a rapid flight over the fragmented landscape that lies beyond the Second Law. We then highlight the connections that emerge from the recent work presented in this volume. Finally we summarise these connections in a tentative road map that also highlights some directions for future research.

R. C. Dewar (✉)
Research School of Biology, The Australian National University, Canberra, ACT 0200, Australia
e-mail: roderick.dewar@anu.edu.au

C. H. Lineweaver
Research School of Astronomy and Astrophysics and Research School of Earth Sciences, The Australian National University, Canberra, ACT 0200, Australia

R. K. Niven
School of Engineering and Information Technology, The University of New South Wales at ADFA, Canberra, ACT 2600, Australia

K. Regenauer-Lieb
School of Earth and Environment, The University of Western Sydney and CSIRO Earth Science and Resource Engineering, Crawley, WA 6009, Australia

R. C. Dewar et al. (eds.), *Beyond the Second Law*,
Understanding Complex Systems, DOI: 10.1007/978-3-642-40154-1_1,

1.1 The Challenge: Understanding and Predicting Non-equilibrium Behaviour

Non-equilibrium,[1] dissipative systems abound in nature. Examples span the biological and physical worlds, and cover a vast range of scales: from biomolecular motors, living cells and organisms to ecosystems and the biosphere; from turbulent fluids and plasmas to hurricanes and planetary climates; from growing crystals and avalanches to earthquakes; from cooling coffee cups to economies and societies; from stars and supernovae to clusters of galaxies and beyond.

A characteristic feature of all open, non-equilibrium systems is that they import energy and matter from their surroundings in one form and re-export it in a more degraded (higher entropy) form. A sheared viscous fluid driven out of thermodynamic equilibrium by the external input of kinetic energy eventually dissipates and expels that energy to its environment as heat; the Earth absorbs short-wave radiation at solar temperatures and re-emits it to space as long-wave radiation at terrestrial temperatures; living organisms use the chemical free energy ultimately derived from photons to grow and survive, eventually dissipating it to their environment as heat and carbon dioxide.

In association with these exchanges of energy and matter, spatial gradients in temperature and chemical concentration are set up and maintained, both internally and between the system and its environment. The patterns of flows and their associated gradients self-organize into intricate dynamical structures that continually transport and transform energy and mass into higher entropy forms: thus emerge plant vascular systems, food webs, river networks, and turbulent eddies such as Jupiter's Red Spot and the convective cells on the Sun's surface. Idealised systems in equilibrium with their surroundings exhibit no flows or gradients; they appear static, structureless, lifeless. In stark contrast, non-equilibrium systems, even purely physical ones, appear to be alive in a sense that perhaps even defines life itself, at least thermodynamically [1].

In view of their ubiquity in nature, understanding and predicting the behaviour of non-equilibrium systems lies at the heart of many questions of fundamental and practical importance, from the origin of a low entropy universe and the evolution of life, to the development of nanotechnology and the prediction of climate change. What is life? And what are the general requirements for its emergence on Earth and elsewhere? What determines the rate at which the universe tends towards thermodynamic equilibrium? How is the functioning of nanoscale devices affected by molecular-scale fluctuations in energy and mass flow? How will the large-scale flows of energy and mass that characterise Earth's climate respond to increased atmospheric greenhouse gas concentrations?

Answering such questions has been a long-standing scientific challenge, largely because the scientific principles and tools required to understand and predict

---

[1] *Equilibrium* is used here in the thermodynamic sense, and not in the dynamic sense of stationarity.

non-equilibrium behaviour have been lacking. In many cases we may not know the underlying equations of motion exactly (especially the case in biology); with only the conservation laws (of energy, mass, momentum and/or charge) as guiding principles, there remains a large number of possible behaviours to choose from. Even when the underlying equations of motion are known (more or less) exactly—for example, the Navier–Stokes equation of fluid mechanics[2]—computational limitations may restrict our ability to solve them. One response to this challenge is to exploit the fact that the macroscopic behaviour of complex, non-equilibrium systems represents the emergent outcome of a large number of microscopic degrees of freedom. Some of those underlying degrees of freedom may behave as 'noise' that averages out at macroscopic scales. This offers the possibility of predicting the emergent macroscopic behaviour from the laws of thermodynamics. We might then hope to understand the behaviour of non-equilibrium systems statistically, in terms of the average, collective behaviour of a large number of individual degrees of freedom. Here again, however, traditional thermodynamics gives us little to go on.

The First Law of Thermodynamics only gives us energy conservation, while the Second Law is qualitative—it tells us only the direction in which an isolated non-equilibrium system will evolve on the average: towards the state of equilibrium, in which the system's thermodynamic entropy adopts its largest value subject to any constraints on it. The Second Law thus implies that, on average, the total thermodynamic entropy $S_{tot} = S_{sys} + S_{env}$ of an isolated system consisting of an open non-equilibrium subsystem (sys) plus its environment (env) will not decrease (i.e. $dS_{tot}/dt \geq 0$). In particular, if the open subsystem is in a steady state (i.e. $dS_{sys}/dt = 0$), then on average the entropy of the environment (S_{env}) will not decrease (i.e. $dS_{env}/dt \geq 0$). As noted above, this behaviour is evident in the observed tendency of open systems to re-export energy and matter to their environment in a higher entropy form than that in which they receive it.

Crucially, however, the Second Law is mute on two counts. Firstly, it does not predict the actual value of dS_{tot}/dt (i.e. the average rate at which S_{tot} increases). Secondly, as the mathematical physicist James Clerk Maxwell was one of the first to appreciate [2], the Second Law is statistical in character, rather than being a dynamical law. It is a statement about the average behaviour of isolated systems. However, it does not tell us the probability of statistical fluctuations in energy and mass flow for which, at least momentarily and locally, $dS_{tot}/dt < 0$, as when (for example) a group of gas molecules happens to move collectively from a region of low concentration to a region of higher concentration. And yet knowing these quantities—the average rate of entropy increase, and the probability of entropy-decreasing fluctuations—is central to answering some of the fundamental and practical questions mentioned above. Beyond the Second Law, the behaviour of entropy production becomes a key focus of study.

---

[2] Strictly speaking, the Navier–Stokes equation is only approximate; the (linear) expression for the stress tensor is only valid close to equilibrium.

The aim of this introductory chapter is to orient and guide the reader of this book. We begin with a rapid flight over the landscape of non-equilibrium principles that have been proposed beyond the Second Law (Sect. 1.2). For now, that landscape is still forming; it remains a rather fragmented one and we highlight some of the challenges one encounters in trying to negotiate it (Sect. 1.3). The key challenge is to find connections within this landscape, to construct bridges between previously isolated islands. In Sect. 1.4 we highlight some of the connections suggested to us by the recent work presented in this volume. Summarizing these in Sect. 1.5, we offer a tentative road map of the current landscape, as well as possible directions for future research.

Given the current state of play, we attempt no more than a partial synthesis here—partial in both perspective and scope. Thus Sects. 1.4 and 1.5 present one particular view of this exciting area of science, and where it might go next. It does not represent a consensus view of the contributing chapter authors, as will be clear from the diversity of perspectives this book brings together. And even at that, it does not pretend to paint a complete picture. Nevertheless, we hope this tentative road map will encourage the reader to develop his or her own vision of the landscape beyond the Second Law, and of the most fruitful paths to explore within it.

1.2 Beyond the Second Law: The Search for New Principles

Our main aim here is to give a brief overview of the landscape of non-equilibrium principles that have been proposed beyond the Second Law. Discussion of the key challenges in negotiating this landscape (ambiguities of meaning etc.) is deferred to Sect. 1.3.

1.2.1 Paltridge's MaxEP, the Fluctuation Theorem …

Within the last few decades, significant progress has been made towards developing and applying new principles of thermodynamics for non-equilibrium systems that go beyond the Second Law. With regard to the average value of $\mathrm{d}S_{\mathrm{tot}}/\mathrm{d}t$ and the probability of entropy-decreasing fluctuations, two key concepts have emerged: respectively, the principle of Maximum Entropy Production (MaxEP) and the Fluctuation Theorem (FT).

MaxEP is often stated verbally as a sort of codicil to the Second Law, according to which it is asserted that an open system adopts the stationary state ($\mathrm{d}S_{\mathrm{sys}}/\mathrm{d}t = 0$) in which $\mathrm{d}S_{\mathrm{tot}}/\mathrm{d}t = \mathrm{d}S_{\mathrm{env}}/\mathrm{d}t$ attains its largest value possible within the constraints acting on the system. That is, a stationary open subsystem plus its environment not only tends to equilibrium ($\mathrm{d}S_{\mathrm{tot}}/\mathrm{d}t = \mathrm{d}S_{\mathrm{env}}/\mathrm{d}t \geq 0$) but, it is claimed, does so as fast as possible (maximum $\mathrm{d}S_{\mathrm{env}}/\mathrm{d}t$) subject to any constraints.

In the seminal work of Garth Paltridge [3–5] in the 1970s and 1980s, MaxEP was applied to simple steady-state energy balance models of Earth's climate. Maximizing the entropy production associated with material heat transport in the atmosphere and oceans produced realistic predictions of the stationary latitudinal profiles of surface temperature, cloud fraction and equator-to-pole material heat transport. Somewhat surprisingly, this success was achieved when the maximization was subject to the sole constraint[3] of global energy balance, in the absence of any dynamical information such as planetary rotation rate.

Paltridge's MaxEP principle selects one among several climate states compatible with global energy balance [3–6]. It is the archetype for analogous MaxEP principles constrained only by global mass balance that have been applied with similar success to other non-equilibrium selection problems (e.g. crystal growth morphology, macromolecular evolution, plant growth strategies) [7–13]. For brevity, in the following we will refer to these collectively as 'Paltridge's MaxEP'—i.e. MaxEP principles in which the key constraints are global energy and/or mass balance. Despite these successes, the theoretical basis for Paltridge's MaxEP has remained elusive and this has hampered its acceptance by the wider scientific community.

The Fluctuation Theorem (FT) [14–16] concerns the probabilities of trajectories and their time reverse in microscopic phase space. Roughly speaking, the FT states that the probability of observing an entropy change $-d$ relative to that of an entropy change $+d$ over a given time period is exponentially small in d. Since d is an extensive quantity in both space and time (i.e. the entropy change increases with both the size of the system and the time period), the FT implies that macroscopic decreases in entropy, although possible, are extremely rare. In contrast, we expect to see frequent entropy-decreasing fluctuations in small (e.g. nanoscale) systems observed over short periods. Significantly, the FT also implies the Second Law inequality, i.e. the ensemble average[4] of d is non-negative.

1.2.2 … and other Principles

Prior to Paltridge's MaxEP principle, several earlier non-equilibrium principles had also been proposed, involving entropy production or dissipation in one guise or another (see e.g. the excellent review in [13]). A selection of these are summarised in Table 1.1: they include Onsager's MaxEP principle [17, 18], Prigogine's minimum entropy production (MinEP) theorem [19], Kohler's MaxEP principle in statistical transport theory [20, 21], and Ziegler's MaxEP principle for

[3] However, Paltridge's energy balance model still contained a number of *ad hoc* assumptions and parameterizations (see Herbert and Paillard Chap. 9).

[4] The ensemble average is over the probability distribution of microscopic trajectories in phase space (see Sect. 1.4.1).

Table 1.1 A fragmented landscape. A selection of different dissipation- and entropy-related variational principles $H(y|C)$, defined in terms of the function that is maximised (H), the variables being optimised (y), the constraints (C), and the key prediction. The table entries above and below the dashed line describe principles that are conjectured to apply, respectively, close to and far from equilibrium (linear and non-linear regimes). Σ_i denotes a sum over $i = 1, \ldots n$ (this may be generalised to continuous systems). Abbreviations: *EP* entropy production; *GCM* general circulation model; *KE* kinetic energy; *LBE* linearized Boltzmann equation; *RTE* radiative transport equation; *UBT* upper bound theory of fluid turbulence

Variational principle	Maximised function, H	Variables, y	Constraints, C	Key prediction
Onsager MaxEP [17, 18]	$\Sigma_i J_i X_i - {}^1\!/_2 \Sigma_{ij} R_{ij} J_i J_j$	Fluxes[1] $J_i (i = 1, \ldots n)$	Fixed forces: $X_i = X_i^* \ (i = 1, \ldots n)$	Linear flux-force relations: $J_i = \Sigma_j R_{ij}^{-1} X_j^*$
Prigogine MinEP [19]	$-\Sigma_{ij} R_{ij}^{-1} X_i X_j$	Free forces $X_i (i = k+1, \ldots n)$	Fixed forces: $X_i = X_i^* (i = 1, \ldots k)$; Linear flux-force: $X_i = \Sigma_j R_{ij} J_j$	$J_i = 0 \ (i = k+1, \ldots n)$
Kohler MaxEP (solution to LBE) [20, 21]	EP of molecular collisions $\sigma_{coll}(f)$ (source term in continuity equation for entropy $s = -\int f \ln f d\mathbf{v}$)	One-particle velocity ($\mathbf{v}$) distribution function $f(\mathbf{v})$	Fixed temperature and concentration fields; $\sigma_{coll}(f)$ = entropy export (steady-state condition)	Stationary $f(\mathbf{v})$ near equilibrium
- - - -	- - - -	- - - -	- - - -	- - - -
Radiative MinEP (solution to RTE) (Chap. 12)	$-\sigma_{tot}(I_\nu)$, where $\sigma_{tot}(I_\nu)$ = total (radiation + matter) EP	Radiation intensity, I_ν	Various moment constraints on I_ν	Radiation transport coefficients
Ziegler MaxEP [22]	Dissipation $\sigma(J)$ (assumed to be a known function of the fluxes J_i)	Fluxes[1] $J_i (i = 1, \ldots n)$	Fixed forces: $X_i = X_i^* \ (i = 1, \ldots n)$; $\sigma(J) = \Sigma_i J_i X_i^*$	Non-linear flux-force relations: $X_i^* \propto \partial\sigma(J)/\partial J_i$ (orthogonality)
Paltridge MaxEP [3–13]	Various EP-related functions of the form $\Sigma_i J_i X_i$	J_i and $X_i (i = 1, \ldots n)$	Steady-state energy/mass balance (neither X_i nor J_i need be fixed)	Stationary J_i and X_i (non-linear regime)

(continued)

Table 1.1 (continued)

Variational principle	Maximised function, H	Variables, y	Constraints, C	Key prediction
Malkus UBT [23–28]	Various dissipation-related functions of turbulent flow (including KE dissipation)	Mean velocity field	A restricted set of spatial integrals of the Navier–Stokes equation	Mean stationary velocity field (non-linear regime)
Max KE dissipation [29]	Dissipation of KE in a GCM	GCM parameters	Steady-state GCM dynamics	Dynamical climate features
Boltzmann-Gibbs-Jaynes MaxEnt [30–34]	Relative entropy $-\Sigma_i p_i \ln(p_i/q_i)$	Posterior probability of outcome i, p_i	Available or relevant information: $\Sigma_i p_i f_{ik} = F_k (k = 1, \ldots m)$; Prior probability q_i of outcome i	Most likely sampling distribution p_i

[1] The Onsager and Ziegler MaxEP principles can also be formulated in the space of forces $(y = X)$ subject to fixed fluxes $(J = J^*)$

dissipative materials [22]. Also, starting in the 1950s, several variational principles for fluid turbulence were developed by Malkus and others, based on maximising various dissipation-like functions of the flow [23–28]—an approach known as the Upper Bound Theory (UBT) of fluid turbulence. Table 1.1 also includes three other variational principles: a variant of Kohler's principle applied to radiative transport, in which entropy production is minimized rather than maximized (Christen and Kassubek Chap. 12); a principle of maximum kinetic energy (KE) dissipation, suggested by recent climate simulations using a General Circulation Model (GCM) [29], which is also one of the principles emerging from UBT; and the Boltzmann-Gibbs-Jaynes Maximum Entropy (MaxEnt) algorithm [30–34].

Anticipating the discussion in Sect. 1.3, the landscape presented by these principles is a fragmented one. In order to compare and contrast the elements of this landscape, Table 1.1 describes each principle in terms of the dissipation- or entropy-related function H that is maximized, the variables (y) being optimised, and the constraints (C). Key predictions of each principle are given in the last column.

1.2.2.1 Onsager's MaxEP, Prigogine's MinEP

Onsager's original motivation was to establish a theoretical framework for the development of near-equilibrium thermodynamics [17, 18]. Specifically, Onsager's MaxEP principle may be used to derive the near-equilibrium, linear 'constitutive relations' between generalised thermodynamic fluxes J_i and forces X_i, i.e. $J_i = \Sigma_j L_{ij} X_j$ (generalisations of the laws of Fick and Ohm, for example), where the matrix of coupling coefficients[5] is symmetric (i.e. $L_{ij} = L_{ji}$, also known as reciprocity). Prigogine's principle [19] assumes linear flux-force relations as a starting point, and some of the forces are then relaxed: it describes the behaviour of the entropy production, given by $\Sigma_{ij} L_{ij} X_i X_j$, 'when we let go of some of the leads' [21].

1.2.2.2 Kohler's MaxEP, Radiative MinEP

In a separate context, Kohler [20] established a mathematical variational principle to solve the linearised Boltzmann equation (LBE) describing the statistical transport properties of a rarified gas. Subsequently, Ziman [21] recast the Boltzmann equation in the language of thermodynamic fluxes and forces and showed Kohler's principle to be mathematically equivalent to Onsager's MaxEP principle. This suggested to Ziman that Kohler's principle was not just a convenient mathematical trick but had the following physical interpretation: the entropy production of molecular collisions is maximized subject to fixed thermodynamic forces (e.g. temperature and concentration fields), and to the steady-state condition

[5] Here L_{ij} is the inverse of the matrix R_{ij} in Table 1.1.

that the internal entropy production is balanced by dissipation of heat into the environment.

For the problem of radiative transfer in gases or plasmas, the relevant principle appears to be one of MinEP rather than MaxEP (Christen and Kassubek, Chap. 12 and references therein; see also Niven and Noack, Chap. 7). Moreover, when the radiative transfer equation is considered as a LBE, the linearisation is exact because photons do not interact with each other, so that the solution is valid for radiation that is arbitrarily far from thermal equilibrium.

1.2.2.3 Ziegler's MaxEP

The original motivation behind Ziegler's MaxEP principle [22] was to derive the non-linear constitutive relation between generalised forces X_i and fluxes J_i (e.g. stress–strain relations) in dissipative materials far from equilibrium. What are the fluxes given the forces (and vice versa)? As Table 1.1 indicates, the key prediction of Ziegler's MaxEP (subject to fixed generalised forces $X_i = X_i{}^*$ and the constraint $\sigma(J) = \Sigma_i J_i X_i{}^*$) is a constitutive relation that satisfies an orthogonality condition (OC), according to which the generalised force X^* (considered as a vector with components $X_i{}^*$) lies in the direction normal to the contours of $\sigma(J)$ in flux space.[6] Ziegler originally derived the OC using a geometrical argument, based on the assumption that the vector X can be derived solely from properties of the scalar dissipation function $\sigma(J)$; the existence and nature of the function $\sigma(J)$ were also assumptions (Houlsby, Chap. 4).

Ziegler noted the equivalence of the OC to a variational principle (Ziegler's MaxEP)—i.e. maximizing $\sigma(J)$ with respect to J under the constraints in Table 1.1—as a possibly more general thermodynamic basis for the OC. And yet a fundamental basis for the assumptions underlying either derivation of Ziegler's OC (geometrical or variational) has yet to be established; moreover, a direct experimental test of the OC has yet to be derived [35]. In practice, therefore, the OC has been adopted as a working hypothesis for classifying different theoretical behaviours of dissipative materials.

1.2.2.4 Upper Bound Theory of Fluid Turbulence, Maximum KE Dissipation

The UBT of fluid turbulence was developed by Malkus and others [23–28] to predict the mean turbulent velocity field, by maximizing various dissipation-related functionals of the flow. Initially these took the form of maximum transport principles (maximum heat flow, maximum momentum transport) [23–25, 27].

---

[6] An equivalent orthogonality condition for the direction of the generalised fluxes J in force space can be stated in terms of the contours of $\sigma(X)$.

Later, Kerswell [26] analysed a more general family of functionals related to KE dissipation by the mean and fluctuating components of the flow. Maximum KE dissipation by the mean flow was also proposed more recently by Malkus [28].

Crucially, the maximization was subject to a restricted number of dynamical constraints, obtained as integral properties of the Navier-Stokes equation (e.g. global power balance and horizontal mean momentum balance within a horizontally sheared fluid layer) rather than the full dynamics. The UBT is thus analogous in spirit to Paltridge's MaxEP: i.e. select one of many possible stationary states compatible with a restricted number of constraints representing the relevant physics on macroscopic scales, the rest being treated as 'noise'. However, one key difference is that UBT includes some dynamical information (momentum balance) in addition to global energy balance; another key difference is in the nature of the extremized function (e.g. viscous dissipation of KE [28] rather than thermal dissipation [3–6]).

Intriguingly, simulations using the FAMOUS GCM [29] also showed that key dynamical features of Earth's climate were close to a maximum of KE dissipation (Table 1.1).

1.2.2.5 Boltzmann-Gibbs-Jaynes Maximum Entropy

Finally we have the Boltzmann-Gibbs-Jaynes principle of Maximum Entropy (MaxEnt) [30–34]. This principle stands somewhat apart from the others in Table 1.1, both conceptually and in practice (see Dewar and Maritan, Chap. 3; Niven and Noack, Chap. 7). MaxEnt predicts a probability distribution p_i over microscopic outomes i, from which macroscopic quantities may be predicted as averages over p_i. The maximized function H is the relative entropy (or negative Kullback–Leibler divergence) of p_i and a prior distribution q_i; H reduces to the Shannon entropy when q_i is uniform. The maximization is subject to constraints on certain moments of p_i (representing available or relevant physical information), as well as the specified prior probabilities q_i. MaxEnt has several interpretations (Chaps. 3 and 7, and references therein). One fairly concrete interpretation of the MaxEnt distribution is that it corresponds to the most likely frequency distribution of outcomes that would be observed in a long sequence of independent observations of a system that is subject to the given constraints; MaxEnt also has an information-theoretical interpretation as the 'least-informative' p_i [32, 33].

MaxEnt has a long history, starting with Boltzmann's discovery that MaxEnt expresses the asymptotic behaviour of multinomial probabilities [30], and the early development of equilibrium statistical mechanics by Gibbs [31]. The later reappearance of $-\Sigma_\mathrm{i} p_\mathrm{i} \ln p_\mathrm{i}$ (Shannon entropy) in the development of information theory [36, 37], as a measure of missing information, led Jaynes to see MaxEnt as a general method of statistical inference from incomplete information [32, 33]. In view of its general nature, Jaynes promoted MaxEnt as a theoretical framework for non-equilibrium as well as equilibrium statistical mechanics. When applied to non-equilibrium systems, MaxEnt leads to non-linear flux-force relationships that automatically satisfy Onsager reciprocity and reduce to linear form in the near-equilibrium limit [38–40]. Although MaxEnt provides a foundation for equilibrium and non-equilibrium thermodynamics, its physical interpretation remains a subject of debate (e.g. [41]) that has, like Paltridge's MaxEP, hampered its wider acceptance.

1.3 A Fragmented Landscape

After this rapid tour, the student would be forgiven for being confused by the sheer number and variety of entropy production-related principles, as well as by the diverse ways in which they have been applied to non-equilibrium systems. One sees a fragmented landscape of principles and applications, and faces three key difficulties in negotiating it.

1.3.1 Different Histories

One difficulty is historical: the above theoretical principles (Table 1.1) were developed at different times, more or less independently of one other. Some theoretical links between the earlier variational principles (Onsager, Prigogine, Ziegler) have been identified [13]. For example, Ziegler's MaxEP principle reduces to Onsager's in the near-equilibrium limit, while Prigogine's is a corollary of Onsager's that involves additional constraints. However, the links (if any) between Paltridge's MaxEP, the Fluctuation Theorem, Kohler's MaxEP, radiative MinEP, UBT, maximum KE dissipation and MaxEnt (and between these and Ziegler's principle) have remained obscure.

1.3.2 Different Meanings

A second difficulty, and one that compounds the first, is semantic. The terms *entropy production* or *dissipation* are defined and used by different workers in different ways, creating ample room for confusion. Some approaches take entropy production as a given function of thermodynamic fluxes and forces [3–9, 11, 12, 17–19, 22–28], while others define entropy production from an underlying microscopic picture [10, 14–16, 21, 42, 43]. Moreover it does not help that Onsager called his MaxEP principle 'least dissipation of energy', or that Paltridge originally called his principle 'minimum entropy exchange' before resorting to MaxEP!

Yet further scope for confusion arises in the context of extremal principles. For example, from the name alone one might conclude that Paltridge's MaxEP principle contradicts Prigogine's principle of MinEP, whereas these two extremal principles refer to quite different situations. Paltridge's MaxEP principle (e.g. as applied to climate systems [3–6]) is a selection principle between different far-from-equilibrium stationary states. In contrast, Prigogine's principle describes the non-stationary behaviour of the entropy production of near-equilibrium systems[7] when a subset of the thermodynamic force constraints is relaxed; it says only that the unique stationary state has lower entropy production than any non-stationary state, and does not provide a selection principle in situations where there are multiple stationary states (for a further critique of Prigogine's MinEP, see [44]).

To have any chance of making sense of the landscape, one must look beyond the semantics and identify three key aspects (Table 1.1) of each extremal principle, which we denote by $H(y|C)$: (1) which entropy production or dissipation function (H) is being maximised? (2) with respect to which variable(s) (y)? and (3) subject to which constraint(s) (C)? Unless extremal principles are clearly stated in this way, the potential for confusing apples with pears is essentially infinite.

1.3.3 Lack of Foundations

A third related difficulty in negotiating the current landscape lies in the somewhat *ad hoc* way in which, for example, Paltridge's MaxEP has been applied in practice, with many aspects open to ambiguity. Which entropy production function (H) is to be maximised? In some discussions of the physical interpretation of H, the system entropy $S_{sys} = \int s_{sys} dV$ (V = system volume) is treated as a physical quantity obeying a local continuity equation ($\partial s_{sys}/\partial t = -\nabla \cdot \boldsymbol{j}_s + \sigma$ with entropy flux $\boldsymbol{j}_s$ and local entropy production rate σ). In a stationary state all the entropy production $\int \sigma dV (\geq 0)$ within the system is exported to the environment, and by MaxEP is then meant maximum $\int \sigma dV$. However, missing from this interpretation of MaxEP is a clear definition of σ itself! In the definition of σ for flow systems, Niven and Noack (Chap. 7) also reveal some fundamental problems related to decomposition of the flow into mean and fluctuating parts (the entropy production closure problem).

Thus it was only through a process of trial and error that Paltridge [3] stumbled upon maximisation of the entropy production associated with thermal dissipation by the equator-to-pole heat transport in the oceans and atmosphere,[8] rather than the global entropy production associated with short- and long-wave radiative exchange at the top of the atmosphere. In contrast, as noted above, simulations

[7] The regime of linear force-flux relations.

[8] In Paltridge's zonally-averaged climate model [4], thermal dissipation is given by $\sigma = \Sigma_i J_i X_i$ where J_i = material heat transport between meridional zones i and $i + 1$, and $X_i = 1/T_{i+1} - 1/T_i$ is the corresponding inverse temperature difference.

using the FAMOUS GCM [29] showed that key dynamical climate features were close to a maximum of KE dissipation rather than thermal dissipation.

A similar ambiguity about the choice of dissipation function to maximise afflicts the UBT in fluid turbulence; for example, from among a general family of dissipation functionals Kerswell [26] was unable to identify a universal one whose maximization applies to all flow problems. In all of these cases, the main source of ambiguity in the choice of extremized function is the lack of a rigorous theoretical foundation for Paltridge's MaxEP and the UBT. The theoretical basis of Kohler's MaxEP is on firmer ground, as it may be proved as the mathematical solution to the linearized Boltzmann equation [13, 20, 21]. Overall, however, what is lacking is a selection principle for selection principles!

The debate on the theoretical basis of MaxEP extends to whether MaxEP is even a physical principle at all [e.g. 13, 41]. Another source of ambiguity lies in the choice of appropriate constraints, a limitation that applies also to MaxEnt. Moreover, by its very nature as a generic inference algorithm, applications of MaxEnt also require a choice to be made for the set of outcomes i whose probabilities p_i are to be predicted (Table 1.1); i may represent microscopic paths in phase space, macroscopic fluxes defined within the system and/or on its boundary, or indeed any quantity of which we have incomplete information (see Chaps. 3 and 7, and references therein).

Finally, so far we have been referring to near-equilibrium and far-from-equilibrium principles (Table 1.1) when we have not defined what we mean by distance from equilibrium. If by near-equilibrium systems we mean systems obeying linear constitutive relations (Table 1.1 and footnote 7), then the definition is circular. Of course, entropy production itself is a measure of distance from equilibrium (since it vanishes in equilibrium) but, as we have seen, there are many definitions of entropy production!

1.4 Making Connections

Here we try to pull together some of the common threads running through this book, with entropy production playing a key unifying role. An attempt will be made to synthesise these connections in Sect. 1.5. In anticipation, the reader is referred to the tentative road map shown in Fig. 1.1.

1.4.1 The Fluctuation Theorem, MaxEnt, and a Generic MaxEP Principle

We have seen that the plethora of non-equilibrium extremal principles in Table 1.1 involve various definitions of entropy production and dissipation. Is there a

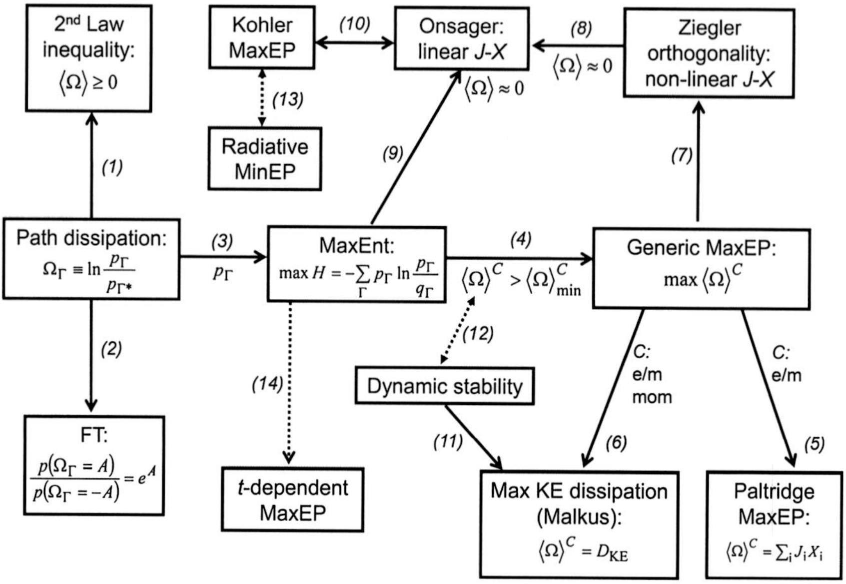

Fig. 1.1 The Second Law and beyond: a tentative road map of theoretical links (*solid arrows*, roads *1–11*) and future research directions (*dashed arrows*, roads *12–14*). See text for notation

fundamental entropy production- or dissipation-like quantity on which we can build a more coherent picture? Here we suggest there is, defined at the level of microscopic trajectories in phase space, a definition that links the Fluctuation Theorem, MaxEnt and MaxEP.

Specifically, let p_Γ denote the probability that the system follows microscopic trajectory (or path) Γ in phase space,[9] and let Γ^* denote the time-reverse of Γ. In equilibrium we expect that $p_\Gamma = p_{\Gamma^*}$ (no net fluxes on average); however, when a system is driven out of equilibrium by an external force (e.g. non-uniform heating on the boundary), the odds are changed in favour of a particular average flow direction, and then we expect that $p_\Gamma \neq p_{\Gamma^*}$. The trajectory-dependent quantity Ω_Γ (called the *dissipation function* in Reid et al., Chap. 2) defined by

$$\Omega_\Gamma = \ln \frac{p_\Gamma}{p_{\Gamma^*}} \tag{1.1}$$

emerges as a central concept in both the Fluctuation Theorem (Reid et al., Chap. 2) and a proposed generic MaxEP principle derived from MaxEnt (Dewar and

[9] The notation is simplified here for clarity. In terms of the notation of Chap. 2, for example, $p_\Gamma = p(\mathbf{\Gamma}(0), 0)$ is the probability of observing the system in an infinitesmal region around $\mathbf{\Gamma}(0)$ at time $t = 0$, where $\mathbf{\Gamma}(t)$ denotes the phase space vector at time t, and $p_{\Gamma^*} = p(\mathbf{\Gamma}^*(\tau), 0)$ where $\mathbf{\Gamma}^*(\tau)$ is obtained from $\mathbf{\Gamma}(\tau)$ by reversing all the particle velocities.

Maritan, Chap. 3). Two key results follow as mathematical consequences of the above definition[10]:

$$\frac{p(\Omega_\Gamma = A)}{p(\Omega_\Gamma = -A)} = e^A \tag{1.2}$$

and

$$\langle\Omega\rangle = \sum\nolimits_\Gamma p_\Gamma \ln\frac{p_\Gamma}{p_{\Gamma*}} \geq 0, \tag{1.3}$$

with equality if and only if $p_\Gamma = p_{\Gamma*}$; the sum in Eq. (1.3) extends over all microscopic trajectories (i.e. it includes both Γ and Γ*). Equation (1.2) is the Fluctuation Theorem; it implies that trajectories with negative Ω_Γ are exponentially less likely to be observed than trajectories with positive Ω_Γ. In Eq. (1.3), $\langle\Omega\rangle$ is the Kullback-Leibler divergence of p_Γ and $p_{\Gamma*}$; it is a measure of how different p_Γ and $p_{\Gamma*}$ are. In other words, $\langle\Omega\rangle$ is a measure of time-reversal symmetry breaking (or *irreversibility* in the language of Chap. 3)—and therein lies its fundamental character, as well as the key to its eventual physical interpretation as a thermodynamic entropy production. Equation (1.3) states that the average value of Ω_Γ is non-negative; this result, known as the Second Law inequality, is a mathematical consequence of Gibbs' inequality. The value $\langle\Omega\rangle = 0$ (i.e. $p_\Gamma = p_{\Gamma*}$) corresponds to thermodynamic equilibrium, underpinning the interpretation of $\langle\Omega\rangle$ as a measure of distance from equilibrium.

Clearly, the physical consequences of these purely mathematical results can only emerge when some additional physical information is built into p_Γ. In Chap. 2, this is done by deriving p_Γ from a microscopic model of the underlying molecular dynamics. In Chap. 3, p_Γ is derived[11] from MaxEnt constrained by a restricted subset of the underlying dynamics, representing those key aspects of the dynamics that are known (or assumed) to be relevant on macroscopic scales. Equation (1.3) is also consistent (up to a factor 2) with the definition of entropy production that emerges from the MaxEnt analysis of flow systems in Chap. 7, in which constraints on various mean flow rates are imposed.

In each case, whether p_Γ (or $p(f)$, see footnote 11) is derived from molecular dynamics (Chap. 2) or macroscopic flux constraints (Chaps. 3 and 7), $\langle\Omega\rangle$ may be interpreted physically as a *generalised entropy production*, defined for systems arbitrarily far from equilibrium. The FT and Second Law inequality then become statements about physically-meaningful dissipation-like quantities. The generic

[10] If $\delta(i,j)$ denotes the Kronecker delta function [0 if $i \neq j$, 1 if $i = j$], Eq. (1.2) follows from $p(\Omega_\Gamma = A) = \Sigma_\Gamma p_\Gamma \delta(\Omega_\Gamma, A) =$ [change of variable] $\Sigma_\Gamma p_{\Gamma*}\delta(\Omega_{\Gamma*}, A) =$ (from Eq. 1.1) $\Sigma_\Gamma p_\Gamma \exp(-\Omega_\Gamma)\delta(-\Omega_\Gamma, A) = e^A \Sigma_\Gamma p_\Gamma \delta(\Omega_\Gamma, -A) = e^A p(\Omega_\Gamma = -A)$. Equation (1.3) follows from Gibbs' inequality: $-\Sigma_i p_i \ln p_i \leq -\Sigma_i p_i \ln q_i$ for any probability distributions p_i and q_i, with equality if and only if $p_i = q_i$ for all i.

[11] In Chap. 3, MaxEnt is used to construct $p(f)$, the probability distribution of macroscopic fluxes f, rather than p_Γ; the formalism can be re-expressed in terms of p_Γ as shown in [42].

MaxEP principle derived from the MaxEnt argument given in Chap. 3 then supplements the Second Law inequality with the statement that not only is $\langle\Omega\rangle$ non-negative but $\langle\Omega\rangle$ takes on its maximum value attainable under the imposed constraints.[12]

1.4.2 Which Entropy Production is Extremised? An Emerging Pattern

The majority of applications described in Part III of this book pertain to MaxEP and its variants. In the absence of an accepted theoretical basis for MaxEP, a pragmatic approach has seemed to be the only way forward. By applying MaxEP in many different ways, using different candidate entropy production functions, and seeing in which situations it appears to work in practice, one might gain insights into its theoretical basis.

Using simple energy balance models (EBMs) that avoid some of the *ad hoc* assumptions and parameterizations of Paltridge's EBM [3–5], Herbert and Paillard (Chap. 9) and Fukumura and Ozawa (Chap. 11) provide further evidence that the broad *thermal* characteristics of some planetary climates (e.g. latitudinal profiles of surface temperature and meridional heat flows) may be reproduced with reasonable accuracy by maximizing the thermal entropy production associated with the material transport of heat across temperature gradients, subject only to the constraint of global energy balance. These results do not invoke any dynamical constraints, but are nevertheless conditional on there being sufficient mass to sustain advective heat transport (Chap. 11), a conclusion also reached in [45]. Subject to this proviso, the implication then is that fluids with different dynamical properties (e.g. different viscosities), but subject to the same global energy balance constraint, will self-organize their velocity fields to achieve the same overall optimal pattern of heat flow.

In contrast, in order to predict dynamical characteristics that depend explicitly on the *velocity field* (including the velocity field itself), a principle of maximum KE dissipation appears to take precedence. In climatology, this result is suggested by a dynamic sensitivity analysis of FAMOUS, a GCM of intermediate complexity [29]; in horizontal shear turbulence, maximum KE dissipation also accurately reproduces the mean velocity profile over a large range of forcing conditions [28]; the same principle also emerges as an accurate predictor of the steady orientation of a body settling in a viscous fluid (Vaidya, Chap. 13). Yet another principle, maximum heat flow, appears to govern the stability of stationary convective states

---

12 Specifically (see Chap. 3), when the non-equilibrium driving force is such that $\langle\Omega\rangle^C > \langle\Omega\rangle^C_{min}$, then MaxEnt implies $\langle\Omega\rangle = \langle\Omega\rangle_{max}$. Here C denotes a restricted set of stationarity constraints, the nature of which determines the physical nature of $\langle\Omega\rangle^C$ as an entropy production or dissipation functional; $\langle\Omega\rangle^C_{min}$ and $\langle\Omega\rangle^C_{max}$ are the lower and upper bounds on $\langle\Omega\rangle^C$.

in lattice-Boltzmann simulations of Rayleigh-Bénard convection (Weaver et al., Chap. 14); under fixed-temperature boundary conditions this is equivalent to maximum thermal entropy production.

From this pragmatic approach, therefore, the empirical and numerical evidence appears to suggest that there is no universal entropy production functional that is maximized in all problems. However, from a variety of different applications two key principles have emerged—maximum thermal dissipation and maximum KE dissipation—as a guide to constructing a more fundamental theory. One would like such a theory to tell us a *priori* which functional to maximize. What insights do existing theoretical approaches provide? For example, can we unify at least some of the extremal functions in Table 1.1 in terms of the generalised entropy production $\langle\Omega\rangle$ discussed in Sect. 1.4.1?

1.4.3 MaxEP and Dynamic Stability: Emerging Theories

Several chapters in the theoretical section of this book (Part II), and studies elsewhere, suggest that MaxEP and dynamic stability are closely related. For example, Kleidon et al. (Chap. 8) propose a maximum power principle (equivalent to maximum KE dissipation) for the selection of different flow structures in a simple model of Earth system KE; they use a heuristic dynamical stability argument to suggest that stationary states of maximum KE dissipation are preferred because they are the most stable. Ozawa and Shimokawa (Chap. 6) come to a similar conclusion by deriving a necessary condition for the local KE dissipation rate of a convective fluid to increase over time. In an earlier study, Malkus [28] derived maximum global KE dissipation by the mean flow as a necessary condition for the stability of stationary states in horizontal shear turbulence.

Further numerical evidence that MaxEP and dynamic stability are intimately linked emerges from the stability analysis of crystal growth morphologies (Martyushev, Chap. 20), which shows that the coexistence of two growth morphologies occurs when their respective entropy productions (defined in terms of the local velocity of the crystal surface) are equal. Simulations of Rayleigh-Bénard convection (Weaver et al., Chap. 14) indicate that stationary convective states of maximum heat transport are the most stable (cf. [25]).

A common feature of the analyses linking dynamical stability to maximum KE dissipation is their incorporation of momentum balance[13] as a constraint, in addition to the global energy balance constraint under which Paltridge maximized thermal dissipation. This suggests that there might be a link between the choice of extremized function and the choice of constraints, and, further, that dynamical stability underlies that link.

[13] In Chap. 8 and [28], only a spatially-averaged momentum balance constraint is imposed, rather than the full Navier–Stokes equation.

Dewar and Maritan (Chap. 3) propose such a connection, based on the application of MaxEnt to non-equilibrium systems under generic dynamical constraints. Here, the physical nature of the generalised entropy production $\langle\Omega\rangle$ indeed depends on the choice of constraints. Specifically, when only global energy (or mass) balance is imposed, MaxEnt predicts that $\langle\Omega\rangle$ is the entropy production by heat (or mass) flow, consistent with the extremised function in Paltridge's MaxEP. In horizontal shear turbulence, when the additional (spatially-averaged) momentum constraint of Malkus [28] is imposed, MaxEnt predicts that $\langle\Omega\rangle$ is KE dissipation by the mean flow, consistent with [28]. Moreover, dynamical stability plays a crucial role here: when the non-equilibrium driving force is sufficient to make the stationary state of minimum $\langle\Omega\rangle$ dynamically unstable (see footnote 12 and Chap. 3), MaxEnt predicts that $\langle\Omega\rangle$ adopts its maximum value.

The suggestion here is that there may indeed exist a universal entropy production functional that is maximized in all problems—in the form of $\langle\Omega\rangle$ defined by Eq. (1.3)—but that $\langle\Omega\rangle$ manifests itself as thermodynamic entropy production in different ways (e.g. thermal dissipation, KE dissipation) according to the constraints on the system. We might think of this as the result of different constraints confining the dissipation of free energy to different degrees of freedom, as described by different thermodynamic dissipation functions.

1.4.4 MaxEnt and Ziegler's MaxEP

What is the link, if any, between Ziegler's MaxEP and the other extremal principles in Table 1.1? As we noted in Sect. 1.2, Ziegler's MaxEP lacks a rigorous theoretical basis. Given the proposed argument for a MaxEnt basis of MaxEP as a generic stationary state selection principle (Chap. 3), it is natural to examine whether MaxEnt might also provide a basis for Ziegler's MaxEP (or equivalently Ziegler's orthogonality condition, see Table 1.1).

Such a link was established in [43] for systems close to equilibrium.[14] Specifically, under fixed mean fluxes $J_i{}^*$, MaxEnt yields an orthogonality condition between the associated Lagrange multipliers λ_i and $J_i{}^*$, i.e. $J_i{}^* \propto \partial\sigma(\lambda)/\partial\lambda_i$. Here $\sigma(\lambda) = \langle\Omega\rangle$ is the dissipation function defined by Eq. (1.3), which according to MaxEnt can be expressed as a function of either J^* or λ. Equivalently, we can consider λ^* as given and derive the orthogonality condition $\lambda_i{}^* \propto \partial\sigma(J)/\partial J_i$. If we then identify λ_i with the thermodynamic 'force' conjugate to J_i, these results are identical to Ziegler's orthogonality condition in, respectively, X-space and J-space (Table 1.1). Whereas Ziegler's MaxEP assumes the functional form of $\sigma(J)$ a priori, this emerges from MaxEnt a posteriori.

Thus, close to equilibrium, Ziegler's orthogonality condition (OC) characterises the MaxEnt relation between flux constraints and their Lagrange multipliers. The fact

[14] The restriction of the analysis in [43] to near-equilibrium systems was pointed out in [46, 47].

that Ziegler's OC can also be derived from a separate maximization principle (Ziegler's MaxEP principle, Table 1.1) may perhaps be a 'red herring' for two reasons. Firstly, it is known that the generic relation between MaxEnt Lagrange multipliers and constraints can be solved mathematically as a variational principle [48]. Secondly, the max/min character of Ziegler's OC depends on the nature of the auxilliary constraints on λ; for example, $\sigma(\lambda)$ has a minimum with respect to variations in λ restricted to the plane $\Sigma_i \lambda_i J_i^* =$ constant (cf. Ziegler's MaxEP, Table 1.1). Applications to dissipative materials and land-atmosphere energy exchange are discussed, respectively, by Houlsby (Chap. 4) and Wang et al. (Chap. 16).

The equivalence of Ziegler's MaxEP and MaxEnt subject to given fluxes J^* is also apparent in the fact that both lead to linear flux-force relations close to equilibrium [13, 38–40] (see also Seleznev and Martyushev, Chap. 5). Moreover, the equivalence may be more general: in the derivation of a generic MaxEP principle from MaxEnt (Dewar and Maritan, Chap. 3), Ziegler's OC emerges as a property of MaxEP stationary states arbitrarily far from equilibrium.

1.4.5 The Physical Interpretation of MaxEP

The suggestion, then, is that MaxEnt offers a common theoretical framework that links at least some of the non-equilibrium extremal principles[15] in Table 1.1 (see also Fig. 1.1). If this is correct, the question of the physical significance of MaxEP (Paltridge, UBT, Ziegler ...) boils down to that of MaxEnt itself.

Mathematically, MaxEnt is an algorithm that constructs a probability distribution p_i over some set of outcomes i subject to given constraints C (usually a restricted subset of the full underlying dynamics). The MaxEnt probability distribution p_i coincides with the most likely frequency distribution of outcomes that would be observed in an infinitely long sequence of independent samples, provided we have correctly identified the relevant constraints C that apply during the experiment (see Chap. 3 and references therein; also Chap. 7).

This implies that MaxEnt (hence MaxEP) can be used to answer two complementary questions: Given the constraints, what is the most likely system behaviour? Or, given the observed system behaviour, what are the key constraints governing it? Therefore we can use MaxEP in two ways—as a statistical selection principle or a method for inferring the relevant constraints—depending on which question we are asking. The latter question is less straightforward to answer than the former. It requires a trial and error approach in which by comparing MaxEnt predictions and observations we eventually home in on the relevant constraints (which usually comprise some restricted subset of the full dynamics, e.g. global energy balance, global momentum balance). In the former we can interpret

[15] For an alternative perspective, see Chap. 5.

MaxEnt as a physical (statistical) selection principle, just as we do the Second Law.

While it may be tempting to interpret MaxEP as a dynamical principle that reflects the evolution of a system towards the most stable stationary state (as suggested in Sect. 1.4.3), we should recall Maxwell's insight that the Second Law is statistical in nature, and not a dynamical principle [2]. Likewise, it may be more insightful to interpret stability arguments for MaxEP, based on a restricted subset of the full dynamics, in a statistical sense, as describing the most likely behaviour under those constraints.

Haff (Chap. 21) discusses the interplay between MaxEP and constraints in the context of biological and technological evolution. Evolutionary 'hang-ups' [49] may be short-term internal constraints that reflect slow degrees of freedom (e.g. the long abiotic phase of Earth's evolution, or the energy that is temporarily stranded in fossil fuels). Some of those constraints may relax over longer timescales. Thus, while predicting the current evolutionary state may not be straightforward, MaxEP may more readily tell us where evolution is heading once only a few easily-identifiable external constraints remain (e.g. global energy or mass balance). Dobovišek et al. (Chap. 19) investigate the extent to which the evolution of enzyme kinetics accords with MaxEP and MaxEnt constrained by mass balance. Lineweaver (Chap. 22) explores the question of how MaxEP may relate to the rate of evolution of the universe towards a state of thermodynamic equilibrium.

1.5 Towards a Synthesis

Figure 1.1 depicts a tentative road map of the theoretical links (solid arrows, roads *1-11*) suggested in Sect. 1.4, and in addition some directions for future research (dashed arrows, roads *12-14*). We emphasise again that this particular view of the landscape beyond the Second Law is a partial one, based mainly on the material presented in Chaps. 2–4, 6, 8, 9, 11–13, 18 and 19. Section 1.5.2 highlights some of the viewpoints and issues not featured in Fig. 1.1.

1.5.1 A Tentative Road Map

As indicated by roads *1-3*, the path probability p_Γ and path dissipation function Ω_Γ (Reid et al., Chap. 2) play a unifying role by linking the Second Law inequality, the Fluctuation Theorem (FT) and Maximum Entropy (MaxEnt). Road *4* links MaxEnt to a generic MaxEP principle (Dewar and Maritan, Chap. 3), in which the path relative entropy ($H = -\Sigma_\Gamma p_\Gamma \ln p_\Gamma / q_\Gamma$) is maximized with respect to p_Γ, subject to a restricted set of stationarity constraints (C) together with the criterion that the state of minimum dissipation is dynamically unstable, $\langle\Omega\rangle^C > \langle\Omega\rangle^C_{min}$. As indicated by the superscript, the physical nature of the dissipation function $\langle\Omega\rangle^C$

(i.e. its interpretation as a thermodynamic entropy production) depends on the nature of C. MaxEnt implies MaxEP, i.e. $\langle \Omega \rangle^{C}$ is maximized with respect to the stationary states compatible with C. When C represents global energy and/or mass balance (e/m, road *5*), we recover Paltridge's MaxEP (cf. Chaps. 9, 11, 19) in which $\langle \Omega \rangle^{C}$ takes the form of flux times force (e.g. heat flux times inverse temperature gradient; or mass flux times chemical affinity). When C includes global power balance *and* a spatially-averaged momentum balance constraint in horizontal shear turbulence [28] (e/m, mom, road *6*), we recover maximum KE dissipation (cf. Chaps. 6, 8, 13) in which $\langle \Omega \rangle^{C}$ is the KE dissipation associated with the mean velocity field (D_{KE}, Fig. 1.1).

The generic MaxEP principle (Chap. 3) also leads to Ziegler's orthogonality condition (non-linear constitutive relations, cf. Chap. 4) for systems arbitrarily far from equilibrium (road *7*). This reduces to linear flux-force relationships (road *8*) in the near-equilibrium limit, $\langle \Omega \rangle \approx 0$, a result that can also be obtained directly from MaxEnt (road *9*, [38–40]). As shown by Ziman [21], Kohler's MaxEP follows rigorously as the mathematical solution to the linearized Boltzmann equation for gas transport; and when expressed in the language of thermodynamic fluxes and forces, it is equivalent to Onsager's MaxEP (road *10*).

In a separate thread, Malkus's dynamical stability analysis [28], involving the same constraints as road *6*, also leads to maximum KE dissipation by the mean flow (road *11*). This raises the possibility of interpreting $\langle \Omega \rangle^{C}$ as a fundamental statistical measure of dynamical stability (road *12*). Intuitively, this might reflect the fact that maximizing $\langle \Omega \rangle^{C}$ ensures that p_{Γ} and $p_{\Gamma *}$ are maximally different, so that entropy-decreasing trajectories that would destabilise the stationary state of maximum entropy production are also maximally improbable. Dynamical stability might also be understood statistically through the 'maximum caliber' interpretation of MaxEnt [39, 44], in which the predicted macroscopic path of a non-equilibrium system is representative of the largest number of plausible microscopic paths. This echoes the interpretation by Malkus [25] that the statistical stability of maximum-dissipation turbulent states reflects the high local density of flow solutions in phase space.

Another direction for future study is the link, if any, between Kohler's MaxEP and radiative MinEP (road *13*). With appropriate constraints, both principles offer solutions to transport problems that can be expressed mathematically as a linearized Boltzmann equation (for mass and radiation, respectively). As far as we are aware, however, a mathematical derivation of radiative MinEP that follows the same lines as the mathematical derivation of Kohler's MaxEP [e.g. 13, 20] has yet to be given explicitly (cf. Chap. 12).

Finally, can MaxEnt provide a theoretical basis for a non-stationary version of MaxEP? (road *14*). A time-dependent formulation of MaxEP is explored by Vallino et al. (Chap. 18) in the context of biogeochemistry; see also [39, 50]. If such a principle could be established, entropy production might be to macroscopic dynamics what the Lagrangian functional is to microscopic dynamics (cf. Chap. 5, Sect. 5.5).

1.5.2 Other Perspectives and Open Questions

So far we have highlighted the role of MaxEnt as a common theoretical framework for some of the principles in Table 1.1. Its potential in this regard stems largely from its generic nature—in all problems the same principle is applied (maximum relative entropy); only the nature of the outcomes i and constraints C differs between problems. Fig. 1.1 offers a tentative road map centred on MaxEnt subject to the specific constraints applied by Dewar and Maritan in Chap. 3 (stationarity, dynamic instability of the MinEP state). In Chap. 7, Niven and Noack apply MaxEnt to flow systems subject to constraints on mean flow rates. There, the interplay between changes in entropy within and outside the system can be described in terms of changes in a potential function, $\Phi = -\ln Z$, where Z is the partition function. By analogy with equilibrium thermodynamics, a principle of minimum[16] Φ for open, stationary systems is proposed, with Φ the non-equilibrium analogue of the Planck potential (or free energy); from this, principles of MaxEP or MinEP might then arise depending on the particular constraints on the system under study. Seleznev and Martyushev (Chap. 5) proposes that MaxEP is an independent physical principle whose theoretical foundation does not rely on MaxEnt at all.

Yoshida and Kawazura (Chap. 15) examine the link between entropy production and stability in a turbulent fluid-plasma system. Using a simple low-dimensional dynamical model, they find that whether the thermal entropy production of the (stable) non-linear stationary state is larger or smaller than the (unstable) linear stationary state depends on the system connectivity (series vs. parallel) and the type of forcing (flux-driven vs. force-driven). Analyses of pipe flow systems show a similar dependence of the relative size of the entropy production rates of laminar versus turbulent flow on the type of forcing [51–54]. Do these results challenge emerging theories suggesting that the most stable states always have the largest entropy production (Chaps. 3, 6, 8 and [28])? Or do the latter theories only apply to selection of one among several non-linear stationary states, and not to selection between one linear state and one non-linear state (Chap. 15)? Alternatively, do these results provide evidence for a principle of minimum Φ, analogous to the minimum free energy principle of equilibrium thermodynamics, that might reduce to MaxEP or MinEP under different circumstances (Chap. 7)?

Boshi et al. (Chap. 10) analyse a GCM incorporating the ice-albedo feedback to show that dynamical transitions between the *snowball* and *warm* stationary climate states are characterised thermodynamically by signature variations in the effective Carnot efficiency of the climate considered as a heat engine. Herbert et al. [49] used a simpler energy balance model to demonstrate a close analogy between the relative stability of snowball and warm states and their thermal entropy production rates, suggesting MaxEP as the relevant selection criterion. A MaxEP principle also appears to govern selection between crystal growth morphologies

[16] The minimum is along a path in the space of flux states.

(Martyushev, Chap. 20). How do these results relate to the theoretical landscape of Fig. 1.1, or to the MaxEP/MinEP dichotomy described above?

The emergence of maximum heat transport as characteristic of stable states in a lattice-Boltmann simulation of Rayleigh-Bénard convection (Weaver et al., Chap. 14) echoes the earlier maximum transport principles of UBT [23]. A subsequent stability analysis by Malkus [25] suggested that, for very high Rayleigh number flows, maximum momentum transport takes precedence over maximum heat transport. How are these results to be reconciled? Does maximum heat transport take precedence at lower Rayleigh numbers?

Clearly, there is much climbing to be done before we reach a consensus view on the theoretical basis of the various MaxEP/MinEP principles described in this chapter. Conquering that lofty peak will require a close interplay between bottom-up and top-down modelling approaches—numerical simulations (GCMs, lattice-Boltzmann models, molecular dynamics simulations), dynamical stability analyses, and variational methods. However, model analyses are not sufficient. Ultimately the models must be confronted with observational data, as exemplified by several contributions to this volume (e.g. Chaps. 11–13, 16, 18, 19). Together, these approaches will continue to provide a fertile testing-ground for emerging theories of MaxEP and other non-equilibrium principles beyond the Second Law.

References

1. Schrödinger, E.: What is life? (With mind and matter and autobiographical sketches). CUP, Cambridge (1992)
2. Maxwell, J.C.: Letter to John William Strutt (1870). In: PM Harman (ed) The scientific letters and papers of James Clerk Maxwell. CUP, Cambridge UK **2**, 582–583 (1990)
3. Paltridge, G.W.: Global dynamics and climate-a system of minimum entropy exchange. Q. J. Roy. Meteorol. Soc. **101**, 475–484 (1975)
4. Paltridge, G.W.: The steady-state format of global climate. Q. J. Roy. Meteorol. Soc. **104**, 927–945 (1978)
5. Paltridge, G.W.: Thermodynamic dissipation and the global climate system. Q. J. Roy. Meteorol. Soc. **107**, 531–547 (1981)
6. Lorenz, R.D., Lunine, J.I., Withers, P.G., McKay, C.P.: Titan, mars and earth: entropy production by latitudinal heat transport. Geophys. Res. Lett. **28**, 415–418 (2001)
7. Hill, A.: Entropy production as a selection rule between different growth morphologies. Nature **348**, 426–428 (1990)
8. Martyushev, L.M., Serebrennikov, S.V.: Morphological stability of a crystal with respect to arbitrary boundary perturbation. Tech. Phys. Lett. **32**, 614–617 (2006)
9. Juretić, D., Županović, P.: Photosynthetic models with maximum entropy production in irreversible charge transfer steps. Comp. Biol. Chem. **27**, 541–553 (2003)
10. Dewar, R.C., Juretić, D., Županović, P.: The functional design of the rotary enzyme ATP synthase is consistent with maximum entropy production. Chem. Phys. Lett. **430**, 177–182 (2006)
11. Dewar, R.C.: Maximum entropy production and plant optimization theories. Phil. Trans. R. Soc. B **365**, 1429–1435 (2010)

12. Franklin, O., Johansson J., Dewar, R.C. Dieckmann, U., McMurtrie, R.E., Brännström, Å, Dybzinski, R.: Modeling carbon allocation in trees: a search for principles. Tree Physiol. **32**, 648–666 (2012)
13. Martyushev, L.M., Seleznev, V.D.: Maximum entropy production principle in physics, chemistry and biology. Phys. Rep. **426**, 1–45 (2006)
14. Evans, D.J., Searle, D.J.: The fluctuation theorem. Adv. Phys. **51**, 1529–1585 (2005)
15. Seifert, U.: Stochastic thermodynamics: principles and perspectives. Eur. Phys. J. B **64**, 423–431 (2008)
16. Sevick, E.M., Prabhakar, R., Williams, S.R., Searles, D.J.: Fluctuation theorems. Ann. Rev. Phys. Chem. **59**, 603–633 (2008)
17. Onsager, L.: Reciprocal relations in irreversible processes I. Phys. Rev. **37**, 405–426 (1931)
18. Onsager, L.: Reciprocal relations in irreversible processes II. Phys. Rev. **38**, 2265–2279 (1931)
19. Prigogine, I.: Introduction to thermodynamics of irreversible processes. Wiley, New York (1967)
20. Kohler, M.: Behandlung von Nichtgleichgewichtsvorgängen mit Hilfe eines Extremalprinzips. Z. Physik. **124**, 772–789 (1948)
21. Ziman, J.M.: The general variational principle of transport theory. Can. J. Phys. **34**, 1256–1273 (1956)
22. Ziegler, H.: An Introduction to Thermomechanics. North-Holland, Amsterdam (1983)
23. Malkus, W.V.R.: The heat transport and spectrum of thermal turbulence. Proc. R. Soc. **225**, 196–212 (1954)
24. Malkus, W.V.R.: Outline of a theory for turbulent shear flow. J. Fluid Mech. **1**, 521–539 (1956)
25. Malkus, W.V.R.: Statistical stability criteria for turbulent flow. Phys. Fluids **8**, 1582–1587 (1996)
26. Kerswell, R.R.: Upper bounds on general dissipation functionals in turbulent shear flows: revisiting the 'efficiency' functional. J. Fluid Mech. **461**, 239–275 (2002)
27. Ozawa, H., Shimokawa, S., Sakuma, H.: Thermodynamics of fluid turbulence: a unified approach to the maximum transport properties. Phys. Rev. E **64**, 026303 (2001)
28. Malkus, W.V.R.: Borders of disorders: in turbulent channel flow. J. Fluid Mech. **489**, 185–198 (2003)
29. Pascale, S., Gregory, J.M., Ambaum, M.H.P., Tailleux, R.: A parametric sensitivity study of entropy production and kinetic energy dissipation using the FAMOUS AOGCM. Clim. Dyn. **38**, 1211–1227 (2012)
30. Boltzmann, L.: Über die Beziehung zwischen dem zweiten Hauptsatze der mechanischen Wärmetheorie und der Wahrscheinlichkeitsrechnung respektive den Sätzen über das Wärmegleichgewicht. Wien. Ber. **76**, 373–435 (1877)
31. Gibbs, J.W.: Elementary principles of statistical mechanics. Ox Bow Press, Woodridge (1981). Reprinted
32. Jaynes, E.T.: Information theory and statistical mechanics. Phys. Rev. **106**, 620–630 (1957)
33. Jaynes, E.T.: Information theory and statistical mechanics II. Phys. Rev. **108**, 171–190 (1957)
34. Jaynes, E.T.: Probability Theory: The Logic of Science. In: Bretthorst, G.L. (ed.). CUP, Cambridge (2003)
35. Houlsby, G.T., Puzrin, A.M.: Principles of hyperplasticity. Springer, London (2006)
36. Shannon, C.E.: A mathematical theory of communication. Bell Sys. Tech. J. **27**, 379–423 and 623–656 (1948)
37. Shannon, C.E., Weaver, W.: The mathematical theory of communication. University of Illinois Press, Urbana (1949)
38. Grandy, W.T. Jr.: Foundations of Statistical Mechanics. Volume II: Nonequilibrium Phenomena. D. Reidel, Dordrecht (1987)
39. Grandy, W.T. Jr.: Entropy and the time evolution of macroscopic systems. International series of monographs on physics, vol. 141, Oxford University Press, Oxford (2008)

40. Niven, R.K.: Steady state of a dissipative flow-controlled system and the maximum entropy production principle. Phys. Rev. E **80**, 021113 (2009)
41. Dewar, R.C.: Maximum entropy production as an inference algorithm that translates physical assumptions into macroscopic predictions: Don't shoot the messenger. Entropy **11**, 931–944 (2009)
42. Dewar, R.C.: Information theory explanation of the fluctuation theorem, maximum entropy production and self-organized criticality in non-equilibrium stationary states. J. Phys. A: Math. Gen. **36**, 631–641 (2003)
43. Dewar, R.C.: Maximum entropy production and the fluctuation theorem. J. Phys. A: Math. Gen. **38**, L371–L381 (2005)
44. Jaynes, E.T.: The minimum entropy production principle. Ann. Rev. Phys. Chem. **31**, 579–601 (1980)
45. Jupp, T.E., Cox, P.M.: MEP and planetary climates: insights from a two-box climate model containing atmospheric dynamics. Phil. Trans. R. Soc. B. **365**, 1355–1365 (2010)
46. Bruers, S.A.: Discussion on maximum entropy production and information theory. J. Phys. A: Math. Theor. **40**, 7441–7450 (2007)
47. Grinstein, G., Linsker, R.: Comments on a derivation and application of the 'maximum entropy production' principle. J. Phys. A: Math. Theor. **40**, 9717–9720 (2007)
48. Agmon, N., Alhassid, Y., Levine, R.D.: An algorithm for finding the distribution of maximal entropy. J. Comput. Phys. **30**, 250–258 (1979)
49. Dyson, F.J.: Energy in the universe. Sci. Amer. **225**, 51–59 (1971)
50. Herbert, C., Paillard, D., Dubrulle, B.: Entropy production and multiple equilibria: the case of the ice-albedo feedback. Earth Syst. Dynam. **2**, 13–23 (2011)
51. Thomas, T.Y.: Qualitative analysis of the flow of fluids in pipes. Am. J. Math. **64**, 754–767 (1942)
52. Paulus, D.M., Gaggioli, R.A.: Some observations of entropy extrema in fluid flow. Energy **29**, 2487–2500 (2004)
53. Martyushev, L.M.: Some interesting consequences of the maximum entropy production principle. J. Exper. Theor. Phys. **104**, 651–654 (2007)
54. Niven, R.K.: Simultaneous extrema in the entropy production for steady-state fluid flow in parallel pipes. J. Non-Equil. Thermodyn. **35**, 347–378 (2010)

Part II
Theoretical Perspectives on Entropy Production

Chapter 2
The Dissipation Function: Its Relationship to Entropy Production, Theorems for Nonequilibrium Systems and Observations on Its Extrema

James C. Reid, Sarah J. Brookes, Denis J. Evans and Debra J. Searles

Abstract In this chapter we introduce the dissipation function, and discuss the behaviour of its extrema. The dissipation function allows the reversibility of a nonequilibrium process to be quantified for systems arbitrarily close to or far from equilibrium. For a system out of equilibrium, the average dissipation over a period, t, will be positive. For field driven flow in the thermodynamic and small field limits, the dissipation function becomes proportional to the rate of entropy production from linear irreversible thermodynamics. It can therefore be considered as an entropy-like quantity that remains useful far from equilibrium and for relaxation processes. The dissipation function also appears in three important theorems in nonequilibrium statistical mechanics: the fluctuation theorem, the dissipation theorem and the relaxation theorem. In this chapter we introduce the dissipation function and the theorems, and show how they quantify the emergence of irreversible behaviour in perturbed, steady state, and relaxing nonequilibrium systems. We also examine the behaviour of the dissipation function in terms of the extrema of the function using numerical and analytical approaches.

J. C. Reid (✉) · D. J. Searles
Australian Institute of Bioengineering and Nanotechnology and School of Chemistry and Molecular Biosciences, The University of Queensland, Brisbane Qld 4072, Australia
e-mail: uqjreid9@uq.edu.au

D. J. Searles
e-mail: d.bernhardt@uq.edu.au

S. J. Brookes · D. J. Searles
Queensland Micro- and Nanotechnology Centre, School of Biomolecular and Physical Sciences, Griffith University, Brisbane Qld 4072, Australia
e-mail: s.brookes@griffith.edu.au

D. J. Evans
Research School of Chemistry, Australian National University, Canberra, ACT 0200, Australia
e-mail: denevans@mac.com

R. C. Dewar et al. (eds.), *Beyond the Second Law*,
Understanding Complex Systems, DOI: 10.1007/978-3-642-40154-1_2,

2.1 Introduction

The treatment of thermodynamic systems can be considered to be split between systems that are in or near equilibrium, and nonequilibrium systems. Equilibrium systems are well quantified by a variety of state functions, such as entropy, temperature and free energy, that are independent of the moment to moment behaviour of the system. Nonequilibrium systems can exhibit a variety of behaviours including relaxation, ageing, various meta-stable states, and steady states. Even in a steady state, they are generally more difficult to classify as many of the basic state functions, including the temperature and entropy, are undefined for far from equilibrium states and the nonequilibrium distribution function of a steady state system is fractal and non-analytic.

As discussed recently [1], the Gibbs' entropy,

$$S_G(t) \equiv -k_B \int d\boldsymbol{\Gamma} f(\boldsymbol{\Gamma},t)\ln(f(\boldsymbol{\Gamma},t)) \equiv -k_B\langle \ln(f(\boldsymbol{\Gamma},t))\rangle_{f(\boldsymbol{\Gamma},t)}, \tag{2.1}$$

is not useful for the description of the relaxation of nonequilibrium Hamiltonian systems or for steady states because it is constant in the first case and divergent in the latter. Here the notation $\langle\ldots\rangle_{f(\boldsymbol{\Gamma},t)}$ is an ensemble average with respect to the distribution function $f(\boldsymbol{\Gamma},t)$. Close to equilibrium and in the thermodynamic limit, the spontaneous entropy production rate from nonlinear irreversible thermodynamics [2] can be well defined and is a useful quantity. However, far from equilibrium where temperature is not well defined a new quantity needs to be considered. Recently we have shown that the dissipation function, Ω, is a powerful quantity which reduces to the spontaneous entropy production rate at small fields for field driven flow [3]. As discussed below, it also satisfies the inequality, $\langle\Omega_t\rangle > 0$ for nonequilibrium systems and therefore it is considered an entropy-like quantity or a generalisation of the entropy for far from equilibrium systems.

In this chapter we will first define the dissipation function and give an overview of some recently derived theorems where dissipation plays a key role. We will also consider the extremal behaviour of the dissipation function by referring to some results from previously published work.

2.2 The Dissipation Function and the Fluctuation Theorem

The dissipation function quantifies the thermodynamic reversibility of a trajectory. It compares the probability of observing an arbitrary system trajectory with the probability of observing the time reverse of that trajectory (its conjugate anti-trajectory) in the same ensemble of trajectories [3]:

$$\Omega_t(\boldsymbol{\Gamma}(0)) = \ln\frac{P(\boldsymbol{\Gamma}(0),0)}{P(\boldsymbol{\Gamma}^*(t),0)}. \tag{2.2}$$

Here $\boldsymbol{\Gamma} \equiv \{q_1, p_1, \ldots q_N, p_N\}$ is the phase space vector of the system which corresponds to a system trajectory, $\Omega_t(\boldsymbol{\Gamma}(0))$ is the total dissipation (or time integral of the dissipation) for a trajectory originating at $\boldsymbol{\Gamma}(0)$ and evolving for a time t, $P(\boldsymbol{\Gamma}(0), 0) = f(\boldsymbol{\Gamma}(0), 0)d\boldsymbol{\Gamma}(0)$ is the probability of observing a system in an infinitesimal region around $\boldsymbol{\Gamma}(0)$ in the initial system distribution with distribution function $f(\boldsymbol{\Gamma}(0), 0)$, and $\boldsymbol{\Gamma}^*(t)$ is the result of applying a time reversal map to $\boldsymbol{\Gamma}(t)$. It can therefore be written in expanded form as,

$$\Omega_t(\boldsymbol{\Gamma}(0)) = \ln \frac{f(\boldsymbol{\Gamma}(0), 0)d\boldsymbol{\Gamma}(0)}{f(\boldsymbol{\Gamma}^*(t), 0)d\boldsymbol{\Gamma}^*(t)}. \tag{2.3}$$

While the expression above is for a deterministic system, the dissipation function can be defined for a general dynamic system [4], and has been applied to quantum and stochastic systems in addition to deterministic ones [5–7].

The time integral of the dissipation is positive when the observed system trajectory, starting at $\boldsymbol{\Gamma}(0)$, is more probable than the conjugate trajectory, starting at $\boldsymbol{\Gamma}(t)$, and negative when the observed trajectory is less likely than its conjugate. In order for the dissipation function to be well defined for a system we need the conjugate trajectory to exist for every possible trajectory in the accessible phase space of the system,[1] which for a deterministic system requires ergodic consistency (that the volume of phase space occupied at time t is congruent with or a subset of the volume at time 0) and time reversal symmetric (any time dependent external parameters are even around $t/2$). We can also define an instantaneous dissipation function:

$$\Omega(\boldsymbol{\Gamma}(t)) = \frac{d\Omega_t(\boldsymbol{\Gamma}(0))}{dt}. \tag{2.4}$$

For clarity of notation, we note that $\Omega_t(\boldsymbol{\Gamma}(0)) = \int_0^t \Omega(\boldsymbol{\Gamma}(s))ds$.

The dissipation function evaluates the relative reversibility of a single observation of a system. To understand the overall system behaviour, we need to look at the distribution of the dissipation function. The fluctuation theorem does this by considering the relative probability of observing processes that have positive and negative total dissipation in nonequilibrium systems. This theorem was first derived by Evans and Searles [3, 8], and results in the relationship:

$$\frac{P(\Omega_t = A \pm dA)}{P(\Omega_t = -A \pm dA)} = e^A, \tag{2.5}$$

where $P(\Omega_t = A \pm dA)$ is the probability of observing a trajectory with a dissipation total infinitesimally close to A.

The fluctuation theorem was derived to solve one of the fundamental paradoxes of statistical mechanics, Lochsmidt's paradox. In 1876, Lochsmidt pointed out a fundamental problem associated with nonequilibrium thermodynamics: the

[1] Note that this is often trivially satisfied in a stochastic system.

macroscopic behaviour of a system is irreversible, as embodied in the second law, but the microscopic motion of all of the individual components is fully time reversible. Therefore for any system change, the opposite system change must also be possible. Even at that time, a resolution to the paradox was recognised by some. As noted by Maxwell [9]:

> The truth of the second law is... a statistical, not a mathematical, truth, for it depends on the fact that the bodies we deal with consist of millions of molecules...
>
> Hence the second law of thermodynamics is continually being violated, and that to a considerable extent, in any sufficiently small group of molecules belonging to a real body.

and Boltzmann [10]:

> ...as soon as one looks at bodies of such small dimension that they contain only a very few molecules, the validity of this theorem [the Second Law] must cease.

This means that the second law of thermodynamics is the limiting result of a statistical effect where the probability of observing the behaviour predicted from thermodynamics becomes more and more likely as the system size grows. However for sufficiently small systems monitored for short periods, we would expect to observe both types of behaviour. The fluctuation theorem quantifies this result.

The fluctuation theorem is an exact expression that applies to any system for which the dissipation function is well defined. It tells us that positive dissipation is exponentially more likely to be observed than negative dissipation. Furthermore, consideration of the ensemble *average* of the dissipation function, that is the average over all the available initial points, must be positive, $\langle \Omega_t \rangle \geq 0$, a result known as the second law inequality [11]. These two results expand the generally irreversible behaviour of macroscopic thermodynamic systems into a finite size regime where the second law only holds probabilistically, as demonstrated by a variety of experiments on systems as diverse and optically trapped colloids, torsional pendulums, and electric circuits [12–14].

It is instructive to consider a simple, nonequilibrium, thermostatted system of volume, V consisting of charged particles driven by a field, F_e. In this case, $\Omega_t = \beta V F_e \int_0^t J_c(s)ds$ where $\beta = 1/(k_B T)$, T is the temperature that the system would relax to in the absence of the field (i.e. the temperature of the surroundings), and J_c is the current density in the direction of the field. Writing the time-average of the current density along a trajectory as $\bar{J}_{c,t} = \frac{1}{t}\int_0^t J_c(s)ds$, the fluctuation relation can then be stated:

$$\frac{P(\bar{J}_{c,t} = A \pm dA)}{P(\bar{J}_{c,t} = -A \pm dA)} = e^{A\beta F_e V t}, \tag{2.6}$$

From this equation, we can see that as the system size or time of observation is increased, the relative probability of observing positive to negative current density increases exponentially so the current density has a definite sign and the second law of thermodynamics is retrieved. Furthermore, $\langle \bar{J}_{c,t} \rangle \geq 0$. In obtaining these results, nothing is assumed about the form of the distribution of current density (it does not have to be Gaussian), and their application is not restricted to this special case of a field driven system, but is very widely applicable. The conditions of ergodic consistency and microscopic time reversibility (that is reversibility of the equations of motion of the particles in the system) are all that are required. Furthermore, in the weak field limit, the rate of entropy production, $\dot{S}$, is given from linear irreversible thermodynamics as $\dot{S} \equiv \sum \langle J_i \rangle V X_i / T$ where the sum is over the product of all conjugate thermodynamic fluxes, J_i and thermodynamics forces, X_i divided by the temperature of the system, T. Clearly the dissipation here is related to this: $\lim_{F_e \to 0} \dot{S}(t) = k_B \langle \Omega(t) \rangle$. The difference at high fields is because the temperature that appears in the dissipation function is that which the system would relax to if the fields were removed rather than any non-equilibrium system temperature observed with the field on.[2] The change in entropy for a process will be similarly related to the time-integral of the dissipation $\lim_{F_e \to 0} \Delta S = k_B \langle \Omega_t \rangle$.

The second law of thermodynamics is often stated in terms of the thermodynamic entropy and this is equivalent to the dissipation in some cases. However we note that the argument of the fluctuation theorem and second law inequality is the dissipation function and these results apply widely: to field driven, boundary driven and relaxation processes arbitrarily close to, or far from, equilibrium. Therefore we argue that away from equilibrium the dissipation function is the appropriate property to consider. It is a well defined, unambiguous quantity for deterministic nonequilibrium systems, given by Eq. (2.3).

2.3 The Dissipation Theorem and the Relaxation Theorem

The dissipation function appears as the central argument of the fluctuation theorem and the second law inequality. Dissipation is also important in quantifying a range of nonequilibrium behaviours, including appearing as the argument of exact expressions for nonlinear response (the dissipation theorem [16]) and relaxation towards equilibrium (the relaxation theorem [17, 18]).

From the definition of the dissipation function, it can be shown that the time-evolution of the phase space distribution function is given by [16]:[3]

$$f(\boldsymbol{\Gamma}, t) = \exp\left(- \int_0^{-t} \Omega(\boldsymbol{\Gamma}(s)) ds \right) f(\boldsymbol{\Gamma}, 0) \qquad (2.7)$$

---

[2] Outside of equilibrium, microscopic temperature expressions are ill defined. Often expressions such as the kinetic temperature (the equipartition expression in momenta) or configurational temperature (a similar expression in position) are used, however these expressions only correspond to the temperature of the system, and each other, at equilibrium [15].

[3] Equation (2.7) applies to systems with no field or a constant field. For the case of a time-dependent field see [19].

The time argument in $f(\boldsymbol{\Gamma}, t)$ represents the time over which the distribution function has evolved from the initial distribution function. Using this result, the dissipation theorem enables the time evolution of ensemble averages of arbitrary phase variables to be calculated for systems that are arbitrarily close or far from equilibrium. In its transient time-correlation form it is written [16]:

$$\langle B(\boldsymbol{\Gamma}(t))\rangle = \langle B(\boldsymbol{\Gamma}(0))\rangle + \int_0^t \langle \Omega(\boldsymbol{\Gamma}(0))B(\boldsymbol{\Gamma}(s))\rangle ds, \tag{2.8}$$

where $B(\boldsymbol{\Gamma})$ is a phase function (i.e. a function whose value can be obtained from the instantaneous value of $\boldsymbol{\Gamma}$) and $\Omega(\boldsymbol{\Gamma}(0))$ is the instantaneous dissipation at time 0. It was first derived in [16], however a simpler derivation is presented in [20].[4] While it may initially appear redundant, as the ensemble average of the quantity of interest can always be measured directly using the same experiment as the dissipation theorem, in some systems, such as weakly driven ones, the dissipation average converges more quickly than the direct average [22–25]. It also proves useful in understanding the response of a system. This relationship is exact at arbitrary field, and therefore it gives the nonlinear response of a system to an external field. However, it also applies to nonequilibrium systems where there is no external field, but there is a change to the system, such as a change in temperature, removal of an applied pressure gradient etc.

The question of how a system relaxes towards equilibrium, and how to determine the functional form of the equilibrium distribution function can also be answered by considering the dissipation. Applying the dissipation theorem to a relaxing system, using the second law inequality and assuming the property of T-mixing [26] (that is the decay of time-correlations in transients), it has been proven that a system will relax to a unique equilibrium state. This result is referred to as the relaxation theorem [17, 18]. If a system that is initially out of equilibrium has a well defined dissipation function and time decay of correlations then the relaxation theorem predicts that [17, 18]:

- The instantaneous dissipation will be zero for a system in equilibrium ($\Omega(\boldsymbol{\Gamma}(t)) = 0$).
- The state where the instantaneous dissipation remains 0 with time is unique and is the canonical distribution for a thermostatted system [17, 18] and the microcanonical distribution for a constant energy system [17, 18].
- The ensemble average of the instantaneous dissipation will go to zero as the relaxing system approaches equilibrium ($\lim_{t\to\infty}\langle \Omega(\boldsymbol{\Gamma}(t))\rangle = 0$).
- The ensemble average of the time-integrated dissipation will always be positive, following the second law inequality ($\langle \Omega_t\rangle > 0$), and will reach a limiting value $\langle \Omega_\infty\rangle$ at long times. Therefore if the conditions on an equilibrium system are changed, the system will relax to a new equilibrium state and the total dissipation during the process will be finite and positive.

4 Note that a special case of this relation was derived much earlier, see [21] for details.

The relaxation theorem does not however say that the average *instantaneous* dissipation function must be greater than zero at all times, accommodating non-monotonic relaxation [27].

We can consider applying Eq. (2.7) to the case where a system initially in equilibrium state 1 with distribution function $f_1(\boldsymbol{\Gamma}) = f(\boldsymbol{\Gamma}, 0)$ is subject to a change in conditions and allowed to relax to a new equilibrium state 2 with distribution function $f_2(\boldsymbol{\Gamma}) = \lim_{t\to\infty} f(\boldsymbol{\Gamma}, t)$. As mentioned previously, the change in Gibbs' entropy, $-k_B\langle \ln f_2\rangle_{f_2} + k_B\langle \ln f_1\rangle_{f_1}$ is zero if the dynamics are Hamiltonian and will be equal to the heat removed from the system divided by the thermodynamic temperature of the underlying equilibrium state, with a non-Hamiltonian thermostat [1]. Taking the ensemble average with respect to the *initial* distribution function, for the left and right hand side of Eq. (2.7), it can be shown that:[5]

$$k_B\langle \Omega_\infty\rangle_{f_1} = k_B\langle \ln f_2\rangle_{f_1} - k_B\langle \ln f_1\rangle_{f_1} \tag{2.9}$$

In obtaining this result, we have made use of the fact that the equilibrium distributions do not change when the sign of the momentum is changed [16]. This looks similar to the change in Gibbs' entropy, however the ensemble averages are *both* with respect to the distribution function of the *initial* state. This clearly shows the difference between the change in the Gibbs' entropy and the total dissipation.

The importance of dissipation in all these results shows that apart from being an entropy-like quantity for field driven flow, it is an important general property for all nonequilibrium systems.

2.4 Extrema of the Dissipation Function

The dissipation function is similar to the entropy production, and while not directly connected to a state function, the various fluctuation theorems provide exact, nonequilibrium relations that are fundamentally important to describing non-equilibrium systems. Given the similarity between the two functions, it is interesting to consider whether a phenomenological theory such as MaxEnt (maximum entropy) for equilibrium systems, and MaxEP (maximum entropy production rate) for nonequilibrium systems can be applied to the dissipation. Several papers have considered the links between entropy production, maximum entropy production

---

[5] We note that $-\int_0^{-t} \Omega(\boldsymbol{\Gamma}(s))ds = \int_0^t \Omega(\boldsymbol{\Gamma}^*(s))ds$ if the dynamics are time reversible and that $\int \Omega(\boldsymbol{\Gamma}^*, s) f(\boldsymbol{\Gamma}) d\boldsymbol{\Gamma} = \int \Omega(\boldsymbol{\Gamma}^*, s) f(\boldsymbol{\Gamma}^*) d\boldsymbol{\Gamma}^* = \langle \Omega(s)\rangle$ since the probability of observing ensemble members is constant ($f(\boldsymbol{\Gamma})d\boldsymbol{\Gamma} = f(\boldsymbol{\Gamma}^*)d\boldsymbol{\Gamma}^*$, see [1]). We assume that the system eventually relaxes to an equilibrium state, $f_2(\boldsymbol{\Gamma}) = \lim_{t\to\infty} \ln f(\boldsymbol{\Gamma}, t)$. From Eq. (2.7), $\lim_{t\to\infty} \ln f(\boldsymbol{\Gamma}, t) = \lim_{t\to\infty}(-\int_0^{-t} \Omega(\boldsymbol{\Gamma}(s))ds + \ln f(\boldsymbol{\Gamma}, 0))$, which can be expressed $\ln f_2(\boldsymbol{\Gamma}) = \int_0^\infty \Omega(\boldsymbol{\Gamma}^*(s))ds + \ln f_1(\boldsymbol{\Gamma})$. Then taking the ensemble average with respect to the initial distribution function we obtain $\langle \ln f_2\rangle_{f_1} = \langle \Omega_\infty\rangle_{f_1} + \langle \ln f_1\rangle_{f_1}$.

rates, fluctuation theorems and nonlinear response [19, 28–33]. As noted by Williams and Evans [19], MaxEP cannot be applied rigorously to nonequilibrium systems in general, as the distribution function at any time (including in the steady state) is not just a function of the dissipation at that time. However it might provide a good approximation in some cases.

A system that relaxes towards equilibrium can do so in one of two ways: conformally and non-conformally.[6] From the relaxation theorem, it can be shown that a system that behaves conformally relaxes monotonically to equilibrium at t_{eq}.[7] Given the bounds on the instantaneous dissipation function, $\langle \Omega(t_{eq}) \rangle = \langle \Omega(0) \rangle = 0$, we expect it to peak at some time before returning to zero. The total dissipation will therefore reach a maximum at equilibrium, $\langle \Omega_{t_{eq}} \rangle > \langle \Omega_{t_{eq}-\Delta t} \rangle$, $\forall \Delta t < t_{eq}$. That is, the ensemble of systems will move through a number of different states (with different distribution functions and average values of the average instantaneous dissipation) as it evolves towards the equilibrium state. However, because the total dissipation is maximised when the system reaches equilibrium, we can use Eq. (2.9) to note that the final stable state will have a greater value of $\langle \ln f(\boldsymbol{\Gamma}, t) \rangle_{f_1}$ than any other state it passes through. This is a special case where it is clear that this function is a maximum compared with other states that it passes through. If a system relaxes non-conformally then there is no such extremal result. This is the most common form of relaxation in nature, and is the one we will study using numerical simulations.

For systems that reach a nonequilibrium steady state, qualitatively different behaviour is expected. From the relaxation theorem and the second law inequality, we know that if a field is applied to a system that is initially at equilibrium the total dissipation will be positive at all times after application of a field. If a steady state is ultimately reached, this implies that the ensemble average of the instantaneous dissipation function in the steady state must be positive, and that the average of the total dissipation function will approach infinity at long times. Furthermore, considering Eq. (2.8) with $B = \Omega$, if the time autocorrelation function $\langle \Omega(\boldsymbol{\Gamma}(0)) \Omega(\boldsymbol{\Gamma}(t)) \rangle$ decays monotonically with time, then the value of the ensemble average of the instantaneous dissipation function, $\langle \Omega(t) \rangle$ will be higher when it reaches its steady state than for any another state it passes through. Therefore the system is in the state that maximises the dissipation function (rate of entropy production). Again this is a special case, so we use numerical simulations to study the behaviour of the instantaneous dissipation function more generally; examining whether it is a maximum in the steady state or if a transient state has a higher average instantaneous dissipation value.

---

[6] A conformal system relaxes such that the nonequilibrium distribution is of the form $(f(\boldsymbol{\Gamma}, t) = \exp(-\beta H(\boldsymbol{\Gamma}) + \lambda(t) g(\boldsymbol{\Gamma}))/Z, \forall t)$ and the deviation function, g, is a constant over the relaxation.

[7] Strictly a system relaxes as time tends towards infinity, but in practice at t_{eq} the system has relaxed.

2.4.1 Relaxing System: Trapped Particle

We choose a system with non-monotonic relaxation based on an optical trapping experiment called the capture experiment [12]. In this model a particle is bound by a harmonic potential to a point in space within a two dimensional fluid surrounded by thermostatted walls [27]. At the beginning of the experiment we begin with the test particle in equilibrium with a trapping constant of k_0, then at time $t = 0^+$ we discontinuously change the trap constant to k_t, and allow the system to relax to equilibrium with k_t. The system contains a number of energy storage modes including the optical trap of the particle, the harmonic binding of the walls, and an integral feedback thermostat; all of which can phase shift the response in dissipation and cause non-monotonic relaxation.

To study this system nonequilibrium molecular dynamics simulations (NEMD) were performed using a 4th order Runge–Kutta algorithm with $k_0 = 2$, $k_t = 8$, and therefore it is expected that the particle will move closer to the trap on average.[8] From the equations of motion for the system:

$$\dot{\mathbf{q}}_i(t) = \mathbf{p}_i(t)/m, \tag{2.10}$$

$$\dot{\mathbf{p}}_i(t) = \mathbf{F}_{I,i}(t) - \delta_{1,i} k_t \mathbf{q}_i(t) - S_w \alpha(t) \mathbf{p}_i(t) + S_w \mathbf{F}_{w,i}(t), \tag{2.11}$$

$$\dot{\alpha}(t) = \frac{3k_B}{Q}(T_k(t) - T), \tag{2.12}$$

and from our definition of the dissipation function, Eq. (2.2), we can derive a dissipation function of the form:

$$\Omega_t(\boldsymbol{\Gamma}(0)) = \frac{\beta(k_0 - k_t)}{2}(\mathbf{q}_1^2(t) - \mathbf{q}_1^2(0)). \tag{2.13}$$

Here i is the particle index, m is the mass of the particles, δ is the Kronecker delta, S_w is the thermostat switch that is 1 for the wall particles and 0 for the fluid and trapped particles, $\mathbf{F}_{I,i}$ is the intermolecular force acting on the particle, $\mathbf{F}_{w,i}$ is the harmonic force constraining the wall particles to their positions, α is the thermostat multiplier that constrains the momenta of the system, Q is the thermal mass of the thermostat, T is the thermostat target temperature, and T_k is the kinetic temperature of the walls, $\beta = 1/k_B T$, $\mathbf{q}_1$ is the position of the trapped particle relative to the harmonic trap centre. The particles interact via the Weeks, Chandler and Andersen (WCA) potential (i.e. a Lennard-Jones potential that is truncated at the potential energy minimum) [34].

From this we can fully describe the behaviour of the dissipation: if the initial trapping constant is greater than the final trapping constant, $k_0 > k_t$, then the function will be positive when the trapped particle finishes further from the trap

---

[8] Simulation parameters: 50 Fluid particles, 22 Wall particles, T = 1, $\rho = 0.3$, 100,000 trajectories.

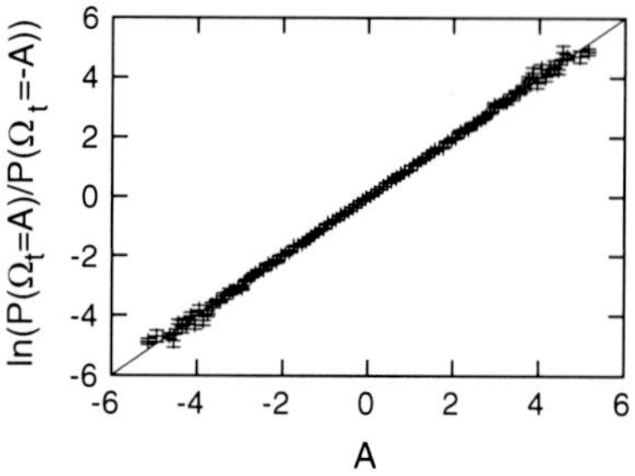

Fig. 2.1 Plot of the logarithm of both sides of Eq. (2.5) at the end of the simulation, with standard error bars, and a line that shows the expected relation. Reprinted with permission from Reid et al. [27]. Copyright 2012 American Institute of Physics

centre than it started and negative when it is closer, and if the initial trapping constant is less than the final trapping constant, $k_t > k_0$, then the function will be positive when the trapped particle finishes closer to the trap centre than it started and negative when it is further away. However, there is always a possibility of observing a negative value for the total dissipation in this system. Finally the instantaneous form of the dissipation function for this system is easily derived to be:

$$\Omega(\boldsymbol{\Gamma}(t)) = \beta(k_0 - k_t)\mathbf{q}_1(t)\dot{\mathbf{q}}_1(t). \tag{2.14}$$

Again it can be seen that its sign is connected to the difference in trapping constants. For example, when the initial trapping constant is greater than the final trapping constant, $k_0 > k_t$, then the function will be positive when the particle is travelling away from the trap centre.

We will first examine the fluctuation theorems for this system to show that it is consistent with their predictions. In Fig. 2.1 we plot the logarithm of both sides of Eq. (2.5) at the end of the simulation, and see extremely good agreement with the straight line of slope one predicted by the fluctuation theorem. To test the dissipation theorem, we can take advantage of the fact that the instantaneous dissipation function is a phase function and can be the argument of the theorem, $B(\boldsymbol{\Gamma}) = \Omega(\boldsymbol{\Gamma})$ in Eq. (2.8). In Fig. 2.2 we see extremely close agreement between the average as directly measured, and the average as calculated from the dissipation theorem. We also observe that the average of the instantaneous dissipation function goes to zero as predicted by the relaxation theorem at long times (Fig. 2.2).

Looking at the extrema, the average of the instantaneous dissipation function against time in Fig. 2.2 is reminiscent of an underdamped harmonic oscillator; this means that the maximum and minimum are reached in the first ‘cycle’, before going to zero at long times. In Fig. 2.3 we plot the average dissipation function against time. It appears that the average of the total dissipation reaches a maxima at equilibrium. This suggests that the non-conformal modes relax more quickly

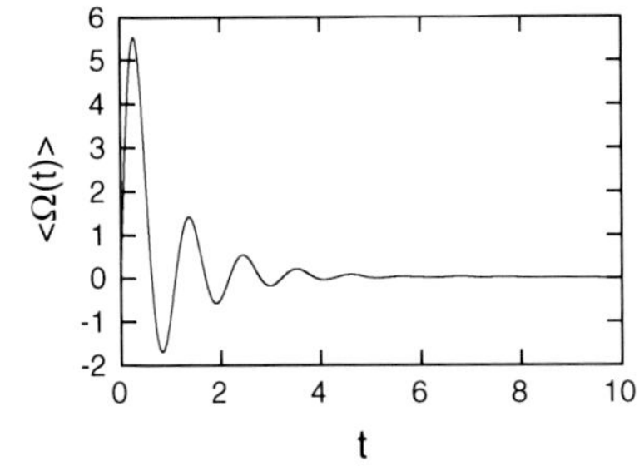

Fig. 2.2 Plot of the average instantaneous dissipation ($\Omega(\boldsymbol{\Gamma}(t))$, –), and the dissipation theorem estimate of the average instantaneous dissipation ($\theta(\Omega, t)$ = RHS of Eq. (2.8), - - -), with time. Note that these results are indistinguishable at this scale. Also note how the function approaches zero at long time as predicted by the relaxation theorem. Reprinted with permission from Reid et al. [27]. Copyright 2012 American Institute of Physics

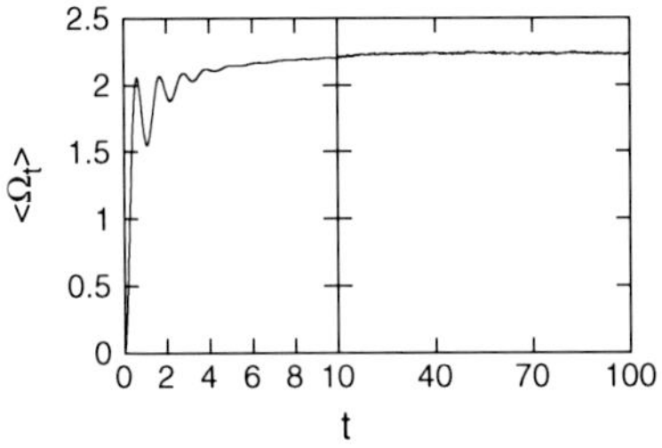

Fig. 2.3 Plot of the average dissipation ($\langle \Omega_t(\boldsymbol{\Gamma}(0)) \rangle$, —) with time over two time scales. Note how the trend of the function is to increase in spite of the oscillations. Reprinted with permission from Reid et al. [27]. Copyright 2012 American Institute of Physics. Note that it has been modified to show two time scales

than the conformal modes, leading to an overall relaxation to the state with maximum Ω_t. This is an interesting result, and bears further study to determine if it is a universal feature of relaxing systems, or merely a feature of this system.

2.4.2 Transient and Steady State System: Poiseuille Flow

Here we examine a simple system, thermostatted Poiseuille flow, that produces a nonequilibrium steady state. A fluid of identical particles in equilibrium is subjected to a field at time $t = 0^+$ that exerts a constant force on the fluid particles in the positive x-direction. At short times, the system will exhibit a transient state as particles begin to accelerate in the appropriate direction, while at long times the particles will have velocities that are mediated by their interactions with the walls of the system, and a steady state will be reached with a maximum streaming velocity in the center of the system, and a minimum streaming velocity proximate to the walls.

We study this system using nonequilibrium molecular dynamics simulations.[9] The equations of motion for the N particles are given by:

$$\dot{\mathbf{q}}_{\mathrm{i}}(t) = \mathbf{p}_i(t)/m, \tag{2.15}$$

$$\dot{\mathbf{p}}_i(t) = \mathbf{F}_{I,i}(t) + S_f F_e \mathbf{i} - S_w \alpha(t)\mathbf{p}_i(t) + S_w \mathbf{F}_{w,i}(t), \tag{2.16}$$

$$\alpha(t) = \frac{\sum S_w(\mathbf{F}_{I,i} + \mathbf{F}_{w,i}).\mathbf{p}_i}{\sum S_w \mathbf{p}_i.\mathbf{p}_i}. \tag{2.17}$$

F_e is the field which acts in the x-direction on each of the N_f fluid particles and S_f is the field switch that is 1 for the fluid particles and 0 for the wall particles. As in the previous example, a WCA interparticle potential is used. In order to better observe non-linear behaviour in this system, a polymer melt consisting of chains of 4 particles bound by a Finite Extensible Nonlinear Elastic Potential (FENE) [35] was studied in addition to studying an atomistic fluid. This adds an additional term into the interparticle force, $\mathbf{F}_{I,i}$, due to interactions of the particle with its neighbours in the chain. We can again derive the dissipation using Eq. (2.2),

$$\Omega(\boldsymbol{\Gamma}(t)) = \beta F_e \sum_{i=1}^{N_f} \frac{p_{xi}(t)}{m}. \tag{2.18}$$

This dissipation function is extensive with the number of fluid particles in the field. Particles give a positive contribution to the dissipation function when they move in the same direction as the field, and a negative dissipation they move in the opposite direction to the field, and the dissipation function can be expressed in terms of a particle current, $J_x = \sum_{i=1}^{N_f} \dot{q}_{xi}/N_f$, as $\Omega(t) = \beta J_x(t) N_f F_e$.

In Fig. 2.4 we plot the logarithm of both sides of Eq. (2.5) at the end of the simulation, and see extremely good agreement with the straight line of slope one predicted by the fluctuation theorem. To test the dissipation theorem for this system, we use the particle current J_x instead of the instantaneous dissipation as they are proportional, but the particle current is a more familiar property. The second law inequality predicts that $\langle \Omega_t \rangle > 0$ and hence $\langle J_{x,t} \rangle > 0$ at all times for a nonequilibrium system, however it says nothing about $\langle \Omega_x(t) \rangle$ or $\langle J_x(t) \rangle$. Indeed it is well known that $\langle \Omega(t) \rangle$ can be negative (see [36]). The results presented in Fig. 2.5 show that $\langle J_x(t) \rangle$, and hence the average instantaneous dissipation function, is positive at all times. It would be interesting to determine if this will always be the case for systems with a constant field. In Fig. 2.5 we also observe the numerical utility of the dissipation theorem: at high field the agreement between the directly calculated current and that obtained from the dissipation theorem is

[9] Simulation parameters: 64 Fluid particles, 64 Wall particles, T = 1, $\rho = 0.8$, 100,000 Trajectories.

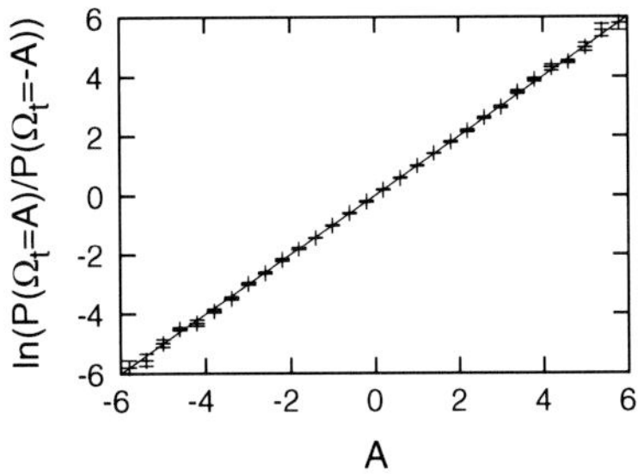

Fig. 2.4 Plot of the logarithm of both sides of Eq. (2.5) at the end of the simulation, with standard error bars, and a line that shows the expected relation. Reprinted from [24], with modifications to formatting. Published under licence in Journal of Physics: Conference Series by IOP Publishing Ltd

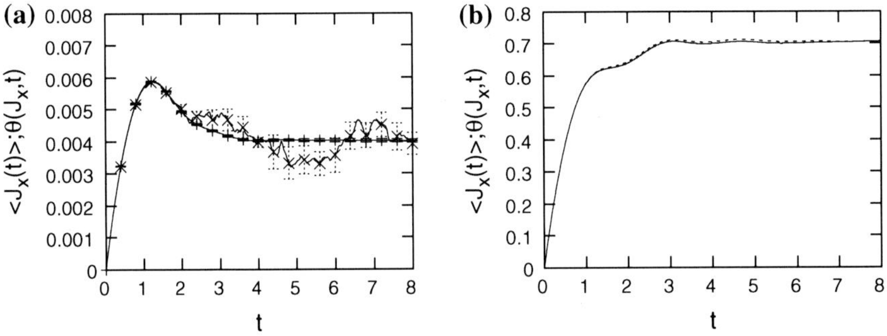

Fig. 2.5 Plot of the average particle current $\langle J_x(t)\rangle$, - - -), and the dissipation theorem estimate of the average particle current ($\theta(J_x, t)$ = RHS of Eq. (2.8), —), with time for a 4-atom polymer system at two field strengths. Note how the dissipation theorem gives better convergence at low field for this system. Reprinted from [24], with modifications to formatting. Published under licence in Journal of Physics: Conference Series by IOP Publishing Ltd. **a** $F_e = 0.01$. **b** $F_e = 1.00$

similar as in the case of the capture experiment; but at low field the dissipation theorem does a much better job of determining the behaviour of the particle current. In Fig. 2.6 we plot the average of the particle current normalised by the field against time for various field strengths. It is notable that the dissipation average peaks before dropping to a steady state for most field values. Therefore it is clear that the instantaneous dissipation is not a maximum in the steady state—the system evolves through an unstable maximum. However for strong fields in the polymer melt we reach a value that is higher than the transient response. This is consistent with the arguments of Williams and Evans [19] where they suggest that under some conditions, (large fields in this case), the state with the maximum dissipation (or rate of "entropy" production) might be the steady state that is observed. From these results we infer that the linear response of the system has a transient peak, but the non-linear response increases with time. In Fig. 2.7, we

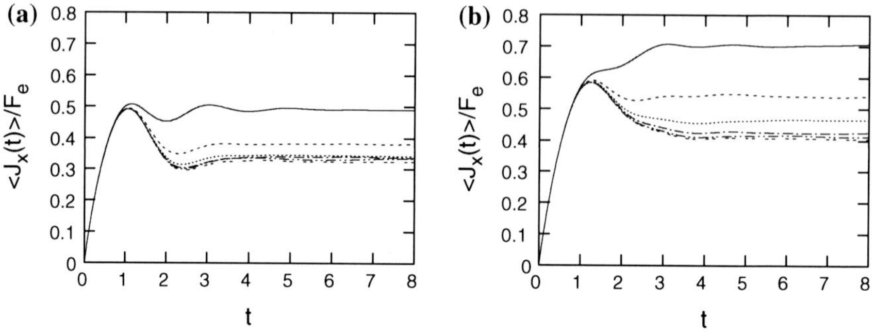

Fig. 2.6 Plot of the average particle current divided by the field ($\langle J_x \rangle / F_e$) for an atomic fluid and a melt of 4-atom long polymers. The field strengths of the lines are, from lowest to highest at $t = 8$, ($F_e = 0.01$, - - - -), ($F_e = 0.1$, — .. –), ($F_e = 0.2$, – . –), ($F_e = 0.3$, ···), ($F_e = 0.5$,- - -), ($F_e = 1$, —). Note that as the field increases, the value of the asymptotic particle current rises relative to the peak of the transient response, and for the polymer melt, the high field asymptotic value exceeds the transient peak. Reprinted from [24]; **a** is a rescale of Fig. 2a from [24], and **b** is a modified version of Fig. 6b. Published under licence in Journal of Physics: Conference Series by IOP Publishing Ltd. **a** Atomic fluid. **b** 4-Atom polymer melt

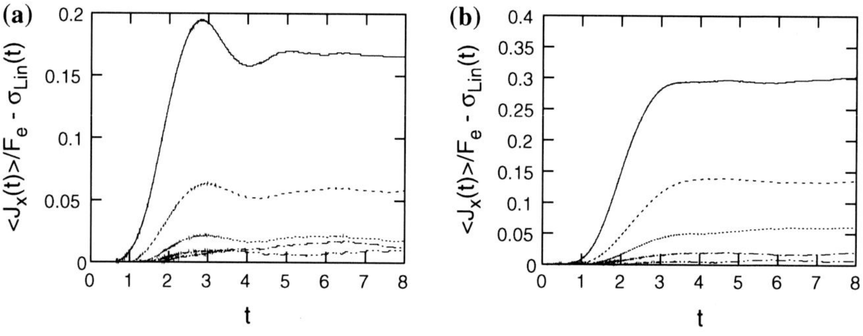

Fig. 2.7 Plot of the non-linear response in the particle current ($\langle J_x \rangle / F_e - \sigma_{Lin}(t)$, $\sigma_{Lin}(t) = \langle J_x(F_e = 0.01) \rangle / 0.01$) for an atomic fluid and a melt of 4-atom long polymers. The field strengths of the lines are, from lowest to highest at $t = 8$, ($F_e = 0.01$, - - - -), ($F_e = 0.1$, —.. –), ($F_e = 0.2$– . –), ($F_e = 0.3$, ···), ($F_e = 0.5$, - - -), ($F_e = 1$,—). Note that the difference in peak to asymptotic particle current is 0.17 for the atomic system and 0.18 for the polymer melt. **a** Atomic fluid. **b** 4-Atom polymer melt

subtract the smallest field response (as a good approximation for the linear response), and see that this is indeed the case for the polymer melt. Since the instantaneous dissipation function settles to a steady value as the steady state is reached, the time integral of the dissipation will be positive at all times (as determined from the second law inequality), but will diverge to infinity with time.

2.5 Discussion and Conclusion

The dissipation function is like entropy production in that it can be used to indicate whether a system's behaviour is in accord with the Second "Law" of Thermodynamics. Positive entropy production or dissipation is taken to be in accord with the Second "Law". Negative dissipation is against the Second "Law" and is seen by the fluctuation theorems as being unlikely. In most text books negative entropy production is seen as not only against the Second "Law" but is expressly impossible.

Close to equilibrium in large systems, where entropy production can be defined, the average dissipation is equal to the entropy production. However unlike entropy production, the dissipation function is well defined arbitrarily near or far from equilibrium and for systems of arbitrary size. Furthermore it obeys a number of exact theorems: the Fluctuation Theorem, the Second "Law" Inequality and the various Relaxation Theorems. Dissipation is also the fundamental quantity required for all linear and nonlinear response theory both for driven and relaxing systems. Dissipation is the central quantity of nonequilibrium statistical mechanics. It also gives for the first time a definition of equilibrium. An equilibrium phase space distribution has the property that the dissipation is identically zero everywhere in the ostensible phase space.

The most profound aspect of these theorems concerning the role played away from equilibrium by dissipation is that the logical status of thermodynamics has changed. The "laws" of thermodynamics were termed "laws" because, apart from the first law, they were deemed unprovable from the laws of mechanics. Indeed the Second "Law" was thought for over 100 years to be in conflict with the time reversible laws of classical and quantum mechanics. All this changed with the first proof of a fluctuation theorem in 1994.

In the present chapter we have presented numerical results that demonstrate many of the various properties of the dissipation, and also provide insight into its temporal extrema and evolution for relaxing systems and driven T-mixing systems that must have a unique steady state.

We found that while the time integrated dissipation appears to approach a maximum for relaxation to equilibrium, in thermostatted, relaxing systems the instantaneous dissipation function has a peak in the transient response, subsequently decaying to zero at equilibrium. For driven thermostatted T-mixing systems the average instantaneous dissipation often passes through one or more unstable maxima before reaching a stable final time independent state—a nonequilibrium steady state. This suggests that a theorem of maximum instantaneous dissipation for steady states does not generally apply.

MaxEP is a theory focused on identifying which steady state a system will select if multiple steady states are possible. In order to provide further information on the behaviour of the dissipation, we can consider a simple system where multiple steady state solutions are known to exist: the heat flow in a one-dimensional lattice [37]. Here two possible steady states exist which depend on the initial conditions: a soliton and a diffusive heat flow. The state with maximum dissipation should stay the same at all field strengths, however at low fields, the soliton does not form whereas at high fields it does. This is not consistent with the state of maximum dissipation being the one that it observed. It would be of interest to study this system in more detail.

Our results might be interpreted as being in contradiction with MaxEP. However this may not be the case. We note that the irreversibility defined in Chap. 3, "A theoretical basis of maximum entropy production", is equivalent to the time integral of the dissipation function given by Eq. (2.2) (see discussion in Sect. 3.5.2). However, the results that we have presented are not necessarily in conflict with current understanding of MaxEP (see Chap. 3) because MaxEP requires the external forcing to be sufficiently large that low entropy states are unstable and because the system is also subject to various constraints. It is clear that the results will depend on the constraints imposed, and therefore the problem can be reformulated as a problem in identification of the appropriate constraints. Ideally, a fundamental maximal principle would justify *a priori* all constraints that are required, but a theoretical basis for this has not yet been established. Indeed this might not be possible. However, as discussed above, it may be possible to determine under what conditions an extremum principle will apply through systematic analysis of systems at a microscopic level.

As MaxEP is useful at very extreme forcing, it will be difficult to study this numerically at a microscopic level for the types of system we have considered here. This is because most systems that generate multiple steady states, such as convection or turbulent flow, are computationally expensive as they require huge numbers of particles and strong fields or both. These instabilities are hydrodynamic rather than molecular in origin and they require large system sizes to increase the Reynolds numbers above various turbulent thresholds. Traditionally, turbulence has rarely been studied using molecular dynamics (for early work see [38, 39]). However modern computing facilities are making this more accessible (see [40, 41]) and it would be interesting to carry out simulations of these systems.

If some objective way can be found to select these constraints, the MaxEP formalism, if applied exactly, could still be impractical. This can be seen for the derivation of the dissipation theorem for driven systems from an extremum principle. The problem in this case was that the number of constraints required in order to obtain the exact answer, was in fact infinite [32, 42].

Whether MaxEP can be established as a useful approximation is an open question. We still need a more objective way of selecting the necessary constraints.

References

1. Evans, D.J., Williams, S.R., Searles, D.J.: J. Chem. Phys. **135**, 194107 (2011)
2. de Groot, S., Mazur, P.: Non-Equilibrium Thermodynamics. Dover Books on Physics. Dover Publications, NY (1984)

3. Evans, D.J., Searles, D.J.: Adv. Phys. **51**, 1529 (2002)
4. Reid, J.C., Sevick, E.M., Evans, D.J.: Europhys. Lett. **72**, 726 (2005)
5. Kurchan, J.: J. Phys. A: Math. Gen. **31**, 3719 (1998)
6. Lebowitz, J.L., Spohn, H.: J. Stat. Phys. **95**, 333 (1999)
7. Monnai, T.: Phys. Rev. E **72**, 027102 (2005)
8. Evans, D.J., Searles, D.J.: Phys. Rev. E **50**, 1645 (1994)
9. Maxwell, J.: Nature **17**, 278 (1878)
10. Broda, E., Gay, L.: Ludwig Boltzmann: man, physicist, philosopher. Ox Bow Press, Woodbridge (1983). Translating: L. Boltzmann, Re-joinder to the Heat Theoretical Considerations of Mr E. Zermelo (1896)
11. Searles, D.J., Evans, D.J.: Aust. J. Chem. **57**, 1119 (2004)
12. Carberry, D.M., Reid, J.C., Wang, G.M., Sevick, E.M., Searles, D.J., Evans, D.J.: Phys. Rev. Lett. **92**, 140601 (2004)
13. Joubaud, S., Garnier, N.B., Ciliberto, S.: J. Stat. Mech: Theory Exp. **2007**, P09018 (2007)
14. Garnier, N., Ciliberto, S.: Phys. Rev. E **71**, 060101 (2005)
15. Jepps, O.G., Ayton, G., Evans, D.J.: Phys. Rev. E **62**, 4757 (2000)
16. Evans, D.J., Searles, D.J., Williams, S.R.: J. Chem. Phys. **128**, 014504 (2008)
17. Evans, D.J., Searles, D.J., Williams, S.R.: J. Stat. Mech: Theory Exp. **2009**, P07029 (2009)
18. Evans, D.J., Searles, D.J., Williams, S.R.: Diffusion fundamentals III. In: Chmelik, C., Kanellopoulos, N., Karger, J., Theodorou, D. (eds.), pp. 367–374. Leipziger Universitats Verlag, Leipzig (2009)
19. Williams, S.R., Evans, D.J.: Phys. Rev. E **78**, 021119 (2008)
20. Evans, D.J., Williams, S.R., Searles, D,J.: Nonlinear dynamics of nanosystems. In: Radons, G., Rumpf, B., Schuster, H. (eds.), pp. 84–86. Wiley-VCH, NJ (2010)
21. Evans, D.J., Morriss, G.P.: Statistical Mechanics of Nonequilibrium Liquids. Cambridge Unviersity Press, Cambridge (2008)
22. Todd, B.D.: Phys. Rev. E **56**, 6723 (1997)
23. Desgranges, C., Delhommelle, J.: Mol. Simul. **35**(5), 405 (2009)
24. Brookes, S.J., Reid, J.C., Evans, D.J., Searles, D.J.: J. Phys: Conf. Ser. **297**, 012017 (2011)
25. Hartkamp, R., Bernardi, S., Todd, B.D.: J. Chem. Phys. **136**, 064105 (2012)
26. Evans, D.J., Searles, D.J., Williams, S.R.: J. Chem. Phys. **132**, 024501 (2010)
27. Reid, J.C., Evans, D.J., Searles, D.J.: J. Chem. Phys. **136**, 021101 (2012)
28. Dewar, R.: J. Phys. A: Math. Gen. **36**, 631 (2003)
29. Dewar, R.C.: Entropy **11**, 931 (2009)
30. Dewar, R.C.: J. Phys. A: Math. Gen. **38**, L371 (2005)
31. Hoover, W.G.: J. Stat. Phys. **42**, 587 (1986)
32. Evans, D.J.: Phys. Rev. A **32**, 2923 (1985)
33. Lorenz, R.: Science **299**, 837 (2003)
34. Weeks, J.D., Chandler, D., Andersen, H.C.: J. Chem. Phys. **54**, 5237 (1971)
35. Kröger, M.: Phys. Rep. **390**, 453 (2004)
36. Williams, S.R., Evans, D.J., Mittag, E.: C.R. Phys. **8**, 620 (2007)
37. Zhang, F., Isbister, D.J., Evans, D.J.: Phys. Rev. E **64**, 021102 (2001)
38. Cui, S.T., Evans, D.J.: Mol. Simul. **9**, 179 (1992)
39. Meiburg, E.: Phys. Fluids **29**, 3107 (1986)
40. Dzwinel, W., Alda, W., Pogoda, M., Yuen, D.: Physica D **137**, 157 (2000)
41. Kadau, K., Germann, T.C., Hadjiconstantinou, N.G., Lomdahl, P.S., Dimonte, G., Holian, B.L., Alder, B.J.: Proc. Natl. Acad. Sci. U.S.A. **101**, 5851 (2004)
42. Hoover, W.G.: Lecture Notes in Physics 258, Molecular Dynamics. Springer, London (1986)

Chapter 3
A Theoretical Basis for Maximum Entropy Production

Roderick C. Dewar and Amos Maritan

Abstract Maximum entropy production (MaxEP) is a conjectured selection criterion for the stationary states of non-equilibrium systems. In the absence of a firm theoretical basis, MaxEP has largely been applied in an *ad hoc* manner. Consequently its apparent successes remain something of a curiosity while the interpretation of its apparent failures is fraught with ambiguity. Here we show how Jaynes' maximum entropy (MaxEnt) formulation of statistical mechanics provides a theoretical basis for MaxEP which answers two outstanding questions that have so far hampered its wider application: What do the apparent successes and failures of MaxEP actually mean physically? And what is the appropriate entropy production that is maximized in any given problem? As illustrative examples, we show how MaxEnt underpins previous applications of MaxEP to planetary climates and fluid turbulence. We also discuss the relationship of MaxEP to the fluctuation theorem and Ziegler's maximum dissipation principle.

3.1 MaxEP: What Does It Mean and How Do We Use It?

The conjecture of maximum entropy production (MaxEP) as a selection criterion for non-equilibrium stationary states has shown some promising successes in studies of, for example, planetary climates [1, 2], fluid turbulence [3, 4], crystal growth morphology [5, 6], biological evolution and adaptation [7–10], and earthquake dynamics [11]. The practical significance of MaxEP is that it appears to

R. C. Dewar (✉)
Research School of Biology, The Australian National University,
Canberra, ACT 0200, Australia
e-mail: roderick.dewar@anu.edu.au

A. Maritan
Department of Physics G. Galilei, University of Padua, Via Marzolo 8, 35131 Padua, Italy

R. C. Dewar et al. (eds.), *Beyond the Second Law*,
Understanding Complex Systems, DOI: 10.1007/978-3-642-40154-1_3,

be able to make realistic predictions of non-equilibrium stationary states on the basis of a limited set of dynamical constraints, without having to solve the underlying equations of motion in their full complexity. Potentially MaxEP is a non-equilibrium selection criterion of some generality [12]. However, without a theoretical basis to underpin MaxEP, its wider application has been hampered by two unresolved questions, one conceptual, the other practical: What is the physical interpretation of the apparent successes of MaxEP? And what is the entropy production (EP) function to be maximized in any given problem? In the absence of answers to these two questions, to date MaxEP has been applied in a largely *ad hoc* manner and its successes remain something of an unexplained curiosity.

For example, the early successes of MaxEP using 1-D zonally-averaged energy balance models of Earth's climate [1, 2] were obtained by maximizing the material EP associated with meridional heat transport in the atmosphere and oceans, even though radiative EP is numerically by far the dominant contribution to global EP [13]. More recent studies suggest that the appropriate choice of EP function depends on the type of climate model used (e.g. energy balance vs. resolved fluid dynamics). For example, in a study of FAMOUS, a 3-D atmosphere–ocean general circulation model (GCM) [14], no maximum in the material EP was found. Instead, realistic climate features were obtained at a maximum in the EP associated with kinetic energy dissipation. Thus no universal EP function seems to apply to all climate models. However, to date only an a posteriori justification for the choice of EP function has seemed possible.

A similar arbitrariness pervades a close relative of MaxEP, namely, the upper bound approach to fluid turbulence [3, 4, 15], in which mean flow properties are derived by maximizing various dissipation-related functionals of the flow. Kerswell [15] analysed a family of dissipation functionals of the form $f = \mathcal{D}(\mathcal{D}_v/\mathcal{D}_m)^n$ and $f = \mathcal{D}_m(\mathcal{D}_v/\mathcal{D}_m)^n$ $(n \geq 0)$ involving the total dissipation $\mathcal{D}$, dissipation in the mean flow $\mathcal{D}_m$, and dissipation in the fluctuating flow $\mathcal{D}_v$. Within this family Kerswell was unable to identify a universal dissipation functional that applies to all flow problems, and he concluded somewhat pessimistically [15]:

> The challenge therefore remains to find a functional whose optimization over a tractably reduced set of dynamical constraints leads to the emergence of realistic optimal velocity fields. Unfortunately, it remains unclear how to construct such a functional beyond intelligent guessing.

In the absence of a firm theoretical basis for MaxEP, an equally important uncertainty afflicts the interpretation of apparent failures of MaxEP. In an analysis of simple dynamical food web models, Meysman and Bruers [16] found that transitions between steady states under an increase in the external driving force (chemical potential gradient in resources) did not always lead to a higher EP, contrary to what some verbal statements of MaxEP might suggest. A similar result was found for transitions between the linear and non-linear branches in a simple phenomenological model of heat flow in a plasma/fluid system ([17]; see also Yoshida and Kawazura, Chap. 15). These results might appear to signal the limits of MaxEP as a general physical principle, and yet without a theoretical basis it remains unclear if and how MaxEP relates to the behaviour of such simple dynamical models with only a few degrees of freedom. There is also the question of whether MaxEP is a physical principle at all [18]. The interpretation of apparent failures of MaxEP is thus fraught with ambiguity.

What is needed, but is currently lacking, is a theoretical foundation for MaxEP in the form of some underlying principle that underpins and guides its practical application. In particular, one would like such a foundation to tell us which EP function to maximize, and whether this function is universal or specific to each problem. It might also clarify where it would be inappropriate to apply MaxEP. Only then might we begin to understand what the apparent successes and failures of MaxEP are actually telling us (if anything) about the real world.

In this chapter we show how Jaynes' maximum entropy (MaxEnt) formulation of statistical mechanics [19] provides the theoretical basis for a generic MaxEP principle. The origins of MaxEnt go back to Gibbs' formulation of equilibrium statistical mechanics [20] which, following the work of Shannon [21, 22], was developed further in the context of information theory by Jaynes [19, 23] and others [24], and extended to non-equilibrium systems [25]. The MaxEnt basis for MaxEP proposed here answers the conceptual and practical questions that have plagued previous applications of MaxEP as a non-equilibrium selection criterion: what does it mean physically, and what is the EP function that is maximized in any given problem?

The chapter is organized as follows. In Sect. 3.2 we discuss the interpretation of MaxEnt and its role in statistical mechanics. This provides the conceptual foundation for Sect. 3.3, in which we show that MaxEnt implies a generic MaxEP principle, and we derive a general expression for the EP function that is maximized. Section 3.3 also briefly outlines how the present theory overcomes the problems of earlier attempts to derive MaxEP from MaxEnt [18, 26–31]. Section 3.4 illustrates the general theory with two examples: planetary climates [1, 2] and fluid turbulence [4, 15]. Section 3.5 links MaxEP to two other non-equilibrium results involving EP: the fluctuation theorem ([32–34]; see also Reid et al., Chap. 2) and Ziegler's maximum dissipation principle ([35]; see also Houlsby, Chap. 4; Seleznev and Martyushev, Chap. 5). We end with an outlook on some challenges for the future (Sect. 3.6).

3.2 The Role of MaxEnt in Statistical Mechanics: The Messenger, Not the Message

In Sect. 3.3 we will derive a generic MaxEP principle from MaxEnt. In order to understand what the apparent successes and failures of MaxEP actually mean physically, we therefore need to understand MaxEnt and the role it plays in statistical mechanics. Here we discuss MaxEnt from two alternative viewpoints: as an inference algorithm in the context of information theory, and as a statistical selection principle in the context of experimental frequencies. However, the key role of MaxEnt in statistical mechanics—translating given dynamical constraints into macroscopic predictions—does not depend on which viewpoint one adopts. The discussion here is based on [18].

3.2.1 MaxEnt as an Inference Algorithm

The practical goal of statistical mechanics (equilibrium or non-equilibrium) is to make predictions of macroscopic behaviour on the basis of a relatively small number of dynamical assumptions and constraints C, without having to solve the underlying equations of motion in their full complexity. That such a goal is attainable is suggested by the empirical observation that, for systems with sufficient degrees of freedom, macroscopic behaviour can be accurately reproduced experimentally through the control of a relatively small number of macroscopic parameters. This implies that those microscopic details that are not under experimental control must be largely irrelevant to the experimental result, and behave as 'noise'. The interpretation of this microscopic irrelevance is that the overwhelming majority of possible microscopic histories which the system could follow under the conditions of the experiment look the same on macroscopic scales [19]—so it does not matter which of these microscopic histories is the one actually followed. The constraints C represent that subset of the full dynamics which is relevant for predicting the macroscopic behaviour under investigation. Two practical challenges then are how to identify the relevant dynamics C in any given problem, and how to make macroscopic predictions from them.

This is where MaxEnt comes in. In its information theory context, MaxEnt is a general inference method for updating a prior distribution q_i to a posterior distribution p_i in the light of new information [23–25]. MaxEnt consists of maximizing the relative entropy (negative Kullback–Leibler divergence)

$$H(p\|q) = -\sum_i p_i \ln\frac{p_i}{q_i} \tag{3.1}$$

with respect to p_i, subject to the new information in the form of various moment constraints on p_i (and normalisation of p_i). In information terms, $-H(p\|q)$ may be interpreted as the information gained by using p_i instead of q_i. MaxEnt is the unique inference algorithm which satisfies a set of four axioms (uniqueness, invariance under co-ordinate transformations, system independence, subset independence) based on the fundamental requirement that the resulting p_i should not depend on how the new information C is taken into account [24].

In its application to statistical mechanics, the new information C represents the dynamical constraints from which macroscopic predictions are to be made; the prior q_i then describes what is known (if anything) about the system in the absence of constraints C. We denote the MaxEnt probability distribution by p_i*.

By maximizing $H(p\|q)$ (i.e. minimizing the information gain) with respect to p_i subject to C, any other distribution $p_i{}' \neq p_i{}^*$ compatible with C necessarily encodes more information than just C. In this way MaxEnt ensures that p_i*, and the macroscopic predictions obtained as expectation values over p_i*, reflect the constraints C *and no other assumptions*. This is the key feature of MaxEnt that underpins its role in statistical mechanics [18].

Specifically, we make an initial guess for the relevant dynamics C, then use MaxEnt to make macroscopic predictions from that guess (via expectation values over p_i*). Disagreement between predictions and experiment informs a new guess and so on until, by trial and error, acceptable agreement is reached. MaxEnt does not tell us a priori which dynamical constraints to apply, and therefore it makes no claim as to the success of its predictions. Instead, its role is simply to ensure that the macroscopic predictions faithfully reflect the assumed constraints C and no other assumptions. Only then can we be sure in the end that we have correctly identified the relevant dynamics.

From this viewpoint, falsification of MaxEnt is meaningless because MaxEnt is not a physical principle; it is an inference algorithm that passively translates the information from physical constraints into macroscopic predictions. It is the assumed set[1] of relevant constraints C that is being tested, not MaxEnt. MaxEnt is the messenger, not the message [18].

3.2.2 MaxEnt as a Statistical Selection Principle

An alternative interpretation of the MaxEnt distribution p_i* as the frequency of experimental observations of i can be given, by considering a sequence of N independent experimental observations (or trials) of the system [36]. The prior probability that in such a sequence the outcome i is observed n_i times is given by the multinomial distribution $Q(\{n_i\}|\{q_i\},N) = N!\prod_i q_i^{n_i}/n_i!$ where q_i is the prior probability of outcome i and $\sum_i n_i = N$. We then have the asymptotic result

$$\lim_{N\to\infty}\frac{1}{N}\ln Q(\{n_i\}|\{q_i\},N) = H(p\|q) \tag{3.2}$$

where $p_i = n_i/N$ is the relative frequency of outcome i. By maximising both sides of (3.2) with respect to p_i subject to moment constraints of the form $\sum_i p_i a_{ik} = A_k\ (k = 1\ldots m)$ and $\sum_i p_i = 1$, we see that p_i*, which in the context of information theory represents the state of knowledge about the outcome of a single trial based on the information $\{A_k\}$, coincides numerically with the most probable frequency distribution n_i*/N obtained in an infinitely long sequence of independent trials.

[1] The constraints themselves may be perfectly accurate (e.g. global energy balance). In that case it is the completeness of the chosen set that is being tested.

The validity of this correspondence rests on the obvious but crucial requirement that *the same constraints* $\{A_k\}$ *also apply during the experiment.*

The mathematical correspondence described by Eq. (3.2) often appears in statistical mechanics textbooks as a combinatorial justification for MaxEnt within the more restricted context of a system of N independent, distinguishable particles or entities. Then p_i represents the fraction of particles in single-particle state i. However, this does not mean that MaxEnt is only valid for large ($N \to \infty$) systems of independent, distinguishable particles (cf. [37, 38]). When N refers instead to the number of independent (and distinguishable) observations, Eq. (3.2) is valid regardless of the number and distinguishability of the entities making up the system, because i then refers to the state of an arbitrary physical system. In particular, Eq. (3.2) applies equally to equilibrium and non-equilibrium systems; in the latter case i might represent, for example, the microscopic trajectory of a system over some time interval (e.g. [26]).

Equation (3.2) thus provides a combinatorial justification of MaxEnt for arbitrary physical systems, complementary to its justification from information theory [24]. MaxEnt can then be viewed also as a statistical selection principle, but only in the conditional sense that it predicts the most likely frequency distribution of experimental outcomes *provided we have correctly identified the relevant dynamical constraints that apply during the experiment.* But here again MaxEnt does not tell us a priori what those constraints are.

3.3 MaxEP from MaxEnt

In this section we show how MaxEnt provides a theoretical basis for MaxEP as a selection criterion for the stationary states of non-equilibrium systems subject to a given restricted set of dynamical constraints.

As discussed in the previous section, the constraints represent the assumed relevant dynamics (i.e. a subset of the full dynamics). We therefore begin by specifying a restricted set of stationarity conditions (e.g. global energy balance). For a weakly-driven (near-equilibrium) system, typically there is only one non-equilibrium state compatible with these stationarity conditions. We call this the *basal state*. One example would be the laminar flow solution in shear turbulence (see Sect. 3.4). Another example would be heat transport by conduction in a plane horizontal layer of fluid heated from below (Rayleigh-Bénard cell). In that case, there is simply no room for MaxEP (or any other selection criterion) to operate. Therefore we introduce an additional condition that the basal state is dynamically unstable, describing a strongly-forced (far-from-equilibrium) system. Then there are typically many non-equilibrium states compatible with the imposed stationarity conditions.

Note that, in principle, if we were to integrate the underlying equations of motion (e.g. the Navier-Stokes equation) in their full complexity, generically only one stationary state would be predicted and we would not need a selection criterion at all. Here, however, we are trying to bypass that complexity by imposing only a restricted number of stationarity conditions, which then allows a multiplicity of solutions.

This then is the situation where we seek a selection criterion among different non-equilibrium states. In the example of shear turbulence to be discussed in Sect. 3.4, for instance, this situation corresponds to Reynolds numbers larger than the critical value for the onset of turbulence, when the basal state of laminar flow is dynamically unstable. In the Rayleigh-Bénard cell, it corresponds to Rayleigh numbers larger than the critical value for the onset of convection, when the conductive state is dynamically unstable. We then want to select one from among the many flow solutions allowed when we apply only a restricted set of stationarity conditions rather than the full dynamics.

To apply the dynamical instability condition, we introduce an information-theoretical measure of the distance from equilibrium, or *irreversibility* I, defined in terms of the relative probabilities of forward and reverse fluxes. Dynamical instability is then introduced via the strict inequality constraint $I > I_{\min}$, where $I_{\min}$ is the minimal irreversibility of the basal state. We then show from MaxEnt that I adopts its maximum possible value under the stationarity constraints. With I then interpreted as thermodynamic EP, this result yields a MaxEP selection criterion. This derivation of MaxEP shows how the physical nature of the EP function that is maximised depends on the nature of the applied constraints.

This section ends with a brief outline of the key differences between this and previous attempts to derive MaxEP from MaxEnt.

3.3.1 Setting up the Constraints

Non-equilibrium stationary states are characterised by the presence of macroscopic fluxes (e.g. of energy, mass or momentum) within the system, and between the system and its environment. Here we denote the instantaneous value of these fluxes by the vector $\boldsymbol{f}$, which may be finite (e.g. heat fluxes in a discrete box model of Earth's climate, [1, 2], to be discussed in Sect. 3.4.1) or infinite dimensional (e.g. the 3-D velocity field in fluid turbulence, [3, 4, 15], as in Sect. 3.4.2). The flux vector $\boldsymbol{f}$ (e.g. heat flux, see Sect. 3.4.1) may be related to some local density ρ (e.g. internal energy density) through the continuity equation $\partial\rho/\partial t = -\nabla\cdot \boldsymbol{f} + h$ (or its discrete equivalent) where h is a local source (e.g. radiative heating rate per unit volume); alternatively, the components of $\boldsymbol{f}$ might themselves be identified with local densities (e.g. fluid momentum density, whose local rate of change obeys the Navier-Stokes equation, see Sect. 3.4.2). The macroscopic state of the system is then described by $\boldsymbol{f}$ (and/or ρ), and stationary states are defined in a statistical sense by the condition that the expectation value of $\partial\rho/\partial t$ over the probability density function (p.d.f.) $p(\rho)$ vanishes; non-equilibrium states are characterised by a non-vanishing expectation value of $\boldsymbol{f}$ over the p.d.f. $p(\boldsymbol{f})$.

Here we use MaxEnt to construct $p(f)$; that is, we maximise the relative entropy

$$H = -\int p(f) \ln \frac{p(f)}{q(f)} \mathrm{d}f \tag{3.3}$$

with respect to $p(f)$, subject to given dynamical constraints C, where $q(f)$ is a prior p.d.f. By Gibbs' inequality, $H \leq 0$ with equality if and only if $p(f) = q(f)$. Since C represents non-equilibrium forcing conditions, it is natural to choose a prior p.d.f. with the symmetry property $q(-f) = q(f)$; this describes the equilibrium (zero-flux) state $\boldsymbol{F} = \int q(f) f \mathrm{d}f = 0$. Although in general the detailed form of the prior depends on the underlying space of microscopic outcomes [23], here the symmetry property of $q(f)$ is all we will need for the following analysis, which is therefore quite general. The constraints C (i.e. the assumed relevant dynamics) typically involve a restricted number of stationarity conditions. In order to keep the formalism general, these constraints are written in the generic form

$$\int p(f)\, \varphi_m(f)\, \mathrm{d}f = 0 \tag{3.4}$$

where $\varphi_m(f)$ are given function(al)s of the fluxes, and the constraint label m may be discrete or continuous. For example, (3.4) might describe steady-state global energy, mass or momentum balance (as illustrated later in Sect. 3.4). The right-hand side of (3.4) can be set to zero without loss of generality by suitably defining the function $\varphi_m(f)$. We also have the normalization constraint

$$\int p(f)\, \mathrm{d}f = 1 \tag{3.5}$$

As discussed above, in addition to (3.4) and (3.5), we introduce the condition that the non-equilibrium forcing is sufficiently strong that the basal stationary state is dynamically unstable. To enforce this instability condition we introduce the *irreversibility* defined by the Kullback–Leibler (KL) divergence of $p(f)$ and $p(-f)$, i.e. $I = \int p(f) \ln\{p(f)/p(-f)\} \mathrm{d}f$. By Gibbs' inequality, $I \geq 0$ with equality if and only if $p(f) = p(-f)$, so that $I = 0$ corresponds to the equilibrium state $\boldsymbol{F} = \boldsymbol{0}$. The irreversibility I is thus a natural information-theoretic measure of the distance from equilibrium, or time-reversal symmetry breaking, since it measures the extent to which $p(f)$ differs from $p(-f)$.

The basal state is the non-equilibrium state characterised by minimal irreversibility, $I = I_{\min}(C) > 0$, the value of which depends on the stationarity conditions C of Eq. (3.4); dynamic instability of the basal state is then enforced by imposing the strict inequality $I > I_{\min}(C)$. Given $I > I_{\min}(C)$, the question naturally arises as to how large I can be. We assume there is some upper bound $I_{\max}(C)$ also determined by the stationarity conditions (3.4), so that $I \leq I_{\max}(C)$. Physically this is reasonable; under a finite non-equilibrium driving force the system cannot be an infinite distance from equilibrium. Mathematically, this requires that for any instantaneous flux state f that has non-zero probability, the reverse flux state $-f$ also has non-zero probability (i.e. $p(f) > 0 \Rightarrow p(-f) > 0$). In the following, it proves convenient to apply MaxEnt subject to $I \leq I_0$, where I_0 represents a trial estimate of $I_{max}(C)$, and then to take the limit $I_0 \to I_{max}(C)$. We therefore impose the lower- and upper-bound inequality constraints on I as follows:

$$I_{min}(C) < I \equiv \int p(f) \ln \frac{p(f)}{p(-f)} df \leq I_0 \tag{3.6}$$

Finally we introduce a trial mean flux $\boldsymbol{F}$ (which is subsequently relaxed) via the auxiliary constraint

$$\int p(f) f df = \boldsymbol{F} \tag{3.7}$$

It might seem artificial to introduce (3.7) as an additional constraint when $\boldsymbol{F}$ is ultimately determined (selected) by the constraints C, rather than being an independent constraint in its own right. The motivation for introducing $\boldsymbol{F}$ in (3.7) is that $\boldsymbol{F}$ then represents a trial estimate of the actual fluxes $\boldsymbol{F}(C)$ selected under C. Introducing $\boldsymbol{F}$ in this way allows us to establish an extremal principle whereby $\boldsymbol{F}(C)$ is determined by varying the trial solution $\boldsymbol{F}$. The approach here is analogous to the way in which equilibrium variational principles (e.g. minimum free energy) can be derived from MaxEnt by enlarging the set of fixed macroscopic variables X to include one or more free unconstrained variables Y, then maximizing $S = H_{max}(X, Y)$ with respect to Y with X held fixed.

From its definition we intuitively expect the irreversibility I to be closely related to thermodynamic EP. However, we emphasise that the link between I and thermodynamic EP as a function of generalized forces and fluxes emerges only after we have applied MaxEnt under the conditions (3.4)–(3.7), as shown in Sect. 3.3.3. This is analogous to the MaxEnt formalism of equilibrium statistical mechanics, in which only after it is maximized subject to equilibrium constraints is the Shannon entropy identified with thermodynamic entropy [23].

3.3.2 The MaxEnt Solution

The MaxEnt solution for $p(f)$ under constraints (3.4)–(3.7) is given implicitly by

$$p(f)^* = \frac{q(f)}{Z} \exp\left\{\boldsymbol{\lambda} \cdot f + \boldsymbol{\alpha} \cdot \boldsymbol{\varphi}(f) - \mu\left(d(f) - e^{-d(f)}\right)\right\} \tag{3.8}$$

where $d(f) = \ln\{p(f)/p(-f)\}, \boldsymbol{\varphi}(f)$ denotes the vector with components $\varphi_m(f), Z = Z(\boldsymbol{\lambda}, \boldsymbol{\alpha}, \mu)$ is a normalisation factor (partition function) that enforces (3.5), and $\boldsymbol{\lambda}, \boldsymbol{\alpha}$ and μ are Lagrange multipliers for (3.7), (3.4) and the upper-bound inequality constraint (3.6) respectively. The maximised relative entropy is

$$S(\boldsymbol{F}, I_0, C) \equiv H_{max} = \ln Z(\boldsymbol{\lambda}, \boldsymbol{\alpha}, \mu) - \boldsymbol{\lambda} \cdot \boldsymbol{F} + \mu(I_0 - 1) \tag{3.9}$$

where we have used (3.4) and the identity $\int p(f)\exp\{-d(f)\}df = 1$. Each of the Lagrange multipliers $\boldsymbol{\lambda}, \boldsymbol{\alpha}$ and μ are functions of the trial values $\boldsymbol{F}$ and I_0; they also depend on the functional form of $\boldsymbol{\varphi}(f)$ as indicated by the argument C on the l.h.s. of (3.9). Under small variations in $\boldsymbol{F}$ and I_0 that satisfy (3.4), the change in S is

$$\delta S = -\boldsymbol{\lambda} \cdot \delta F + \mu \delta I_0 \tag{3.10}$$

From the Karush–Kuhn–Tucker conditions for optimization under inequality constraints [39], $\mu = 0$ when the upper bound inequality constraint in (3.6) is inactive ($I < I_0$), whereas $\mu > 0$ when the inequality constraint is active ($I = I_0$). The sign of μ in the latter case can be understood from the fact that $\mu = \partial S/\partial I_0$ and that as I_0 increases, the feasibility space under which H is maximized becomes larger so $S = H_{max}$ cannot decrease. We also note that the case $\mu > 0$ requires $I_0 < I_{max}(C)$, i.e. the trial value I_0 is an underestimate[2] of the actual upper bound $I_{max}(C)$.

The next step is to maximize $S(\boldsymbol{F}, I_0, C)$ with respect to the trial flux solution $\boldsymbol{F}$ with I_0 held fixed. The two cases $I < I_0$ and $I = I_0$ must be treated separately. For $\mu = 0$ ($I < I_0$) the upper-bound constraint on I is inactive, and the solution that maximizes S is the basal stationary state $I = I_{min}(C)$ because this is the state closest to the equilibrium state $S = I = 0$ where S has a global maximum and I has a global minimum. However this solution is excluded by the dynamic instability condition $I > I_{min}(C)$. Therefore the only case consistent with the stationarity conditions C and $I > I_{min}(C)$ is the case $\mu > 0$ ($I = I_0$).

Confining our attention now to the case $I = I_0$, the final step is to take the limit $I_0 \to I_{max}(C)^-$ where the superscript reminds us that the limit is taken from below. Since $\mu > 0$ for $I_0 \le I_{max}(C)$, while $\mu = 0$ for $I_0 > I_{max}(C)$, continuity of the function $\mu(I_0)$ implies

$$\lim_{I_0 \to I_{max}(C)^-} \mu(I_0) = 0^+ \tag{3.11}$$

In summary, in the absence of the dynamic instability condition $I > I_{min}(C)$, MaxEnt predicts the basal state $I = I_{min}(C)$, i.e. minimal irreversibility. When the basal state is excluded by the dynamical instability condition $I > I_{min}(C)$, MaxEnt predicts a p.d.f. $p(f)$ for which $I = I_{max}(C)^-$, i.e. maximal irreversibility. The latter solution is characterised by $\mu = 0^+$, which from (3.8) then yields the final result:

$$p(f)^* = \frac{q(f)}{Z} \exp\{\boldsymbol{\lambda} \cdot f + \boldsymbol{\alpha} \cdot \boldsymbol{\varphi}(f)\} \tag{3.12}$$

There are no MaxEnt solutions intermediate between $I = I_{min}(C)$ and $I = I_{max}(C)$.

---

[2] If instead we had $I_0 > I_{max}(C)$ then, since $I \le I_{max}(C)$ by definition, we would have $I < I_0$ and so $\mu = 0$.

3.3.3 Identifying the Entropy Production Functional

From (3.12) and $q(-f) = q(f)$, the irreversibility $I = \int p(f)\ln\{p(f)/p(-f)\}\mathrm{d}f$ is given as a functional of $\boldsymbol{F}$ by

$$I(\boldsymbol{F}) = 2\boldsymbol{\lambda}(\boldsymbol{F})\cdot\boldsymbol{F} + 2\boldsymbol{\alpha}(\boldsymbol{F})\cdot\boldsymbol{\Phi}^{\mathrm{A}}(\boldsymbol{F}) \tag{3.13}$$

where

$$\boldsymbol{\Phi}^{\mathrm{A}}(\boldsymbol{F}) \equiv \frac{1}{2}\int p(f)\{\boldsymbol{\varphi}(f) - \boldsymbol{\varphi}(-f)\}\mathrm{d}f \tag{3.14}$$

is the expectation value of the anti-symmetric part of $\boldsymbol{\varphi}(f)$. When $\varphi_m(-f) = \pm\varphi_m(f)$ for all m, i.e. when each constraint function $\varphi_m(f)$ is either a symmetric (S) or an anti-symmetric (AS) function of f, Eq. (3.4) implies $\boldsymbol{\Phi}^{\mathrm{A}} = \mathbf{0}$ so that

$$I(\boldsymbol{F}) = 2\boldsymbol{\lambda}(\boldsymbol{F})\cdot\boldsymbol{F} \quad (\mathrm{S/AS}). \tag{3.15}$$

Alternatively, if not all of the $\varphi_m(f)$ are symmetric or anti-symmetric with respect to f (i.e. no pure symmetry, NS), then $\boldsymbol{\Phi}^{\mathrm{A}} \neq \mathbf{0}$ and in principle the more general form (3.13) applies. However in this case it can be shown that the MaxEnt solution implies[3] $\boldsymbol{\lambda} = \mathbf{0}$ so that

$$I(\boldsymbol{F}) = 2\boldsymbol{\alpha}(\boldsymbol{F})\cdot\boldsymbol{\Phi}^{\mathrm{A}}(\boldsymbol{F}) \quad (\mathrm{NS}). \tag{3.16}$$

In summary, for non-equilibrium stationary systems under given constraints C for which the basal state $I = I_{\min}(C)$ is dynamically unstable, MaxEnt is equivalent to an extremal selection criterion of MaxEP:

$$\boldsymbol{F}(C) = \arg\max_{\boldsymbol{F}|C} I(\boldsymbol{F}) \tag{3.17}$$

i.e. $\boldsymbol{F}(C)$ is the value of $\boldsymbol{F}$ that maximizes $I(\boldsymbol{F})$ under the constraints C, where the EP functional $I(\boldsymbol{F})$ is identified as

$$I(\boldsymbol{F}) = \begin{Bmatrix} 2\boldsymbol{\lambda}(\boldsymbol{F})\cdot\boldsymbol{F} & (\mathrm{S/AS}) \\ 2\boldsymbol{\alpha}(\boldsymbol{F})\cdot\boldsymbol{\Phi}^{\mathrm{A}}(\boldsymbol{F}) & (\mathrm{NS}) \end{Bmatrix}. \tag{3.18}$$

Equation (3.18) shows there is no universal EP functional that applies to all problems. The functional $I(\boldsymbol{F})$ to be maximized depends on the specific form and symmetry properties of the constraint functionals $\varphi_m(f)$ describing the assumed relevant dynamics C. This conclusion is analogous to what one finds in the MaxEnt formalism of equilibrium statistical mechanics, where the expression for

---

[3] With $\mu = 0^+$ (Eq. 3.11), $S = S(\boldsymbol{\lambda}, \boldsymbol{\alpha})$ is a functional of $\boldsymbol{\lambda}$ and $\boldsymbol{\alpha}$ alone. We have $S(\boldsymbol{\lambda}, \boldsymbol{\alpha}) \leq S(\boldsymbol{\lambda} = \mathbf{0}, \boldsymbol{\alpha})$ because removing a constraint—in this case, the auxiliary constraint (3.7)—cannot lead to a decrease in S. Therefore the MaxEnt solution for $p(f)$ in the case $\boldsymbol{\Phi}^{\mathrm{A}} \neq \mathbf{0}$ has $\boldsymbol{\lambda} = \mathbf{0}$. Note that we cannot set $\boldsymbol{\lambda} = \mathbf{0}$ in the case $\boldsymbol{\Phi}^{\mathrm{A}} = \mathbf{0}$ (i.e. the case S/AS) because (3.13) would then give $I(\boldsymbol{F}) = 0$ which contravenes $I(\boldsymbol{F}) > I_{\min}(C)$; in this case $I(\boldsymbol{F})$ is given by (3.15).

the thermodynamic entropy (interpreted as the maximized Shannon entropy) in terms of extensive and intensive variables depends on the nature of the equilibrium constraints, i.e. there is no universal expression for thermodynamic entropy independent of the constraints (e.g. [23], Eq. 11.59 therein).

Note here that we have *defined* thermodynamic EP as the function $I(\boldsymbol{F})$ which emerges naturally from MaxEnt subject to given non-equilibrium constraints. We have therefore avoided the semantic trap of starting with some pre-conceived notion of thermodynamic EP (e.g. as the time derivative of some physically defined entropy) and then tried to show that this is a maximum. That our definition of thermodynamic EP turns out to be consistent with previously-defined dissipation functions is illustrated by the examples discussed in Sect. 3.4.

3.3.4 Comparison with Previous Attempts to Derive MaxEP from MaxEnt

In [26, 27], two different derivations of MaxEP from MaxEnt were attempted, based on maximizing the Shannon entropy of the probability distribution of microscopic trajectories, subject to physical constraints. Subsequently, some technical limitations of these derivations were noted [28, 29].

The first derivation [26] introduced an *ad hoc* assumption that the Shannon path entropy is an increasing function of EP, so that MaxEnt would imply MaxEP. However Bruers [28] presented a counter-example to that assumption, implying MinEP not MaxEP. A key constraint missing from [26] but included here is the condition $I > I_{\min}(C)$ that the basal state is dynamically unstable. Without this condition MaxEnt indeed predicts the basal state $I = I_{\min}(C)$ (i.e. MinEP), as noted in [28]. With the condition $I > I_{\min}(C)$ MaxEnt leads to $I = I_{\max}(C)$ (i.e. MaxEP).

In the second derivation [27], a Gaussian approximation for $p(f)$ was used to derive an orthogonality condition for EP (first proposed by Ziegler [35]), from which a version of MaxEP was then derived. But as noted in [28, 29], the specific argument used in [27] is only valid close to equilibrium. The present theory overcomes this limitation; the constraints (3.4)–(3.7) and the Karush–Kuhn–Tucker conditions leading from MaxEnt to MaxEP apply regardless of how far the system is driven from equilibrium.

Niven [30] proposed a derivation of MaxEP from MaxEnt subject to prescribed fluxes, analogous to the auxiliary constraint (3.7) on the trial fluxes $\boldsymbol{F}$ in the present theory (see also Niven and Noack, Chap. 7). As one possibility among several, he showed that MaxEP for stationary states follows if one assumes there is a net increase in EP during the dynamic transition from one macrostate to another. However, the precise conditions under which this assumption is valid were not specified, so this approach sets out the landscape of possible extremal principles (which include MinEP) rather than a methodology for deciding which one applies in any given problem. Virgo [31] gave a heuristic MaxEnt-based argument for

MaxEP that assumes a relation between thermodynamic entropy at a given time t_1 and the Shannon entropy constrained by information at some earlier time t_0, the validity of which remains to be shown; this approach is somewhat tentative and difficult to formalize.

The two key ingredients missing from these previous studies, but included here, are (1) the explicit representation of the relevant dynamics in the form of the stationarity conditions (3.4), from which the EP functional can be determined via (3.18); and (2) the lower bound $I > I_{\min}(C)$ enforcing dynamic instability of the basal state, leading to MaxEP. By itself the auxiliary constraint (3.7) on $\boldsymbol{F}$ is not enough; rather, its role is to introduce a trial solution whose subsequent variation to maximise $S(\boldsymbol{F}, I_0, C)$ in (3.9) leads to MaxEP as given by (3.17).

3.4 Two Illustrative Examples

In this section we illustrate the above derivation of MaxEP for the two cases described by (3.18). The first case (S/AS) is illustrated by the application of MaxEP to simple energy balance models of planetary climates. This example illustrates the generic case in which the assumed relevant dynamics consists of a single global energy or mass balance constraint (see Chap. 1, Fig. 1.1, road 5). The second case (NS) is illustrated by the upper bound approach to fluid turbulence, in which several stationarity conditions of the form (3.4) are derived from spatial integrals of the Navier-Stokes equation, reflecting both energy and momentum balance (Fig. 1.1, road 6). These two examples illustrate how MaxEnt provides a theoretical justification for previous *ad hoc* applications of MaxEP to planetary climates [1, 2] and fluid turbulence [4, 15].

3.4.1 Zonal Energy Balance Models of Meridional Heat Flow in Planetary Climates

In [1, 2] MaxEP was applied to simple 1-D zonally-averaged energy balance models of the meridional heat flow in planetary climates. In these models the assumed relevant dynamics C consisted of the prescribed mean annual short-wave (SW) irradiance $F_{\mathrm{SW},i}\downarrow$ into each latitudinal zone i, and the global radiative balance between total incoming SW and total outgoing long-wave (LW) radiative fluxes, $\sum_i F_{\mathrm{LW},i}\uparrow = \sum_i F_{\mathrm{SW},i}\downarrow$.

In the derivation of MaxEP from MaxEnt (Sect. 3.3), the basal state of minimum irreversiblity is assumed to be dynamically unstable, and this helps us to identify the appropriate flux f in any given problem. In the present case the appropriate identification is $f = \{f_i\}$ where f_i is the meridional heat flux from zone $i - 1$ to zone i. The basal state then corresponds to zero meridional heat flow

$\boldsymbol{F} = \boldsymbol{0}$ ($I = I_{\min} \approx 0$ since we are ignoring heat transport by diffusion), i.e. the basal state is the state of radiative equilibrium where $F_{\mathrm{LW},i}\uparrow = F_{\mathrm{SW},i}\downarrow$ in each zone. With this identification, the condition $I > I_{\min}(C)$ means that differential radiative heating is strong enough to set in motion meridional heat flow between the equator and poles.

Denoting the heat convergence into zone i by $\Delta F_i = F_i - F_{i+1}$, global radiative balance is equivalent to $\sum_i \Delta F_i = 0$ which may be written in the form (3.4) ($m = 1$), i.e. $\int p(f)\varphi_1(f)\mathrm{d}f = 0$ with $\varphi_1(f) = \sum_i \Delta f_i$. Since here the constraint functional $\varphi_1(f)$ is anti-symmetric in f (i.e. the case S/AS), from (3.15) the appropriate EP functional is

$$I(\boldsymbol{F}) = 2\boldsymbol{\lambda}(\boldsymbol{F}) \cdot \boldsymbol{F} = 2\sum\nolimits_i \lambda_i F_i \propto -\sum\nolimits_i F_i \Delta\left(\frac{1}{T_i}\right) = \sum\nolimits_i \left(\frac{\Delta F_i}{T_i}\right). \quad (3.19)$$

Here we have used local heat balance to identify λ_i (up to a multiplicative constant) with the gradient in inverse temperature $-\Delta(1/T_i) = 1/T_{i+1} - 1/T_i$ (see Appendix A). The function (3.19), derived here from MaxEnt, agrees with the choice of EP function maximized in [1, 2].

In the light of its theoretical basis in MaxEnt, we can now reinterpret previous applications of MaxEP to planetary climates [1, 2]. Maximizing (3.19) predicts meridional heat fluxes that faithfully reflect the constraint of global energy balance, and no other stationarity constraints. Associated predictions for the latitudinal distributions of cloud cover and surface temperature are close to observations [1, 2]. Although in Paltridge's 1978 paper [1] the predicted meridional flux itself was unrealistically low, this was subsequently rectified using a more realistic definition of the temperature T_i appearing in (3.19) [40]. The predictive successes of MaxEP[4] here indicates that global energy balance captures the relevant dynamics governing these large-scale thermal features of Earth's climate.

3.4.2 Shear Turbulence

The upper bound approach to predicting the mean features of turbulent flow has a history going back to the 1950s [4, 15, 41]. In order to bypass the enormous computational challenge of numerically solving the Navier-Stokes equation (NSE), the upper bound approach maximizes various dissipation-like functionals of the flow subject to a restricted number of dynamical constraints based on spatial integrals of the NSE. Here we discuss and reinterpret some recent applications of the upper bound approach to turbulent shear flow [4, 15] in terms of the present MaxEnt derivation of MaxEP. We largely follow the notation of [15].

[4] Paltridge [1] also maximized the vertical flux of latent and sensible heat from ground to atmosphere in each zone. However, this additional constraint does not alter the MaxEnt derivation of (3.19) as the appropriate EP function for predicting the meridional flux.

3.4.2.1 Plane Couette Flow

First we consider plane Couette flow (PCF; see [15], Sect. 2.1) in which a fluid layer is horizontally sheared between two parallel infinite plates at $z = \pm\frac{1}{2}d$ moving at constant velocities $\mp\frac{1}{2}V_0\hat{\boldsymbol{x}}$ respectively. Measuring distances and velocities in units of d and ν/d respectively (ν = kinematic viscosity), the ith component of the NSE in dimensionless form is

$$NS_i \equiv \frac{\partial u_i}{\partial t} + u_j \partial_j u_i + \partial_i p - \nabla^2 u_i = 0 \tag{3.20}$$

(sum over repeated index $j = x, y, z$) where $\boldsymbol{u}$ is the velocity field and p is the pressure field. We also have the incompressibility condition $\partial_j u_j = 0$ and the boundary condition $u = \mp\frac{1}{2}Re\hat{\boldsymbol{x}}$ for $z = \pm\frac{1}{2}$, where $Re = V_0 d/\nu$ is the Reynolds number.

In [4, 15] the two spatial integrals of the NSE considered were horizontal mean momentum balance at each z and global power balance:

$$\overline{NS_x} = 0 \qquad -\frac{1}{2} \le z \le +\frac{1}{2} \tag{3.21}$$

$$\langle u_i NS_i \rangle = 0. \tag{3.22}$$

As in [15], the overbar and angular brackets denote, respectively, a horizontal $(x - y)$ average and a global $(x - y - z)$ average.

Here we will also introduce double vertical bars $\|\ \|$ to denote an ensemble average, i.e. an expectation value over $p(\boldsymbol{u})$, the probability density function of the instantaneous velocity field $\boldsymbol{u}$. Here $p(\boldsymbol{u})$ describes both thermal and hydrodynamical fluctuations. The Reynolds decomposition of the velocity field into mean and fluctuating components is $\boldsymbol{u} = U(z)\hat{\boldsymbol{x}} + \boldsymbol{v}$, with $\|\boldsymbol{v}\| = 0$. By invoking statistical stationarity (i.e. $\|\partial u_i/\partial t\| = 0$), and the incompressibility and boundary conditions, the ensemble averages of (3.21) and (3.22) may be written in the form (see [15], Eqs. 2.4 and 2.8)

$$\|\partial_z \overline{u_x} - \overline{v_x v_z} + \langle v_x v_z \rangle + Re\| = 0 \qquad -\frac{1}{2} \le z \le +\frac{1}{2} \tag{3.23}$$

$$\left\|\langle \partial_j u_i \partial_j u_i \rangle - Re^2 - Re\langle v_x v_z \rangle\right\| = 0. \tag{3.24}$$

In (3.24) the terms $\mathcal{D} \equiv \left\|\langle \partial_j u_i \partial_j u_i \rangle\right\|$, $\mathcal{D}_{lam,PCF} \equiv Re^2$ and $\mathcal{D}_{turb} \equiv \|Re\langle v_x v_z \rangle\|$ are, respectively, the ensemble-average total dissipation rate (per unit mass) of the fluid, the laminar dissipation and the enhancement in dissipation due to turbulence (all in units of ν^3/d^4). The sum of $\mathcal{D}_{lam,PCF}$ and $\mathcal{D}_{turb}$ is equal to the total work done on the fluid by the moving plates, and (3.24) states that this is equal to $\mathcal{D}$, the total dissipation of kinetic energy to heat. Two other relevant dissipation functionals are the ensemble-average dissipations in the mean and fluctuating velocity fields, given, respectively, by $\mathcal{D}_m \equiv \langle U'^2 \rangle$ (where $U' = \mathrm{d}U(z)/\mathrm{d}z$) and $\mathcal{D}_v \equiv \left\|\langle \partial_j v_i \partial_j v_i \rangle\right\|$, with the Reynolds decomposition implying $\mathcal{D} = \mathcal{D}_m + \mathcal{D}_v$.

To make the connection with the present MaxEnt theory of MaxEP, as anticipated above we identify $\boldsymbol{f}$ with the instantaneous velocity field $\boldsymbol{u}$. Then the stationarity conditions (3.23) and (3.24) are of the general form (3.4) with $\varphi_{1,z}(\boldsymbol{u}) = \partial_z \overline{u}_x - \overline{v_x v_z} + \langle v_x v_z \rangle + Re$ and $\varphi_2(\boldsymbol{u}) = \langle \partial_j u_i \partial_j u_i \rangle - Re^2 - Re\langle v_x v_z \rangle$, for which the corresponding Lagrange multipliers will be denoted by $\alpha_1(z)$ and α_2. We identify the basal state with the laminar solution $\boldsymbol{u}_{lam} = U_{lam}(z)\hat{\boldsymbol{x}} = -Rez\hat{\boldsymbol{x}}$. As we will see, the minimal irreversibility $I_{\min}(C)$ is then just the laminar dissipation rate $\mathcal{D}_{lam,PCF} = Re^2$. The dynamical instability condition $I > I_{\min}(C)$ implies that Re exceeds the critical Reynolds number for the onset of turbulent flow.

It follows[5] that $\Phi_1^A(z) = U'(z)$ and $\Phi_2^A = 0$ (i.e. case NS), so from (3.16) the appropriate EP functional for PCF is given by

$$I_{PCF} = 2 \int_{-1/2}^{+1/2} \alpha_1(z)\, U'(z) \mathrm{d}z. \tag{3.25}$$

As shown in Appendix B, we can make the identification $\alpha_1(z) \propto U'(z)$, so that

$$I_{PCF} \propto \langle U'^2 \rangle = \mathcal{D}_m. \tag{3.26}$$

Therefore, MaxEnt subject to constraints (3.23) and (3.24) is equivalent to maximizing $\mathcal{D}_m$, the dissipation in the mean velocity field. From (3.26), the minimum value of I_{PCF} is then (up to a multiplicative constant) $I_{\min,PCF} = \langle {U_{lam}}'^2 \rangle = \mathcal{D}_{lam,PCF} = Re^2$, as advertised.

3.4.2.2 Plane Poiseuille Flow

A similar analysis can be done for plane Poiseuille flow (PPF, see [15], Sect. 2.2), where the plates are stationary and the fluid is forced by a constant pressure gradient $\rho\nu^2 A/d^3$ (ρ = fluid density) in the $\hat{\boldsymbol{x}}$ direction. The dimensionless NSE is now

$$NS_i \equiv \frac{\partial u_i}{\partial t} + u_j \partial_j u_i + \partial_i p - A\delta_{i,x} - \nabla^2 u_i = 0 \tag{3.27}$$

which is supplemented by the incompressibility condition and the non-slip condition $\boldsymbol{u} = \boldsymbol{0}$ on the plates. As in PCF we set $\boldsymbol{f} = \boldsymbol{u}$. The basal state is again the laminar solution, which for PPF is given by $\boldsymbol{u}_{lam} = U_{lam}(z)\hat{\boldsymbol{x}} = \frac{1}{8}A(1 - 4z^2)\hat{\boldsymbol{x}}$. The PPF analogues of (3.23) and (3.24) are given by (see [15], Eqs. 2.11 and 2.17)

$$\| \partial_z \overline{u_x} - \overline{v_x v_z} + \langle v_x v_z \rangle + Az \| = 0 \qquad -\frac{1}{2} \leq z \leq +\frac{1}{2} \tag{3.28}$$

---

5 Note that under $\boldsymbol{u} \to -\boldsymbol{u}, \boldsymbol{v} \to -2U(z)\hat{\boldsymbol{x}} - \boldsymbol{v}$ so that $v_x v_z \to v_x v_z + 2U(z)v_z$, and since $\|v_z\| = 0$ the terms involving $v_x v_z$ do not contribute to $\Phi_1^A(z)$ or Φ_2^A.

$$\left\| \left\langle \partial_j u_i \partial_j u_i \right\rangle - \frac{16}{3} Re^2 - 8Re \langle z v_x v_z \rangle \right\| = 0 \tag{3.29}$$

where the Reynolds number for PPF is $Re = 3\langle U \rangle / 2 = \frac{1}{8}A - 3\langle z v_x v_z \rangle / 2$ (see [15], Eq. 2.12). In (3.28) we can set $\|\langle v_x v_z \rangle\| = 0$ assuming symmetry of $U(z)$ about the mid-plane $z = 0$, while in (3.29) $\mathcal{D}_{lam,PPF} \equiv 16Re^2/3$ is the laminar dissipation for PPF. Then the stationarity conditions (3.28) and (3.29) can be rewritten in the form (3.4) with $\varphi_{1,z}(\boldsymbol{u}) = \partial_z \bar{u}_x - \overline{v_x v_z} + Az$ and $\varphi_2(\boldsymbol{u}) = \left\langle \partial_j u_i \partial_j u_i \right\rangle - 16Re^2/3 - 8Re\langle z v_x v_z \rangle$. Therefore $\Phi_1^A(z) = U'(z)$ and $\Phi_2^A = 0$, as for PCF. Again we can make the identification $\alpha_1(z) \propto U'(z)$ (Appendix B) so that the appropriate EP functional is

$$I_{PPF} \propto \left\langle U'^2 \right\rangle = \mathcal{D}_m \tag{3.30}$$

Thus, for both PCF and PPF, the dissipation in the mean velocity field, $\mathcal{D}_m$, emerges from MaxEnt as the EP functional appropriate to the stationarity constraints (3.23)–(3.24) and (3.28)–(3.29), respectively.

3.4.2.3 The Upper Bound Approach to Turbulence Reinterpreted as MaxEnt

$\mathcal{D}_m$ was indeed one of the dissipation functionals considered by Kerswell [15] in his application of the upper bound approach to PCF and PPF subject to the stationarity conditions (3.23)–(3.24) and (3.28)–(3.29), respectively. In fact, among the family of dissipation functionals he considered, $\mathcal{D}_m$ was identified as the only candidate whose maximization was capable of predicting a vanishing interior mean shear (i.e. no velocity defect law) in the optimal flow as $Re \to \infty$. This asymptotic condition is currently believed to hold in real flows for PCF but not for PPF. From this Kerswell [15] then concluded that $\mathcal{D}_m$ cannot be a dissipation functional for both PCF and PPF and (as quoted in the Introduction) that it remains unclear how to construct such a universal functional ‘beyond intelligent guessing’.

However, the present MaxEnt theory of MaxEP suggests an alternative conclusion. MaxEnt implies that $\mathcal{D}_m$ is indeed the appropriate EP functional for both PCF and PPF, i.e. that maximization of $\mathcal{D}_m$ predicts mean velocity fields that faithfully reflect the stationarity conditions (3.23)–(3.24) or (3.28)–(3.29), and no other stationarity conditions. The disagreement in the case of PPF between the predicted and observed interior shear as $Re \to \infty$ then signals that some relevant dynamics is missing from the assumed constraint set C.

A clue to this missing physics was given in a subsequent study by Malkus [4], who applied the upper bound approach to PPF subject to conditions (3.28)–(3.29), with *two additional stability criteria*: (1) Rayleigh's criterion that $\mathrm{d}^2U(z)/\mathrm{d}z^2$ does not change sign across the entire channel, and (2) the boundary layer is marginally stable, which leads to the determination of a smallest spatial scale of momentum transport. As Malkus [4] showed, maximization of $\mathcal{D}_m$ under these additional constraints (which do not change the MaxEnt-derived EP functional $\mathcal{D}_m$) reproduces to a remarkable extent observed flow features such as the logarithmic region and the velocity defect law across the entire flow profile and over a large range of Reynolds numbers (e.g. see [4], Fig. 2.2). Thus, previous results obtained from the upper bound approach to fluid turbulence can be reinterpreted as the playing out of statistical mechanics under successive guesses for the relevant dynamics (Sect. 3.2).

Interestingly, Malkus [4, 42] was also led to consider maximization of $\mathcal{D}_m$ through considerations of dynamical stability. He determined a sufficiency condition for a solution of the NSE to be dynamically unstable, leading to maximum $\mathcal{D}_m$ as a necessary condition for reproducibility of the mean flow at a given *Re*. The fact that this instability argument, which is specific to the stationarity conditions derived from (3.21)–(3.22), leads to the same EP functional as MaxEnt (under the same constraints) supports our interpretation of the lower-bound inequality criterion $I > I_{\min}(C)$ as a dynamical instability condition.

We recall that it is the EP associated with kinetic energy dissipation, rather than turbulent heat transport, which also appears to be the appropriate EP functional whose maximization best describes the dynamic behaviour of the FAMOUS 3-D atmosphere-ocean GCM [14]. In view of the way $\mathcal{D}_m$ emerges here from MaxEnt as the appropriate EP functional in the case of horizontal shear turbulence subject to constraints (3.23)–(3.24), this suggests that global energy balance alone is not sufficient to characterise the relevant dynamics of FAMOUS, and that some additional momentum constraint—analogous to (3.24)—is required. This is to be expected when the dynamic behaviour under investigation depends on the velocity field, not just heat transport.

3.5 Relation to Other Entropy Production Principles

In this penultimate section we discuss the relation between MaxEP, the fluctuation theorem, and Ziegler's orthogonality/maximum dissipation principle (see also Chap. 1, Fig. 1.1).

3.5.1 The Fluctuation Theorem

The irreversibility I defined in (3.6) is related to the *dissipation function* Ω_t ([32–34], see also Chap. 2). For a microscopic trajectory over time interval $(0, \tau)$, the dissipation function is defined by $\Omega_\tau(\mathbf{\Gamma}(0)) = \ln\{p(\mathbf{\Gamma}(0), 0)/p(\mathbf{\Gamma}^*(\tau), 0)\}$ where $\mathbf{\Gamma}(t)$ is the phase space vector of the system at time t, $p(\mathbf{\Gamma}(0), 0)$ is the probability of observing the system in an infinitesmal region around $\mathbf{\Gamma}(0)$ at time $t = 0$, and $\mathbf{\Gamma}^*(\tau)$ is the phase space vector obtained from $\mathbf{\Gamma}(\tau)$ by reversing all the particle velocities. In fact $I = \langle\Omega_\tau\rangle$ if we interpret the flux vector f as a time-average over $(0, \tau)$. This equality reflects the fact that, like $\langle\Omega_\tau\rangle$, I can be written as a sum over phase space trajectories, in which the density of trajectories is symmetrical under time reversal. Like $\Omega_\tau, d(f) = \ln\{p(f)/p(-f)\}$ then obeys the fluctuation theorem[6] $p(d)/p(-d) = e^d$.

We can in fact recast the present formalism in terms of $p(\mathbf{\Gamma}(0), 0)$ instead of $p(f)$, including the expression for the Shannon entropy H, as was done in [19]. This establishes a direct connection between MaxEP and the fluctuation theorem, through the dissipation function $I = \langle\Omega_\tau\rangle$. Thus the irreversibility I [the KL divergence of $p(f)$ and $p(-f)$], the EP function $I(\boldsymbol{F})$ [the MaxEnt-predicted value of I) and the trajectory-based dissipation function $\Omega_\tau(\mathbf{\Gamma}(0)]$ are intimately related.

3.5.2 Ziegler's Orthogonality/Maximum Dissipation Principle

The present MaxEnt derivation of MaxEP also sheds new light on the connection between MaxEP and Ziegler's maximum dissipation (MaxD) principle ([35], also Chaps. 4 and 5). Dewar [27] found that, close to equilibrium [29], the MaxEnt-derived irreversibility I obeys an orthogonality property first noted by Ziegler [35], which can be expressed in variational form (i.e. MaxD). However, MaxD is not the same as the MaxEP selection criterion established here. MaxD emerges when MaxEnt is applied to $p(f)$ in the special case where $\boldsymbol{F}$ is known [27], and expresses in variational form the constitutive relation between $\boldsymbol{F}$ and the corresponding Lagrange multiplier $\boldsymbol{\lambda}$ established by MaxEnt in that special case. Unlike MaxEP, MaxD does not select between different possible $\boldsymbol{F}$. Nevertheless, we can establish a more general link between MaxEP and Ziegler's orthogonality property as follows.

If there are several possible flux vectors $\boldsymbol{F}$ satisfying the upper bound constraint $I = I_{\max}(C)$ then, in accordance with MaxEnt, the one with the largest value of S is to be selected. Assuming a continuum of such vectors $\boldsymbol{F}$, the variation in S is given from (3.10) by $\delta S = -\boldsymbol{\lambda} \cdot \delta\boldsymbol{F}$. Since $\delta\boldsymbol{F}$ is confined to the surface $I(\boldsymbol{F}) = I_{\max}(C)$ in $\boldsymbol{F}$-space, and since $\delta S = 0$ at the MaxEnt solution $\boldsymbol{F} = \boldsymbol{F}(C)$, the corresponding Lagrange multiplier vector $\boldsymbol{\lambda}(C)$ is normal to the surface $I(\boldsymbol{F}) = I_{\max}(C)$. This is just the statement of Ziegler's orthogonality property, which would therefore appear to hold for systems arbitrarily far from equilibrium (cf. [27, 29]) (see Chap. 1, Fig. 1.1, road 7).

[6] In fact with d so defined, the fluctuation theorem is a purely mathematical result (see Chap. 1, Footnote 12). As we have seen, the physical interpretation of $\langle d\rangle$ as entropy production emerges in MaxEnt through the physical constraints that determine $p(f)$.

3.6 Conclusion and Outlook

The present theory suggests that MaxEP, like MaxEnt from which it has been derived, is at heart not a physical principle after all.[7] In statistical mechanics, MaxEP plays the role of an inference algorithm that faithfully translates the assumed relevant dynamics C into macroscopic predictions. Disagreement between MaxEP predictions and observations signals an inadequate constraint set rather than a failure of MaxEP *per se*. MaxEP successes indicate that we have identified the relevant dynamics that govern the phenomena under study. Complementing this view, MaxEP (like MaxEnt) can also be viewed as a statistical principle that selects the most likely state under given dynamical constraints. However, a key challenge lies in choosing the relevant constraints.

Thus while the present theory may serve to replace 'intelligent guessing' [15] of the EP function with an algorithm (Eq. 3.18), it appears that intelligent guessing is still required in the choice of constraints. One possible approach to systematizing this choice might be to use renormalization group techniques (e.g. [43]) to explicitly construct the relevant dynamics on different spatial and temporal scales, by successive coarse-graining of the microscopic equations of motion. However, the technical challenges are formidable.

Nevertheless the present theory may help to underpin and guide the application of MaxEP to a wider range of non-equilibrium systems. Moreover, it may indicate where it might be inappropriate to apply MaxEP. For example, it raises a question over the analysis of EP in deterministic models with few degrees of freedom, in which the most stable steady state is already determined by the low-dimensional dynamics; it is then unclear what role MaxEP (or any selection principle) has to play. In contrast, MaxEP appears to describe the relative stability of multiple steady states in models with many degrees of freedom (e.g. [44]; see also Chaps. 6, 8, 13, 14).

Thus another challenge is to clarify the theoretical relationship between maximum irreversibility and dynamic stability (Fig. 1.1, road 11), a link suggested here by the fact that MaxEnt/MaxEP predicts the same dissipation functional as Malkus's instability criterion in shear turbulence [4, 42].

Finally, we tentatively suggest that the MaxEnt derivation of MaxEP (i.e. maximal time-reversal symmetry breaking) might be applicable to other symmetries. Thus, if the set of possible physical outcomes x is closed under the transformation $x \to Tx$ (e.g. here $x = f$, T = time reversal under which $f \to -f$) then the KL divergence $I = \int p(x)\ln\{p(x)/p(Tx)\}$ is a measure of T-symmetry breaking. Perhaps MaxEnt then implies maximal symmetry breaking more generally.

[7] For an alternative perspective, see Chap. 5. See also Chap. 1, Sect. 1.4.5.

Acknowledgments Financial support from the Isaac Newton Institute for Mathematical Sciences (to RD; Programme CLP: *Mathematical and Statistical Approaches to Climate Modelling and Prediction*, Aug–Dec 2010) and Fondazione Cariparo (to AM) is gratefully acknowledged. RD thanks the participants of the annual MaxEP workshops (2003–2011) for many stimulating discussions and encouragement.

Appendix A: Identifying the Lagrange Multiplier λ (Sect. 3.4.1)

In Sect. 3.4.1 (zonal energy balance models), the MaxEnt p.d.f. is $p(f) \propto e^A$ where $A = \sum_i (\alpha\Delta f_i + \lambda_i f_i)$, where f_i is the meridional heat flux from zone $i - 1$ to zone i, $\Delta f_i = f_i - f_{i+1}$ is the heat convergence into zone i, and α and λ_i are the Lagrange multipliers associated with the global energy balance constraint $\sum_i \Delta F_i = 0$ and the trial flux F_i, respectively. Here we show that λ_i may be identified with the gradient in inverse temperature $-\Delta(1/T_i) = 1/T_{i+1} - 1/T_i$. Consider local heat balance over some finite time interval $(0, \tau)$. Let f_i and Δf_i denote time-averaged rates over $(0, \tau)$, and let $u_i(t)$ denote the heat content of zone i at time t. Local heat balance then gives $\{u_i(\tau) - u_i(0)\}/\tau = \Delta f_i - r_i$ where $r_i = f_{\mathrm{LW},i}\uparrow - F_{\mathrm{SW},i}\downarrow$ is the net radiation leaving zone i, with ensemble average $R_i = \Delta F_i$. By introducing constraints on the ensemble averages of $u_i(\tau)$ in each zone, for which the corresponding Lagrange multipliers may be identified with[8] $-1/T_i$ we then have $A = -\sum_i u_i(\tau)/T_i + \sum_i (\alpha\Delta f_i + \lambda_i f_i)$ which from heat balance can be rewritten as $A = -\frac{1}{2}\sum_i \{u_i(0) + u_i(\tau)\}/T_i + \sum_i \Delta f_i\left(\alpha - \frac{1}{2}\tau/T_i\right) + \frac{1}{2}\tau\sum_i r_i/T_i + \sum_i \lambda_i f_i$. Summing by parts reduces this to $A = -\frac{1}{2}\sum_i \{u_i(0) + u_i(\tau)\}/T_i + \sum_i f_i\left(\lambda_i + \frac{1}{2}\tau\Delta(1/T_i)\right) + \frac{1}{2}\tau\sum_i r_i/T_i$. Since knowledge of the radiation fluxes R_i is equivalent to knowledge of the heat fluxes F_i, the second term in A is redundant in the presence of the third term, implying $\lambda_i = -\frac{1}{2}\tau\Delta(1/T_i)$ as we wished to show. Substituting this into Eq. (3.19) for the mean entropy production then gives $I = 2\sum_i \lambda_i F_i \propto -\sum_i F_i\Delta(1/T_i) = \sum_i \Delta F_i/T_i$.

Appendix B: Identifying the Lagrange Multiplier $\alpha_1(z)$ (Sect. 3.4.2)

In Sect. 3.4.2.1 (shear turbulence in plane Couette flow), the MaxEnt p.d.f. is $p(f) \propto e^A$ where $A = \langle \alpha_1(z)(\partial_z\bar{u}_x - \overline{v_x v_z} + \langle v_x v_z\rangle + Re)\rangle + \alpha_2(\langle\partial_j u_i \partial_j u_i\rangle - Re^2 - Re\langle v_x v_z\rangle)$. Ignoring the fluctuation terms involving $v_x v_z$ and setting $\boldsymbol{u} = U(z)\hat{\boldsymbol{x}}$ (mean-field approximation) yields $A = \int dz\{\alpha_1(z)U'(z) + \alpha_2 U'(z)^2\}$ where we have dropped the constant terms involving Re and Re^2. The action A is stationary with respect to $U(z)$ when $\alpha_1(z) + 2\alpha_2 U'(z) = c$ with c constant. Substituting this

---

8 Here we assume units such that Boltzmann's constant equals 1.

into Eq. (3.25) for the mean entropy production gives $I_{PCF} = 2\int \mathrm{d}z\alpha_1(z)\ U'(z) \propto \int \mathrm{d}zU'(z)^2$ (Eq. 3.26), where we have dropped the constant term $c\int \mathrm{d}z\ U'(z) = -cRe$. Thus effectively we can set $\alpha_1(z) \propto U'(z)$ in (3.25). A similar argument applies to plane Poiseuille flow, leading to the same expression for I_{PPF} (Eq. 3.30).

References

1. Paltridge, G.W.: Global dynamics and climate—a system of minimum entropy exchange. Q. J. Roy. Meteorol. Soc. **101**, 475–484 (1975); The steady-state format of global climate. *Ibid.* **104**, 927–945 (1978); Thermodynamic dissipation and the global climate system. *Ibid.* **107**, 531–547 (1981)
2. Lorenz, R.D., Lunine, J.I., Withers, P.G., McKay, C.P.: Titan, mars and earth: entropy production by latitudinal heat transport. Geophys. Res. Lett. **28**, 415–418 (2001)
3. Ozawa, H., Shimokawa, S., Sakuma, H.: Thermodynamics of fluid turbulence: a unified approach to the maximum transport properties. Phys. Rev. E **64**, 026303 (2001)
4. Malkus, W.V.R.: Borders of disorders: in turbulent channel flow. J. Fluid Mech. **489**, 185–198 (2003)
5. Hill, A.: Entropy production as a selection rule between different growth morphologies. Nature **348**, 426–428 (1990)
6. Martyushev, L.M., Serebrennikov, S.V.: Morphological stability of a crystal with respect to arbitrary boundary perturbation. Tech. Phys. Lett. **32**, 614–617 (2006)
7. Juretić, D., Županović, P.: Photosynthetic models with maximum entropy production in irreversible charge transfer steps. Comp. Biol. Chem. **27**, 541–553 (2003)
8. Dewar, R.C., Juretić, D., Županović, P.: The functional design of the rotary enzyme ATP synthase is consistent with maximum entropy production. Chem. Phys. Lett. **430**, 177–182 (2006)
9. Dewar, R.C.: Maximum entropy production and plant optimization theories. Phil. Trans. R. Soc. B **365**, 1429–1435 (2010)
10. Franklin, O., Johansson, J., Dewar, R.C., Dieckmann, U., McMurtrie, R.E., Brännström, Å., Dybzinski, R.: Modeling carbon allocation in trees: a search for principles. Tree Physiol. **32**, 648–666 (2012)
11. Main, I.G., Naylor, M.: Maximum entropy production and earthquake dynamics. Geophys. Res. Lett. **35**, L19311 (2008)
12. Martyushev, L.M., Seleznev, V.D.: Maximum entropy production principle in physics, chemistry and biology. Phys. Rep. **426**, 1–45 (2006)
13. Essex, C.: Radiation and the irreversible thermodynamics of climate. J. Atmos. Sci. **41**, 1985–1991 (1984)
14. Pascale, S., Gregory, J.M., Ambaum, M.H.P., Tailleux, R.: A parametric sensitivity study of entropy production and kinetic energy dissipation using the FAMOUS AOGCM. Clim. Dyn. **38**, 1211–1227 (2012)
15. Kerswell, R.R.: Upper bounds on general dissipation functionals in turbulent shear flows: revisiting the 'efficiency' functional. J. Fluid Mech. **461**, 239–275 (2002)
16. Meysman, F.J.R., Bruers, S.: Ecosystem functioning and maximum entropy production: a quantitative test of hypotheses. Phil. Trans. R. Soc. B **365**, 1405–1416 (2010)
17. Kawazura, Y., Yoshida, Z.: Entropy production rate in a flux-driven self-organizing system. Phys. Rev. E **82**(066403), 1–8 (2010)
18. Dewar, R.C.: Maximum entropy production as an inference algorithm that translates physical assumptions into macroscopic predictions: Don't shoot the messenger. Entropy **11**, 931–944 (2009)

19. Jaynes, E.T.: Information theory and statistical mechanics. Phys. Rev. **106**, 620–630 (1957); Information theory and statistical mechanics II. *Ibid.* **108**, 171–190 (1957)
20. Gibbs, J.W.: Elementary Principles of Statistical Mechanics. Ox Bow Press, Woodridge (1981). (Reprinted)
21. Shannon, C.E.: A mathematical theory of communication. Bell Sys. Tech. J. **27**, 379–423, 623–656 (1948)
22. Shannon, C.E., Weaver, W.: The mathematical theory of communication. University of Illinois Press, Urbana (1949)
23. Jaynes, E.T.: Probability Theory: the Logic of Science. Bretthorst, G.L. (ed.). CUP, Cambridge (2003)
24. Shore, J.E., Johnson, R.W.: Axiomatic derivation of the principle of maximum entropy and the principle of minimum cross-entropy. IEEE Trans. Info. Theor. **26**, 26–37 (1980)
25. Jaynes, E.T.: Where do we stand on maximum entropy? In: Levine, R.D., Tribus, M. (eds.) The maximum entropy formalism, pp. 15–118. MIT Press, Cambridge (1978)
26. Dewar, R.C.: Information theory explanation of the fluctuation theorem, maximum entropy production and self-organized criticality in non-equilibrium stationary states. J. Phys. A: Math. Gen. **36**, 631–641 (2003)
27. Dewar, R.C.: Maximum entropy production and the fluctuation theorem. J. Phys. A: Math. Gen. **38**, L371–L381 (2005)
28. Bruers, S.A.: Discussion on maximum entropy production and information theory. J. Phys. A: Math. Theor. **40**, 7441–7450 (2007)
29. Grinstein, G., Linsker, R.: Comments on a derivation and application of the 'maximum entropy production' principle. J. Phys. A: Math. Theor. **40**, 9717–9720 (2007)
30. Niven, R.K.: Steady state of a dissipative flow-controlled system and the maximum entropy production principle. Phys. Rev. E **80**, 021113 (2009)
31. Virgo, N.: From maximum entropy to maximum entropy production: a new approach. Entropy **12**, 107–126 (2010)
32. Evans, D.J., Searle, D.J.: The fluctuation theorem. Adv. Phys. **51**, 1529–1585 (2005)
33. Seifert, U.: Stochastic thermodynamics: principles and perspectives. Eur. Phys. J. B **64**, 423–431 (2008)
34. Sevick, E.M., Prabhakar, R., Williams, S.R., Searles, D.J.: Fluctuation theorems. Ann Rev. Phys. Chem. **59**, 603–633 (2008)
35. Ziegler, H.: An introduction to thermomechanics. North-Holland, Amsterdam (1983)
36. Jaynes, E.T.: Information theory and statistical mechanics. In: Ford, K. (ed.) Statistical Physics, pp. 181–218. Benjamin, New York (1963)
37. Niven, R.K.: Non-asymptotic thermodynamic ensembles. Europhys. Lett. **86**, 20010 (2009)
38. Niven, R.K., Grendar, M.: Generalized classical, quantum and intermediate statistics and the Pólya urn model. Phys. Lett. A **373**, 621–626 (2009)
39. Kuhn, H.W., Tucker, A.W.: Nonlinear programming. In: Neyman, J. (ed.) Proceedings of the 2nd Berkeley Symposium on Mathematical Statistics and Probability, pp 481–492. UCA, Berkeley (1951)
40. Paltridge, G.W., Farquhar, G.D., Cuntz, M.: Maximum entropy production, cloud feedback, and climate change. Geophys. Res. Lett. **34**, L14708 (2007)
41. Malkus, W.V.R.: The heat transport and spectrum of thermal turbulence. Proc. R. Soc. **225**, 196–212 (1954); Outline of a theory for turbulent shear flow. J. Fluid Mech. **1**, 521–539 (1956)
42. Malkus, W.V.R.: Statistical stability criteria for turbulent flow. Phys. Fluids **8**, 1582–1587 (1996)
43. Giles, M.J.: Turbulence renormalization group calculations using statistical mechanics methods. Phys. Fluid. **6**, 595–604 (1994)
44. Shimokawa, S., Ozawa, H.: On the thermodynamics of the oceanic general circulation: Irreversible transition to a state with higher rate of entropy production. Q. J. Roy. Meteorol. Soc. **128**, 2115–2128 (2002)

Chapter 4
Dissipation Rate Functions, Pseudopotentials, Potentials and Yield Surfaces

Guy T. Houlsby

Abstract This chapter is about the role of the dissipation rate function—and other functions derived from it—in determining the constitutive behaviour of dissipative materials. It consists of a discussion of some general theory, followed by examples. We address the class of materials for which knowledge of the functional form of the dissipation rate function supplies the complete constitutive response, without recourse to further assumptions. A careful distinction is drawn between functions that are true potentials and those that are pseudopotentials (defined in the chapter), in order to clarify some aspects of terminology that the Author has elsewhere found confusing. In plasticity theory the intimate relationship between the dissipation rate function and the yield surface is explored. The chapter is illustrated by examples of simple one- and two-dimensional conceptual models, as well as full continuum models. Both rate independent (plastic) and rate-dependent (viscous, or viscoplastic) models are addressed.

List of Symbols

Symbol Meaning

Symbol	Meaning
Roman Symbols	
$d(x, v)$	Dissipation rate function
$d^*(x, \chi)$	Dissipation rate function (alternative functional form)
$I_X(\chi)$	Indicator function of a set X
k	Plastic strength
n	Order of homogeneous dissipation rate function in velocities
$N_X(\chi)$	Normal cone to a convex set X
r	Reference velocity
R	Reference force

G. T. Houlsby (✉)
Department of Engineering Science, University of Oxford, Parks Road, Oxford OX1 3PJ, UK
e-mail: guy.houlsby@eng.ox.ac.uk

R. C. Dewar et al. (eds.), *Beyond the Second Law*,
Understanding Complex Systems, DOI: 10.1007/978-3-642-40154-1_4,

v	Generalised velocity (of state variable)
x	State variable
$w(x, \chi)$	Flow potential
$y(x, \chi)$	Yield function
$\bar{y}(x, \chi)$	Canonical form of yield function
$z(x, v)$	Force potential

Greek Symbols

$\gamma_X(\chi)$	Gauge or Minkowski function of a convex set X
θ	Absolute temperature
λ	Scalar multiplier
μ	Viscosity
Λ	Lagrangian multiplier
τ	Dummy variable in integral transform
χ	Generalized force

Mathematical Symbolism

$$S(x) = \begin{cases} -1, & x<0 \\ \in [-1,1], & x=0 \\ +1, & x>0 \end{cases}$$ Generalised signum function

$$\langle x \rangle = \begin{cases} 0, & x<0 \\ x, & x \geq 0 \end{cases}$$ Macaulay brackets

$\langle x, y \rangle$ Inner product

4.1 General Theory

In thermodynamic approaches to the mechanics of dissipative materials, the dissipation rate is usually considered as a non-negative function of some **generalised velocities**, which we shall denote here as v (which may be scalar, vectorial or tensorial in character). It is important to note that the generalised velocities may belong to two major classes: firstly there are the rates of internal variables (e.g. the plastic strain rate) and secondly there are fluxes of various quantities, which are driven by spatial gradients of associated variables (e.g. the flux of electrical charge driven by the voltage gradient). For many applications it is important to make a distinction between these two very different sources of dissipation, one associated with a temporal gradient and the other associated with a spatial gradient, but for our purposes here it is sufficient to treat the two classes together, and use the term generalised velocity either for a rate of change or for a flux.

Dissipation only occurs in the presence of non-zero generalised velocity, and it is a small step (often taken implicitly) to assume therefore that the dissipation rate can be expressed as a function of these velocities. We need to recognise though that even the existence of a dissipation rate function is an assumption. For instance, more generally we could allow the dissipation rate to be a **functional** ("function of a function") of the entire time-history of the velocities. However, it is conventional to assume that the dissipation rate can be written simply as a non-negative function of the velocities and of certain state variables, which we denote by x (which also may be scalar, vectorial or tensorial in character). Thus $d = d(x, v) \geq 0$. The effects of the history of the material are effectively captured entirely within the current values of the state variables, and need not be treated by resort to functionals of the history. This is the key feature of "thermodynamics with internal variables", TIV, otherwise known as generalised thermodynamics (see for example [7]), as opposed to the so-called "rational mechanics", which makes much use of functionals (see for example [9]).

As the entropy production is simply d/θ, where θ is the absolute temperature, the assumption of the existence of a dissipation rate function is precisely equivalent to the assumption of a function for the entropy production.

The question that is now posed is whether knowledge of the functional form of $d(x, v)$ furnishes us with additional information, possibly even complete information, about the constitutive behaviour of the material. Compare, for instance, with the classical thermodynamics of fluids. In that case (in the absence of dissipation), knowledge of the internal energy as a function of appropriate state variables is sufficient to define the entire constitutive behaviour. Can a similar objective be achieved for dissipative materials?

The next step is usually to identify a set of **generalised forces** χ which are work-conjugate to the generalised velocities, and such that $d = \langle \chi, v \rangle$ (where $\langle \chi, v \rangle$ denotes an appropriate inner product). Different authors make different assumptions about the nature of χ. For instance in some texts if v represents the plastic strains then χ is explicitly identified with the stresses. In other texts (e.g. [10]), χ is simply identified with **generalised stresses**, not necessarily identical to the true stresses. These distinctions need not concern us at this stage: it is sufficient that we consider a set of generalised forces of unspecified nature. However, what is common to all the approaches is that the key to determining the constitutive behaviour is the establishment of the relationship between v and χ. Once this relationship is known, the constitutive behaviour is fully determined. Clearly the problem can be cast in one of two explicit forms, either in terms of establishing the velocities v as a function of the forces χ, or in the inverse form of determining the forces χ in terms of the velocities v. Implicit forms are also possible.

One possibility is that the expression $d = d(x, v)$ is merely used to calculate the value of the dissipation rate (and check that it is non-negative) whilst introducing independently some other specification of the constitutive behaviour; for instance the velocities may be expressed as explicit functions of the forces (or *vice versa*). However, we are interested here in the case where the constitutive behaviour can itself be deduced from $d(x, v)$ without the need to introduce additional assumptions. If this objective can be achieved it has a very particular advantage: the constitutive response can be defined through the knowledge of just the scalar function $d(x, v)$. Contrast this with the case where the velocities are expressed as functions of the forces—in that case, if there are n velocities, then n equations have to be specified, and the dissipation rate then calculated to check that it is non-negative for all possible conditions.

Perhaps the simplest argument for derivation of the forces from the dissipation rate was advanced by Ziegler [10] (see his Chap. 14, Sect. 14.3), but his reasoning may have earlier origins. The argument goes broadly as follows. If the forces χ can be determined from the velocities v and knowledge of $d(x, v)$, then the direction (in a generalised sense) of χ can only be given by a direction that can be obtained solely from knowledge of $d(x, v)$. Ziegler argues that the only such direction is $\partial d/\partial v$. One might argue that if, for instance, v and χ are finite-dimensional vectors, then χ could be in the same direction as v, but Ziegler dismisses this argument by appealing to the fact that v should properly be definable in terms of some curvilinear coordinate system in which a proper distinction is necessary between covariant and contravariant tensors. In this case, if v is contravariant, then χ is covariant, and it is meaningless to speak of them being in the same direction. By a similar argument Ziegler dismisses the possibility that the direction of χ could be determined from contracted inner products involving higher derivatives of $d(x, v)$. The same argument could be advanced in more modern terminology in terms of vectors and one-forms. Note, however, that Ziegler's entire argument is based on the premise that χ **can** be derived from knowledge of v and $d(x, v)$, and some would argue that this itself is an unjustified assumption.

Ziegler thus argues that $\chi = \lambda \frac{\partial d}{\partial v}$, where λ is some yet-to-be-determined scalar. However, simply by noting that $d = \langle \chi, v \rangle = \lambda \langle \frac{\partial d}{\partial v}, v \rangle$, one can immediately derive $\lambda = \dfrac{d}{\langle \frac{\partial d}{\partial v}, v \rangle}$ and therefore $\chi = \dfrac{d}{\langle \frac{\partial d}{\partial v}, v \rangle} \dfrac{\partial d}{\partial v}$. Thus we have achieved our objective of determining χ as a function of v entirely from knowledge of $d(x, v)$. This result is known as **Ziegler's orthogonality condition**: χ is orthogonal (normal) to a level set of $d(x, v)$.

Orthogonality can be deduced from other assumptions, or alternatively it can be interpreted in different ways. For instance one can pose the problem as one of maximal dissipation rate, or equivalently maximal entropy production. This is an appealing notion, in that the Second Law requires that the dissipation rate be non-negative, and it is a logical extension to strengthen this to a statement that (subject to any relevant constraints) the dissipation rate should be maximal. The main problem comes in the treatment of the constraints, and many authors simply acknowledge the issue and then hastily avoid dealing with it in any detail. The treatment of constraints within the context of a general boundary problem is not trivial. However, the maximal dissipation rate concept can be posed in the following way: what are the velocities that, for given χ, maximise the dissipation rate, subject just to the side condition that $d = \langle \chi, v \rangle$? We proceed by first forming an augmented Lagrangian function $d' = d + \Lambda(d - \langle \chi, v \rangle)$ in which Λ is a

Lagrangian multiplier, and then maximising with respect to v by writing $\frac{\partial d'}{\partial v} = \frac{\partial d}{\partial v} + \Lambda\left(\frac{\partial d}{\partial v} - \chi\right) = 0$.

Rearrangement gives $\chi = \frac{1+\Lambda}{\Lambda}\frac{\partial d}{\partial v}$, the same result as before for the direction of χ, with $\lambda = \frac{1+\Lambda}{\Lambda}$. We then follow the same argument as before to establish the value of the multiplier $\lambda = \frac{1+\Lambda}{\Lambda} = \dfrac{d}{\left\langle \frac{\partial d}{\partial v}, v \right\rangle}$.

It is appealing that orthogonality can be derived either from geometric arguments about the direction of χ, or from a principle of maximal dissipation rate. Either way, orthogonality serves as the tool to extract more information from the dissipation rate function, and allows knowledge of the function to provide us with the entire constitutive response. In fact this is such a useful result that it could be advanced as an independent argument for orthogonality: a constitutive model employing orthogonality requires only the specification of a scalar function, as opposed to a set of tensorial relationships. It could be argued that the structure of a constitutive law based on orthogonality is so much simpler than many other structures that one should explore these materials first, and only reject orthogonality if it proves inadequate for describing real materials (of course, this comes close to advancing an argument based on the aesthetics of the theory, but we avoid that pitfall, and prefer an argument simply based on pragmatism).

4.1.1 Materials Obeying Orthogonality

The remainder of this chapter is devoted to materials for which orthogonality applies. Note that these materials are often referred to as "standard materials" by authors such as Lemaitre and Chaboche [5], Maugin [6, 7].

When χ can be derived from $d(x, v)$ by an expression of the form $\chi = \lambda\frac{\partial d}{\partial v}$, and more specifically from $\chi = \dfrac{d}{\left\langle \frac{\partial d}{\partial v}, v \right\rangle}\dfrac{\partial d}{\partial v}$, then we say that $d(x, v)$ acts as a **pseudo-potential** for χ. By pseudopotential we mean that the partial derivative of $d(x, v)$ provides the direction, but not the magnitude, of χ. Contrast this with the situation where, if some function z exists such that $z = z(x, v)$ and $\chi = \partial z/\partial v$, then $z = z(x, v)$ is a **true potential** (or just **potential**) for χ, in that it supplies directly both direction and magnitude of χ. There are numerous mathematical advantages if we can identify such a true potential. Most notably we can then exploit the mathematics of the Legendre transform (or more generally the Fenchel dual) to obtain from $z = z(x, v)$ another potential $w(x, \chi)$ such that $z + w = \langle \chi, v \rangle = d$, and $v = \partial w/\partial \chi$. Thus either of the two fundamental forms of the constitutive relationships can then be derived.

How can $z = z(x, v)$ be derived from $d(x, v)$? First of all we can observe that if $d(x, v)$ is homogeneous and of order n in v, then from Euler's theorem $\left\langle \frac{\partial d}{\partial v}, v \right\rangle = nd$ and $\lambda = \dfrac{d}{\left\langle \frac{\partial d}{\partial v}, v \right\rangle} = \dfrac{1}{n}$. This is an important and useful special case. For rate-independent (plastic) materials $d(x, v)$ is homogeneous and of degree 1 in v, so $\lambda = 1$,

$\chi = \partial d/\partial v$ and we immediately identify that $z(x, v) = d(x, v)$. We return to this case later. For a linear viscous material then $d(x, v)$ is homogeneous and of degree 2 in v, so $\lambda = \frac{1}{2}$, $\chi = \frac{1}{2}\frac{\partial d}{\partial v}$ and we can immediately identify $z(x, v) = \frac{1}{2}d(x, v)$. For any homogeneous function of degree n, it follows that $z(x, v) = \frac{1}{n}d(x, v)$. For any function that can be expressed as the sum of homogeneous functions $d(x, v) = \sum d_n(x, v)$ where $d_n(x, v)$ is homogeneous of degree n in v, it follows that $z(x, v) = \sum \frac{1}{n} d_n(x, v)$.

Houlsby and Puzrin [4] show that more generally the true potential can be derived from the integral transform $z(x, v) = \int_0^1 \frac{d(x, \tau v)}{\tau} d\tau$. The inverse process is of course much more straightforward: $d(x, v) = \langle \chi, v \rangle = \langle \frac{\partial z}{\partial v}, v \rangle$.

To avoid confusion with the dissipation rate, which is a pseudopotential, we follow Houlsby and Puzrin and call the function $z(x, v)$ the **force potential** (as it acts as a potential for the determination of the generalised force χ). We call $w(x, \chi)$ the **flow potential** as it acts as a potential that determines the "flow" v. Confusingly, Maugin [7] refers to both of these functions as the dissipation potential (in his Sect. 5.3, Maugin's D is our z, and his $D*$ is our w). This ambiguous terminology has also been used elsewhere.

Now let us consider the case where the dissipation rate can alternatively be expressed as a function of the forces $d = d^* = d^*(x, \chi)$. Note, however, that for the very important case of rate-independent plasticity this is not possible. The question now arises as to how the velocities can be derived from the dissipation rate function. Following exactly Ziegler's argument above, but with the roles of force and velocity interchanged, one can deduce $v = \lambda^* \frac{\partial d^*}{\partial \chi}$, where λ^* is an undetermined scalar. Then by noting that $d^* = \langle \chi, v \rangle = \lambda^* \left\langle \chi, \frac{\partial d^*}{\partial \chi} \right\rangle$, one can immediately derive $\lambda^* = \dfrac{d^*}{\left\langle \chi, \frac{\partial d^*}{\partial \chi} \right\rangle}$ and therefore $v = \dfrac{d^*}{\left\langle \chi, \frac{\partial d^*}{\partial \chi} \right\rangle} \dfrac{\partial d^*}{\partial \chi}$. Thus $d^*(x, \chi)$ is a pseudopotential for v. Clearly we can relate $d^*(x, \chi)$ and $w(x, \chi)$, and a similar process to that used before yields the result $w(x, \chi) = \int_0^1 \frac{d^*(x, \tau\chi)}{\tau} d\tau$. Again the inverse process is of course much more straightforward: $d^*(x, \chi) = \langle \chi, v \rangle = \left\langle \chi, \frac{\partial w}{\partial \chi} \right\rangle$.

4.2 Rate-independent Plastic Materials

4.2.1 Legendre Transformations when the Dissipation Rate is First Order in Velocities

We now turn to the special case where d is homogeneous and first order in v, which arises in the extremely important problem of rate-independent plasticity. In this case we have $\langle \frac{\partial d}{\partial v}, v \rangle = d$ and $\lambda = 1$, so that $z = d$ and $d(x, v)$ is a true potential for

χ, not just a pseudopotential. However, it also follows that $w = d - z = 0$, so that we are faced with the fact that the flow potential is always identically zero. This special case is extensively treated in the Appendix to Collins and Houlsby [1], and by Houlsby and Puzrin [4]. They observe that in this special case it is possible to derive a function of the forces that is identically zero when yield occurs and that this function acts as a pseudopotential for the velocities: $y = y(x, \chi) = 0$, $v = \lambda' \frac{\partial y}{\partial \chi}$. This function is none other than the yield function (but in generalised force space, not true stress space) and the fact that it acts as a pseudopotential means that there is "associated flow" in generalised force space. An important observation, pursued by Houlsby [2] and in later publications, is that this associated flow does not carry over to true stress space in the case of "frictional" behaviour, so this approach to plasticity is able to accommodate non-association in conventional plasticity terms. ("Frictional" behaviour is defined in this context as the explicit dependence of the dissipation rate on the true stresses. The dependence usually takes the form that the dissipation rate is proportional to the mean stress.)

4.2.2 Alternative Approach: Fenchel Duals

The Legendre transform of a homogeneous first order function can be treated as a special case as above, but it is more usefully treated using the language of **convex analysis**, in which the Legendre transform is a special case of the Fenchel dual. Houlsby and Puzrin [4] provide (in their Appendix D) an introduction to convex analysis in this context. The main restriction is that we have to confine our attention to dissipation rate functions that are convex, but this does not prove to be excessively arduous, as virtually every dissipation rate function that might be of interest satisfies this criterion.

The essential results from convex analysis are as follows. The homogeneous first order dissipation rate function $d(x, v)$ can be interpreted as a **support function** of a convex set X in the space of χ. This convex set is none other than the set of accessible values of χ, and so its boundary is the yield surface in generalised stress space. The Fenchel dual of the dissipation rate function is the **indicator function** $I_X(x, \chi)$ of this set—a function of χ that is zero if χ is a member of the set X and $+\infty$ if it is not. Whilst the indicator of the set of course defines the yield surface, it is not in a convenient mathematical form. A much more convenient function to define is the **gauge function** $\gamma_X(x, \chi)$ of the set (also called the Minkowski distance function), which is a homogeneous first order function that is zero at the origin of generalized force space, and equal to unity at the boundary of the set. The gauge function is formally the **polar** of the support function (in this case the dissipation rate function). The restriction that the set of accessible generalised stress states must include the origin of generalized force space is also not one that causes difficulty.

The gauge function can be used to define a conventional yield function $\bar{y}(x, \chi) = \gamma_X(x, \chi) - 1$ which is zero at yield, negative for accessible generalised

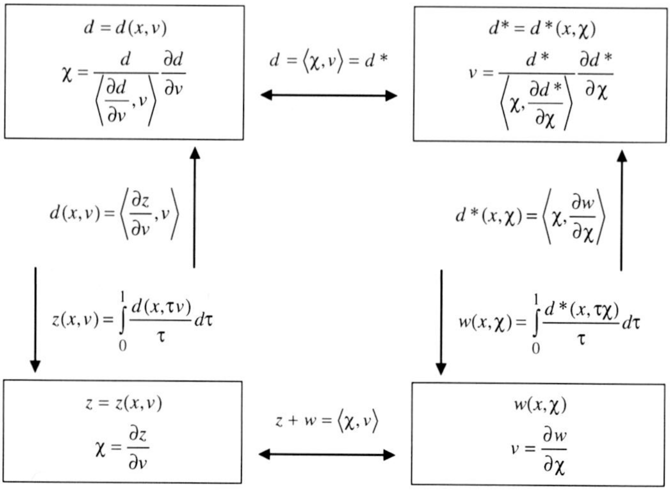

Fig. 4.1 Summary of functions and their relationships

stress states and positive for inaccessible generalised stress states. We call this the **canonical yield function**, as it is a preferred form of the yield function, and use the notation $\bar{y}(x,\chi)$ to distinguish it from other possible forms of the yield function $y(x,\chi)$. Strictly though it is not necessary to introduce the yield function as a separate function, as the gauge could serve our entire purpose here. We introduce the yield function principally for comparison with conventional practice in plasticity theory.

The generalised stresses are given by the **subdifferential** of the dissipation rate function $\chi \in \partial_v d(x,v)$, where we introduce the notation ∂_v to indicate the subdifferential with respect to the variable v. Alternatively the velocities are given by the subdifferential of the indicator function $v \in \partial_\chi \mathrm{I}_\mathrm{X}(x,\chi) = N_\mathrm{X}(x,\chi)$ where $N_\mathrm{X}(x,\chi)$ is the **normal cone** of the set X. It can easily be shown that at yield $\gamma_\mathrm{X}(x,\chi) = 1$ and that the normals to the indicator function and the gauge function are in the same direction, so that one can also write $v \in \lambda' \partial_\chi \gamma_\mathrm{X}(x,\chi)$. If $\gamma_\mathrm{X}(x,\chi)$ is smooth (which is often but not always the case), then this of course simply reduces to $v = \lambda' \frac{\partial \gamma_\mathrm{X}}{\partial \chi}$. We can then note that $d = \langle \chi, v \rangle = \lambda' \left\langle \chi, \frac{\partial \gamma_\mathrm{X}}{\partial \chi} \right\rangle$, but as $\gamma_\mathrm{X}(x,\chi)$ is by definition first order in χ, it follows that $\left\langle \chi, \frac{\partial \gamma_\mathrm{X}}{\partial \chi} \right\rangle = \gamma_\mathrm{X}$, and furthermore at yield we have $\gamma_\mathrm{X} = 1$, so that $\lambda' = d$. Thus the multiplier has a particularly simple physical interpretation. Exactly the same result applies if alternatively we use the canonical yield function form and define $v = \lambda' \frac{\partial \bar{y}}{\partial \chi}$.

A summary of the relationships between the different functions, and the expressions derived from them, is given in Fig. 4.1.

4.3 Examples

4.3.1 One Dimensional Examples

We first consider some simple examples of analogues for materials in one dimension, in which both the velocity v and the generalised force χ are scalar. The inner product $\langle \chi, v \rangle$ in this case is simply χv. Such models do not allow exploration of the shapes of yield surfaces, but allow us to explore the implications of some basic functional forms of d, v, w and d^*. We begin with linear viscosity, and go on to treat rate-independence as a special case of non-linearity.

Note as an aside that in the one-dimensional case, because $d = \chi v = d^*$, then trivially $\chi = d/v$ or $v = d^*/\chi$ furnishes the constitutive response. However, as these trivial results do not carry over to higher dimensionality we do not pursue them, but instead use the one dimensional examples to illustrate results that do generalise to higher dimensionality.

4.3.1.1 Example 1.1: A Linear Viscous Material

Consider $d = \mu v^2$, which is a homogeneous expression of degree $n = 2$. The parameter μ is a viscous constant (for consistency with later results we could write $\mu = k/r$ where k is a strength parameter and r a reference velocity, or even $\mu = \kappa R/r$ where κ is a dimensionless strength parameter and R a reference force). Clearly $\frac{\partial d}{\partial v} = 2\mu v$, the inner product $\left\langle \frac{\partial d}{\partial v}, v \right\rangle = \frac{\partial d}{\partial v} v = 2\mu v^2 = 2d$ and $\lambda = \frac{d}{\left(\frac{\partial d}{\partial v} v\right)} = \frac{1}{2}$. We therefore calculate $\chi = \lambda \frac{\partial d}{\partial v} = \mu v$ and thus the constitutive behaviour is determined. This material behaviour corresponds of course to the well-known Newtonian fluid with a constant viscosity.

Alternatively we could calculate (using either the special procedure or the particular formula for a homogeneous function) $z = \frac{1}{2}\mu v^2$ and immediately derive $\chi = \lambda \frac{\partial d}{\partial v} = \mu v$.

A further alternative is to take the Legendre transform and determine $w = d - z = \frac{1}{2}\mu v^2$. Substituting $\chi = \mu v$ to convert this to a function of the forces one obtains $w = \frac{1}{2\mu}\chi^2$. Differentiation then immediately leads to $v = \frac{\partial w}{\partial \chi} = \frac{\chi}{\mu}$.

Finally, one could calculate $d^* = \frac{\chi^2}{\mu}$. Straightforward manipulation gives $\lambda^* = \frac{1}{2}$, from which we obtain $v = \frac{1}{2}\frac{\partial d^*}{\partial \chi} = \frac{\chi}{\mu}$.

Note very importantly that, applying the above procedures, it is only necessary to know any one of d, z, w and d^* in order to define the entire constitutive response, and (if necessary) to derive all the other functions.

4.3.1.2 Example 1.2: A Non-linear Viscous Material

Now consider $d = kr(|v|/r)^n$ which is a homogeneous expression of degree n. Note that we now use k and r so that all parameters have simple dimensionality. We have to introduce the absolute value of v in order to cope with negative velocities and non-integer values of n. We can derive $\frac{\partial d}{\partial v} = nk\left(\frac{|v|}{r}\right)^{n-1} S(v)$, where $S(x)$ is a generalised signum function (see Notation), so that $\frac{\partial d}{\partial v} v = nkr\left(\frac{|v|}{r}\right)^n = nd$, and $\lambda = \dfrac{d}{\left(\frac{\partial d}{\partial v} v\right)} = \dfrac{1}{n}$. We can therefore calculate $\chi = \frac{1}{n}\frac{\partial d}{\partial v} = k\left(\frac{|v|}{r}\right)^{n-1} S(v)$. The behaviour that this describes is that of a non-Newtonian fluid in which the viscosity is a function of shear rate. If $n > 1$ the fluid is "shear thickening", and the apparent viscosity increases with shear rate. If $n < 1$ the fluid is "shear thinning", with the apparent viscosity decreasing with shear rate. Fluids that exhibit either of these properties are encountered in practice.

Alternatively we could calculate (again using either the special procedure or the particular formula for a homogeneous function) $z = \frac{1}{n} d = \frac{kr}{n}\left(\frac{|v|}{r}\right)^n$, and immediately derive $\chi = \frac{\partial z}{\partial v} = k\left(\frac{|v|}{r}\right)^{n-1} S(v)$.

Taking the Legendre transform we determine $w = d - z = kr\left(\frac{n-1}{n}\right)\left(\frac{|v|}{r}\right)^n$, and recalling that $\chi = k\left(\frac{|v|}{r}\right)^{n-1} S(v)$ or alternatively $\frac{|v|}{r} = \left(\frac{|\chi|}{k}\right)^{1/(n-1)}$ we can convert w to a function of the forces as $w = kr\left(\frac{n-1}{n}\right)\left(\frac{|\chi|}{k}\right)^{n/(n-1)}$. Differentiation confirms the result that $v = \frac{\partial w}{\partial \chi} = r\left(\frac{|\chi|}{k}\right)^{1/(n-1)} S(\chi)$.

Finally, one could calculate $d^* = kr\left(\frac{|\chi|}{k}\right)^{n/(n-1)}$. Straightforward manipulation gives $\lambda^* = \frac{n-1}{n}$, from which we obtain as before $v = \lambda^* \frac{\partial d^*}{\partial \chi} = r\left(\frac{|\chi|}{k}\right)^{1/(n-1)} S(\chi)$.

Again it is only necessary to know any one of d, z, w and d^* to define the entire constitutive response.

4.3.1.3 Example 1.3: Rate Independent Plastic Behavior

We now consider the special case of the above when $n = 1$, and $d = k|v|$. We can immediately derive $\partial d/\partial v = kS(v)$, so that $\frac{\partial d}{\partial v} v = kS(v)v = k|v| = d$ and $\lambda = \dfrac{d}{\left(\frac{\partial d}{\partial v} v\right)} = 1$. We therefore have simply $\chi = \lambda \frac{\partial d}{\partial v} = kS(v)$. The physical interpretation of this expression is clearly that either $v = 0$ and χ is undetermined (but $|\chi| \le k$), or that $v \ne 0$ and $\chi = \pm k$. This corresponds to simple rigid-plastic behaviour. Such a model would serve, for example as a first approximation to the behaviour of a ductile steel in a one-dimensional tension test.

In this case the dissipation rate is a true potential for the force, not merely a pseudopotential, and one is justified in calling it a dissipation rate potential.

Alternatively we could calculate (again using either the special procedure or the particular formula for a homogeneous function) $z = d = k|v|$, and immediately derive the force from $\chi = \partial z/\partial v = kS(v)$.

If we try to form the Legendre transform, we note that $w = d - z = 0$, so the flow potential is identically zero. As noted above, this case is better pursued using the language of convex analysis, but let us first simply explore it using *ad hoc* procedures.

Here we note that, if v is non-zero, then $\chi = \pm k$, so that we could write an expression for a "yield function" $y = |\chi| - k = 0$. There are numerous alternative ways we could choose to write the yield function, for instance $y = \chi^2 - k^2 = 0$, $y = \frac{\chi^2}{k^2} - 1 = 0$, but the particular form of the canonical yield function (the preferred form as defined above) is $\bar{y} = \frac{|\chi|}{k} - 1 = 0$.

Provided that v is non-zero, then $\bar{y}(\chi) = 0$ must be satisfied. Thus, from the existence of a dissipation rate potential, we have established the existence of a yield surface.

We observe that we can write $v = \lambda' \frac{\partial \bar{y}}{\partial \chi} = \lambda' \frac{S(\chi)}{k}$ so that $\bar{y}$ is a pseudopotential for the velocity. In fact in this very simple perfectly plastic model λ' is not further determined, and can simply take an arbitrary magnitude. Note, however, that $\chi v = \lambda' \frac{S(\chi)}{k} \chi = \lambda' \frac{|\chi|}{k} = \lambda'$, so that $\lambda' = d$, which is an example of the general result for the canonical yield function.

Now reconsider this case within the formalism of convex analysis. First we note that the indicator function of the set X of accessible states of χ is formally the Fenchel dual of the dissipation rate function (which in turn can be identified with the support function of the set). Thus $\mathrm{I_X}(\chi) = \sup_v(\chi v - d) = \sup_v(\chi v - k|v|)$. It can readily be established that if $|\chi| \leq k$, $\mathrm{I_X}(\chi) = 0$, whilst if $|\chi| > k$, $\mathrm{I_X}(\chi) = +\infty$. We can write this as $\mathrm{I_X}(\chi) = \mathrm{I}_{[-1,1]}(\chi/k)$. Although this establishes clearly the set of accessible force states as $\mathrm{X} = \{\chi \mid -1 \leq \chi/k \leq 1\} = \{\chi \mid |\chi| \leq k\}$, it is an inconvenient mathematical form.

More useful is to identify the gauge function, which is the polar of the dissipation rate, $\gamma_{\mathrm{X}}(\chi) = \sup_{0 \neq v}\left(\frac{\chi v}{k|v|}\right)$, and clearly $\gamma_{\mathrm{X}}(\chi) = \frac{|\chi|}{k}$. This immediately identifies the canonical yield function as $\bar{y}(\chi) = \gamma_{\mathrm{X}}(\chi) - 1 = \frac{|\chi|}{k} - 1$.

We now explore some more complex forms of the ways that rates can enter the dissipation process. First of all we look at the case of a compound dissipation rate function.

4.3.1.4 Example 1.4: A Compound Dissipation Rate Function—Plasticity and Viscosity

Consider the dissipation rate function $d = k|v| + \mu v^2$. The easiest way to proceed is to use either the special procedure (for a function that can be expressed as a sum of homogeneous functions) or the general procedure to derive $z = k|v| + \frac{\mu}{2}v^2$. This can immediately be used to derive $\chi = \partial z/\partial v = kS(v) + \mu v$. A little thought reveals that χ and v must have the same sign, and that if v is non-zero $|\chi| > k$. Conversely if $|\chi| \leq k$ then $v = 0$. We can therefore invert the expression for χ to obtain $v = \frac{\langle|\chi|-k\rangle}{\mu}S(\chi)$.

The material behaviour described by this model is that of the well-known "Bingham fluid", which is a good approximation to many fluids, e.g. colloidal systems. If the viscous term is relatively small and the plasticity term dominates then this model also serves as the starting point for plastic materials that exhibit a mild rate-dependence, which they commonly do.

Taking the Legendre transform we have $w = d - z = \chi v - k|v| - \frac{\mu}{2}v^2$. After substituting the solution for v above, we can show that $w = \frac{\langle|\chi|-k\rangle^2}{2\mu}$. It follows of course that $v = \frac{\partial w}{\partial \chi} = \frac{\langle|\chi|-k\rangle}{\mu}S(\chi)$.

It is unlikely that one would need to know the form of d^*, but for completeness we note that this can be derived as $d^* = \chi v = \frac{\langle|\chi|-k\rangle|\chi|}{\mu}$.

4.3.1.5 Example 1.5: A Complex Dissipation Rate Function—Rate Process Theory

All the above have involved relatively trivial functional forms for the dissipation rate. We now illustrate a case where the dissipation rate takes a slightly more complex form, and in particular one that cannot be decomposed as a finite sum of homogeneous functions.

Consider $d = \mu r v \sinh^{-1}(v/r)$, where $\mu = k/r$ is a viscous constant and r is a constant that has the dimensions of the velocity. We first determine the force potential by means of the integral transform:

$$z(v) = \int_0^1 \frac{d(\tau v)}{\tau}d\tau = \int_0^1 \mu r v \sinh^{-1}\left(\frac{\tau v}{r}\right)d\tau$$

$$= \mu r v\left[\tau \sinh^{-1}\left(\frac{\tau v}{r}\right) - \sqrt{\tau^2 + \frac{r^2}{v^2}}\right]_0^1 = \mu r\left(v \sinh^{-1}\left(\frac{v}{r}\right) + r - \sqrt{v^2 + r^2}\right)$$

Note if $\left(y = \sinh^{-1}\left(\frac{x}{a}\right)\right)$ then $\left(\frac{dy}{dx} = \frac{1}{\sqrt{a^2+x^2}}\right)$.

It is then straightforward to obtain $\chi = \partial z/\partial v = \mu r \sinh^{-1}(v/r)$, which can of course be inverted to $v = r\sinh\left(\frac{\chi}{\mu r}\right)$.

We can form the Legendre transform $w = d - z = \mu r\left(\sqrt{v^2 + r^2} - r\right)$, and substitute the above solution for v to obtain $w = \mu r^2\left(\cosh\left(\frac{\chi}{\mu r}\right) - 1\right)$. Differentiation easily confirms the result $v = \frac{\partial w}{\partial \chi} = r\sinh\left(\frac{\chi}{\mu r}\right)$.

Again it would be unusual to require the functional form of d^*, but we can note that it is $d^* = r\chi\sinh\left(\frac{\chi}{\mu r}\right)$.

The above results are of interest because the structure of the response $v = r\sinh\left(\frac{\chi}{\mu r}\right)$ is representative of **thermally activated** processes, also called **rate process theory**. At low generalised forces $\sinh\left(\frac{\chi}{\mu r}\right) \approx \frac{\chi}{\mu r}$ and $v \approx \frac{\chi}{\mu}$, so that the response approaches linear viscous behaviour. At high generalised forces $\sinh\left(\frac{\chi}{\mu r}\right) \approx \frac{1}{2}\exp\left(\frac{\chi}{\mu r}\right)$ and $v \approx \frac{r}{2}\exp\left(\frac{\chi}{\mu r}\right)$ or $\chi = \mu r \log\left(\frac{2v}{r}\right)$, which is typical of certain processes in which force increases approximately with the logarithm of velocity. Materials such as soils exhibit rate-dependence of this type (although the strength also has a significant plastic component).

4.3.2 Two-dimensional Examples

We now extend the above models to consider some two-dimensional examples. These give insight into the way the functions determine the ratios between the forces or velocities. In the case of plasticity they also illustrate the derivation of the yield surface. It is useful to consider some simple cases with two velocity components $v = (v_1, v_2)$ (it may be useful think of these as shear strain rates) and corresponding forces $\chi = (\chi_1, \chi_2)$ (which may be thought of as shear stresses). The inner product $\langle \chi, v\rangle$ is simply the dot product $\chi_1 v_1 + \chi_2 v_2$.

4.3.2.1 Example 2.1: A Linear Viscous Material (Based on Example 1.1)

Consider $d = \mu\left(v_1^2 + v_2^2\right)$, which is a homogeneous expression of degree $n = 2$ with the parameter μ a viscous constant. Clearly:

$$\frac{\partial d}{\partial v_1} = 2\mu v_1 \quad \text{and} \quad \frac{\partial d}{\partial v_2} = 2\mu v_2.$$

The inner product $\frac{\partial d}{\partial v} v$ is given by $\frac{\partial d}{\partial v_1} v_1 + \frac{\partial d}{\partial v_2} v_2 = 2\mu\left(v_1^2 + v_2^2\right) = 2d$, so that $\lambda = \dfrac{d}{\left(\frac{\partial d}{\partial v} v\right)} = \dfrac{1}{2}$. We therefore calculate:

$$\chi_1 = \frac{1}{2}\frac{\partial d}{\partial v_1} = \mu v_1 \quad \text{and} \quad \chi_2 = \frac{1}{2}\frac{\partial d}{\partial v_2} = \mu v_2.$$

Thus the entire constitutive behaviour is determined. Importantly, we find that the forces are in the same ratio as the velocities: $\frac{\chi_1}{\chi_2} = \frac{v_1}{v_2}$. This represents well the behaviour of most fluids under combined shear stresses.

Alternatively we could calculate (using either the special procedure or the particular formula for a homogeneous function) $z = \frac{1}{2}d = \frac{\mu}{2}\left(v_1^2 + v_2^2\right)$ and immediately derive:

$$\chi_1 = \frac{\partial z}{\partial v_1} = \mu v_1 \quad \text{and} \quad \chi_2 = \frac{\partial z}{\partial v_2} = \mu v_2.$$

A further alternative is to take the Legendre transform and determine $w = d - z = \frac{\mu}{2}\left(v_1^2 + v_2^2\right)$. Substituting $\chi_1 = \mu v_1$, $\chi_2 = \mu v_2$ to convert this to a function of the forces one obtains $w = \frac{1}{2\mu}\left(\chi_1^2 + \chi_2^2\right)$. Differentiation then immediately leads to

$$v_1 = \frac{\partial w}{\partial \chi_1} = \frac{\chi_1}{\mu} \quad \text{and} \quad v_2 = \frac{\partial w}{\partial \chi_2} = \frac{\chi_2}{\mu}.$$

Finally, one could calculate $d^* = \frac{1}{\mu}\left(\chi_1^2 + \chi_2^2\right)$. Straightforward manipulation gives $\lambda^* = \frac{1}{2}$, from which we obtain

$$v_1 = \frac{1}{2}\frac{\partial d^*}{\partial \chi_1} = \frac{\chi_1}{\mu} \quad \text{and} \quad v_2 = \frac{1}{2}\frac{\partial d^*}{\partial \chi_2} = \frac{\chi_2}{\mu}.$$

As before in the 1-D case, it is only necessary to know any one of d, z, w and d^* to define the entire constitutive response, and (if necessary) to derive all the other functions.

4.3.2.2 Example 2.2: A Non-linear Viscous Material (Based on Example 1.2)

Now consider $d = kr\left(\frac{v_1^2+v_2^2}{r^2}\right)^{n/2}$, which is a homogeneous expression of degree n. Clearly:

$$\frac{\partial d}{\partial v_1} = \frac{nk}{r} v_1 \left(\frac{v_1^2 + v_2^2}{r^2}\right)^{(n/2)-1} \quad \text{and} \quad \frac{\partial d}{\partial v_2} = \frac{nk}{r} v_2 \left(\frac{v_1^2 + v_2^2}{r^2}\right)^{(n/2)-1}.$$

The inner product is $\frac{\partial d}{\partial v_1} v_1 + \frac{\partial d}{\partial v_2} v_2 = nkr\left(\frac{v_1^2+v_2^2}{r^2}\right)^{n/2} = nd$, so that $\lambda = \frac{d}{\left(\frac{\partial d}{\partial v} v\right)} = \frac{1}{n}$. We therefore calculate:

$$\chi_1 = \frac{1}{n}\frac{\partial d}{\partial v_1} = \frac{k}{r}v_1\left(\frac{v_1^2+v_2^2}{r^2}\right)^{(n/2)-1} \quad \text{and} \quad \chi_1 = \frac{1}{n}\frac{\partial d}{\partial v_2} = \frac{k}{r}v_2\left(\frac{v_1^2+v_2^2}{r^2}\right)^{(n/2)-1}.$$

Note that, as for the linear model, the forces are in the same ratio as the velocities. As in Example 1.2, shear thickening or thinning behaviour can be described depending on the value of n.

Alternatively we could calculate (again using either the special procedure or the particular formula for a homogeneous function) $z = \frac{1}{n}d = \frac{kr}{n}\left(\frac{v_1^2+v_2^2}{r^2}\right)^{n/2}$ and immediately derive the same expressions as above for v_1 and v_2 by differentiation of z.

Taking the Legendre transform we determine $w = d - z = kr\left(\frac{n-1}{n}\right)\left(\frac{v_1^2+v_2^2}{r^2}\right)^{n/2}$. Observing that $\frac{\chi_1^2+\chi_2^2}{k^2} = \left(\frac{v_1^2+v_2^2}{r^2}\right)^{n-1}$, we can convert w to a function of the forces as:

$$w = kr\left(\frac{n-1}{n}\right)\left(\frac{\chi_1^2+\chi_2^2}{k^2}\right)^{n/2(n-1)}.$$

Differentiation then immediately leads to

$$v_1 = \frac{\partial w}{\partial \chi_1} = \frac{r}{k}\chi_1\left(\frac{\chi_1^2+\chi_2^2}{k^2}\right)^{(2-n)/2(n-1)}$$

and similarly for v_2. It follows that $\frac{v_1^2+v_2^2}{r^2} = \left(\frac{\chi_1^2+\chi_2^2}{k^2}\right)^{1/(n-1)}$, so it can easily be verified that the constitutive behaviour is the same as derived from d or z.

Finally, one could calculate $d^* = kr\left(\frac{\chi_1^2+\chi_2^2}{k^2}\right)^{n/2(n-1)}$. Straightforward manipulation gives $\lambda^* = \frac{n-1}{n}$, from which we obtain the same expressions as above for the velocities from $v_1 = \frac{n-1}{n}\frac{\partial d^*}{\partial \chi_1}$ and similarly for v_2.

Once again it is only necessary to know any one of d, z, w and d^* to define the entire constitutive response.

4.3.2.3 Example 2.3: Rate Independent Plastic Behaviour (Based on Example 1.3)

We now consider the special case of the above when $n = 1$, and $d = k\sqrt{v_1^2+v_2^2}$. We can immediately derive

$$\frac{\partial d}{\partial v_1} = \frac{kv_1}{\sqrt{v_1^2+v_2^2}}$$

and similarly for v_2. We obtain $\lambda = 1$ and so

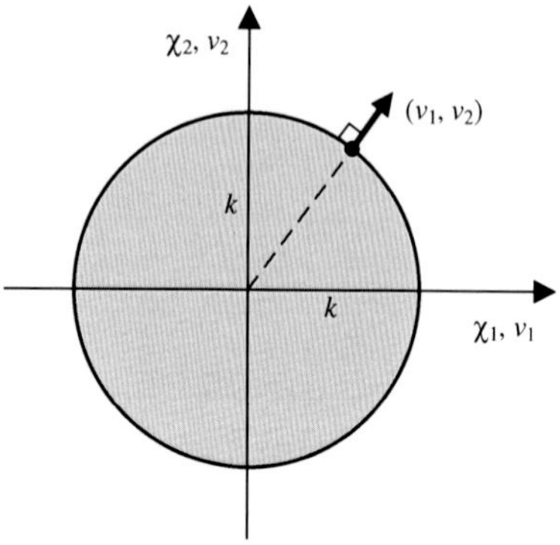

Fig. 4.2 Yield surface and flow vectors for isotropic model

$$\chi_1 = \frac{\partial d}{\partial v_1} = \frac{k v_1}{\sqrt{v_1^2 + v_2^2}}$$

and similarly for χ_2. In this case the dissipation rate is a true potential for the forces, not merely a pseudopotential.

Alternatively we could calculate $z = d$, and immediately derive the same expressions for χ_1 and χ_2.

If we try to form the Legendre transform, we note that $w = d - z = 0$, so the flow potential is identically zero. This case is better pursued using convex analysis, but again we first explore it using *ad hoc* procedures.

Here we note that, squaring and adding expressions for χ_1 and χ_2 one obtains $\chi_1^2 + \chi_2^2 = k^2$. We rearrange this into the form

$$\bar{y} = \frac{\sqrt{\chi_1^2 + \chi_2^2}}{k} - 1 = 0$$

and we call $\bar{y}(\chi_1, \chi_2)$ the canonical yield function. (Canonical because although the yield function can be written in many different ways, but this is a preferred form). The yield surface is shown in Fig. 4.2. Provided that (v_1, v_2) is non-zero, then $\bar{y}(\chi_1, \chi_2) = 0$ must be satisfied. Thus, from the existence of a dissipation rate potential, we have proven the existence of a yield surface.

As for the linear and non-linear viscous models, we can observe that once more the forces are in the same ratio as the velocities. However, we can also observe in Fig. 4.2 that (because the yield surface is circular and centered on the origin), it also follows that the velocity vector is normal to the yield surface. As we shall see below in Example 2.4, it is this feature of normality that is much more fundamental. In anisotropic models the forces are not necessarily in the same proportion as the velocities.

We observe that $v_1 = \lambda' \frac{\partial \bar{y}}{\partial \chi_1} = \frac{\lambda'}{k} \frac{\chi_1}{\sqrt{\chi_1^2 + \chi_2^2}} = \frac{\lambda' \chi_1}{k^2}$ so that $\bar{y}$ is a pseudopotential for the velocities. Note that one can readily show that $\lambda' = d$, an example of the general result for the canonical yield function.

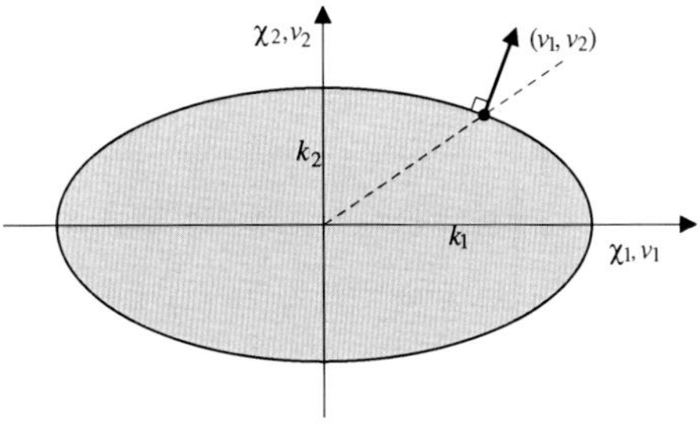

Fig. 4.3 Yield surface and flow vectors for anisotropic model

Now reconsider this case within the formalism of convex analysis. First we note that the indicator function of the set X of accessible states of (χ_1, χ_2) is the Fenchel dual of the dissipation rate function (support function). Thus $I_X(\chi_1, \chi_2) = \sup_{v_1, v_2}(\chi_1 v_1 + \chi_2 v_2 - d) = \sup_{v_1, v_2}\left(\chi_1 v_1 + \chi_2 v_2 - k\sqrt{v_1^2 + v_2^2}\right)$. It can be established that if $\sqrt{\chi_1^2 + \chi_2^2} \leq k$, $I_X(\chi_1, \chi_2) = 0$, whilst if $\sqrt{\chi_1^2 + \chi_2^2} > k$, $I_X(\chi_1, \chi_2) = +\infty$. Although this establishes clearly the set of accessible force states as $X = \left\{ (\chi_1, \chi_2) \mid \sqrt{\chi_1^2 + \chi_2^2} \leq k \right\}$, it is an inconvenient mathematical form.

More useful is to identify the gauge function, which is the polar of the dissipation rate, $\gamma_X(\chi_1, \chi_2) = \sup_{0 \neq v_1, v_2}\left(\frac{\chi_1 v_1 + \chi_2 v_2}{k\sqrt{v_1^2 + v_2^2}}\right)$, and with some manipulation one can show that $\gamma_X(\chi_1, \chi_2) = \frac{\sqrt{\chi_1^2 + \chi_2^2}}{k}$. This immediately identifies the canonical yield function as $\bar{y}(\chi_1, \chi_2) = \gamma_X(\chi_1, \chi_2) - 1 = \frac{\sqrt{\chi_1^2 + \chi_2^2}}{k} - 1$.

4.3.2.4 Example 2.4: Anisotropic Plasticity (Based on Example 2.3)

It was observed that in for Example 2.3 that the velocities are simply "in the same direction" as the forces, as $v_1 = \frac{\lambda' \chi_1}{k^2}$ and $v_2 = \frac{\lambda' \chi_2}{k^2}$, so that $\frac{v_1}{v_2} = \frac{\chi_1}{\chi_2}$ (see Fig. 4.2). However, this is just a special case. As a counterexample consider the anisotropic material represented by $d = \sqrt{k_1^2 v_1^2 + k_2^2 v_2^2}$. It is straightforward to derive $\chi_1 = \frac{k_1^2 v_1}{\sqrt{k_1^2 v_1^2 + k_2^2 v_2^2}}$, $\chi_2 = \frac{k_2^2 v_2}{\sqrt{k_1^2 v_1^2 + k_2^2 v_2^2}}$, leading to $\frac{v_1}{v_2} = \frac{k_2^2}{k_1^2}\frac{\chi_1}{\chi_2}$, so that the velocities are not in the same direction as the forces, see Fig. 4.3. The velocities are, however, normal to the canonical yield surface, which is $\bar{y}(\chi_1, \chi_2) = \sqrt{\frac{\chi_1^2}{k_1^2} + \frac{\chi_2^2}{k_2^2}} - 1 = 0$, and this normality is common to both Figs. 4.2 and 4.3. It is widely observed experimentally that plastic strain rates for non-frictional materials obey this "normality" criterion. A number of important theoretical results follow (as pursued by many authors in the 1950s, with Drucker in particular making numerous notable contributions). For instance, the upper and lower bound theorems of plasticity theory

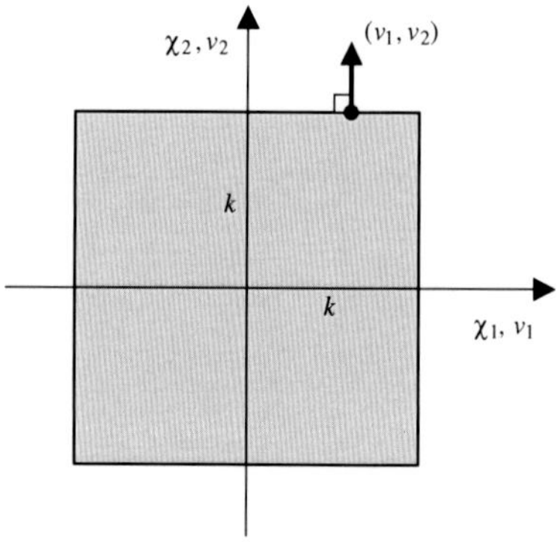

Fig. 4.4 "Square" yield surface

can be proven for materials that exhibit normality. These allow structural collapse loads to be bounded from both above and below, in many cases leading to exact solutions.

4.3.2.5 Example 2.5: A "Square" Yield Surface

The shape of the yield surface depends on the way that the velocities are combined in the dissipation rate function. We introduce the following as a somewhat artificial example. However, it has some relationship with "Tresca" plasticity. We demonstrate that a multilinear yield surface is related to a dissipation rate function which consists of independent additive terms. Consider the function $d = k(|v_1| + |v_2|)$. We can immediately derive:

$$\frac{\partial d}{\partial v_1} = kS(v_1)$$

and similarly for v_2. We obtain $\lambda = 1$ and so

$$\chi_1 = \frac{\partial d}{\partial v_1} = kS(v_1)$$

and similarly for χ_2. The gauge function is given by $\gamma_X(\chi_1, \chi_2) = \sup\limits_{0 \neq v_1, v_2} \left(\frac{\chi_1 v_1 + \chi_2 v_2}{k(|v_1| + |v_2|)} \right)$. With some manipulation one can show that $\gamma_X(\chi_1, \chi_2) = \frac{\max(|\chi_1|, |\chi_2|)}{k}$ and the canonical yield surface is therefore $\bar{y}(\chi_1, \chi_2) = \gamma_X(\chi_1, \chi_2) - 1 = \frac{\max(|\chi_1|, |\chi_2|)}{k} - 1$, a "square" yield surface, Fig. 4.4.

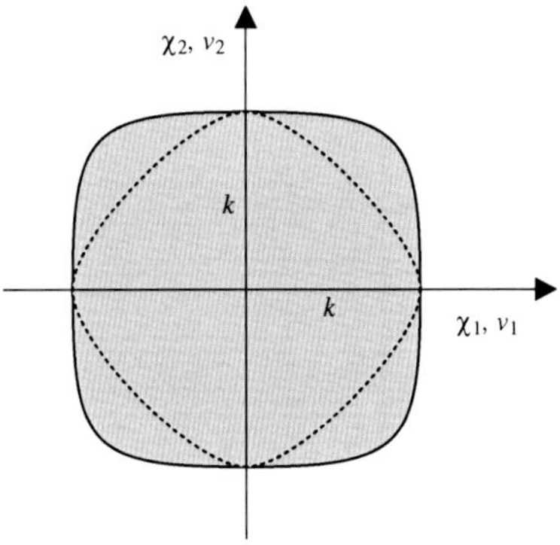

Fig. 4.5 Superelliptical yield surface

4.3.2.6 Example 2.6: More General Shapes of Yield Surface

More generally, consider the function $d = k(|v_1|^m + |v_2|^m)^{1/m}$, $1 \le m < \infty$. We can derive

$$\frac{\partial d}{\partial v_1} = k|v_1|^{m-1}S(v_1)(|v_1|^m + |v_2|^m)^{(1-m)/m}$$

and similarly for v_2. We obtain $\lambda = 1$ and so

$$\chi_1 = \frac{\partial d}{\partial v_1} = k|v_1|^{m-1}S(v_1)(|v_1|^m + |v_2|^m)^{(1-m)/m}$$

and similarly for χ_2. Adopting the *ad hoc* approach to derivation of the yield surface, we observe that, by raising appropriate powers of χ_1 and χ_2 and adding, one obtains $\left(\frac{|\chi_1|}{k}\right)^{m'} + \left(\frac{|\chi_2|}{k}\right)^{m'} = 1$, where $m' = m/(m-1)$. We can rearrange this into the canonical form of the yield surface:

$$\bar{y} = \frac{\left(|\chi_1|^{m'} + |\chi_2|^{m'}\right)^{1/m'}}{k} - 1 = 0.$$

In general this is a "superelliptical" shaped yield surface. Figure 4.5 shows two such surfaces—the solid line for the case $m < 2$ and the dotted line for $m > 2$. For $m = 1$ the surface becomes square (Fig. 4.4), for $m = 2$ it is the circular surface (Fig. 4.2) and for $m \to \infty$ (Example 2.6) it becomes the diamond-shaped surface $\bar{y} = \frac{|\chi_1| + |\chi_2|}{k} - 1 = 0$ (Fig. 4.6). In this last case the dissipation rate function can be written as $d = k\max(|v_1|, |v_2|)$. The Author is, however, not aware of practical applications of this sort of model.

Fig. 4.6 "Diamond-shaped" yield surface

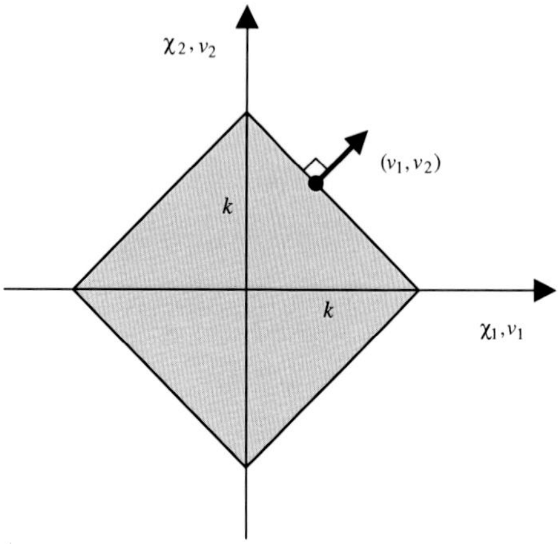

4.3.2.7 Example 2.7: An Asymmetric Yield Surface

Yield surfaces with more specialised application can be derived from more complex dissipation rate functions. The following model, for instance, serves as the starting point for the modelling of the behaviour of soft clays in the discipline of soil mechanics. Consider the dissipation rate function $d = k_1 v_1 + \sqrt{k_1^2 v_1^2 + k_2^2 v_2^2}$, which gives $\chi_1 = k_1 + \frac{k_1^2 v_1}{\sqrt{k_1^2 v_1^2 + k_2^2 v_2^2}}$, $\chi_2 = \frac{k_2^2 v_2}{\sqrt{k_1^2 v_1^2 + k_2^2 v_2^2}}$, leading immediately to a yield surface $y(\chi_1, \chi_2) = \sqrt{\frac{(\chi_1 - k_1)^2}{k_1^2} + \frac{\chi_2^2}{k_2^2}} - 1 = 0$, which is shown in Fig. 4.7. This can be reduced to canonical form as $\bar{y}(\chi_1, \chi_2) = \frac{(\chi_1/k_1)^2 + (\chi_2/k_2)^2}{2\chi_1/k_1} - 1 = 0$. (Note that, in spite of the lack of the modulus sign on the first term in the dissipation rate function, the dissipation rate function is never negative, as the magnitude of the second term, which is always positive for non-zero velocities, is never less than the magnitude of the first.) Yield surfaces of the form shown in Fig. 4.7 find an important application in soil mechanics, in which v_1 would play the role of the volumetric strain rate and v_2 the role of the deviatoric strain rate. In this context the elliptical yield locus, passing through the origin, is identified as the "Modified Cam-Clay" yield surface [8]. Note that there are alternative approaches for the definition of this yield surface which involve use of combined isotropic and kinematic hardening. To emphasise the link with soil mechanics in which the (p,q) notation is used for the stresses we could rewrite the dissipation rate as $d = \frac{p_c}{2}\left(v_p + \sqrt{v_p^2 + M^2 v_q^2}\right)$ and the yield surface (in non-canonical form) as $y = \chi_p(\chi_p - p_c) + \frac{\chi_q^2}{M^2} = 0$. As the generalized stresses are in this case equal to the true stresses, this also leads to the yield surface $y = p(p - p_c) + \frac{q^2}{M^2} = 0$, which is familiar in soil mechanics.

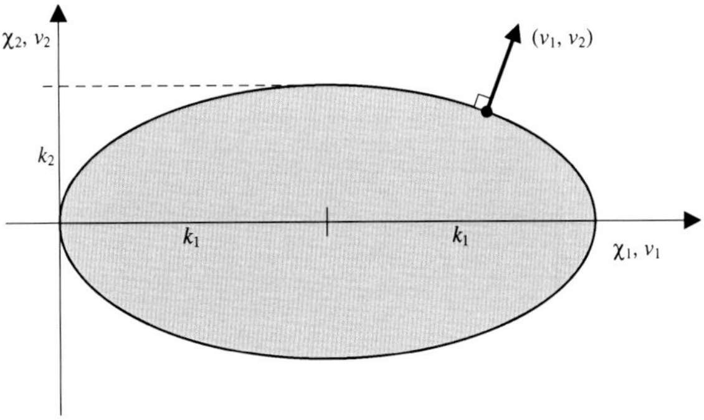

Fig. 4.7 Asymmetric yield surface

Table 4.1 Summary of forms of functions for rate-dependent models

Example	z	w
1.1: Linear viscosity	$\frac{\mu}{2}v^2$	$\frac{1}{2\mu}\chi^2$
1.2: Nonlinear viscosity	$\frac{kr}{n}\left(\frac{\lvert v\rvert}{r}\right)^n$	$kr\left(\frac{n-1}{n}\right)\left(\frac{\lvert\chi\rvert}{k}\right)^{\frac{n}{n-1}}$
1.4: Visco-plastic	$k\lvert v\rvert + \frac{\mu}{2}v^2$	$\frac{\langle\lvert\chi\rvert-k\rangle^2}{2\mu}$
1.5: Rate process theory	$\mu r\left(v\sinh^{-1}\left(\frac{v}{r}\right) + r - \sqrt{v^2+r^2}\right)$	$\mu r^2\left(\cosh\left(\frac{\chi}{\mu r}\right) - 1\right)$
2.1: Linear viscosity	$\frac{\mu}{2}\left(v_1^2+v_2^2\right)$	$\frac{1}{2\mu}\left(\chi_1^2+\chi_2^2\right)$
2.2: Nonlinear viscosity	$\frac{kr}{n}\left(\frac{v_1^2+v_2^2}{r^2}\right)^{\frac{n}{2}}$	$kr\left(\frac{n-1}{n}\right)\left(\frac{\chi_1^2+\chi_2^2}{k^2}\right)^{\frac{n}{2(n-1)}}$

Summaries of the key functions used in the above examples are given in Tables 4.1 and 4.2.

4.3.3 Advanced Considerations in Continuum Mechanics

Each of the above models serves as the starting point for a full continuum model, which we express here using the subscript tensor notation. In order to avoid repetition of the development we simply show in Table 4.3 how the dissipation rate functions for the simpler conceptual models map into the full continuum models. We introduce the deviator of the velocity tensor $v'_{ij} = v_{ij} - \frac{1}{3}v_{kk}\delta_{ij}$, recognising that in most models the dissipation rate depends only on the deviatoric strain rate. In the development of such models it is usual also to introduce the incompressibility constraint $v_{kk} = 0$. See [3, 4] for discussion of constraints in this context.

If any of the parameters defining the dimensions of the yield surfaces in the above models (for example k, k_1, k_2) is a function of any of the state parameters x, and furthermore $\dot{x} = \dot{x}(v)$ (where $\dot{x}(v)$ must be a homogeneous first order function

Table 4.2 Summary of forms of functions for rate-independent (plastic) models

Example	$d = z$	$\bar{y} = 0$
1.3: Plasticity	$k\lvert v\rvert$	$\frac{\lvert\chi\rvert}{k} - 1$
2.3: Circular surface	$k\sqrt{v_1^2 + v_2^2}$	$\frac{\sqrt{\chi_1^2+\chi_2^2}}{k} - 1$
2.4: Anisotropy	$\sqrt{k_1^2 v_1^2 + k_2^2 v_2^2}$	$\sqrt{\frac{\chi_1^2}{k_1^2} + \frac{\chi_2^2}{k_2^2}} - 1$
2.5: Square surface	$k(\lvert v_1\rvert + \lvert v_2\rvert)$	$\frac{\max(\lvert\chi_1\rvert, \lvert\chi_2\rvert)}{k} - 1$
2.6: Superelliptical surface	$k(\lvert v_1\rvert^m + \lvert v_2\rvert^m)^{1/m}$	$\frac{\left(\lvert\chi_1\rvert^{m'} + \lvert\chi_2\rvert^{m'}\right)^{1/m'}}{k} - 1$
2.6: Diamond-shaped surface	$k\max(\lvert v_1\rvert, \lvert v_2\rvert)$	$\frac{\lvert\chi_1\rvert + \lvert\chi_2\rvert}{k} - 1$
2.7: Asymmetric surface	$k_1 v_1 + \sqrt{k_1^2 v_1^2 + k_2^2 v_2^2}$	$\frac{(\chi_1/k_1)^2 + (\chi_2/k_2)^2}{2\chi_1/k_1} - 1$

Table 4.3 Development of dissipation rate functions for 1-D, 2-D and continuum models

Example	1-D	2-D	Continuum
Linear viscosity	μv^2 (Ex. 1.1)	$\mu(v_1^2 + v_2^2)$ (Ex. 2.1)	$\mu v'_{ij} v'_{ji}$
Nonlinear viscosity	$kr\left(\frac{\lvert v\rvert}{r}\right)^n$ (Ex. 1.2)	$kr\left(\frac{v_1^2+v_2^2}{r^2}\right)^{n/2}$ (Ex. 2.2)	$\mu\left(\frac{v'_{ij} v'_{ji}}{r^2}\right)^{n/2}$
Plasticity	$k\lvert v\rvert$ (Ex. 1.3)	$k\sqrt{v_1^2 + v_2^2}$ (Ex. 2.3)	$k\sqrt{v'_{ij} v'_{ji}}$
Visco-plastic	$k\lvert v\rvert + \mu v^2$ (Ex. 1.4)		$k\sqrt{v'_{ij} v'_{ji}} + \mu v'_{ij} v'_{ji}$
Rate process theory	$\mu r v \sinh^{-1}\left(\frac{v}{r}\right)$ (Ex. 1.5)		$\mu r\sqrt{v'_{ij} v'_{ji}} \sinh^{-1}\left(\frac{\sqrt{v'_{ij} v'_{ji}}}{r}\right)$
Critical state family		$k_1 v_1 + \sqrt{k_1^2 v_1^2 + k_2^2 v_2^2}$ (Ex. 2.7) or $\frac{p_c}{2}\left(v_p + \sqrt{v_p^2 + M^2 v_q^2}\right)$	$\frac{p_c}{2}\left(v_{ii} + \sqrt{v_{ii} v_{jj} + \frac{2}{3} M^2 v'_{ij} v'_{ji}}\right)$

of v), then the models involve some form of strain hardening, with the specific details depending on the forms of the relevant functions. For instance, if we modify the model in Example 2.3 so that $d = (k_0 + k_1 x)\sqrt{v_1^2 + v_2^2}$ and $\dot{x} = \sqrt{v_1^2 + v_2^2}$, then the resulting model involves a simple linear hardening with strain.

Houlsby and Puzrin [4] discuss other ways to develop hardening models, depending on the functional form of free energy functions in the "hyperplasticity" formulation that they define. In general, kinematic hardening is most naturally described by an approach in which extra terms are introduced in the energy functions, whilst isotropic hardening is most naturally described by the approach outlined above in which the dissipation rate function is altered.

Houlsby and Puzrin also pursue the fact that orthogonality does not necessarily imply "normality" or "associated flow" in the sense that is usually meant in classical plasticity theory: i.e. that the plastic strain vector is normal to the yield surface in true stress space. In order to address this issue it is necessary to make careful distinctions between the true stress and the generalised stress, and such issues lie outside the scope of this chapter.

References

1. Collins, I.F., Houlsby, G.T.: Application of thermomechanical principles to the modelling of geotechnical materials. Proc. R. Soc. Lond. Ser. A **453**, 1975–2001 (1997)
2. Houlsby, G.T.: A study of plasticity theory and its applicability to soils. PhD Thesis, Cambridge University (1981)
3. Houlsby, G.T., Puzrin, A.M.: A thermomechanical framework for constitutive models for rate-independent dissipative materials. Int. J. Plast **16**(9), 1017–1047 (2000)
4. Houlsby, G.T., Puzrin, A.M.: Principles of hyperplasticity. Springer, London (2006)
5. Lemaitre, J., Chaboche, J.-L.: Mechanics of solid materials. Cambridge University Press (1990)
6. Maugin, G.A.: The thermodynamics of plasticity and fracture. Cambridge University Press (1992)
7. Maugin, G.A.: The thermodynamics of nonlinear irreversible behaviours. World Scientific, Singapore (1999)
8. Roscoe, K.H., Burland, J.B.: On the generalized behavior of 'wet' clay. In: Heyman, J., Leckie, F.A. (eds.) Engineering Plasticity. Cambridge University Press, pp. 535–609 (1968)
9. Truesdell, C.: A first course in rational continuum mechanics. Academic Press, New York (1977)
10. Ziegler, H.: An Introduction to thermomechanics. North Holland, Amsterdam (1977). (2nd edition 1983)

Chapter 5
Fluctuations, Trajectory Entropy and Ziegler's Maximum Entropy Production Principle

Vladimir D. Seleznev and Leonid M. Martyushev

Abstract This chapter discusses two current interpretations of the maximum entropy production principle—as a physical principle and as an inference procedure. A simple model of relaxation of an isolated system towards equilibrium is considered for this purpose.

Table of Notation

Symbol Meaning

Roman Symbols

t	Time
$A_i(t)$	Set of macroscopic state variables
A_i^{eq}	Set of macroscopic state variables in equilibrium
$W(\boldsymbol{\alpha})$	Probability of finding the equilibrium system in macrostate $\boldsymbol{\alpha}$
$P(\boldsymbol{\alpha}_0\|\boldsymbol{\alpha}, \tau)$	Conditional probability that, given the system is in initial state $\boldsymbol{\alpha}_0$, it will be in state α after time τ
$S(\boldsymbol{\alpha})$	Entropy in state $\boldsymbol{\alpha}$
$S_{\mathrm{tr}}(\boldsymbol{\alpha}_0\|\boldsymbol{\alpha}, \tau)$	Trajectory entropy that, given the system is in initial state $\boldsymbol{\alpha}_0$, it will be in state $\boldsymbol{\alpha}$ after time τ
$\Delta S(\boldsymbol{\alpha})$	$S(\boldsymbol{\alpha}) - S(0)$
$\boldsymbol{X}_i$	Thermodynamic forces
$\boldsymbol{J}_i$	Thermodynamic fluxes
L_{ij}	Kinetic coefficients

V. D. Seleznev · L. M. Martyushev (✉)
Ural Federal University, 19 Mira Street, Ekaterinburg, Russia 620002,
e-mail: leonidmartyushev@gmail.com

L. M. Martyushev
Institute of Industrial Ecology, Russian Academy of Sciences, 20 S. Kovalevskaya Street, Ekaterinburg, Russia 620219,

R. C. Dewar et al. (eds.), *Beyond the Second Law*,
Understanding Complex Systems, DOI: 10.1007/978-3-642-40154-1_5,

Greek Symbols

α_i	$=A_i(t) - A_i^{eq}$
$\boldsymbol{\alpha}$	Macroscopic (non-equilibrium) state vector with components α_i
τ	Observation time
$\Gamma(\boldsymbol{\alpha})$	Number of microscopic states realizing macrostate $\boldsymbol{\alpha}$
$\Gamma(\boldsymbol{\alpha}_0 \vert \boldsymbol{\alpha}, \tau)$	Number of microscopic trajectories realizing the transition from $\boldsymbol{\alpha}_0$ to $\boldsymbol{\alpha}$ during time τ
Ω, Ξ	Normalization constants
σ	Entropy production (or entropy production density for the model under consideration)
β_{ij}	Coefficient that is inversely proportional to the variance of $\boldsymbol{\alpha}$ relative to its equilibrium value ($\boldsymbol{\alpha} = 0$)
γ_{ij}	Coefficient that is inversely proportional to the variance of $\boldsymbol{\alpha}$ relative to its average (most probable) value $\alpha^*(\boldsymbol{\alpha}_0, \tau)$ during the transition from $\boldsymbol{\alpha}_0$ during time τ
γ_{ij}^0	$=\gamma_{ij}\tau$

We consider the relaxation of an isolated system towards equilibrium, assuming detailed balance and Onsager's fluctuation approximation. Small deviations from equilibrium are quantified in terms of two state variables. For this system, expressions are obtained for both the dependence of trajectory entropy on random thermodynamic fluxes and the dependence of entropy production on the most probable thermodynamic fluxes. Onsager's linear relations are obtained from this model using two methods: maximization of trajectory entropy and Ziegler's maximization of entropy production. We discuss two current interpretations of the maximum entropy production principle—as a physical principle and as an inference procedure.

5.1 Introduction

There are several recent examples of the successful application of the maximum entropy production (MaxEP) principle in physics (kinetic theory of gases, hydrodynamics, theories of crystallization and radiation etc.), biology, and chemistry (see [1–3] and Part III of this volume). This has led naturally to an interest in the theoretical basis of MaxEP and its relation to other principles [1–12].

MaxEP has been considered as a natural generalization of the second law of thermodynamics, starting from the work by Kohler and Ziman [13, 14] that used MaxEP to solve the Boltzmann equation. While the second law states that the entropy of an isolated non-equilibrium system increases, MaxEP states that this increase occurs at the maximum possible rate. Moreover, while the second law leads to the basic thermodynamics relations and treatment of phase transitions for equilibrium (quasistatic) processes, MaxEP has led to the basic laws of nonequilibrium thermodynamics and nonequilibrium (kinetic) phase transitions [15–17]. On this basis, MaxEP has been interpreted as an independent entropic principle that complements and generalises the second law. Such an interpretation was proposed in particular by Ziegler, who independently developed a version of MaxEP involving entropy production variation subject to fixed thermodynamic forces, leading to an expression for thermodynamic fluxes (in particular, Onsager's linear relations [1, 15, 16]). As a result, many researchers have considered MaxEP to be a new and important principle of the physics of nonequilibrium processes; this viewpoint has been developed either from statistical (kinetic) [13, 14] or thermodynamic considerations [15, 16].

However, a somewhat different viewpoint on MaxEP has also been developed. The Boltzmann–Gibbs entropy of a macroscopic state is a measure of the number of microscopic states that realize this macroscopic state. In equilibrium, the number of microscopic states is a maximum, corresponding to a maximum in the entropy. For certain models, this microstate interpretation of entropy represents a statistical justification of the second law of thermodynamics. Later, Shannon and Jaynes generalized this view of entropy: by analogy with the Boltzmann–Gibbs entropy, they introduced the so-called information entropy applicable to the description of objects of any nature [18, 19]. Maximization of information entropy (also known in the literature as MaxEnt) is then a general method for determining the probability of a particular state of the system. Subsequently this approach was applied to the relationship between the probability of a nonequilibrium process and the number of microscopic trajectories realizing it, so that maximization of information entropy (a measure of the number of microscopic trajectories) defines the most probable macroscopic evolution of a nonequilibrium system (see e.g. [4–11], [20–25]).

This approach has led to the suggestion that, for non-equilibrium processes, maximization of information entropy associated with microscopic trajectories leads to MaxEP. While this suggestion has yet to be established rigorously, it is very promising for understanding and illustrating (using specific models) the microscopic interpretation of MaxEP and its relation to other results in non-equilibrium physics. This approach has given rise to the idea that, if MaxEP is a simple consequence of Jaynes' information entropy maximization, then it is simply "an inference algorithm that translates physical assumptions into macroscopic predictions" [4–11], rather than a physical law or principle. For brevity, we will refer to this as the informational (or MaxEnt) interpretation of MaxEP, as opposed to the above statistical and thermodynamic interpretation.

In order to obtain a better understanding of these two interpretations of MaxEP, and to compare them, it would be desirable to find a problem to which both approaches can be applied independently. In this chapter, our objective is to consider an elementary transfer problem using these two contrasting approaches—Ziegler's MaxEP formalism (thermodynamic) and maximization of trajectory entropy (informational)—to determine the similarities and differences between them.

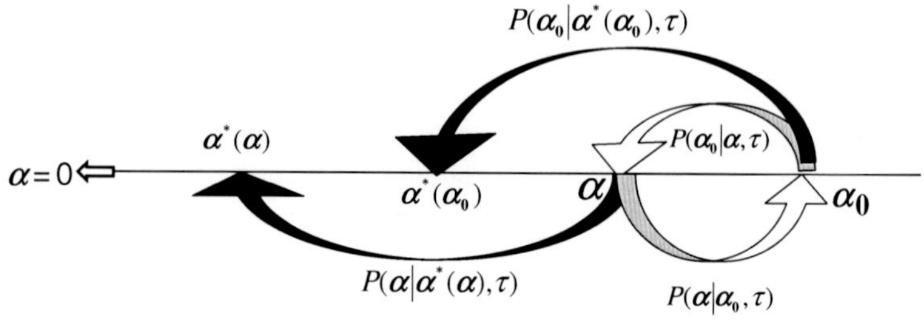

Fig. 5.1 Illustrating the transitions between states of the system discussed in the text. The equilibrium state is $\boldsymbol{\alpha} = 0$. $P(\boldsymbol{\alpha}|\boldsymbol{\alpha}_0, \tau) < P(\boldsymbol{\alpha}|\boldsymbol{\alpha}^*(\boldsymbol{\alpha}), \tau) = \Xi = P(\boldsymbol{\alpha}_0|\boldsymbol{\alpha}^*(\boldsymbol{\alpha}_0), \tau) > P(\boldsymbol{\alpha}_0|\boldsymbol{\alpha}, \tau)$

The remainder of the chapter is structured as follows. Section 5.2 introduces the problem, the basic model assumptions and equations. Sections 5.3 and 5.4 consider the problem through the trajectory entropy maximization method and Ziegler's procedure, respectively. Concluding Sect. 5.5 summarizes the main results, and briefly discusses the different interpretations of MaxEP.

5.2 Onsager's Model and Linear Thermodynamic Relations

We consider an isolated system with possible fluctuations (of molecular speed distribution, etc.). An arbitrary non-equilibrium macroscopic state of the system will be described by the set of state variables $A_i(t)$ that acquire the values A_i^{eq} in equilibrium. We denote the difference between the state variables and their equilibrium values by $\alpha_{\mathrm{i}} = A_i(t) - A_i^{eq}$. For brevity we introduce the state vector $\boldsymbol{\alpha}$ with components α_{i}. We assume that the observation time τ is much smaller than the time of relaxation to equilibrium, i.e. we consider the instantaneous response of the system in a non-equilibrium state.

Associated with macroscopic state $\boldsymbol{\alpha}$ is the number of microscopic states $\Gamma(\boldsymbol{\alpha})$ which, as is known [26, 27], is a maximum in equilibrium. As usual, we assume that all microscopic states corresponding to $\boldsymbol{\alpha}$ are equiprobable. We define the probability of finding an equilibrium system in the state $\boldsymbol{\alpha}$ as

$$W(\boldsymbol{\alpha}) \propto \Gamma(\boldsymbol{\alpha}) \tag{5.1}$$

The process of relaxation to equilibrium will be characterized by the conditional probability that, given the system is initially in state $\boldsymbol{\alpha}_0$, it will be in state $\boldsymbol{\alpha}$ after time τ [27]. Here we denote this probability by $P(\boldsymbol{\alpha}_0|\boldsymbol{\alpha}, \tau)$ (Fig. 5.1). We shall assume that this conditional transition probability is proportional to the number of microscopic trajectories from state $\boldsymbol{\alpha}_0$ to state $\boldsymbol{\alpha}$ during time τ, i.e.

$$P(\boldsymbol{\alpha}_0|\boldsymbol{\alpha}, \tau) \propto \Gamma(\boldsymbol{\alpha}_0|\boldsymbol{\alpha}, \tau). \tag{5.2}$$

Let the transition from $\boldsymbol{\alpha}_0$ to $\alpha^*(\boldsymbol{\alpha}_0)$ during time τ be the most probable transition; according to (5.2), this macroscopic transition is associated with the maximum number of microscopic trajectories.

Following [26], we define the entropies in the state $\boldsymbol{\alpha}$ and in the equilibrium state $\boldsymbol{\alpha} = 0$ by $S(\boldsymbol{\alpha}) = \ln\Gamma(\boldsymbol{\alpha})$ and $S(0) = \ln\Gamma(0)$, respectively. Then their difference is

$$\Delta S(\boldsymbol{\alpha}) = S(\boldsymbol{\alpha}) - S(0) = \ln\frac{\Gamma(\boldsymbol{\alpha})}{\Gamma(0)} = \ln\frac{W(\boldsymbol{\alpha})}{W(0)} \tag{5.3}$$

By analogy, we introduce the trajectory entropies $S_{\rm tr}(\boldsymbol{\alpha}_0|\boldsymbol{\alpha}, \tau) = \ln\Gamma(\boldsymbol{\alpha}_0|\boldsymbol{\alpha}, \tau)$ and $S_{\rm tr}(\boldsymbol{\alpha}_0|\alpha^*(\boldsymbol{\alpha}_0), \tau) = \ln\Gamma(\boldsymbol{\alpha}_0|\alpha^*(\boldsymbol{\alpha}_0), \tau)$, so that their difference is

$$\begin{aligned} \Delta S_{tr}(\alpha_0|\boldsymbol{\alpha}, \tau) &= S_{tr}(\alpha_0|\boldsymbol{\alpha}, \tau) - S_{tr}(\alpha_0|\alpha^*(\alpha_0), \tau) \\ &= \ln\frac{\Gamma(\alpha_0|\boldsymbol{\alpha}, \tau)}{\Gamma(\alpha_0|\alpha^*(\alpha_0), \tau)} = \ln\frac{P(\alpha_0|\boldsymbol{\alpha}, \tau)}{P(\alpha_0|\alpha^*(\alpha_0), \tau)} \end{aligned} \tag{5.4}$$

As a result

$$P(\boldsymbol{\alpha}_0|\boldsymbol{\alpha}, \tau) = P(\boldsymbol{\alpha}_0|\alpha^*(\alpha_0), \tau) \cdot \mathbf{e}^{\Delta S_{tr}(\alpha_0|\boldsymbol{\alpha}, \tau)} \tag{5.5}$$

In the following, we will consider processes for which $P(\boldsymbol{\alpha}_0|\alpha^*(\boldsymbol{\alpha}_0), \tau)$ is independent of $\boldsymbol{\alpha}_0$. Such an approximation is quite common [27, 28].

In equilibrium, the number of system transitions in the forward $\boldsymbol{\alpha}_0 \to \boldsymbol{\alpha}$ and reverse $\boldsymbol{\alpha} \to \boldsymbol{\alpha}_0$ directions during time τ should be equal. It can then be shown [27, 29, 30] that the condition of detailed balance holds, i.e.

$$W(\alpha_0)P(\alpha_0|\boldsymbol{\alpha}, \tau) = W(\boldsymbol{\alpha})P(\boldsymbol{\alpha}|\alpha_0, \tau). \tag{5.6}$$

Condition (5.6) holds for fluctuations of an equilibrium system [27, 29, 30]. Following the classic work of Onsager [29, 30] and the monograph [27] where this approach is described in its most complete form, we suppose (Onsager's hypothesis) that this detailed balance condition also holds for conditions close to equilibrium. In other words, the evolution of an equilibrium system from initial state $\boldsymbol{\alpha}_0$ will be similar to the evolution of a specifically prepared close-to-equilibrium system brought to the same state $\boldsymbol{\alpha}_0$ and then left to spontaneously relax.

Furthermore, for notational simplicity, here we consider that $\boldsymbol{\alpha}$ has only two components; the results below may be generalized to any number of components. According to (5.6), we have:

$$\frac{W(\alpha_1, \alpha_2)}{W(\alpha_{10}, \alpha_{20})} = \frac{P(\alpha_{10}, \alpha_{20}|\alpha_1, \alpha_2, \tau)}{P(\alpha_1, \alpha_2|\alpha_{10}, \alpha_{20}, \tau)}. \tag{5.7}$$

Using (5.3)–(5.5), Eq. (5.7) can also be written in the form

$$\Delta S(\boldsymbol{\alpha}) - \Delta S(\alpha_0) = \Delta S_{tr}(\alpha_0|\boldsymbol{\alpha}, \tau) - \Delta S_{tr}(\boldsymbol{\alpha}|\alpha_0, \tau). \tag{5.8}$$

In Onsager's classic study [29, 30], the condition of detailed balance, together with the assumption that the mean change of $\boldsymbol{\alpha}$ during time τ is linearly related to $\boldsymbol{\alpha}$ itself, leads to the so-called reciprocal relations and to Gaussian forms for $W(\boldsymbol{\alpha})$ and $P(\boldsymbol{\alpha}_0|\boldsymbol{\alpha}, \tau)$. Here we consider an inverse problem: the Gaussian forms for $W(\boldsymbol{\alpha})$ and $P(\boldsymbol{\alpha}_0|\boldsymbol{\alpha}, \tau)$ are postulated and the linear relations for the change of $\boldsymbol{\alpha}$ are obtained using the principle of detailed balance. As we will show in Sects. 5.3 and 5.4, this statement of the problem provides the shortest route to expressions for the trajectory entropy, the entropy production, and other quantities required for the objective set out in the Introduction.

Thus, we assume only small deviations from equilibrium. In this case, the Gaussian distribution is a frequently used approximation for the probability of deviations from equilibrium [26, 27]:

$$W(\boldsymbol{\alpha}) = \Omega \exp(-\beta_{11}\alpha_1^2 - \beta_{22}\alpha_2^2 - 2\,\beta_{12}\alpha_1\alpha_2), \tag{5.9}$$

and for the trajectory probability [27–30]:

$$P(\boldsymbol{\alpha}_0|\boldsymbol{\alpha}, \tau) = \Xi \cdot \exp\{-\gamma_{11}(\alpha_1^*(\boldsymbol{\alpha}_0, \tau) - \alpha_1)^2) \\ - \gamma_{22}(\alpha_2^*(\boldsymbol{\alpha}_0, \tau) - \alpha_2)^2 - 2\gamma_{12}(\alpha_1^*(\boldsymbol{\alpha}_0, \tau) - \alpha_1)(\alpha_2^*(\boldsymbol{\alpha}_0, \tau) - \alpha_2)\} \tag{5.10}$$

where Ω, Ξ are normalization constants (independent of $\boldsymbol{\alpha}$, $\boldsymbol{\alpha}_0$, but dependent on A_i^{eq}); β_{ij} is a coefficient that is inversely proportional to the variance of $\boldsymbol{\alpha}$ relative to its equilibrium value ($\boldsymbol{\alpha} = 0$); and γ_{ij} is a coefficient[1] that is inversely proportional to the variance of $\boldsymbol{\alpha}$ relative to its average (most probable) value $\alpha_i^*(\boldsymbol{\alpha}_0, \tau)$ during the transition from state $\boldsymbol{\alpha}_0$ during time τ. It should be noted that, within this approximation, these variances are assumed independent of $\boldsymbol{\alpha}_0$, $\boldsymbol{\alpha}$ [27–30]. It should also be emphasized that β_{ij} does not dependent on τ (because it characterizes fluctuations in the equilibrium state); in contrast, γ_{ij} increases with decreasing τ (for $\tau \to 0$, $\alpha_i^* \to \alpha_{i0}$ and the distribution tends to a delta function). We make the simple assumption that $\gamma_{ij} = \gamma_{ij}^0/\tau$ [27–30], where γ_{ij}^0 is some constant.

By substituting (5.9) and (5.10) into (5.7), we obtain

$$\frac{\exp(-\beta_{11}\alpha_1^2 - \beta_{22}\alpha_2^2 - 2\beta_{12}\alpha_1\alpha_2)}{\exp(-\beta_{11}\alpha_{10}^2 - \beta_{22}\alpha_{20}^2 - 2\beta_{12}\alpha_{10}\alpha_{20})} \\ = \frac{\exp\left(-\gamma_{11}(\alpha_1^*(\boldsymbol{\alpha}_0, \tau) - \alpha_1)^2 - \gamma_{22}(\alpha_2^*(\boldsymbol{\alpha}_0, \tau) - \alpha_2)^2 - 2\gamma_{12}(\alpha_1^*(\boldsymbol{\alpha}_0, \tau) - \alpha_1)(\alpha_2^*(\boldsymbol{\alpha}_0, \tau) - \alpha_2)\right)}{\exp\left(-\gamma_{11}(\alpha_1^*(\boldsymbol{\alpha}, \tau) - \alpha_{10})^2 - \gamma_{22}(\alpha_2^*(\boldsymbol{\alpha}, \tau) - \alpha_{20})^2 - 2\gamma_{12}(\alpha_1^*(\boldsymbol{\alpha}, \tau) - \alpha_{10})(\alpha_2^*(\boldsymbol{\alpha}, \tau) - \alpha_{20})\right)} \tag{5.11}$$

Here $\alpha_i^*(\boldsymbol{\alpha}, \tau)$ is the most probable state in the case of the transition from $\boldsymbol{\alpha}$ during the time τ.

---

[1] According to their definitions [27–29], $\gamma_{ii} > 0$ and $\gamma_{11}\gamma_{22} - \gamma_{12}^2 \geq 0$.

Taking the logarithm of (5.11), we obtain

$$\begin{aligned}&\beta_{11}(\alpha_{10}^2-\alpha_1^2)+\beta_{22}(\alpha_{20}^2-\alpha_2^2)+2\beta_{12}(\alpha_{10}\alpha_{20}-\alpha_1\alpha_2)\\&=\gamma_{11}((\alpha_1^*(\boldsymbol{\alpha},\tau)-\alpha_{10})^2-(\alpha_1^*(\boldsymbol{\alpha}_0,\tau)-\alpha_1)^2)+\gamma_{22}((\alpha_2^*(\boldsymbol{\alpha},\tau)-\alpha_{20})^2-(\alpha_2^*(\boldsymbol{\alpha}_0,\tau)-\alpha_2)^2)\\&\quad+2\gamma_{12}((\alpha_1^*(\boldsymbol{\alpha},\tau)-\alpha_{10})(\alpha_2^*(\boldsymbol{\alpha},\tau)-\alpha_{20})-(\alpha_1^*(\boldsymbol{\alpha}_0,\tau)-\alpha_1)(\alpha_2^*(\boldsymbol{\alpha}_0,\tau)-\alpha_2)).\end{aligned} \tag{5.12}$$

Recalling the definitions of the state entropy and the trajectory entropy introduced above, from (5.8) and (5.12) we then find

$$\Delta S(\boldsymbol{\alpha}_0)=-\beta_{11}\alpha_{10}^2-\beta_{22}\alpha_{20}^2-2\beta_{12}\alpha_{10}\alpha_{20}, \tag{5.13}$$

$$\Delta S(\boldsymbol{\alpha})=-\beta_{11}\alpha_1^2-\beta_{22}\alpha_2^2-2\beta_{12}\alpha_1\alpha_2, \tag{5.14}$$

$$\begin{aligned}\Delta S_{\mathrm{tr}}(\boldsymbol{\alpha}_0|\boldsymbol{\alpha},\tau)=&-\gamma_{11}(\alpha_1^*(\boldsymbol{\alpha}_0,\tau)-\alpha_1)^2-\gamma_{22}(\alpha_2^*(\boldsymbol{\alpha}_0,\tau)-\alpha_2)^2,\\&-2\gamma_{12}(\alpha_1^*(\boldsymbol{\alpha}_0,\tau)-\alpha_1)(\alpha_2^*(\boldsymbol{\alpha}_0,\tau)-\alpha_2))\end{aligned} \tag{5.15}$$

$$\begin{aligned}\Delta S_{\mathrm{tr}}(\boldsymbol{\alpha}|\boldsymbol{\alpha}_0,\tau)=&-\gamma_{11}(\alpha_1^*(\boldsymbol{\alpha},\tau)-\alpha_{10})^2-\gamma_{22}(\alpha_2^*(\boldsymbol{\alpha},\tau)-\alpha_{20})^2.\\&-2\gamma_{12}(\alpha_1^*(\boldsymbol{\alpha},\tau)-\alpha_{10})(\alpha_2^*(\boldsymbol{\alpha},\tau)-\alpha_{20}))\end{aligned} \tag{5.16}$$

In order to find the relationship between the forward and reverse trajectories, let us transform (5.12); for this purpose, we consider a small time interval τ and apply the Taylor expansion:

$$\alpha_i^*(\boldsymbol{\alpha}_0,\tau)=\alpha_{i0}+\tau\left.\frac{\partial\alpha_i^*(\boldsymbol{\alpha}_0,\boldsymbol{\tau})}{\partial\tau}\right|_{\tau=0}+\ldots \tag{5.17}$$

In the following, we apply two consecutive expansions: first in terms of τ near zero, then in terms of α_i near α_{i0}:

$$\begin{aligned}\alpha_i^*(\boldsymbol{\alpha},\tau)&=\alpha_i+\tau\left.\frac{\partial\alpha_i^*(\boldsymbol{\alpha},\boldsymbol{\tau})}{\partial\tau}\right|_{\tau=0}+\ldots\\&=\alpha_i+\tau\left.\frac{\partial\alpha_i^*(\boldsymbol{\alpha_0},\boldsymbol{\tau})}{\partial\tau}\right|_{\tau=0}+\tau\sum_{i=1}^{2}\left.\frac{\partial\alpha_i^*(\boldsymbol{\alpha},\boldsymbol{\tau})}{\partial\tau\,\partial\alpha_i}\right|_{\tau=0,\alpha_i=\alpha_{i0}}(\alpha_i-\alpha_{i0})+\ldots\end{aligned} \tag{5.18}$$

By neglecting the second-order terms ($\propto\tau(\alpha_i-\alpha_{i0})$), we then obtain

$$\alpha_i^*(\boldsymbol{\alpha}_0,\tau)-\alpha_{i0}=\alpha_i^*(\boldsymbol{\alpha},\tau)-\alpha_i. \tag{5.19}$$

We now introduce the quantities $\Delta\alpha_i^*=\alpha_{i0}-\alpha_i^*(\boldsymbol{\alpha}_0,\tau)$, $\Delta\alpha_i=\alpha_{i0}-\alpha_i$ that characterize the most probable and actual (random) change in the state variables $A_i(\tau)$ relative to the original $A_i(0)$ during the time τ. From (5.19), we can then write

$$\begin{aligned}&\alpha_i^*(\boldsymbol{\alpha}_0,\tau)-\alpha_i=\Delta\alpha_i-\Delta\alpha_i^*;\\&\alpha_i^*(\boldsymbol{\alpha},\tau)-\alpha_{i0}=(\alpha_i^*(\boldsymbol{\alpha}_0,\tau)-\alpha_{i0}+\alpha_i)-\alpha_{i0}=-\Delta\alpha_i^*-\Delta\alpha_i\end{aligned} \tag{5.20}$$

Using the last two expressions, Eq. (5.12) can then be written as[2]:

$$\begin{aligned} &2\Delta\alpha_1(\alpha_{10}\beta_{11} + \beta_{12}\alpha_{20}) + 2\Delta\alpha_2(\beta_{22}\alpha_{20} + \beta_{12}\alpha_{10}) \\ &\quad = 4\Delta\alpha_1(\gamma_{11}\Delta\alpha_1^* + \gamma_{12}\Delta\alpha_2^*) + 4\Delta\alpha_2(\gamma_{22}\Delta\alpha_2^* + \gamma_{12}\Delta\alpha_1^*). \end{aligned} \tag{5.21}$$

Since the deviations ($\Delta\alpha_1$, $\Delta\alpha_2$,) from the initial state are independent, Eq. (5.21) implies

$$\begin{aligned} 2(\beta_{11}\alpha_{10} + \beta_{12}\alpha_{20}) &= 4\gamma_{11}\Delta\alpha_1^* + 4\gamma_{12}\Delta\alpha_2^* \\ 2(\beta_{22}\alpha_{20} + \beta_{12}\alpha_{10}) &= 4\gamma_{22}\Delta\alpha_2^* + 4\gamma_{12}\Delta\alpha_1^*. \end{aligned} \tag{5.22}$$

We now introduce a number of important thermodynamic quantities. Since the system is assumed to be isolated, its rate of change of entropy is equal to the entropy production σ [1, 27]. Following [26, 27, 29, 30], we then introduce the thermodynamic forces X_i and fluxes J_i:

$$\sigma = \frac{dS(\boldsymbol{\alpha})}{dt} = \frac{\partial S}{\partial \alpha_1}\frac{d\alpha_1}{dt} + \frac{\partial S}{\partial \alpha_2}\frac{d\alpha_2}{dt} = X_1 J_1 + X_2 J_2, \tag{5.23}$$

where:

$$X_i = -\frac{\partial S}{\partial \alpha_i} \tag{5.24}$$

$$J_i = -\frac{d\alpha_i}{dt} \tag{5.25}$$

According to (5.13) and (5.24), the thermodynamic forces at the initial moment of relaxation to equilibrium are given by

$$\begin{aligned} X_1 &= 2(\beta_{11}\alpha_{10} + \beta_{12}\alpha_{20}) \\ X_2 &= 2(\beta_{22}\alpha_{20} + \beta_{12}\alpha_{10}) \end{aligned} \tag{5.26}$$

Using (5.25) (5.26),[3] and the relation between γ_{ij} and γ_{ij}^0, expression (5.22) can be rewritten in the form

$$\begin{aligned} X_1 &= 4\gamma_{11}^0 J_1^* + 4\gamma_{12}^0 J_2^* \\ X_2 &= 4\gamma_{12}^0 J_1^* + 4\gamma_{22}^0 J_2^*, \end{aligned} \tag{5.27}$$

or, by inversion,

$$\begin{aligned} J_1^* &= L_{11}X_1 + L_{12}X_2 \\ J_2^* &= L_{21}X_1 + L_{22}X_2, \end{aligned} \tag{5.28}$$

---

[2] For small τ values: $\beta_{ij}\Delta\alpha_i\Delta\alpha_j \ll \gamma_{ij}\Delta\alpha_i\Delta\alpha_j^*$ (because $\gamma_{ij} = \gamma_{ij}^0/\tau$).

[3] For small τ values: $d\alpha_i^*/dt \approx -\Delta\alpha_i^*/\tau$. The minus sign arises from the fact that $\Delta\alpha_i^*$ is the difference between the initial and final value.

where the kinetic coefficients are given by

$$\begin{aligned} L_{11} &= \gamma_{22}^0/4(\gamma_{22}^0\gamma_{11}^0 - (\gamma_{12}^0)^2), \\ L_{12} = L_{21} &= -\gamma_{12}^0/4(\gamma_{22}^0\gamma_{11}^0 - (\gamma_{12}^0)^2), \\ L_{22} &= \gamma_{11}^0/4(\gamma_{11}^0\gamma_{22}^0 - (\gamma_{12}^0)^2). \end{aligned} \tag{5.29}$$

Based on the properties of γ_{ij}^0 (see footnote 1): $L_{ij} > 0$ and $L_{11}L_{22} - L_{12}^2 \geq 0$.

Thus, we have shown that the Gaussian assumption given by (5.9), (5.10) and the principle of detailed balance given by (5.6) lead to Onsager's linear flux-force relations [26, 27, 29, 30]. These relations link the most probable flux in the system with the thermodynamic force during the time interval τ. Our derivation here follows from studying the problem opposite to that considered by Onsager [29, 30].

We conclude this section with a number of useful relations that follow from the above equations. Using Eqs. (5.19), (5.20), we rewrite Eqs. (5.15), (5.16) in the form

$$\begin{aligned} \Delta S_{tr}(\boldsymbol{\alpha_0}|\boldsymbol{\alpha},\tau) = &-\gamma_{11}(\Delta\alpha_1 - \Delta\alpha_1^*)^2 - \gamma_{22}(\Delta\alpha_2 - \Delta\alpha_2^*)^2 \\ &- 2\gamma_{12}(\Delta\alpha_1 - \Delta\alpha_1^*)(\Delta\alpha_2 - \Delta\alpha_2^*) \end{aligned} \tag{5.30}$$

$$\begin{aligned} \Delta S_{tr}(\boldsymbol{\alpha}|\boldsymbol{\alpha}_0,\tau) = &-\gamma_{11}(\Delta\alpha_1 + \Delta\alpha_1^*)^2 - \gamma_{22}(\Delta\alpha_2 + \Delta\alpha_2^*)^2 \\ &- 2\gamma_{12}(\Delta\alpha_1 + \Delta\alpha_1^*)(\Delta\alpha_2 + \Delta\alpha_2^*) \end{aligned} \tag{5.31}$$

According to (5.8) and (5.23), for small τ we have

$$\tau\sigma = S(\boldsymbol{\alpha}) - S(\boldsymbol{\alpha_0}) = \Delta S_{\mathrm{tr}}(\boldsymbol{\alpha_0}|\boldsymbol{\alpha},\tau) - \Delta S_{\mathrm{tr}}(\boldsymbol{\alpha}|\boldsymbol{\alpha_0},\tau) \tag{5.32}$$

Or, according to (5.23)–(5.26),

$$\begin{aligned} \tau\sigma &= X_1\Delta\alpha_1 + X_2\Delta\alpha_2 \\ &= 2\Delta\alpha_1(\alpha_{10}\beta_{11} + \beta_{12}\alpha_{20}) + 2\Delta\alpha_2(\beta_{22}\alpha_{20} + \beta_{12}\alpha_{10}) \end{aligned} \tag{5.33}$$

We are now ready to consider the extremal properties of the trajectory entropy and the entropy production for the above model.

5.3 Information Approach: Maximizing the Trajectory Entropy

When describing nonequilibrium processes using the information approach, first the trajectory entropy is maximized in order to determine the maximum-entropy (MaxEnt) probability distribution over trajectories in phase space. The maximization is subject to certain given (or assumed) constraints (which may be relations between physical quantities that are more or less evident for the process under study). Then MaxEnt distribution function is used to calculate non-equilibrium properties of the process. Importantly, disagreement between predicted nonequilibrium properties and experimental data indicates incorrect constraints only [1, 19]. These constraints are then adjusted and the procedure is repeated. As a result, the information approach is a method of identifying the set of constraints consistent with experimental data.

Thus, such a method is some kind of mathematical device (algorithm). However, this mathematical procedure entails problems. The first problem is that there is too much freedom to choose the constraints and there is no criterion for agreement between predictions and experiment. This "liberty of action" may either yield no desired results at all or allow several solutions that meet the selected criterion but substantially differ in terms of both the constraints and predictions (outside the scope of the selected criterion). The second problem is connected with the choice of the measure of information and, correspondingly, the formula of informational entropy. From the logical viewpoint, there is no best option. There are multiple variants besides the Shannon formula; and many of them prove to be useful in different applications [31].

Here we simply go ahead and maximize the trajectory entropy for the problem under consideration. In the present model, the trajectory entropy as a function of random deviation of α is given by (5.15) or (5.30). The explicit form of this entropy was obtained using a Gaussian form for $P(\boldsymbol{\alpha}_0|\boldsymbol{\alpha}, \tau)$ together with the assumption of detailed balance. Consequently, the trajectory entropy depends not on the distribution function itself but on other variables, and trajectory entropy maximization should lead to relations between these variables. The maximization should be considered as unconstrained because all constraints have already been introduced into the explicit expression for trajectory entropy.

According to Eq.(5.4), the trajectory entropy for the forward trajectory $S_{\text{tr}}(\boldsymbol{\alpha}_0|\boldsymbol{\alpha}, \tau)$ is related to $\Delta S_{\text{tr}}(\boldsymbol{\alpha}_0|\boldsymbol{\alpha}, \tau)$ and to the trajectory entropy for the most probable forward trajectory $S^*_{\text{tr}}(\boldsymbol{\alpha}_0|\alpha^*(\boldsymbol{\alpha}_0), \tau)$ by

$$\frac{\partial S_{tr}(\boldsymbol{\alpha}_0|\boldsymbol{\alpha},\tau)}{\partial \Delta\alpha_i} = \frac{\partial(S^*_{tr}(\boldsymbol{\alpha}_0|\alpha^*(\boldsymbol{\alpha}_0),\tau) + \Delta S_{tr}(\boldsymbol{\alpha}_0|\boldsymbol{\alpha},\tau))}{\partial \Delta\alpha_i} = \frac{\partial \Delta S_{tr}(\boldsymbol{\alpha}_0|\boldsymbol{\alpha},\tau)}{\partial \Delta\alpha_i} \tag{5.34}$$

By substituting the expression for $\Delta S_{\text{tr}}(\boldsymbol{\alpha}_0|\boldsymbol{\alpha}, \tau)$ (Eq. (5.30)), and setting (5.34) equal to zero, it is easily shown that the trajectory entropy maximum leads to the condition $\Delta\alpha_i = \Delta\alpha_i^*$. According to (5.10) and (5.20), the most probable trajectory also satisfies the condition $\Delta\alpha_i = \Delta\alpha_i^*$.

Since the maximum of the trajectory entropy deviation is obtained when $\Delta\alpha_i = \Delta\alpha_i^*$, then using (5.31)–(5.33) we have

$$\begin{aligned}\left.\frac{\partial \Delta S_{tr}(\boldsymbol{\alpha}_0|\boldsymbol{\alpha},\tau)}{\partial \Delta\alpha_i}\right|_{\Delta\alpha_i=\Delta\alpha_i^*} &= \left.\frac{\partial(\tau\sigma + \Delta S_{tr}(\boldsymbol{\alpha}|\boldsymbol{\alpha}_0,\tau))}{\partial \Delta\alpha_i}\right|_{\Delta\alpha_i=\Delta\alpha_i^*} \\ &= \left.\frac{\partial(X_i\Delta\alpha_i + \Delta S_{tr}(\boldsymbol{\alpha}|\boldsymbol{\alpha}_0,\tau))}{\partial \Delta\alpha_i}\right|_{\Delta\alpha_i=\Delta\alpha_i^*} \\ &= X_i - 4\gamma_{ii}\Delta\alpha_i^* - 4\gamma_{ij}\Delta\alpha_j^* = 0,\end{aligned}$$

and hence

$$X_i = 4\gamma_{ii}\Delta\alpha_i^* + 4\gamma_{ij}\Delta\alpha_j^*. \tag{5.35}$$

This expression coincides with Eq. (5.27). Thus, Onsager's linear relations correspond to a maximum in the trajectory entropy. This result indicates that, for the present model, trajectory entropy maximization with a number of constraints leads to a macrotrajectory that satisfies the valid linear relation between thermodynamic fluxes and forces (5.35). This points to the possibility of generalizing the conventional method of equilibrium entropy maximization to non-equilibrium, as proposed by Jaynes and others (see e.g. [4]).

5.4 Thermodynamic Approach: Ziegler's MaxEP

In contrast to the information method (MaxEnt), Ziegler's approach focuses directly on the search for relationships between the most probable quantities. Random variables and their probability distributions lie beyond the scope of this approach. The entropy production is considered to be a known function of thermodynamic fluxes [1, 15, 16]. The relationship between the thermodynamic fluxes and forces is then derived through maximization of entropy production in the space of independent fluxes, subject to fixed thermodynamic forces [1, 15, 16].

In contrast to the informational approach, Ziegler's method is falsifiable (sensu Popper). Entropy production is a physically well-defined macroscopic property of the system connected with energy dissipation to heat. Thermodynamic forces and fluxes also have a clear physical meaning and are measurable in experiments. Consequently, if Ziegler's MaxEP principle yields predictions different from experiment (in particular, concerning the relationship between fluxes and forces), then the principle will have been disproved (or at least its validity will have been limited). One disadvantage of this approach is that, for specific systems, finding the entropy production as a function of fluxes is not always straightforward. There are no standard procedures, each specific case requiring an individual approach. The fact that Ziegler has developed his method only for the systems with a unique correspondence between the thermodynamic force and flux represents another disadvantage.

Formally, Eqs. (5.27) or (5.28) can be derived using Ziegler's procedure [1, 15, 16]. For this purpose, the entropy production was postulated to be a bilinear function of thermodynamic fluxes, and then maximization was carried out subject to fixed forces. In the present model, it is possible to explicitly obtain both the form of the entropy production and the constraint for its maximization. We show this as follows.

Let us rewrite the detailed balance relation (5.32) for the most probable trajectory ($\Delta\alpha_i = \Delta\alpha_i^*$). In this case, $\Delta S_{tr}(\boldsymbol{\alpha_0}|\boldsymbol{\alpha}, \tau) = 0$ [see Eq. (5.30)], and as a result [see Eq. (5.31)], we have

$$\tau\sigma = -\Delta S_{tr}(\boldsymbol{\alpha}|\boldsymbol{\alpha}_0, \tau) = 4\gamma_{11}\Delta\alpha_1^{*2} + 4\gamma_{22}\Delta\alpha_2^{*2} + 8\gamma_{12}\Delta\alpha_1^*\Delta\alpha_2^*$$

Or, using Eqs. (5.25), (5.33), we have

$$X_1\Delta\alpha_1^* + X_2\Delta\alpha_2^* = 4\gamma_{11}\Delta\alpha_1^{*2} + 4\gamma_{22}\Delta\alpha_2^{*2} + 8\gamma_{12}\Delta\alpha_1^*\Delta\alpha_2^*,$$

$$X_1J_1^* + X_2J_2^* = 4\gamma_{11}^0J_1^{*2} + 4\gamma_{22}^0J_2^{*2} + 8\gamma_{12}^0J_1^*J_2^*. \tag{5.36}$$

Clearly, the left-hand side of (5.36) is the entropy production $\sigma(J^*)$ as a function of the thermodynamic forces and fluxes, whereas the entropy production on the right-hand side is written in the terms of thermodynamic fluxes alone. Maximizing $\sigma(J^*) = 4\gamma_{11}^0J_1^{*2} + 4\gamma_{22}^0J_2^{*2} + 8\gamma_{12}^0J_1^*J_2^*$ subject to (5.36) is equivalent to setting

$$\frac{\partial}{\partial J_i^*}\left(\sigma(J^*) - \mu(\sigma(J^*) - X_1J_1^* - X_2J_2^*)\right) = 0, \tag{5.37}$$

(where μ is the Lagrange multiplier), from which it is then easy to obtain Onsager's linear flux-force relations (5.27). See details in [1, 15, 16].

5.5 Conclusion

We have considered the simplest model of a close-to-equilibrium system described by two thermodynamic forces. We assumed the detailed balance condition and Gaussian distributions for the equilibrium and transition probabilities.

Within the scope of this model, we derived an expression for the trajectory entropy as a function of deviations from equilibrium, from which we have shown, for the first time, that maximization of the entropy of microscopic trajectories (MaxEnt) leads to Onsager‘s linear force-flux relations. We also derived an expression for the entropy production as a function of thermodynamic fluxes, as a necessary starting point for Ziegler's MaxEP procedure. The bilinear expression we obtain agrees with that previously postulated by Ziegler and used by him to derive Onsager's linear flux-force relations.

In other words, Onsager's linear flux-force relations can be obtained using at least two independent methods: Ziegler's MaxEP principle, or maximization of trajectory entropy (MaxEnt). Within the scope of the present model, we cannot say that MaxEnt is more general than Ziegler's MaxEP or that the latter follows from MaxEnt (or vice versa). All we can say is that these two different methods[4] lead to the same result, within the near-equilibrium model approximations we made. This model implies linear relationships between fluxes and forces, as shown in

[4] For one approximation, extremization is carried out for a random deviation from the equilibrium, whereas for the other approximation, it is carried out for a thermodynamic flux, i.e. the most probable deviation.

Sect. 5.2. Sections 5.3 and 5.4 show how these relationships can be given a variational interpretation. Here an analogy from mechanics can be drawn, in which Newton's laws of motion can be derived from variational methods (Hamilton, Lagrange).

We conclude with a summary of our view on MaxEP and its two interpretations, as outlined in the Introduction.

1. MaxEP is an important physical principle with empirical support [1–3]. In this interpretation, MaxEP extends the second law of thermodynamics that may be stated as follows: *at each level of description the relationship between the cause and the response of a nonequilibrium system is one that maximizes (under constraints) the entropy production density* [17, 32]. This statement significantly generalizes Ziegler's MaxEP principle, which requires a unique correspondence between the thermodynamic fluxes and forces. This requirement has substantially limited the wider application of Ziegler's principle, in particular to the study of nonequilibrium phase transitions [17, 32].
2. The second law of thermodynamics (i.e. non-negativity of entropy production) identifies a unique direction of time [33]. Time is the most complex and elusive physical concept that still lacks a universally acknowledged definition [33–35]. Despite having learned to measure time, we still fail to understand its nature. In this regard, we suggest that considering MaxEP as a new and important extension of the second law of thermodynamics will increase our understanding of the nature of time, as discussed further in [12, 32].
3. MaxEP is a relatively new principle. Its range of validity has yet to be clearly defined, but it should be based primarily on experiment.[5] MaxEP is best tested using relatively simple experimental systems; in this regard, climate, biological and similar systems may not be ideal for testing (and/or falsifying) MaxEP due to their complexity and ambiguous interpretation (see e.g. criticisms in [36, 37]). We suggest non-equilibrium phase transitions of homogeneous systems as a fruitful experimental testing ground for MaxEP. For example, if the phase with the smallest entropy production is observed to be statistically the most probable phase in the case where a fixed thermodynamic force admits multiple nonequilibrium phases, then MaxEP will have been disproved (or at least its range of validity will have been narrowed).

Microscopic interpretations of MaxEP are certainly important for understanding its range of validity. However, their significance should not be exaggerated. In our view, statistical physics has always played a subordinate role. So, scientists tried to understand and define macroscopic properties that were experimentally discovered and integrally generalized in the thermodynamic postulates (laws) using simple

[5] Mathematical models are absolutely unsuitable for falsification. So, a model is only some more or less crude and often one-sided reflection of some part of a phenomenon, whereas MEPP is the principle reflecting the dissipative properties that are observed in nature rather than in its model.

statistical models (of ideal gas, Markovian processes, etc.). However, disagreement between statistical predictions and thermodynamics has always implied that the statistical model was erroneous, leading to a modified model (whereas the opposite—modification of thermodynamical laws—has never occurred). It is therefore misleading to suppose that the microscopic view on the world somehow explains or proves MaxEP. It is as misleading to suppose that Boltzmann's H-theorem proves the second law of thermodynamics.

4. The interpretation of MaxEP from the viewpoint of Jaynes' information approach (MaxEnt, see e.g. Sect. 5.3) should not be identified with the microscopic interpretation. MaxEnt is a particular approach to the foundation of statistical physics, which has supporters and opponents (see e.g. [38]). The simplicity of the MaxEnt formulation of equilibrium statistical physics is one of its strengths. Its subjective nature has been considered a weakness.

MaxEnt-based attempts to derive MaxEP (considered as a principle of non-equilibrium physics) [4–11] are a very interesting area of study. However, we would like to conclude by raising a number of concerns here.

First, in our view, such attempts still lack rigour, and will require additional assumptions. It is these extra assumptions that will indicate that MaxEP is an independent physical principle.

Second, MaxEnt is a type of microscopic approach. The incompleteness of statistical methods for justifying empirical laws/principle (e.g. MaxEP) is mentioned in point 3 above. However, MaxEnt also has particular limitations of its own. While it is useful as a simple algorithm for obtaining the known (generally accepted) solution,[6] any desired result can be obtained when the solution is unknown (or several solutions are possible).[7] Thus, Jaynes' mathematical procedure (MaxEnt) can be used for obtaining multiple other procedures, but the value of such mathematical exercises becomes, nevertheless, rather doubtful for physics.

In summary, these considerations form the basis of our objection to the informational (as opposed to physical) justification of MaxEP. If MaxEP is a physical principle, then it will be fundamentally distinct from MaxEnt because then MaxEP itself is the key physical constraint, like the first and second law of thermodynamics or charge conservation.

---

[6] Indeed, the researcher's intention to mathematically make the most unprejudiced prediction in the conditions of incomplete information about the system is the essence of this method. Therefore, if a phenomenon is very poorly experimentally studied (i.e. there are insufficient constraints), then anything can be predicted using MaxEnt (i.e. there are no truth criteria). In contrast, when MaxEP or MaxEnt are considered as physical principles, there are far fewer possibilities for drawing arbitrary conclusions. Methods that predict something specific for poorly studied phenomena (from which their falsifiability derives) are especially valuable.

[7] If this cannot be achieved by selecting the constraints, then other kinds of informational entropy can always be used, for example, by Tsallis, Abe, Kullback, and many others [23, 30].

References

1. Martyushev, L.M., Seleznev, V.D.: Maximum entropy production principle in physics, chemistry and biology. Phys. Report **426**(1), 1–45 (2006)
2. Kleidon, A., Lorenz, R.D. (eds.): Non-equilibrium Thermodynamics and the Production of Entropy in Life, Earth, and Beyond. Springer, Heidelberg (2004)
3. Ozawa, H., Ohmura, A., Lorenz, R.D., Pujol, T.: The second law of thermodynamics and the global climate systems—a review of the maximum entropy production principle. Rev. Geophys. **41**(4), 1018–1042 (2003)
4. Dewar, R.: Information theory explanation of the fluctuation theorem, maximum entropy production and self-organized criticality in non-equilibrium stationary state. J. Phys. A: Math. Gen. **36**, 631–641 (2003)
5. Dewar, R.: Maximum entropy production and the fluctuation theorem. J. Phys. A: Math. Gen. **38,** L371–L381 (2005)
6. Grinstein, G., Linsker, R.: Comments on a derivation and application of the 'maximum entropy production' principle. J. Phys. A: Math. Theor. **40**, 9717–9720 (2007)
7. Dewar, R.C.: Maximum entropy production as an inference algorithm that translates physical assumptions into macroscopic predictions: don't shoot the messenger. Entropy **11**, 931–944 (2009)
8. Dewar, R.C., Maritan, A.: The theoretical basis of maximum entropy production. (Chapter in this book)
9. Niven, R.K.: Steady state of a dissipative flow-controlled system and the maximum entropy production principle. Phys. Rev. E **80**, 021113, (15 pp) (2009)
10. Dyke, J., Kleidon, A.: The maximum entropy production principle: its theoretical foundations and applications to the earth system. Entropy **12**, 613–630 (2010)
11. Jones, W.: Variational principles for entropy production and predictive statistical mechanics. J. Phys. A: Math. Gen. **16**, 3629–3635 (1983)
12. Martyushev, L.M.: The maximum entropy production principle: two basic questions. Phil. Trans. R. Soc. B **365**, 1333–1334 (2010)
13. Kohler, M.: Behandlung von Nichtgleichgewichtsvorgängen mit Hilfe eines Extremalprinzips. Z. Physik. **124**, 772–789 (1948)
14. Ziman, J.M.: The general variational principle of transport theory. Can. J. Phys. **34**, 1256–1273 (1956)
15. Ziegler, H.: Some extremum principles in irreversible thermodynamics. In: Sneddon, I.N., Hill, R. (eds.) Progress in Solid Mechanics, vol. 4, pp. 91–193. North-Holland, Amsterdam (1963)
16. Ziegler, H.: An Introduction to Thermomechanics. North-Holland, Amsterdam (1983)
17. Martyushev, L.M., Konovalov, M.S.: Thermodynamic model of nonequilibrium phase transitions. Phys. Rev. E. **84**(1), 011113, (7 pages) (2011)
18. Shannon, C.E.: A mathematical theory of communication. Bell Syst. Tech. J. **27**(379–423), 623–656 (1948)
19. Jaynes, E.T.: Information theory and statistical mechanics. Phys. Rev. **106**(4), 620–630 (1957)
20. Filyukov, A.A., Karpov, V.Ya.: Method of the most probable path of evolution in the theory of stationary irreversible processes. J. Engin. Phys. Thermophys. **13**(6), 416–419 (1967)
21. Monthus, C.: Non-equilibrium steady state: maximization of the Shannon entropy associated with the distribution of dynamical trajectories in the presence of constraints. J. Stat. Mechanics: Theor. Exp. **3**, P03008 (36 pp) (2011)
22. Smith, E.: Large-deviation principle, stochastic effective actions, path entropies, and the structure and meaning of thermodynamic descriptions. Rep. Prog. Phys. **74**, 046601 (38 pp) (2011)
23. Stock, G., Ghosh, K., Dill, K.A.: Maximum Caliber: a variational approach applied to two-state dynamics. J. Chem. Phys. **128**, 194102 (12 pp) (2008)

24. Ge, H., Presse, S., Ghosh, K., Dill, K.A.: Markov processes follow from the principle of maximum caliber. J. Chem. Phys. **136**, 064108 (5 pp) (2012)
25. Ghosh, K., Dill, K.A., Inamdar, M.M., Seitaridou, E., Phillips, R.: Teaching the principles of statistical dynamics. Am. J. Phys. **74**(2), 123–133 (2006)
26. Landau, L.D., Lifshitz E.M.: Statistical Physics, Part 1. vol. 5. Butterworth-Heinemann, Oxford (1980)
27. De Groot, S.R., Mazur, P.: Non-Equilibrium Thermodynamics. North-Holland, Amsterdam (1962)
28. van Kampen, N.G.: Stochastic Processes in Physics and Chemistry. Elsevier, New York (2007)
29. Onsager, L.: Reciprocal Relations in irreversible processes II. Phys. Rev. **38**, 2265–2279 (1953)
30. Onsager, L., Machlup, S.: Fluctuations and irreversible processes. Phys Rev. **91**(6), 1505–1512 (1953)
31. Beck, C.: Generalised information and entropy measures in physics. Contemp. Phys. **50**(4), 495–510 (2009)
32. Martyushev, L.M.: e-print arXiv:1011.4137
33. Reichenbach, H.: The Direction of Time. University of California Press, California (1991)
34. Grunbaum, A.: Philosophical Problems of Space and Time. Knopf, New York (1963)
35. Reichenbach, H.: The Philosophy of Space and Time. Dover, New York (1958)
36. Caldeira, K.: The maximum entropy principle: a critical discussion. An editorial comment. Clim. Change **85**, 267–269 (2007)
37. Goody, R.: Maximum entropy production in climatic theory. J. Atmos. Sci. **64**, 2735–2739 (2007)
38. Lavenda, B.H.: Statistical Physics. A Probabilistic Approach. Wiley, New York (1991)

Chapter 6
The Time Evolution of Entropy Production in Nonlinear Dynamic Systems

Hisashi Ozawa and Shinya Shimokawa

Abstract General characteristics of entropy production in a fluid system are investigated from a thermodynamic viewpoint. A basic expression for entropy production due to irreversible transport of heat or momentum is formulated together with balance equations of energy and momentum in a fluid system. It is shown that entropy production always decreases with time when the system is of a pure diffusion type without advection of heat or momentum. The minimum entropy production (MinEP) property is thus intrinsic to a pure diffusion-type system. However, this MinEP property disappears when the system is subject to advection of heat or momentum due to dynamic motion. When the rate of advection exceeds the rate of diffusion of heat or momentum, entropy production tends to increase over time. The maximum entropy production (MaxEP), suggested as a selection principle for steady states of nonlinear non-equilibrium systems, can therefore be understood as a characteristic feature of systems with dynamic instability. The observed mean state of vertical convection of the atmosphere is consistent with the condition for MaxEP presented in this study.

List of Symbols

Symbol Meaning (SI Units)

Roman Symbols

A	Surface of a system or the Earth (m^2)
c_v	Specific heat at constant volume ($J\ K^{-1}\ kg^{-1}$)
$\mathbf{e}$	Unit vector (–)
F_c	Convective heat flux density (sensible and latent heat) ($J\ m^{-2}\ s^{-1}$)

H. Ozawa (✉)
Hiroshima University, Higashi-Hiroshima 739-8521, Japan
e-mail: hozawa@hiroshima-u.ac.jp

S. Shimokawa
National Research Institute for Earth Science and Disaster Prevention, Tsukuba 305-0006, Japan

R. C. Dewar et al. (eds.), *Beyond the Second Law*,
Understanding Complex Systems, DOI: 10.1007/978-3-642-40154-1_6,

F_r Radiation flux density (J m^{-2} s^{-1})
F_{LW} Longwave radiation flux density (J m^{-2} s^{-1})
F_{SW} Shortwave radiation flux density (J m^{-2} s^{-1})
$\mathbf{J}_i$ Diffusive flux density of *i*-th component
$\mathbf{J}_h$ Diffusive flux density of heat (J m^{-2} s^{-1})
$\mathbf{J}_m$ Diffusive flux density of momentum (kg m^{-2} s^{-1})
k Thermal conductivity (J m^{-1} K^{-1} s^{-1})
L_h Kinetic coefficient for heat diffusion (J m^{-1} K s^{-1})
$\mathbf{n}$ Unit vector normal to system's surface (–)
p Pressure (Pa)
t Time (s)
T Temperature (K)
T_e Effective radiation temperature at the top of the atmosphere (K)
T_r Effective radiation temperature (K)
T_s Surface temperature (K)
T_{sun} Emission temperature of the sun (K)
V Volume of a system (m^3)
$\mathbf{v}$ Velocity (m s^{-1})
$\mathbf{X}_i$ Gradient of intensive variable for *i*-th diffusive flux

Greek Symbols
$\boldsymbol{\delta}$ Unit tensor (–)
κ Thermal diffusivity (m^2 s^{-1})
λ Second viscosity (kg m^{-1} s^{-1})
μ Viscosity (kg s^{-1} m^{-1})
ν Kinematic viscosity (m^2 s^{-1})
$\mathbf{\Pi}$ Viscous stress tensor (Pa)
ρ Density (kg m^{-3})
σ_B Stefan–Boltzmann constant $\approx 5.67 \times 10^{-8}$ (J m^{-2} K^{-4} s^{-1})
$\dot{\sigma}$ Rate of entropy production (J K^{-1} s^{-1})
$\dot{\sigma}_{conv}$ Rate of entropy production due to convective heat flux (J K^{-1} s^{-1})
$\dot{\sigma}_h$ Rate of entropy production due to heat diffusion (J K^{-1} s^{-1})
$\dot{\sigma}_m$ Rate of entropy production due to momentum diffusion (J K^{-1} s^{-1})
$\dot{\sigma}_{rad}$ Rate of entropy production due to absorption of radiation (J K^{-1} s^{-1})
$\dot{\sigma}_{tot}$ Total rate of entropy production in the atmosphere (J K^{-1} s^{-1})

Suffixes to $\dot{\sigma}$
stat Static state with no motion
lam Laminar flow state
adv State with advection

6.1 Introduction

Since an early investigation by Ziegler [1], maximum entropy production (MaxEP) has been suggested as a general thermodynamic property of nonlinear non-equilibrium phenomena, with later studies showing that the MaxEP state is consistent with steady states of a variety of nonlinear phenomena. These include the general circulation of the atmosphere and oceans [2, 3], thermal convection [4], turbulent shear flow [5], climates of other planets [6], oceanic general circulation [7, 8], crystal growth morphology ([9]; Martyushev, this volume) and granular flows [10]. While the underlying physical mechanism is still debated, the MaxEP state is shown to be identical to a state of maximum generation of available energy [11, 12]. Moreover, recent theoretical studies suggest that the MaxEP state is the most probable state that is realized by non-equilibrium systems ([13, 14]; Dewar and Maritan, this volume).

It is known, however, that entropy production in a linear process tends to decrease with time and reach a minimum in a final steady state when a thermodynamic intensive variable (such as temperature) is fixed at the system boundary. This tendency was first suggested for a linear chemical process in a discontinuous system by Prigogine [15], and then extended to the case of a linear diffusion process in a continuous system [16]. Since then, this minimum entropy production (MinEP) principle has become widely known in the field of non-equilibrium thermodynamics. Although a number of attempts have been made to extend this MinEP principle to a general one including nonlinear processes, the results remain controversial and inconclusive (e.g. [17, 18]). In fact, Prigogine [19] explained the situation as:

> It came as a great surprise when it was shown that in systems far from equilibrium the thermodynamic behavior could be quite different—in fact, even *directly opposite* that predicted by the theorem of minimum entropy production.

Sawada [20] pointed out the limitations of the MinEP principle, and instead proposed the MaxEP principle as a general variational principle for nonlinear systems that are far from equilibrium. More recently, Dewar and Maritan (this volume) showed using Jaynes's maximum entropy method that a state of minimum dissipation (MinEP) is selected for a system without dynamic instability, whereas that of maximum dissipation (MaxEP) is selected for a system with dynamic instability. It seems therefore that the existence of dynamic instability plays a key role in determining the behavior of entropy production in nonlinear non-equilibrium systems. However, the nature of the dynamic instability as well as its relation to nonlinearity remains unclear. Moreover, until now, we do not have a reasonable specification of the dynamic conditions under which the MinEP or MaxEP state is realized.

In order to clarify the issues in the phenomena mentioned above, we have investigated the behavior of time evolution of entropy production in a fluid system. Based on a general expression of entropy production and balance equations of energy and momentum, we present a condition under which the MinEP state is realized in the course of time in a system of linear diffusion (Sect. 6.2). We then add nonlinear advection terms in the balance equations, and examine the condition under which the MinEP state becomes unstable and the MaxEP state is realized in the system (Sect. 6.3). We show that the rate of advection of heat or momentum plays an important role in the enhancement of entropy production in a fluid system that possesses dynamic instability. Results obtained from this study are compared with the observed state of vertical atmospheric convection, a typical example of nonlinear dynamic phenomena (Sect. 6.4).

6.2 Linear Diffusion

Let us consider a fluid system in which several irreversible processes take place. These processes can be molecular diffusion of heat under a temperature gradient, molecular diffusion of momentum under a velocity gradient, or diffusion of a chemical component under a gradient of density of the chemical component. All these diffusion processes contribute to an increase in entropy of the whole system consisting of the fluid system and its surroundings. A general expression for the rate of entropy production per unit time by these irreversible processes is given by

$$\dot{\sigma} = \int_V \sum_i \mathbf{J}_i \cdot \mathbf{X}_i \, \mathrm{d}V, \tag{6.1}$$

where $\mathbf{J}_i$ is the i-th diffusive flux density, $\mathbf{X}_i$ is the gradient in the corresponding intensive variable that drives the flux, and the integration is taken over the total volume of the system (e.g. [17]). If the flux density is heat, momentum, or a chemical component, the corresponding intensive variable is temperature ($1/T$), velocity ($-\mathbf{v}/T$), or chemical potential ($-\mu/T$) respectively. It should be noted that the diffusive flux $\mathbf{J}_i$ does not, in principle, include a flux due to advection (i.e. coherent motion of fluid), which is intrinsically a reversible process.[1] However, advection significantly enhances the local gradient of the intensive variable at the moving front, and hence entropy production is also enhanced. We will see how entropy production can change with and without advection.

6.2.1 Heat Diffusion

As the simplest example, let us discuss diffusion of heat under temperature gradient in a fluid system. In this case, Eq. (6.1) is

[1] One can include a reversible flux due to advection in the balance equation of entropy, but it results in no contribution to entropy production after the integration over the whole volume of a fluid system (see, e.g., [21], Sec. 49; [12], Sec. 2.4).

$$\dot{\sigma}_{\mathrm{h}} = \int_V \mathbf{J}_{\mathrm{h}} \cdot \nabla\left(\frac{1}{T}\right) dV = \int_V L_{\mathrm{h}} \left[\nabla\left(\frac{1}{T}\right)\right]^2 \mathrm{d}V, \tag{6.2}$$

where $\mathbf{J}_{\mathrm{h}}$ is the diffusive heat flux density due to heat conduction, T is the temperature and L_{h} is the kinetic coefficient relating the diffusive heat flux and the temperature gradient: $\mathbf{J}_{\mathrm{h}} = L_{\mathrm{h}}\nabla(1/T) = -k\nabla T$, with $k = L_{\mathrm{h}}/T^2$ being the thermal conductivity in Fourier's law. In Eq. (6.2) we have assumed linearity between the diffusive heat flux and the temperature gradient.

We can show that the entropy production due to heat diffusion [Eq. (6.2)] is a monotonically decreasing function of time when the intensive variable (T) is fixed at the boundary of the system and when there is no advective heat transport in the system. Taking thc time derivative of Eq. (6.2), and assuming a constancy of L_{h} in the temperature range of the system ($\mathrm{d}L_{\mathrm{h}}/\mathrm{d}t = 0$), we get

$$\frac{\mathrm{d}\dot{\sigma}_{\mathrm{h}}}{\mathrm{d}t} = 2\int_V L_{\mathrm{h}}\nabla\left(\frac{1}{T}\right) \cdot \frac{\partial}{\partial t}\left[\nabla\left(\frac{1}{T}\right)\right]\mathrm{d}V = 2\int_V \mathbf{J}_{\mathrm{h}} \cdot \nabla\left[\frac{\partial}{\partial t}\left(\frac{1}{T}\right)\right]\mathrm{d}V. \tag{6.3}$$

This expression leads, with integration by parts, to

$$\frac{\mathrm{d}\dot{\sigma}_{\mathrm{h}}}{\mathrm{d}t} = 2\int_A \left[\frac{\partial}{\partial t}\left(\frac{1}{T}\right)\right] \mathbf{J}_{\mathrm{h}} \cdot \mathbf{n}\, \mathrm{d}A - 2\int_V \left[\frac{\partial}{\partial t}\left(\frac{1}{T}\right)\right] \nabla \cdot \mathbf{J}_{\mathrm{h}}\mathrm{d}V, \tag{6.4}$$

where $\mathbf{n}$ is the unit vector normal to the system boundary and directed to outward, and A is the surface bounding the system. The first surface integral varnishes when the temperature is fixed at the boundary (i.e. $\partial T/\partial t = 0$). Using Fourier's law ($\mathbf{J}_{\mathrm{h}} = -k\nabla T$) and assuming the uniformity of k in the system ($\nabla k = 0$), the second volume integral leads to

$$\frac{\mathrm{d}\dot{\sigma}_{\mathrm{h}}}{\mathrm{d}t} = 2\int_V k\,\nabla^2 T \frac{\partial}{\partial t}\left(\frac{1}{T}\right)\mathrm{d}V. \tag{6.5}$$

Equation (6.5) shows that the rate of change of entropy production is a function of the heat diffusion rate ($k\nabla^2 T$) and the rate of change of temperature ($\partial T/\partial t$). The heat diffusion rate is related to the balance equation for internal energy (e.g. [22]) as

$$\rho\frac{\partial}{\partial t}(c_{\mathrm{v}}T) = -\rho\mathbf{v} \cdot \nabla(c_{\mathrm{v}}T) + k\nabla^2 T - p\nabla \cdot \mathbf{v} + \mathbf{\Pi} : \nabla\mathbf{v}, \tag{6.6}$$

where ρ is the fluid density, c_{v} is the specific heat at constant volume, $\mathbf{v}$ is the fluid velocity, p is the pressure and $\mathbf{\Pi}$ is the viscous stress. This equation shows that the rate of temperature increase is caused by the sum of the rates of heat advection, heat diffusion, cooling by volume expansion and viscous heating. Substituting $k\nabla^2 T$ from Eq. (6.6) into Eq. (6.5), and assuming a constancy of c_{v} in the fluid system ($\mathrm{d}c_{\mathrm{v}}/\mathrm{d}t = 0$), we get

$$\frac{\mathrm{d}\dot{\sigma}_{\mathrm{h}}}{\mathrm{d}t} = 2\int_V \left[\rho c_{\mathrm{v}}\frac{\partial T}{\partial t} + \rho c_{\mathrm{v}}\mathbf{v}\cdot\nabla T + p\nabla\cdot\mathbf{v} - \mathbf{\Pi}:\nabla\mathbf{v}\right]\frac{\partial}{\partial t}\left(\frac{1}{T}\right)\mathrm{d}V. \tag{6.7}$$

If we consider a situation with no convective motion ($\mathbf{v} = 0$), Eq. (6.7) reduces to

$$\frac{\mathrm{d}\dot{\sigma}_{\mathrm{h,stat}}}{\mathrm{d}t} = -2\int_V \frac{\rho c_{\mathrm{v}}}{T^2}\left(\frac{\partial T}{\partial t}\right)^2 \mathrm{d}V \le 0, \tag{6.8}$$

where the suffix stat denotes the static state with no motion. The rate of change of entropy production is negative in this static case, because ρ and c_{v} are positive definite. Equation (6.8) shows that entropy production due to pure heat conduction tends to decrease with time, and reaches a minimum in the final steady state ($\partial T/\partial t = 0$) provided there is no convective motion in the fluid. This tendency was first suggested by Prigogine [15], and is called the minimum entropy production (MinEP) principle. While several attempts have been made to extend this principle to a general one including dynamic motion, the results remain controversial and inconclusive [16, 17]. As we shall see in Sect. 6.3.1, when advection due to dynamic motion is nonzero, the local rate of entropy production can either increase or decrease, depending on the rate of heat advection ($\mathbf{v}\cdot\nabla T$); the sign of $\mathrm{d}\dot{\sigma}_{\mathrm{h}}/\mathrm{d}t$ becomes indefinite and even positive in some cases.

6.2.2 Momentum Diffusion

A similar result can be obtained for momentum diffusion due to viscosity under a velocity gradient. Suppose that a viscous fluid with a uniform viscosity is flowing in a system with a constant temperature T. In this case, entropy production due to momentum diffusion is given by

$$\dot{\sigma}_{\mathrm{m}} = \int_V \frac{\mathbf{\Pi}:\nabla\mathbf{v}}{T}\mathrm{d}V. \tag{6.9}$$

Here, the numerator represents the scalar product of the viscous stress tensor and the velocity gradient, and is identical to the heating rate due to viscosity per unit volume per unit time in the fluid. Assuming a linear relation between the viscous stress and the velocity gradient, we can drive the time derivative of the rate of entropy production after a few manipulations[2]:

---

[2] Assuming linearity, $\mathbf{\Pi}:\nabla\mathbf{v} = [2\mu(\nabla\mathbf{v})^{\mathrm{s}} - (2/3)\mu(\nabla\cdot\mathbf{v})\boldsymbol{\delta}]:[(\nabla\mathbf{v})^{\mathrm{s}} + (\nabla\mathbf{v})^{\mathrm{a}}] = 2\mu(\nabla\mathbf{v})^{\mathrm{s}} : (\nabla\mathbf{v})^{\mathrm{s}} - (2/3)\mu(\nabla\cdot\mathbf{v})^2$, with $\boldsymbol{\delta}$ denoting the unit tensor, and $\mathbf{T}^{\mathrm{s}}$ and $\mathbf{T}^{\mathrm{a}}$ denoting symmetric and asymmetric parts of a tensor $\mathbf{T}$. Then, $\partial(\mathbf{\Pi}:\nabla\mathbf{v})/\partial t = 2[2\mu(\nabla\mathbf{v})^{\mathrm{s}} - (2/3)\mu(\nabla\cdot\mathbf{v})\boldsymbol{\delta}]: [\nabla(\partial\mathbf{v}/\partial \mathrm{t})]^{\mathrm{s}} = 2\mathbf{\Pi}:\nabla(\partial\mathbf{v}/\partial t)$.

$$\frac{\mathrm{d}\dot{\sigma}_{\mathrm{m}}}{\mathrm{d}t}=\int\limits_{V}\frac{\partial}{\partial \mathrm{t}}\left(\frac{\boldsymbol{\Pi}:\nabla\mathbf{v}}{T}\right)\mathrm{d}V=2\int\limits_{V}\frac{1}{T}\left[\boldsymbol{\Pi}:\nabla\left(\frac{\partial\mathbf{v}}{\partial t}\right)\right]\mathrm{d}V. \tag{6.10}$$

By a sequence of transformations similar to those from Eq. (6.3) to Eq. (6.5), we get

$$\frac{\mathrm{d}\dot{\sigma}_{\mathrm{m}}}{\mathrm{d}t}=-2\int\limits_{V}\frac{1}{T}\left[\mu\nabla^{2}\mathbf{v}+\frac{\mu}{3}\nabla(\nabla\cdot\mathbf{v})\right]\cdot\left(\frac{\partial\mathbf{v}}{\partial t}\right)\mathrm{d}V, \tag{6.11}$$

where μ is the viscosity of the fluid. Here we have assumed that velocity is fixed at the boundary ($\partial\mathbf{v}/\partial t=0$). The diffusion rate of momentum $[\mu\nabla^{2}\mathbf{v}+\mu\nabla(\nabla\cdot\mathbf{v})/3]$ is related to the balance equation of momentum—the Navier–Stokes equation[3]—as

$$\rho\frac{\partial\mathbf{v}}{\partial t}=-\rho(\mathbf{v}\cdot\nabla)\mathbf{v}-\nabla p+\mu\nabla^{2}\mathbf{v}+\frac{\mu}{3}\nabla(\nabla\cdot\mathbf{v}). \tag{6.12}$$

Substituting Eq. (6.12) into Eq. (6.11) and eliminating the momentum diffusion rate, we get after a few transformations

$$\begin{aligned}\frac{\mathrm{d}\dot{\sigma}_{\mathrm{m}}}{\mathrm{d}t}&=-2\int\limits_{V}\frac{1}{T}\left[\left(\rho\frac{\partial\mathbf{v}}{\partial t}+\rho(\mathbf{v}\cdot\nabla)\mathbf{v}\right)\cdot\left(\frac{\partial\mathbf{v}}{\partial t}\right)-p\frac{\partial}{\partial t}(\nabla\cdot\mathbf{v})\right]\mathrm{d}V\\&\approx-2\int\limits_{V}\frac{1}{T}\left(\rho\frac{\partial\mathbf{v}}{\partial t}+\rho(\mathbf{v}\cdot\nabla)\mathbf{v}\right)\cdot\left(\frac{\partial\mathbf{v}}{\partial t}\right)\mathrm{d}V.\end{aligned} \tag{6.13}$$

Here we have assumed incompressibility ($\nabla\cdot\mathbf{v}=0$) in Eq. (6.13). If we further assume a situation with no advection of momentum, then $(\mathbf{v}\cdot\nabla)\mathbf{v}=0$; that is, there is no velocity gradient along the flow direction, corresponding to a laminar flow in the Stokes approximation.[4] In this specific laminar flow case, we get

$$\frac{\mathrm{d}\dot{\sigma}_{\mathrm{m,lam}}}{\mathrm{d}t}=-2\int\limits_{V}\frac{\rho}{T}\left|\frac{\partial\mathbf{v}}{\partial t}\right|^{2}\mathrm{d}V\leq 0, \tag{6.14}$$

where the suffix lam denotes the laminar flow with no momentum advection. The rate of entropy production in an incompressible laminar flow tends to decrease with time and reach a minimum in the final steady state ($\partial\mathbf{v}/\partial t=0$). This result shows another aspect of MinEP for a laminar flow. In an isothermal condition, this tendency is akin to that of minimum dissipation of kinetic energy in a slow

---

[3] In a general case, the forth term in the right-hand side of Eq. (6.12) should be expressed as a sum of the viscosity μ and the second viscosity λ. Using Stokes' relation ($\lambda=-2\mu/3$), $\mu+\lambda=\mu/3$.

[4] There are a few exceptions. Laminar (or non-turbulent) flow can be realized even with advection of momentum, e.g., in a converging nozzle. However, the flow direction is not parallel in this case, and it may not be regarded as "laminar" in the strict sense of the word.

incompressible steady flow suggested by Helmholtz [23] and Rayleigh [24]. However, as we shall see in Sect. 6.3.2, when advection of momentum is nonzero (i.e. turbulent flow), the sign of $\mathrm{d}\dot{\sigma}_{\mathrm{m}}/\mathrm{d}t$ becomes indefinite, and the entropy production can either decrease or increase depending on the rate of advection determined by the flow pattern produced in the fluid system.

6.3 Nonlinear Advection

We now discuss the effect of advection of heat or momentum on entropy production in a fluid system. The advection process is a typical nonlinear process since it is described as the product of the velocity and gradient of an intensive variable, which is also a function of the velocity. A fundamental difficulty arises from the presence of this nonlinear term in solving the balance equation of energy or momentum [Eq. (6.6) or (6.12)]. Exactly the same difficulty arises from this advection term in solving the equation of entropy production. We do not know, in a deterministic sense, how the rate of entropy production will change once advection becomes a dominant process in the transport of heat or momentum. However, advection of heat or momentum generally increases the local gradient of temperature or velocity at the moving front, which results in an enhancement of entropy production. Here we discuss the conditions under which advection enhances entropy production, using the general equations of entropy production [Eqs. (6.5) and (6.11)] as follows.

6.3.1 *Heat Advection*

Let us go back to the example of entropy production due to heat diffusion. With the presence of convective motion, the MinEP condition [Eq. (6.8)] cannot be justified since it requires $\mathbf{v} = 0$. Even in this case, Eq. (6.5) for the rate of change of entropy production remains valid. Assuming a constancy of c_{v} ($\mathrm{d}c_{\mathrm{v}}/\mathrm{d}t = 0$) in Eq. (6.6), and substituting the rate of change of temperature ($\partial T/\partial t$) into Eq. (6.5), we get

$$\begin{aligned}\frac{\mathrm{d}\dot{\sigma}_{\mathrm{h,adv}}}{\mathrm{d}t} &= -2\int_V \frac{\rho c_{\mathrm{v}}}{T^2}\kappa\nabla^2 T\left(\kappa\nabla^2 T - \mathbf{v}\cdot\nabla T - \frac{p\nabla\cdot\mathbf{v}}{\rho c_{\mathrm{v}}} + \frac{\mathbf{\Pi}:\nabla\mathbf{v}}{\rho c_{\mathrm{v}}}\right)\mathrm{d}V \\ &\approx -2\int_V \frac{\rho c_{\mathrm{v}}}{T^2}\kappa\nabla^2 T(\kappa\nabla^2 T - \mathbf{v}\cdot\nabla T)\,\mathrm{d}V,\end{aligned} \tag{6.15}$$

where the suffix adv denotes the presence of heat advection and $\kappa = k/\rho c_{\mathrm{v}}$ is the thermal diffusivity. The approximation in Eq. (6.15) corresponds to an assumption that the cooling rate by volume expansion ($\nabla\cdot\mathbf{v}$) and the heating rate by viscous dissipation ($\mathbf{\Pi}:\nabla\mathbf{v}$) are negligibly small compared with diffusive heating

($\kappa\nabla^2 T$) and advective cooling ($\mathbf{v}\cdot\nabla T$). Under this assumption, we can get a sufficient condition for the increase of entropy production $\left(\mathrm{d}\dot{\sigma}_{\mathrm{h,adv}}/\mathrm{d}t \geq 0\right)$ as

$$\mathbf{v}\cdot\nabla T \geq \kappa\nabla^2 T \geq 0 \quad \text{or} \quad \mathbf{v}\cdot\nabla T \leq \kappa\nabla^2 T \leq 0 \quad \Rightarrow \quad \frac{\mathrm{d}\dot{\sigma}_{\mathrm{h,adv}}}{\mathrm{d}t} \geq 0. \tag{6.16}$$

Condition (6.16) means that, when advective cooling ($\mathbf{v}\cdot\nabla T$) is greater than diffusive heating ($\kappa\nabla^2 T$), the local temperature decreases further ($\partial T/\partial t \leq 0$) because of Eq. (6.6), and thus entropy production increases because of Eq. (6.5). Alternatively, when advective heating ($-\mathbf{v}\cdot\nabla T > 0$) is greater than diffusive cooling ($-\kappa\nabla^2 T > 0$), the local temperature increases further ($\partial T/\partial t \geq 0$) because of Eq. (6.6), and thus entropy production increases because of Eq. (6.5). These conditions generally hold true during the development of convective motion ($\partial\mathbf{v}/\partial t > 0$) in a fluid system whose Rayleigh number is larger than the critical value for the onset of convection. The rate of entropy production thus tends to increase to a certain maximum value through the development of convective motion with time, as suggested from previous investigations [4, 5]. Moreover, it is known from numerical simulations that a state of convection tends to move to a state with higher rate of entropy production when the system has multiple steady states and the system is subject to external perturbations [7, 8, 25]. These results are consistent with condition (6.16) under which entropy production increases with time through the development of convective motion in a system with dynamic instability. It should be noted that condition (6.16) represents a condition for the increase of entropy production with time, whereas MaxEP has been suggested as a selection principle of the most stable steady state from a set of possible steady states ([12–14, 20]; Dewar and Maritan, this volume). However, when we observe time evolution of a system, the transition to the most stable MaxEP state must occur from a state with lower entropy production, so the two concepts are related through the actual time evolution of the system.[5] Condition (6.16) then shows the actual dynamic process along the evolution towards the MaxEP state inferred from the selection principle.

One can see from condition (6.16) that entropy production can decrease with time when the heat advection rate is smaller than the heat diffusion rate, i.e., $|\mathbf{v}\cdot\nabla T| \leq |\kappa\nabla^2 T|$. Such a situation can be realized in the relaxation period of a convection system towards a steady state, or in a convection system whose boundary temperature is unbounded so that the mean temperature gradient becomes weaker through the development of convective motion. We shall discuss vertical convection of the atmosphere as a typical example in Sect. 6.4. Another such example is thermal convection of a fluid system under fixed heat flux at the boundary. Entropy production as well as the overall temperature contrast at the boundary decreases with the onset of convection in this case (e.g. [26]). A quantitative analysis on the reduction of entropy production using Eq. (6.16) may

[5] The exact correspondence between the time evolution of a system and the probability of states requires an additional assumption, which is related to profound and not yet fully solved problems of the ergodic hypothesis.

therefore be attractive. It should be noted, however, that the decrease of entropy production in this case is not in direct contradiction to the stability criterion of MaxEP, because relative stability of each steady state should in principle be compared under the same boundary forcing condition, i.e., the same temperature contrast at the boundary characterized by the same Rayleigh number.

6.3.2 Momentum Advection

We can obtain a similar result for entropy production due to momentum diffusion. With the presence of advection of momentum $[(\mathbf{v}\cdot\nabla)\mathbf{v} \neq 0]$, the MinEP condition [Eq. (6.14)] cannot be justified. Even in this case, Eq. (6.11) for the rate of change of entropy production remains valid. Assuming incompressibility of fluid and substituting the rate of change of velocity from Eq. (6.12) into Eq. (6.11), we get

$$\frac{\mathrm{d}\dot{\sigma}_{\mathrm{m,adv}}}{\mathrm{d}t} = -2\int_V \frac{\rho}{T}\left(\nu\nabla^2\mathbf{v}\right)\cdot\left(\nu\nabla^2\mathbf{v} - (\mathbf{v}\cdot\nabla)\mathbf{v} - \frac{\nabla p}{\rho}\right)\mathrm{d}V, \tag{6.17}$$

where the suffix adv denotes the presence of momentum advection and $\nu = \mu/\rho$ is the kinematic viscosity. We can then find a sufficient condition for the increase of entropy production $\left(\mathrm{d}\dot{\sigma}_{\mathrm{h,adv}}/\mathrm{d}t \geq 0\right)$ as

$$\left[(\mathbf{v}\cdot\nabla)\mathbf{v} + \frac{\nabla p}{\rho}\right]\cdot\mathbf{e} \geq \left|\nu\nabla^2\mathbf{v}\right| \quad\Rightarrow\quad \frac{\mathrm{d}\dot{\sigma}_{\mathrm{m,adv}}}{\mathrm{d}t} \geq 0, \tag{6.18}$$

where $\mathbf{e} = \nabla^2\mathbf{v}/\left|\nabla^2\mathbf{v}\right|$ is the unit vector in the direction of $\nabla^2\mathbf{v}$.

Condition (6.18) means that, when advective export of momentum $[(\mathbf{v}\cdot\nabla)\mathbf{v}]$ plus pressure deceleration $[\nabla p/\rho]$ in the $\mathbf{e}$ direction is greater than diffusive import of momentum $\left|\nu\nabla^2\mathbf{v}\right|$, the local velocity in that direction decreases further because of Eq. (6.12), and thus entropy production increases because of Eq. (6.11). Alternatively, when advective import of momentum $[-(\mathbf{v}\cdot\nabla)\mathbf{v}]$ plus pressure acceleration $[-\nabla p/\rho]$ in the $-\mathbf{e}$ direction is larger than diffusive export of momentum $\left|\nu\nabla^2\mathbf{v}\right|$, the local velocity increases further because of Eq. (6.12), and thus entropy production increases because of Eq. (6.11). It is known that advection of momentum is negligibly small in laminar flows whereas it is considerably large in turbulent flows. Thus, this condition generally holds true during the development of turbulent motion in a fluid system whose Reynolds number is larger than the critical value for the onset of turbulence. The rate of entropy production thus tends to increase to a maximum value through the development of turbulent motion [5]. Malkus [27] and Busse [28] suggested that the observed mean state of turbulent shear flow corresponds to the state with the maximum rate of momentum transport by turbulent motion. Malkus [29] also showed that velocity profiles estimated from maximum dissipation of kinetic energy due to the mean velocity field and a smallest scale of motion at the system boundary resemble those of observations. Dewar and Maritan (this volume) showed using Jaynes's maximum entropy method that the inferred state is the one with maximum dissipation of kinetic energy due to the mean velocity field in a system with dynamic instability. Since the dissipation rate is proportional to the entropy production rate, these results are consistent with condition (6.18) under which entropy production increases with time towards a maximum value when the system is in a state of dynamic instability. As is the case for thermal convection, when we observe time evolution of such a system, the transition to the most stable MaxEP state must occur from a state with lower entropy production. Condition (6.18) then shows the actual dynamic process along the evolution towards the MaxEP state inferred from the selection principle.

One can also see from this condition (6.18) that entropy production can decrease with time when the momentum advection is less than the rates of diffusion and acceleration by the pressure gradient: $[(\mathbf{v}\cdot\nabla)\mathbf{v} + \nabla p/\rho]\cdot\mathbf{e} \leq |\nu\nabla^2\mathbf{v}|$. Such a condition can be realized in the relaxation period of a turbulent fluid system, or in a fluid system whose boundary velocity is unbounded so that the momentum advection becomes less significant than the sum of momentum diffusion and pressure acceleration. Examples include turbulent shear flow under a fixed shear stress and turbulent pipe flow under a fixed pressure gradient. Entropy production as well as the overall velocity gradient is known to decrease with the onset of turbulence in these cases [30, 31]. Again, notice that the decrease of entropy production in these cases is not in direct contradiction to the stability criterion of MaxEP, because relative stability of each steady state should in principle be compared under the same boundary forcing condition, i.e., the same velocity contrast between the boundary and the interior that is characterized by the same Reynolds number.

It should be noted that the condition [(6.16) or (6.18)] is a *sufficient* condition rather than a *necessary and sufficient* condition for $\mathrm{d}\dot{\sigma}_{\mathrm{adv}}/\mathrm{d}t \geq 0$—the total entropy production can increase even if local entropy production decreases in specific places. Also, this condition does not ensure that the total entropy production is to be a maximum; it describes a condition necessary for entropy production to increase locally. Nevertheless, this condition clearly shows the importance of advection for the behavior of entropy production caused by dynamic instability inherent to a nonlinear fluid system. Without advection, entropy production always decreases to a minimum value with time. With dynamic advection, entropy production can be enhanced to a greater value with a completely different mode of dynamic motion. In what follows, we shall examine a typical example of dynamic phenomena and discuss the behavior of entropy production with the presence of advection, in the light of the above analysis.

6.4 Atmospheric Convection

As a typical example of dynamic phenomena, let us discuss vertical convection of the atmosphere. Figure 6.1a shows a schematic of the global-mean energy balance of the Earth. The Earth absorbs solar shortwave radiation of about

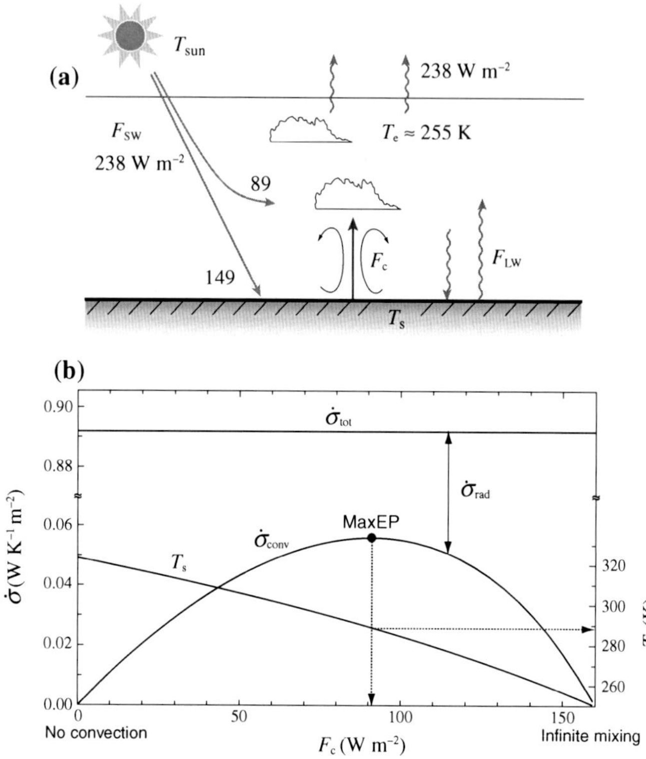

Fig. 6.1 **a** Global-mean energy balance of the earth. Shortwave solar radiation absorbed at the surface ($F_{LW} = 149$ W m^{-2}) is balanced by convective transport of latent and sensible heat (F_c) and by net longwave radiation (F_{LW}). T_s is surface temperature. **b** Entropy production rates per unit surface as a function of the surface convective heat flux F_c, calculated with a vertical grey atmosphere model. The convective entropy production $\dot{\sigma}_{conv}$ reaches a maximum at $F_c = 91$ W m^{-2} and $T_S = 289$ K. In this state the radiative entropy production $\dot{\sigma}_{rad}$ is a minimum (see text for details)

$F_{SW} = 238$ W m^{-2} as the global-mean, and this energy is transferred to various energy transport processes in the system. The energy is eventually emitted back to space by longwave emission at the top of the atmosphere, with an effective temperature $T_e \approx 255$ K according to the Stefan–Boltzmann law ($238 = \sigma_B T_e{}^4$, with σ_B being the Stefan–Boltzmann constant), thereby maintaining global energy balance. Energy balance is also maintained at the Earth's surface: absorbed solar radiation of 149 W m^{-2} is transported upward by convective transport[6] of latent and sensible heat F_c and by net longwave radiation F_{LW}. If convection were more

[6] There is also generation of mechanical energy by volume expansion of the air at the surface. The rate of energy conversion is about 2 W m^{-2}, which thereafter dissipates into heat in the atmosphere. We have included this rate in the convective energy transport considered here.

active (inactive) the surface temperature would decrease (increase). It is known from long-term observations that the convective heat transport is $F_c \approx 100$ W m^{-2} and surface air temperature is $T_s \approx 287$ K for a global-mean state of the present climate.

Entropy production in this climate system is mainly due to irreversible absorption of radiation $\dot{\sigma}_{\text{rad}}$, irreversible heat transport by thermal convection $\dot{\sigma}_{\text{h, conv}}$ and irreversible diffusion of momentum (viscous dissipation of kinetic energy) in the convective atmosphere $\dot{\sigma}_{\text{m, conv}}$:

$$\dot{\sigma}_{\text{tot}} = \dot{\sigma}_{\text{rad}} + \dot{\sigma}_{\text{h, conv}} + \dot{\sigma}_{\text{m, conv}}, \tag{6.19}$$

where $\dot{\sigma}_{\text{tot}}$ is the total rate of entropy production in the climate system. In a steady state, the total rate of entropy production is equal to the rate of entropy discharge into outer space:

$$\dot{\sigma}_{\text{tot}} = F_{\text{SW}} \left(\frac{1}{T_{\text{e}}} - \frac{1}{T_{\text{sun}}} \right) A, \tag{6.20}$$

where T_{sun} is the sun's emission temperature of about 5,760 K and A is the surface area of the Earth. Among these irreversible processes, entropy production due to momentum diffusion $\dot{\sigma}_{\text{m, conv}}$ is smaller than other terms $\dot{\sigma}_{\text{rad}}$ and $\dot{\sigma}_{\text{h, conv}}$ since the viscous dissipation rate is about 2 W m^{-2}, which is one or two orders of magnitude smaller than the convective or radiative energy fluxes in this system (e.g. [12]). This contribution of viscous heating has already been incorporated in the convective heat flux considered here. For a one-dimensional atmosphere we then have

$$\dot{\sigma}_{\text{tot}} \approx \dot{\sigma}_{\text{rad}} + \dot{\sigma}_{\text{conv}} = A \int F_{\text{r}} \frac{\partial}{\partial z} \left(\frac{1}{T_r} \right) \text{d}z + A \int F_{\text{c}} \frac{\partial}{\partial z} \left(\frac{1}{T} \right) \text{d}z, \tag{6.21}$$

where $\dot{\sigma}_{\text{conv}}$ is the entropy production due to convective heat transport, F_{r} is the radiation flux density defined as positive upward, T_{r} is the corresponding radiation effective temperature (mean temperature of the atmospheric medium over a unit optical thickness) and F_{c} is the convective energy flux density defined as positive upward. The transfer of radiant energy in a semi-transparent medium such as the atmosphere is caused by a gradient in the fourth power of temperature (the Planck function) in the medium. In this process, there is no direct advection of radiation due to dynamic motion of the atmosphere. According to the results of Sect. 6.2, we then expect that the radiative entropy production $\dot{\sigma}_{\text{rad}}$ will decrease with time and reach a minimum in the final steady state [Eq. (6.8)]. By contrast, entropy production due to heat diffusion is highly sensitive to advection when the atmosphere is in a state of convective instability. In this case, we can expect that entropy production will evolve with time towards a maximum value when the atmosphere satisfies condition (6.16).

Figure 6.1b shows steady-state total entropy production as a function of convective heat flux F_{c}, calculated with a vertical grey atmosphere model [32].

Longwave and shortwave optical depths are set at 3 and 0.43 respectively, and total shortwave absorption is assumed to be $F_{SW} = 238\ \mathrm{W\ m^{-2}}$ consistent with the observations. The abscissa shows the convective heat flux F_c at the surface, and the ordinate shows corresponding rates of entropy production per unit surface and surface temperature. We can see that, if there is no convection $F_c = 0$, the surface temperature has its maximum value at $T_s = 324$ K and convective entropy production is zero. In this state, radiative entropy production is largest. This static state is unstable against perturbations, however, since the vertical temperature gradient in most of the lower troposphere is larger than the dry adiabatic lapse rate (see, e.g., Fig. 2b in [32]). Convective motion therefore develops in the atmosphere, and the convective heat flux F_c as well as convective entropy production increases. This growth process is consistent with condition (6.16) under which entropy production due to heat diffusion tends to increase with time through the development of convective motion produced by dynamic instability. In the final steady state, the rate of convective entropy production reaches a maximum at $F_c = 91\ \mathrm{W\ m^{-2}}$ and $T_s = 289$ K, in remarkable agreement with observations. In this state, the radiative entropy production is at its minimal value. Further increase in F_c reduces the convective entropy production because the vertical temperature gradient decreases. This situation is also understood from condition (6.16); entropy production decreases when the boundary temperature is unbounded and the mean temperature gradient becomes so small that the advection term becomes less significant than the diffusion term.

6.5 Concluding Remarks

In this chapter, we have discussed some general characteristics of entropy production in a fluid system. We have shown that entropy production always decreases with time when the system is of a pure diffusion type without advection of heat or momentum. Thus, the minimum entropy production (MinEP) property is intrinsic to a system of a pure diffusion type, e.g., heat conduction in a static fluid or momentum diffusion in laminar flow. However, this MinEP property is no longer guaranteed when the system is subject to advection due to dynamic motion. In this case, entropy production increases with time when the rate of advection exceeds the rate of diffusion of the corresponding extensive quantity. The hypothesis of maximum entropy production (MaxEP) suggested as a selection principle for multiple steady states of nonlinear non-equilibrium systems ([1, 5, 12–14, 20]; Dewar and Maritan, this volume) can therefore be seen to be a characteristic feature of systems with nonlinear dynamic instability.

A few remarks may be in order to specify the current situation of the research. As stated in Sect. 6.3.2., the derived condition [(6.16) or (6.18)] is a sufficient condition for the increase of entropy production of a system, but not a necessary and sufficient condition. Thus, total entropy production can increase even if this condition does not hold in local specific places in the system. More rigorous study based on the integral Eqs. [(6.15) or (6.17)] is therefore needed to derive the exact condition for $d\dot{\sigma}_{adv}/dt \geq 0$. Intuitively, the exact condition obtained from the integration over a system should lead to a certain Rayleigh or Reynolds number for the entire system, although our previous attempts could not have been completed yet. This issue thus remains to be a future challenge. It should also be noted that, while we have examined the effects of nonlinear advection terms on entropy production in a fluid system, the constitutive relation between the diffusive flux and the driving force has been assumed to be linear in this study [Eq. (6.2) or (6.10)]. This linear assumption may not hold true, however, when the system is far from equilibrium. In a highly nonlinear regime of a compressible fluid, a discontinuity is known to be produced in the medium, and it propagates as a *shock front* with a supersonic speed. The dynamic process of each shock front is highly complicated, but its net effect is known to enhance entropy production [33]. The dynamic behavior of the shock front is therefore a fascinating topic, and remains a subject of future researches. While the local dynamic process of a nonlinear system is highly intricate because of its nonlinear nature, the ensemble of each process seems to be regulated at a state with a maximum rate of entropy production due to the dynamic processes. In a steady state, the entropy produced by all these irreversible processes is completely discharged into the surrounding system. We therefore suggest that local nonlinear dynamic processes tend to be organized so as to increase entropy in the surrounding system at a possible maximum rate when the system is in a state of nonlinear dynamic instability.

Acknowledgments The authors wish to express their cordial thanks to the organizers of the MaxEP 2011 Workshop in Canberra where the authors' interest in this subject has been stimulated. Valuable comments from two anonymous reviewers are also gratefully acknowledged.

References

1. Ziegler, H.: Zwei Extremalprinzipien der irreversiblen Thermodynamik. Ing. Arch. **30**, 410–416 (1961)
2. Paltridge, G.W.: Global dynamics and climate—a system of minimum entropy exchange. Q. J. Roy. Meteorol. Soc. **101**, 475–484 (1975)
3. Paltridge, G.W.: The steady-state format of global climate. Q. J. Roy. Meteorol. Soc. **104**, 927–945 (1978)
4. Schneider, E.D., Kay, J.J.: Life as a manifestation of the second law of thermodynamics. Math. Comput. Model. **19**, 25–48 (1994)
5. Ozawa, H., Shimokawa, S., Sakuma, H.: Thermodynamics of fluid turbulence: a unified approach to the maximum transport properties. Phys. Rev. E **64**, 026303 (2001)
6. Lorenz, R.D., Lunine, J.I., Withers, P.G., McKay, C.P.: Titan, Mars and Earth: entropy production by latitudinal heat transport. Geophys. Res. Lett. **28**, 415–418 (2001)
7. Shimokawa, S., Ozawa, H.: On the thermodynamics of the oceanic general circulation: irreversible transition to a state with higher rate of entropy production. Q. J. Roy. Meteorol. Soc. **128**, 2115–2128 (2002)
8. Shimokawa, S., Ozawa, H.: Thermodynamics of irreversible transitions in the oceanic general circulation. Geophys. Res. Lett. **34**, L12606 (2007). doi:10.1029/2007GL030208

9. Hill, A.: Entropy production as the selection rule between different growth morphologies. Nature **348**, 426–428 (1990)
10. Nohguchi, Y., Ozawa, H.: On the vortex formation at the moving front of lightweight granular particles. Physica D **238**, 20–26 (2009)
11. Lorenz, E.N.: Generation of available potential energy and the intensity of the general circulation. In: Pfeffer, R.L. (ed.) Dynamics of Climate, pp. 86–92. Pergamon, Oxford (1960)
12. Ozawa, H., Ohmura, A., Lorenz, R.D., Pujol, T.: The second law of thermodynamics and the global climate system: a review of the maximum entropy production principle. Rev. Geophys. **41**, 1018 (2003). doi:10.1029/2002RG000113
13. Dewar, R.: Information theory explanation of the fluctuation theorem, maximum entropy production and self-organized criticality in non-equilibrium stationary states. J. Phys. A **36**, 631–641 (2003)
14. Niven, R.K.: Steady state of a dissipative flow-controlled system and the maximum entropy production principle. Phys. Rev. E **80**, 021113 (2009)
15. Prigogine, I.: Modération et transformations irréversibles des systémes ouverts. Bulletin de la Classe des Sciences. Academie Royale de Belgique **31**, 600–606 (1945)
16. Glansdorff, P., Prigogine, I.: Sur les propriétés différentielles de la production d'entropie. Physica **20**, 773–780 (1954)
17. De Groot, S.R., Mazur, P.: Non-equilibrium Thermodynamics. North-Holland, Amsterdam (1962)
18. Gyarmati, I.: Non-equilibrium Thermodynamics. Field Theory and Variational Principles. Springer, Berlin (1970). [First published in Hungarian, 1967]
19. Prigogine, I.: From Being to Becoming. Time and Complexity in the Physical Sciences. Freeman, San Fransisco (1980)
20. Sawada, Y.: A thermodynamic variational principle in nonlinear non-equilibrium phenomena. Prog. Theor. Phys. **66**, 68–76 (1981)
21. Landau, L.D., Lifshitz, E.M.: Fluid Mechanics, 2nd edn. Pergamon Press, Oxford (1987). [First published in Russian (1944)]
22. Chandrasekhar, S.: Hydrodynamic and Hydromagnetic Stability. Oxford University Press, Oxford (1961)
23. Helmholtz, H.: Zur Theorie der stationären Ströme in reibenden Flüssigkeiten. Verhandlungen des naturhistorisch-medicinischen Vereins zu Heidelberg **5**, 1–7 (1869)
24. Rayleigh, Lord.: On the motion of a viscous fluid. Phil. Mag. **26**, 776–786 (1913)
25. Suzuki, M., Sawada, Y.: Relative stabilities of metastable states of convecting charged-fluid systems by computer simulation. Phys. Rev. A **27**, 478–489 (1983)
26. Kawazura, Y., Yoshida, Z.: Entropy production rate in a flux-driven self-organizing system. Phys. Rev. E **82**, 066403 (2010)
27. Malkus, W.V.R.: Outline of a theory of turbulent shear flow. J. Fluid Mech. **1**, 521–539 (1956)
28. Busse, F.H.: Bounds for turbulent shear flow. J. Fluid Mech. **41**, 219–240 (1970)
29. Malkus, W.V.R.: Borders of disorder: in turbulent channel flow. J. Fluid Mech. **489**, 185–198 (2003)
30. Paulus Jr, D.M., Gaggioli, R.A.: Some observations of entropy extrema in fluid flow. Energy **29**, 2487–2500 (2004)
31. Niven, R.K.: Simultaneous extrema in the entropy production for steady-state fluid flow in parallel pipes. J. Non-Equilib. Thermodyn. **35**, 347–378 (2010)
32. Ozawa, H., Ohmura, A.: Thermodynamics of a global-mean state of the atmosphere —a state of maximum entropy increase. J. Clim. **10**, 441–445 (1997)
33. Heisenberg, W.: Nonlinear problems in physics. Phys. Today **20**(5), 27–33 (1967)

Chapter 7
Control Volume Analysis, Entropy Balance and the Entropy Production in Flow Systems

Robert K. Niven and Bernd R. Noack

Abstract This chapter concerns "control volume analysis", the standard engineering tool for the analysis of flow systems, and its application to entropy balance calculations. Firstly, the principles of control volume analysis are enunciated and applied to flows of conserved quantities (e.g. mass, momentum, energy) through a control volume, giving integral (Reynolds transport theorem) and differential forms of the conservation equations. Several definitions of steady state are discussed. The concept of "entropy" is then established using Jaynes' maximum entropy method, both in general and in equilibrium thermodynamics. The thermodynamic entropy then gives the "entropy production" concept. Equations for the entropy production are then derived for simple, integral and infinitesimal flow systems. Some technical aspects are examined, including discrete and continuum representations of volume elements, the effect of radiation, and the analysis of systems subdivided into compartments. A Reynolds decomposition of the entropy production equation then reveals an "entropy production closure problem" in fluctuating dissipative systems: even at steady state, the entropy production based on mean flow rates and gradients is not necessarily in balance with the outward entropy fluxes based on mean quantities. Finally, a direct analysis of an infinitesimal element by Jaynes' maximum entropy method yields a theoretical framework with which to predict the steady state of a flow system. This is cast in terms of a "minimum flux potential" principle, which reduces, in different circumstances, to maximum or minimum entropy production (MaxEP or MinEP) principles. It is hoped that this chapter inspires others to attain a deeper

R. K. Niven (✉)
School of Engineering and Information Technology, The University of New South Wales at ADFA, Canberra, ACT 2600, Australia
e-mail: r.niven@adfa.edu.au

B. R. Noack
Institut PPRIME, CNRS, Université de Poitiers, ENSMA, CEAT, POITIERS Cedex F-86036, France
e-mail: Bernd.Noack@univ-poitiers.fr

R. C. Dewar et al. (eds.), *Beyond the Second Law*,
Understanding Complex Systems, DOI: 10.1007/978-3-642-40154-1_7,

understanding and higher technical rigour in the calculation and extremisation of the entropy production in flow systems of all types.

List of Symbols

Symbol Meaning (SI Units)

Roman Symbols

A	Area (m^2)
B, b	Conserved quantity ($[B]$); specific (per fluid mass) density ($[B]\,kg^{-1}$)
c_0	Speed of light in vacuum (ms^{-1})
f, $\mathbf{F}$	Generic parameter; generic gradient or driving force (various)
$\mathscr{F}_B$	Bulk flow rate of quantity B ($[B]\,s^{-1}$)
$\mathbf{g}_c$	Specific body force on species c ($N\,kg^{-1} = m\,s^{-2}$)
G, g	Gibbs free energy (J); specific Gibbs free energy ($J\,kg^{-1}$)
$\Delta\tilde{G}_d$	Change in molar Gibbs free energy of reaction d ($J\,mol^{-1}$)
h	Net heat transfer rate by radiation ($J\ s^{-1}\,m^{-6}$)
$\mathfrak{H}$	Generic (information) relative entropy function (–)
I_v, L_v	Energy radiance ($W\ m^{-2}\,s\ sr^{-1}$); entropy radiance ($W\ K^{-1}\ m^{-2}\,s\ sr^{-1}$)
$\mathbf{j}_c$	Molar flux of chemical species c ($mol\,m^{-2}\,s^{-1}$)
$\mathbf{j}_Q$, $\mathbf{j}_E$	Heat flux; energy flux ($J\,m^{-2}\,s^{-1}$)
$\mathbf{j}_S$, $\mathbf{J}_S$	Non-fluid entropy flux; total entropy flux ($J\ K^{-1}\,m^{-2}\,s^{-1}$)
k, k_{SB}	Boltzmann constant ($J\,K^{-1}$); Stefan-Boltzmann constant ($W\ m^{-2}\,K^{-4}$)
$\mathcal{K}$	Steady-state flow constant ($J\,K^{-1}\,m^{-3}\,s^{-1}$)
m	Fluid mass (kg)
$\mathbf{m}$, $\mathbf{n}$	Unit normal to area element; outward unit normal to control surface (–)
M_c	Molar mass of chemical species c ($kg\,mol^{-1}$)
n_c	Molar density of chemical species c ($mol\,kg^{-1}$)
n_i, N	Number of elements (balls) in partition i; total number of elements (–)
p_i, q_i	Inferred probability, prior probability (–)
P	Absolute pressure (Pa)
R	Number of constraints (–)
S, $\hat{S}$, s	Thermodynamic entropy ($J\ K^{-1}$); entropy per volume ($J\ K^{-1}m^{-3}$); specific entropy ($J\ K^{-1}\,kg^{-1}$)
t	Time (s)
T, T_v	Absolute temperature (K); radiative temperature (K)
U, $\hat{U}$, u	Internal energy (J); internal energy per volume ($J\ m^{-3}$); specific internal energy ($J\ kg^{-1}$)
$\mathbf{v}$	Mass-average velocity vector ($m\,s^{-1}$)
V	Volume (m^{-3})
$\mathbf{x}$	Position vector (m)
Z	Partition function (–)

Greek Symbols

γ	Degeneracy of state (−)
ε	Energy level (J)
λ, ζ	Lagrangian multiplier (various)
μ_c	Molar chemical potential of species c (J mol^{-1})
ν	Frequency of radiation (s^{-1})
χ_{cd}	Stoichiometric coefficient of species c in the dth reaction (mol mol^{-1})
$\hat{\dot{\xi}}_c$, $\hat{\dot{\xi}}_d$	Rate per volume of species c; of chemical reaction d (mol m^{-3} s^{-1})
ρ	Fluid density (kg m^{-3})
σ	Amount of thermodynamic entropy produced (J K^{-1})
$\dot{\sigma}$, $\hat{\dot{\sigma}}$, $\breve{\dot{\sigma}}$	Rate of thermodynamic entropy production (J K^{-1} s^{-1}); rate per volume (J K^{-1} m^{-3} s^{-1}); rate per area (J K^{-1}m^{-2} s^{-1})
τ	Viscous stress tensor (Pa)
Φ, ϕ	Potential (negative Massieu) function (−); Planck potential (J K^{-1})
ψ_c	Mass-weighted body force potential on species c (s^{-2})
Ω	Solid angle (sr)
$\breve{\dot{\omega}}$	Rate of entropy production on one side of area (J K^{-1}m^{-2} s^{-1})

Superscripts, Subscripts and Indices

$*$	Stationary state
$+, -$	Final, initial
c, d	Chemical species index, chemical reaction index
$C, \mathcal{C}$	Thermodynamic path index, set of allowable paths
eq, st	Equilibrium system, steady-state system
f, nf, tot	Fluid, non-fluid, total
m, ν	Material, radiative
$i, j, k, \mathbf{i}$	State indices
in, out	In or out of control volume
ℓ, r	Constraint indices
α, β	Compartment indices
κ	Compartment boundary index

Mathematical Symbols

$\bar{f}, \tilde{f}, f'$	Time mean; ensemble mean; fluctuating component
$\dot{f}, \hat{f}, \breve{f}, \tilde{f}$	Per unit time; per unit volume; per unit area; per mole
$\langle f \rangle$	Expectation
$\lfloor f \rfloor$, $(\!| f |\!)$	In-the-mean (product of means) form; mean fluctuating component

7.1 Introduction

Over the past half-century, there has been a growing interest in the analysis of non-equilibrium systems—which by their nature involve flow(s) of one or more quantities—using variational (extremum) principles based on the rate of thermodynamic entropy production and/or allied concepts. These include the maximum dissipation methods first proposed by Helmholtz [1] and Rayleigh [2] and their extension to the upper bound theory of turbulent fluid mechanics [3–5]; Onsager's "minimum dissipation" method [6, 7]; Prigogine's near-equilibrium minimum entropy production (MinEP) theorem [8, 9]; the far-from-equilibrium maximum entropy production (MaxEP) principle advocated by Paltridge [10, 11], Ziegler [12] and others [13–16], the main focus of this book; a MinEP framework for engineering design advocated particularly by Bejan [17]; a MinEP limit on transitions between equilibria [18–20] or steady states [21] respectively in thermodynamic or flow systems; and various minimum and maximum power methods applied to electrical circuits [9, 22–26] and pipe flow networks [27–30]. A broader category of variational technique consists of the maximum relative entropy (MaxEnt) method of Jaynes [31–35], which has seen myriad applications in many fields [36] and has been used in efforts to explain the above MaxEP/MinEP principles [37–41]. Such a zoo of different variational principles provides considerable scope for confusion, especially given their competing claims and partisanship. The entropy concept itself—and in consequence the thermodynamic entropy production—also provides a fertile ground for misunderstanding, which never ceases to yield unexpected traps for beginners and (even) well-established researchers.

In engineering, the method of *control volume analysis* is generally regarded as the most important tool for the analysis of flow systems, underpinning virtually all vehicular, fluid transport, energy generation, manufacturing, civil infrastructure and environmental control systems, and whose basic principles apply to all flows [42–47]. Recently, the authors have been surprised by the lack of appreciation of the control volume method throughout the sciences, even in those disciplines which—one would think—might gain the most from their use. For example, both an "ecosystem" and a "soil" are control volumes, which experience various material and energy flows (inputs and outputs) through their boundaries, and which undergo various internal processes. Their mathematical modelling therefore requires careful control volume analysis. Indeed, although not commonly calculated by engineers, the concept of entropy production itself arises from a control volume analysis of a dissipative system, and can be fruitfully examined from this perspective.

The aim of this chapter is to clarify the basis of the entropy production concept of non-equilibrium thermodynamics—and in consequence its extremisation—using the principles of control volume analysis. In Sect. 7.2, the control volume method and its main results are presented, and applied to flows of various quantities, for both integral and differential forms. Several definitions of steady state are then discussed. In Sect. 7.3.1, we examine the (generic) entropy concept (here labelled $\mathfrak{H}$), which in turn reduces, by a Jaynes' MaxEnt analysis of an equilibrium system, to the thermodynamic entropy S (Sect. 7.3.2). Control volume analysis of the latter (Sect. 7.3.3) enables rigorous definitions of the total thermodynamic entropy production $\dot{\sigma}$ and its local form $\hat{\dot{\sigma}}$. Several special features of the entropy balance are examined, including discrete and continuum representations, radiative effects, compartmentalisation and the definition of steady state. In Sect. 7.3.4, a Reynolds decomposition is used to reveal an "entropy production closure problem", manifested as a discrepancy between the overall mean and mean-of-products components. Finally, in Sect. 7.4 we analyse an infinitesimal control volume by Jaynes' MaxEnt method to directly predict the steady state. This yields a theoretical framework which reduces to (secondary) MaxEP or MinEP principles in different circumstances. The main motivation for this chapter is to inspire others to attain a deeper understanding and higher technical rigour in the calculation and extremisation of the entropy production in flow systems of all types.

7.2 Justification and Principles of Control Volume Analysis

Two Descriptions: Historically, two approaches have been developed for the analysis of flow systems [42–47]:

1. The *Lagrangian description*, which follows the behaviour of individual particles (either molecules or infinitesimal fluid elements) as they move, and so examines individual trajectories within the flow; and
2. The *Eulerian description*, which examines particular points or regions in space through which the flow passes, and so considers the flow field.

The Lagrangian approach has attained a high prominence in physics, giving rise to the field of classical mechanics (e.g. equations of motion, action integrals, principle of least action, Hamiltonian function, Liouville's theorem) and the concept of position-momentum phase space [48]. It also provided the basis of 19th century statistical physics, including Maxwell's velocity distribution, Boltzmann's H-theorem and their successors (including modern lattice-Boltzmann methods) [49], and of 20th century stochastic analyses, such as Markov processes and the Fokker–Planck and Master equations [50]. For all this prominence, however, Lagrangian methods impose considerable computational difficulties and are not widely used in engineering practice, except in specific cases where their use becomes essential (e.g. early re-entry of spacecraft through rarefied gases). Instead, the vast bulk of engineering fluid flow, heat and mass transfer calculations are conducted using the Eulerian description, necessitating a control volume analysis.

Control Volume Analysis: We now introduce the engineering concept of a *control volume* (CV), a geometric region through which one or more fluid(s) can flow, surrounded by a well-defined boundary or *control surface* (CS). The control

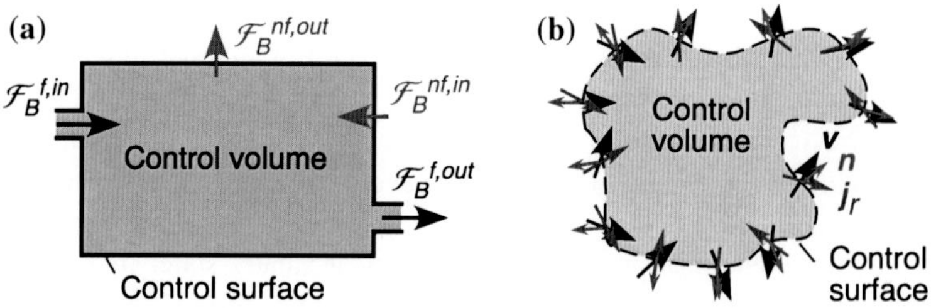

Fig. 7.1 Example control volumes for the analysis of **a** simple (flow rate) and **b** integral (vector flux) flow systems, showing representative fluid and non-fluid flow parameters

volume is assumed to be embedded within a surrounding *environment* (or "rest of the universe") which maintains the flow(s). We also require the concept of a *fluid volume* (FV) (in some references a *material volume* [42] or *system* [45, 47]), an identifiable body of fluid particles (or differential "fluid elements") which moves with time, bounded by its *fluid surface* (FS). We therefore analyse the motion of a fluid volume through a control volume.

Consider the simple fixed, non-deforming control volume shown in Fig. 7.1a, which experiences a discrete set of time-varying flow rate(s) across its control surface, and may also undergo various time-varying rate processes within its volume. We also consider the fluid volume coincident with the control volume at time t, which migrates downstream to a different position at time $t + dt$. For each conserved quantity B (e.g. mass, energy, momentum), the rates of change of B within the fluid and control volumes are connected by the conservation equation [42–47]:

$$\boxed{\frac{DB_{FV(t)}}{Dt} = \frac{\partial B_{CV}}{\partial t} + \mathscr{F}^{out}_{B,f} - \mathscr{F}^{in}_{B,f}} \tag{7.1}$$

where $DB_{FV(t)}/Dt$ is the *substantial*, *material* or *total derivative* of B, denoting its rate of change in motion with the fluid; $\partial B_{CV}/\partial t$ is the rate of change of B within the control volume[1]; and $\mathscr{F}^{out}_{B,f}$ and $\mathscr{F}^{in}_{B,f}$ are respectively the outward and inward flow rates of B due to fluid flow through the control surface.[2] In (7.1), the flow rates only refer to fluid-borne flows; all other flows of B are accounted within the substantial derivative $DB_{FV(t)}/Dt$. Note the "out—in" form of (7.1): in many texts it is written in the opposite sense (often in different notation):

---

[1] Strictly, for a fixed and non-deforming control volume, this should be written dB_{CV}/dt. The partial derivative is adopted to avoid confusion with some authors' use of dB_{CV}/dt to denote the substantial derivative, and for consistency with broader applications to moving control volumes.

[2] In engineering, it is standard practice to designate flow rates by an overdot, here $\dot{B}$. In deference to the different meaning of the overdot in physics, to signify a rate of production within a system, $\mathscr{F}_B$ is used herein for a bulk flow rate of B.

$$\frac{\partial B_{CV}}{\partial t} = \frac{DB_{FV(t)}}{Dt} + \mathscr{F}^{in}_{B,f} - \mathscr{F}^{out}_{B,f} \tag{7.2}$$

but the meaning is identical. If we understand the processes by which B changes within its fluid volume (both internal and external), their rate of change can be equated to $DB_{FV(t)}/Dt$, yielding an overall balance equation for B.

Now consider the more complicated geometry of Fig. 7.1b, in which the flow of B is represented by its time-varying fluid-borne flux $\rho b\mathbf{v}$ (measured in SI units of $[B]\ m^{-2}s^{-1}$) through the control surface, where $\rho(\mathbf{x},t)$ is the fluid density, $b(\mathbf{x},t)$ is the specific (per unit fluid mass) density of B and $\mathbf{v}(\mathbf{x},t)$ is the local (mass-average) velocity, in which $\mathbf{x}$ denotes position and t time. The B balance equation becomes:

$$\frac{DB_{FV(t)}}{Dt} = \frac{\partial B_{CV}}{\partial t} + \oiint_{CS} \rho b\mathbf{v}\cdot\mathbf{n}dA \tag{7.3}$$

where $\mathbf{n}(\mathbf{x} \in \mathrm{CS})$ is the unit normal to the control surface (positive outwards), A is the surface area and $\oiint_{CS}$ denotes integration around the control surface. Expressing $B_{CV} = \iiint_{CV} \rho b dV$, where V is the volume, (7.3) reduces to [42–47]:

$$\boxed{\frac{DB_{FV(t)}}{Dt} = \frac{\partial}{\partial t}\iiint_{CV} \rho b dV + \oiint_{CS} \rho b\mathbf{v}\cdot\mathbf{n}dA} \tag{7.4}$$

Equation (7.4) is known as Reynolds' transport theorem.

Since the control volume used here is stationary and non-deforming, the partial derivative in (7.4) can be brought inside the integral. Furthermore, from Gauss' divergence theorem, $\oiint_{CS} \rho b\mathbf{v}\cdot\mathbf{n}dA = \iiint_{CV} \nabla\cdot(\rho b\mathbf{v})dV$, so (7.4) can be written:

$$\frac{DB_{FV(t)}}{Dt} = \iiint_{CV}\left[\frac{\partial}{\partial t}\rho b + \nabla\cdot(\rho b\mathbf{v})\right]dV \tag{7.5}$$

Also, by integration over mass elements $dm = \rho dV$ of the fluid mass M [46]:

$$\frac{DB_{FV(t)}}{Dt} = \frac{D}{Dt}\iiint_{FV(t)} \rho b dV = \frac{D}{Dt}\int_M b dm = \int_M \frac{Db}{Dt}dm = \iiint_{FV(t)} \rho\frac{Db}{Dt}dV \tag{7.6}$$

using the local substantial derivative $Db/Dt = \partial b/\partial t + \mathbf{v}\cdot\nabla b$.

Equations (7.5)–(7.6) are valid for fluid and control volumes of any size, including infinitesimal volumes dV. It is therefore permissible to equate their integrands, assuming coincident fluid and control volumes in the infinitesimal limit, to give a differential conservation equation for each element dV in the fluid [46, 51]:

Table 7.1 Seven differential balance equations (7.7) for compressible flow (adapted after [51–53])

Property B	b	Balance equation (differential form)
Fluid mass	1	$0=\frac{\partial}{\partial t}\rho+\nabla\cdot(\rho\mathbf{v})$
Species moles	$\frac{N_c}{m}=n_c$	$\rho\frac{Dn_c}{Dt}=\frac{\partial}{\partial t}\rho n_c+\nabla\cdot(\rho n_c\mathbf{v})=-\nabla\cdot\mathbf{j}_c+\hat{\dot{\xi}}_c$
Linear momentum	$\mathbf{v}$	$\rho\frac{D\mathbf{v}}{Dt}=\frac{\partial}{\partial t}(\rho\mathbf{v})+\nabla\cdot(\rho\mathbf{v}\mathbf{v}^{\mathrm{T}})=-\nabla P-\nabla\cdot\boldsymbol{\tau}+\sum_c\rho_c\mathbf{g}_c$
Angular momentum	$\mathbf{x}\times\mathbf{v}$	$\rho\frac{D}{Dt}(\mathbf{x}\times\mathbf{v})=\frac{\partial}{\partial t}\rho(\mathbf{x}\times\mathbf{v})+\nabla\cdot\rho\mathbf{v}(\mathbf{x}\times\mathbf{v})$ $=-\nabla\cdot(\mathbf{x}\times P\boldsymbol{\delta})^{\mathrm{T}}-\nabla\cdot(\mathbf{x}\times\boldsymbol{\tau})^{\mathrm{T}}+\sum_c(\mathbf{x}\times\rho_c\mathbf{g}_c)-\boldsymbol{\varepsilon}:\boldsymbol{\tau}$
Total energy	$e=e_M+u$	$\rho\frac{De}{Dt}=\frac{\partial}{\partial t}(\rho e)+\nabla\cdot(\rho e\mathbf{v})$ $=-\nabla\cdot\mathbf{j}_Q-\nabla\cdot(P\mathbf{v})-\nabla\cdot(\boldsymbol{\tau}\cdot\mathbf{v})-\sum_c M_c\nabla\cdot(\psi_c\mathbf{j}_c)$
Kinetic + potential energy	$e_M=\frac{1}{2}\lvert\mathbf{v}\rvert^2+\psi$	$\rho\frac{De_M}{Dt}=\frac{\partial}{\partial t}(\rho e_M)+\nabla\cdot(\rho e_M\mathbf{v})$ $=-\mathbf{v}\cdot\nabla P-\mathbf{v}\cdot(\nabla\cdot\boldsymbol{\tau})-\sum_c M_c\psi_c\nabla\cdot\mathbf{j}_c$
Internal energy	u	$\rho\frac{Du}{Dt}=\frac{\partial}{\partial t}(\rho u)+\nabla\cdot(\rho u\mathbf{v})$ $=-\nabla\cdot\mathbf{j}_Q-P\nabla\cdot\mathbf{v}-\boldsymbol{\tau}:\nabla\mathbf{v}-\sum_c M_c\mathbf{j}_c\cdot\nabla\psi_c$

Assumptions and relations:
(i) $\rho_c=\rho n_c M_c$, $\sum_c\rho_c\mathbf{v}_c=\rho\mathbf{v}$, $\sum_c n_c M_c=1$, $\sum_c n_c M_c\mathbf{v}_c=\mathbf{v}$, $\sum_c\mathbf{j}_c M_c=0$ and $\mathbf{j}_c=\rho n_c(\mathbf{v}_c-\mathbf{v})$.
(ii) $\hat{\dot{\xi}}_c=\sum_d\chi_{cd}\hat{\dot{\xi}}_d$ and $\sum_c\chi_{cd}=0$. (iii) $\mathbf{g}_c=-\nabla\psi_c$, $\rho\psi=\sum_c\rho_c\psi_c$ and $\sum_c\psi_c\chi_{cd}=0$.

$$\boxed{\rho\frac{Db}{Dt}=\frac{\partial}{\partial t}\rho b+\nabla\cdot(\rho b\mathbf{v})} \tag{7.7}$$

The left-hand term can be further equated to the sum of rates of change of ρb in the infinitesimal fluid volume, due to internal and external processes, giving a local balance equation for B. As with all local formulations, (7.7) employs the *continuum assumption*, in which the system is assumed much larger than the molecular scale, so that its behaviour can be considered continuous even in the infinitesimal limit [42]. Equations (7.4) and (7.7) represent two long-standing traditions of fluid mechanics, integral (global) and local conservation laws, for the analysis of flow systems.

The particular forms of (7.7) for seven physical quantities are listed in Table 7.1. Here "·" is the vector scalar product, ":" is the tensor scalar product, $\top$ is a vector or tensor transpose, $[\boldsymbol{\delta},\boldsymbol{\varepsilon}]$ are the Kronecker delta and third-order permutation tensors; $\left[\rho_c,n_c,M_c,\mathbf{j}_c,\hat{\dot{\xi}}_c\right]$ are respectively the mass density, molar density (molality), molar mass, molar flux and molar rate of production of species c; $[P,\boldsymbol{\tau},\psi]$ are the pressure, stress tensor (positive for compression) and mass-weighted potential; $[\mathbf{g}_c,\psi_c]$ are the specific body force and potential on species c; and $[e,e_M,u,\mathbf{j}_Q]$ are the specific total energy, specific kinetic + potential energy, specific internal energy and heat flux. All fluxes $\mathbf{j}_Q$ and $\mathbf{j}_c$ are measured relative to the local mass-average fluid velocity $\mathbf{v}$. The listed equations are valid for compressible flow under fairly broad assumptions, assuming conservative body forces

$\mathbf{g}_c = -\nabla\psi_c$ on each species c. Other formulations can be derived for different circumstances [52, 54, 55].

Steady State: We now define $\partial B_{CV}/\partial t = 0$ as the *stationary* or *steady state* of a control volume. From (7.4) to (7.5):

$$\frac{\partial B_{CV}}{\partial t} = 0 \quad \Rightarrow \quad \left.\frac{DB_{FV(t)}}{Dt}\right|_{st} = \oiint\limits_{CS} \rho b \mathbf{v} \cdot \mathbf{n} dA = \iiint\limits_{CV} \nabla \cdot (\rho b \mathbf{v})\, dV \tag{7.8}$$

where *st* denotes steady state. We see that at steady state, the internal change of quantity B within the fluid volume is exactly balanced by its flux out of the control surface, and hence its integrated divergence. Similarly, using (7.7) and the definition of divergence [44, 56], we can define the steady state for an infinitesimal element:

$$\frac{\partial}{\partial t}\rho b = 0 \quad \Rightarrow \quad \rho\left.\frac{Db}{Dt}\right|_{st} = \lim_{CV\to 0} \frac{\oiint\limits_{CS} \rho b \mathbf{v} \cdot \mathbf{n} dA}{\iiint\limits_{CV} dV} = \nabla \cdot (\rho b \mathbf{v}) \tag{7.9}$$

Since both $\mathbf{v}(\mathbf{x}, t)$ and $B_{FV(t)}$ (or $b(\mathbf{x}, t)$) are time-dependent, a steady state can involve time-varying fluxes, provided these are exactly balanced by time-varying internal changes. In practice, however, any variability in the fluxes and/or rates will render (7.8)–(7.9) almost impossible to achieve (we could call them a *strict steady state*). It is therefore common in fluid mechanics (but not stated explicitly) to consider the *mean steady state* $\partial\langle B\rangle_{CV}/\partial t = 0$, where $\langle B\rangle$ denotes some mean (stationary first central moment) of B, referred to as a Reynolds average [46, 51, 57, 58]. Usually, $\langle B\rangle$ is equated with the time mean $\overline{B} = \lim_{T\to\infty} T^{-1} \int_0^T B dt$. In some situations, the ensemble mean $\tilde{B} = \lim_{K\to\infty} K^{-1} \sum_{k=1}^{K} B^{(k)}$ is used, where $B^{(k)}$ is the kth realisation of B [46]. For the latter, it is usual practice to invoke the *ergodic hypothesis*, in which the ensemble mean is assumed equivalent to the time mean; this assumption is correct only for certain types of flows. From (7.4) and (7.7):

$$\frac{\partial \langle B\rangle_{CV}}{\partial t} = 0 \Rightarrow \frac{D\langle B\rangle_{FV(t)}}{Dt} = \oiint\limits_{CS} \langle \rho b \mathbf{v}\rangle \cdot \mathbf{n} dA = \iiint\limits_{CV} \nabla \cdot \langle \rho b \mathbf{v}\rangle dV \tag{7.10}$$

$$\frac{\partial}{\partial t}\langle \rho b\rangle = 0 \Rightarrow \left\langle \rho \frac{Db}{Dt} \right\rangle = \nabla \cdot \langle \rho b \mathbf{v}\rangle \tag{7.11}$$

These give much more useful definitions than (7.8)–(7.9).[3] Importantly, since $\langle B\rangle_{CV} = \iiint_{CV} \langle \rho b\rangle dV$ for a stationary control volume, the global and local mean

[3] In consequence, the mean steady state need not be steady! Indeed the Fluctuation Theorem provides a strong argument that, far from equilibrium, it cannot be steady [40].

steady states (7.10)–(7.11) are equivalent, provided both are measured over long time periods. In contrast, the global and local strict steady states (7.8)–(7.9) are not equivalent, except for time-invariant fluxes and internal processes at both global and infinitesimal scales.

Throughout this chapter, the term *equilibrium* is used exclusively in its thermodynamic sense, to indicate the stationary state of a thermodynamic system, while *steady state* (usually qualified) refers to the stationary state of a control volume.

Further Remarks: Control volume analysis thus provides a rigorous framework for the analysis of flow systems, but like all mathematical methods, it holds some traps for beginners. Firstly, it is *essential* that the control volume and its control surface be clearly defined. This almost always requires a schematic diagram. Different control volumes represent different systems (with different steady states) and in general will yield different results. Where is the control surface? Which flows actually pass through the boundary and so must be included? Which flows are internal and so can be neglected? This study also considers only stationary control volumes. A moving and/or deforming control volume may be advantageous in some circumstances, but requires additional care [42–44, 47]. Finally, if a control volume is compartmentalised into sub-volumes, each of which is analysed by balance equations (7.1) or (7.4), the geometry of each compartment must be clearly defined, so that all flows can be identified and attributed to the correct compartments and external or internal boundaries.

7.3 Concept of Entropy

7.3.1 Generic (Information) Entropy

We now turn to the entropy concept, which causes many difficulties but in actual fact is very simple. While many justifications are available, arguably the most profound is the combinatorial basis expounded by Boltzmann and Planck [59, 60], in which we seek the *most probable state* of a probabilistic system. The system is typically represented by an allocation scheme in which N entities (balls) are distributed amongst I categories (boxes), forming individual *microstates* or *configurations* of the system. These are then grouped into observable *macrostates* or *realizations* of the system, specified by the number of balls n_i in each ith box. For distinguishable balls and boxes, the probability of a specified realization is given by the multinomial distribution:

$$\mathbb{P} = Prob(n_1, \ldots, n_I | N, q_1, \ldots, q_I) = N! \prod_{i=1}^{I} \frac{q_i^{n_i}}{n_i!} \tag{7.12}$$

where q_i is the prior or source probability of a ball in the ith box or, in other words, its assigned probability before observation. Seeking the maximum of $\mathbb{P}$ we recognise (as did Boltzmann [59]) that it is easier to maximise $\ln \mathbb{P} = \ln N! + \sum_{i=1}^{I} (n_i \ln q_i - \ln n_i!)$. Introducing the Stirling approximation $\ln N! \approx N \ln N - N$ in the asymptotic limit $N \to \infty$ (or alternatively the Sanov [61] theorem), with some rearrangement we obtain:

$$\mathfrak{H} = \lim_{N \to \infty} \frac{1}{N} \ln \mathbb{P} = - \sum_{i=1}^{I} p_i \ln \frac{p_i}{q_i} \tag{7.13}$$

where we take $p_i = \lim_{N \to \infty} n_i / N$ as the actual (observed or a posteriori) probability of a ball in the ith box. The function $\mathfrak{H}$ is referred to as the *relative entropy* or (negative) Kullback–Leibler function [62]. For equal priors $q_i = I^{-1}$, this simplifies to the *Shannon entropy* [63]:

$$\lim_{N \to \infty} \frac{1}{N} \ln \mathbb{P}_{\text{equal } q_i} \cong \mathfrak{H}_{\text{Sh}} = - \sum_{i=1}^{I} p_i \ln p_i \tag{7.14}$$

modulo a constant. Provided the system is indeed multinomial (7.12), maximising the relative entropy (7.13) (or Shannon entropy (7.14) for equal q_i), subject to any constraints, gives the most asymptotically probable realization of the system.

Adopting this probabilistic (or combinatorial) basis of entropy, we see that Jaynes' MaxEnt method [31–33] can be applied to *any* probabilistic system, not just in thermodynamics. For maximisation, it is necessary to incorporate the normalisation constraint and (usually) R moment constraints, respectively:

$$\sum_{i=1}^{I} p_i = 1, \quad \text{and} \quad \sum_{i=1}^{I} p_i f_{ri} = \langle f_r \rangle, \quad r = 1, \ldots, R, \tag{7.15}$$

where f_{ri} is the ith value of property f_r and $\langle f_r \rangle$ is the expectation of f_{ri}. Applying the calculus of variations, we write the Lagrangian:

$$L = - \sum_{i=1}^{I} p_i \ln \frac{p_i}{q_i} - \lambda_0 \left(\sum_{i=1}^{I} p_i - 1 \right) - \sum_{r=1}^{R} \lambda_r \left(\sum_{i=1}^{I} p_i f_{ri} - \langle f_r \rangle \right) \tag{7.16}$$

where λ_r is the Lagrangian multiplier for the rth constraint. Maximising (7.16) then gives the most probable realization and maximum relative entropy [31–33]:

$$p_i^* = \frac{q_i}{Z} \exp\left(- \sum_{r=1}^{R} \lambda_r f_{ri} \right), \quad \text{with } Z = e^{\lambda_0} = \sum_{i=1}^{I} q_i \exp\left(- \sum_{r=1}^{R} \lambda_r f_{ri} \right) \tag{7.17}$$

$$\mathfrak{H}^* = \ln Z + \sum_{r=1}^{R} \lambda_r \langle \mathrm{f}_r \rangle = -\Phi + \sum_{r=1}^{R} \lambda_r \langle \mathrm{f}_r \rangle \tag{7.18}$$

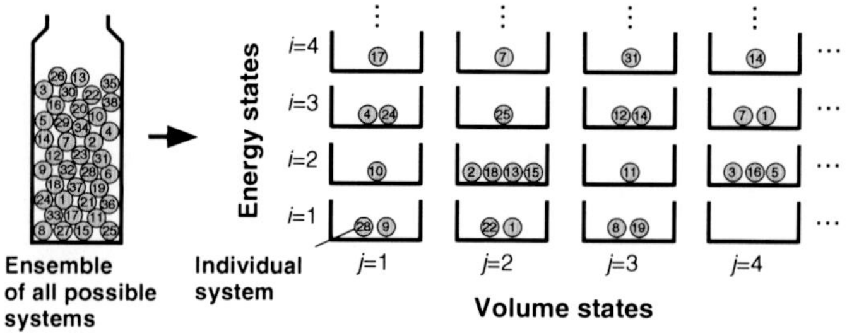

Fig. 7.2 Allocation scheme for the canonical ensemble of equilibrium thermodynamics

where * denotes the inferred state, Z is the partition function and $\Phi = -\ln Z$ is the potential (negative Massieu) function. By further analysis of first and second derivatives under this generic framework, it can be shown that $\mathfrak{H}^*(\langle \mathrm{f}_1 \rangle, \ldots, \langle \mathrm{f}_R \rangle)$ and $\Phi(\lambda_1, \ldots, \lambda_R)$ are Legendre transforms [31–33].

A caveat to the foregoing analysis is that the MaxEnt method is not a method of deductive reasoning, but should instead be viewed as a method of probabilistic inference [31, 33, 39–41, 64]. The distribution inferred by MaxEnt is not necessarily the "most correct" representation, but simply the one which is most probable given the imposed choices of constraints, prior probabilities, state space and the relative entropy function itself. If these assumptions are incomplete or incorrect, the discrepancy will be incorporated in the resulting model. Furthermore, there may be dynamical restrictions which prevent a system from attaining its most probable state. Such phenomenology (metastable states, supersaturated solutions, reaction kinetics, etc.) is well-known in equilibrium thermodynamics and, if necessary, can be handled by the incorporation of additional constraints, restrictions to the state space or additional theoretical apparatus.

7.3.2 Thermodynamic Entropy

The thermodynamic entropy S can now be interpreted as a special case of the generic entropy $\mathfrak{H}$, for a physical system constrained by its *contents* (usually expressed by mean extensive variables). Consider a container of N interacting molecules, for which it is infeasible to examine the allocation of individual molecules to energetic or other states. We therefore consider the *canonical ensemble* of all possible configurations of the system [65–70], in which replicas of the system are allocated to a coupled bivariate classification scheme according to their energy ε_{ij} and volume V_{ij}, where i and j respectively index the discrete energy and volume states of the ensemble. This is illustrated schematically in Fig. 7.2. The probabilities p_{ij} of the ijth energy-volume state of the ensemble are then considered to be constrained by normalisation (7.15), the mean internal energy $U = \sum_{ij} p_{ij} \varepsilon_{ij}$ and mean volume $V = \sum_{ij} p_{ij} V_{ij}$. Adopting the bivariate relative entropy $\mathfrak{H} = -\sum_{ij} p_{ij} \ln\left(p_{ij}/q_{ij}\right)$, the Lagrangian is:

$$L = -\sum_{ij} p_{ij} \ln \frac{p_{ij}}{q_{ij}} - \lambda_0 \left(\sum_{ij} p_{ij} - 1 \right) - \lambda_U \left(\sum_{ij} p_{ij} \varepsilon_{ij} - U \right) - \lambda_V \left(\sum_{ij} p_{ij} V_{ij} - V \right) \tag{7.19}$$

where λ_U and λ_V are Lagrangian multipliers for U and V. Maximisation then yields the most probable realization and maximum relative entropy:

$$p_{ij}^* = \frac{q_{ij}}{Z} \exp(-\lambda_U \varepsilon_{ij} - \lambda_V V_{ij}), \quad \text{with} \quad Z = \sum_{ij} q_{ij} \exp(-\lambda_U \varepsilon_{ij} - \lambda_V V_{ij}) \tag{7.20}$$

$$\mathfrak{H}^* = \ln Z + \lambda_U U + \lambda_V V = -\Phi + \lambda_U U + \lambda_V V \tag{7.21}$$

These are interpreted to represent the inferred or *equilibrium* state of the ensemble [31]. From the empirical body of thermodynamics, or from monotonic considerations, we recognise $\lambda_U = 1/kT$ and $\lambda_V = P/kT$, where k is Boltzmann's constant, T is absolute temperature and P is absolute pressure, while $q_{ij} = \gamma_{ij} / \sum_{ij} \gamma_{ij}$ is commonly expressed in terms of the degeneracy γ_{ij} of the ijth energy-volume level. Furthermore, we can identify $S = k\mathfrak{H}^*$ as the thermodynamic entropy at equilibrium, while $\phi_G = k\Phi = G/T$ is the Planck potential,[4] wherein G is the Gibbs free energy. Equations (7.20)–(7.21) thus provide the core equations of equilibrium thermodynamics [31–35, 69]:

$$p_{ij}^* = \frac{\gamma_{ij}}{\hat{Z}} \exp\left(\frac{-\varepsilon_{ij} - PV_{ij}}{kT} \right), \quad \text{with} \quad \hat{Z} = Z \sum_{ij} \gamma_{ij} = \sum_{ij} \gamma_{ij} \exp\left(\frac{-\varepsilon_{ij} - PV_{ij}}{kT} \right) \tag{7.22}$$

$$S = k \ln Z + \frac{U}{T} + \frac{PV}{T} = -\phi_G + \frac{U}{T} + \frac{PV}{T} \tag{7.23}$$

Further analysis using generalised heat and work concepts [31] gives the differential:

$$\boxed{d\phi_G = -dS + \frac{1}{T} dU + \frac{P}{T} dV} \tag{7.24}$$

Equations (7.22)–(7.24) in turn give a set of derivative relations and Legendre duality between S and ϕ_G [31–36, 69]. Many other formulations are available for different thermodynamic ensembles subject to various constraints [35, 36, 67].

We can now interpret the physical meaning of the potential ϕ_G [39, 69, 72, 73]. Consider a "universe" divided into a system of interest and an external environment. From the second law (7.25), an incremental increase in entropy of the universe can be expressed as a sum of changes within and external to the system

[4] Strictly, Planck used the negative of ϕ_G as his potential function [71, 72].

$dS_{univ} = dS + dS_{ext} \geq 0$. Although dS_{ext} cannot be measured directly, if it alters the system in any way, it must produce a change in its constraints and/or multipliers, hence $dS_{ext} = -\frac{1}{T}dU - \frac{P}{T}dV$, where the negative sign accounts for positive dS_{ext}. Substituting in (7.20), we identify $dS_{univ} = -d\phi_G$. In consequence, if a thermodynamic system can interact with an external environment, its equilibrium state is determined by minimising its Planck potential ϕ_G, thereby maximising the entropy of the universe, rather than by maximising the entropy of the system S alone. For constant T, this reduces to the well-known principle of minimum Gibbs free energy [74].

Minimising ϕ_G requires integration $\Delta\phi_G = \int_{C \in \mathcal{C}} d\phi_G$ over some path C, selected from the set of paths $\mathcal{C}$ with a specified starting point $\phi_{G,0}$ and an endpoint at the minimum potential $\phi_{G,\min}$. Since ϕ_G is a state function, its difference $\Delta\phi_G = \phi_{G,\min} - \phi_{G,0}$ is path-independent, but there may be restrictions on the set of allowable paths $\mathcal{C}$ (e.g., only adiabatic paths or only isobaric paths), causing further restrictions on the minimum potential $\phi_{G,\min}$, or the set of such minima, which can be accessed by the system. Denoting $d\sigma = -(dU + PdV)/T = -dH/T$ as the increment of *entropy produced* by a system, where H is the enthalpy, (7.20) reduces to $d\phi_G = -dS - d\sigma$. Since S and σ are also state functions, the step change can be written as $\Delta\phi_G = -\Delta S - \Delta\sigma$. Minimisation of ϕ_G to give $\Delta\phi_G < 0$ can therefore occur in three ways:

1. By a coupled increase in both S and σ along path C to give $\Delta\phi_G < 0$, hence with $\Delta S > 0$ and $\Delta\sigma > 0$;
2. By a coupled increase in S and decrease in σ along C, hence $\Delta S > 0$ and $\Delta\sigma < 0$, provided that $\Delta S > |\Delta\sigma| > 0$ to ensure $\Delta\phi_G < 0$; or
3. By a couple decrease in S and increase in σ along C, hence $\Delta S < 0$ and $\Delta\sigma > 0$, provided that $\Delta\sigma > |\Delta S| > 0$ to ensure $\Delta\phi_G < 0$.

The choice of scenario is governed by the set of allowable paths $\mathcal{C}$, which controls the flow of various quantities (in this example, heat) through the control surface and hence the competition between dS and $d\sigma$. The first and third scenarios can be interpreted as a constrained maximisation of σ (hence minimisation of H/T) over the set of paths $\mathcal{C}$, while the second can be viewed as a constrained minimisation. Similarly, the first and second scenarios also involve constrained maximisation of S over $\mathcal{C}$, while the third involves its minimisation. This three-fold structure is well established in equilibrium thermodynamics, although is usually presented in terms of the Gibbs free energy rather than the Planck potential [68]. Rather than adopt separate extremum principles for different processes, and to correctly account for changes in entropy within and outside the system, the three scenarios are unified by an overarching minimum Planck potential principle [72, 73], which at constant T reduces, as noted, to that of minimum Gibbs free energy [74].

As will be shown, the above thermodynamically-inspired principle can be established—using the MaxEnt framework—in other, quite different kinds of systems.

7.3.3 Entropy Balance and Entropy Production

Entropy Balance Equations: With the entropy concept in hand, we can now consider the thermodynamic entropy balance in a control volume, such as that shown in Fig. 7.1a. Our first difficulty is that S is not conserved. However, from the second law of thermodynamics, within any closed physical system:

$$dS \geq 0 \tag{7.25}$$

where dS implies a mean differential over a minimum time scale, to allow for brief excursions in the opposite sense. So, despite not being conserved, we can say that S is *preserved*: once created, it cannot be destroyed. In consequence, for an entropically open system—which can exchange entropy with its external environment—(7.1) provides a control volume balance ("*law of preservation*") for S:

$$\frac{DS_{FV(t)}}{Dt} = \frac{\partial S_{CV}}{\partial t} + \mathscr{F}_{S,f}^{out} - \mathscr{F}_{S,f}^{in} \tag{7.26}$$

where $\mathscr{F}_{S,f}^{out}$ and $\mathscr{F}_{S,f}^{in}$ are the outflow and inflow rates of S due to fluid flow through the control surface. The substantial derivative can also be separated, by the de Donder technique, into externally- and internally-driven rates of change of entropy within the fluid volume, giving the overall entropy balance equation (c.f. [75]):

$$\boxed{\frac{DS_{FV(t)}}{Dt} = \frac{\partial S_{CV}}{\partial t} + \mathscr{F}_{S,f}^{out} - \mathscr{F}_{S,f}^{in} = \frac{D_e S_{FV(t)}}{Dt} + \frac{D_i S_{FV(t)}}{Dt}} \tag{7.27}$$

where $D_e S_{FV(t)}/Dt$ represents the rate of change of entropy in the fluid volume due to non-fluid flows (positive inwards), i.e.

$$\frac{D_e S_{FV(t)}}{Dt} = \mathscr{F}_{S,nf}^{in} - \mathscr{F}_{S,nf}^{out} \tag{7.28}$$

Similarly, $D_i S_{FV(t)}/Dt$ denotes the *(rate of) entropy production* in the fluid volume due to internal processes, henceforth labelled $\dot{\sigma}$. The latter serves as a bookkeeping term in (7.27), ensuring that the rate of creation of entropy in the fluid volume satisfies the second law of thermodynamics (7.25):

$$\boxed{\dot{\sigma} = \frac{D_i S_{FV(t)}}{Dt} = \frac{\partial S_{CV}}{\partial t} + \mathscr{F}_{S,tot}^{out} - \mathscr{F}_{S,tot}^{in} \geq 0} \tag{7.29}$$

where $\mathscr{F}_{S,tot} = \mathscr{F}_{S,f} + \mathscr{F}_{S,nf}$ is the total entropy flow rate. Thus, by definition, the rate of entropy production $\dot{\sigma}$ cannot be negative, regardless of whether the newly created entropy is retained in the control volume or exported from it (i.e., independent of the sign of the rate of change of S). Equation (7.29) may therefore be viewed as a powerful manifestation of the second law, applicable to all non-equilibrium systems.

For the integral control volume of Fig. 7.1b, from (7.4):

$$\frac{DS_{FV(t)}}{Dt} = \frac{\partial}{\partial t} \iiint\limits_{CV} \rho s dV + \oiint\limits_{CS} \rho s \mathbf{v} \cdot \mathbf{n} dA \tag{7.30}$$

where s is the specific entropy. From (7.27), this is equal to the internal rate of entropy production in the fluid volume, $\dot{\sigma}$, plus the external rate of input due to non-fluid transport processes, $-\oiint_{FS(t)} \mathbf{j}_S \cdot \mathbf{n} dA$, where $\mathbf{j}_S$ is the non-fluid entropy flux:

$$\frac{DS_{FV(t)}}{Dt} = \frac{D_i S_{FV(t)}}{Dt} + \frac{D_e S_{FV(t)}}{Dt} = \dot{\sigma} - \oiint\limits_{FS(t)} \mathbf{j}_S \cdot \mathbf{n} dA \tag{7.31}$$

Equating (7.30)–(7.31), for coincident fluid and control volumes at time t, gives:

$$\boxed{\dot{\sigma} = \iiint\limits_{CV} \frac{\partial \rho s}{\partial t} dV + \oiint\limits_{\text{coincident } CS \text{ and } FS(t)} [\mathbf{j}_S + \rho s \mathbf{v}] \cdot \mathbf{n} dA = \frac{DS_{FV(t)}}{Dt} + \oiint\limits_{FS(t)} \mathbf{j}_S \cdot \mathbf{n} dA} \tag{7.32}$$

Applying (7.6) and Gauss' theorem then yields:

$$\dot{\sigma} = \iiint\limits_{CV} \left[\frac{\partial}{\partial t} \rho s + \nabla \cdot \mathbf{J}_S\right] dV = \iiint\limits_{FV(t)} \left[\rho \frac{Ds}{Dt} + \nabla \cdot \mathbf{j}_S\right] dV \tag{7.33}$$

where $\mathbf{J}_S = \mathbf{j}_S + \rho s \mathbf{v}$. Finally, subdividing $\dot{\sigma} = \iiint_{CV} \hat{\dot{\sigma}} dV$, where $\hat{\dot{\sigma}}$ is the (*rate of*) *entropy production per unit volume*, and equating integrands (assuming validity at all scales) gives the differential entropy balance equation [9, 52, 53, 76]:

$$\boxed{\hat{\dot{\sigma}} = \frac{\partial}{\partial t} \rho s + \nabla \cdot \mathbf{J}_S = \rho \frac{Ds}{Dt} + \nabla \cdot \mathbf{j}_S \geq 0} \tag{7.34}$$

By a scale invariance argument [9], $\hat{\dot{\sigma}}$ cannot be negative locally (at least over a minimum time scale) at any location, since this would continuously destroy thermodynamic entropy within an identifiable control volume, and so violate the second law of thermodynamics. This is entirely separate to the rate of change of the specific entropy s, which can be positive or negative locally, depending on the sign of the divergence term (i.e. on the local entropy flux out of the element).

Local Entropy Flux and Entropy Production: To reduce (7.34), we seek functional forms of the non-fluid entropy flux $\mathbf{j}_S$ and local entropy production $\hat{\dot{\sigma}}$. For non-radiative processes, the standard approach is to start from the substantial derivative of the specific form of Gibbs' relation (7.23)–(7.24) [8, 52, 53, 76]:

$$\frac{Ds}{Dt} = \frac{D\phi_g}{Dt} + \frac{1}{T} \frac{Du}{Dt} + \frac{P}{T} \frac{D\rho^{-1}}{Dt} \tag{7.35}$$

where $\phi_g = g/T$ is the specific Planck potential and g the specific Gibbs free energy. This adopts the *local equilibrium assumption*, where each infinitesimal element is assumed to be in thermodynamic equilibrium and so can be described by local intensive variables $1/T$, P/T and $\{\mu_c/T\}$, where μ_c is the molar chemical potential of species c. Including the work of chemical diffusion $g = -\sum_c \mu_c n_c$, and substituting for the substantial derivatives of specific volume ρ^{-1}, species molar densities n_c and specific internal energy u (see Table 7.1) gives:

$$\rho \frac{Ds}{Dt} = -\frac{1}{T}\nabla \cdot \mathbf{j}_Q + \sum_c \frac{\mu_c}{T} \nabla \cdot \mathbf{j}_c - \frac{1}{T}\sum_c M_c \mathbf{j}_c \cdot \nabla \psi_c - \frac{1}{T}\tau : \nabla \mathbf{v} - \sum_d \hat{\dot{\xi}}_d \Delta \frac{\tilde{G}_d}{T} \tag{7.36}$$

This is expressed in terms of the molar rate of the dth reaction $\hat{\dot{\xi}}_d = \sum_c \chi_{cd} \hat{\dot{\xi}}_c$ (> 0 if a product) and change in molar Planck potential of the dth reaction, $\Delta\tilde{\phi}_G = \Delta(\tilde{G}_d/T) = \sum_c \chi_{cd}\, \mu_c/T$ (< 0 if spontaneous), where χ_{cd} is the stoichiometric coefficient of species c in the dth reaction. Comparison to (7.34), with some vector calculus, gives the entropy flux and local entropy production [52, 53]:

$$\mathbf{j}_{S,m} = \left(\frac{1}{T}\right)\mathbf{j}_Q - \sum_c \left(\frac{\mu_c}{T}\right)\mathbf{j}_c \tag{7.37}$$

$$\hat{\dot{\sigma}}_m = \mathbf{j}_Q \cdot \nabla\left(\frac{1}{T}\right) - \sum_c \mathbf{j}_c \cdot \left[\nabla\left(\frac{\mu_c}{T}\right) + \frac{M_c \nabla \psi_c}{T}\right] - \frac{\tau : \nabla \mathbf{v}}{T} - \sum_d \hat{\dot{\xi}}_d \Delta \frac{\tilde{G}_d}{T} \tag{7.38}$$

These do not include the effect of radiation, examined in a later section, and so are labelled m to signify the material or thermodynamic component. In generic form, we identify the entropy flux (7.37) as $\mathbf{j}_{S,m} = \sum_r \mathbf{j}_r \lambda_r$, a sum of products of fluxes and conjugate spatial intensive variables selected from $\mathbf{j}_r \in \{\mathbf{j}_Q, \mathbf{j}_c\}$ and $\lambda_r \in \{1/T, -\mu_c/T\}$, while the entropy production (7.38) is $\hat{\dot{\sigma}}_m = \sum_r \mathbf{j}_r \cdot \mathbf{F}_r$, a sum of products of all fluxes or rates and their conjugate gradients or driving forces $\mathbf{F}_r \in \{\nabla(1/T), -\nabla(\mu_c/T), -\nabla\psi_c/T, -\nabla\mathbf{v}/T, -\Delta(\tilde{G}_d/T)\}$ [8, 9, 52]. Usually, $\hat{\dot{\sigma}}_m$ is further simplified—assuming conditions close to thermodynamic equilibrium—using the linear Onsager phenomenological relations and the Curie postulate, to give a bilinear sum of thermodynamic forces [52, 53, 77].

Thermodynamic Representations: Before embarking on further analyses, it is worth scrutinising the physical representation of the bilinear, non-radiative local entropy production (7.38). As evident, it includes two quite different types of physical processes:

Type I Processes: Those which can be represented to occur within an infinitesimal volume element at *local spatial equilibrium* with respect to the spatial intensive

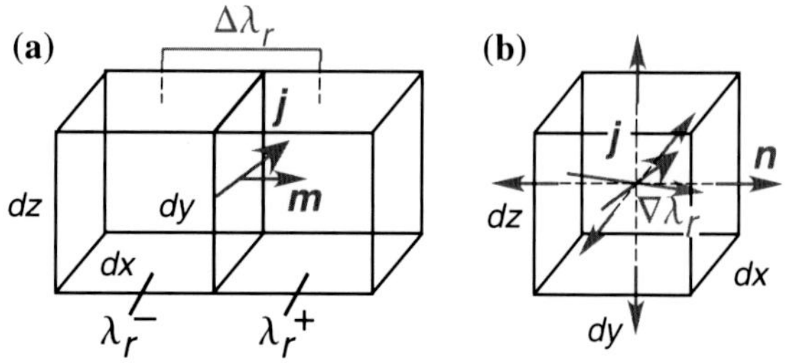

Fig. 7.3 Infinitesimal volume elements for **a** local spatial equilibrium (Type I and II(a)) and **b** continuum (Type II(b)) representations, showing the *r*th flux and its intensive variables

variables $\lambda_r \in \{1/T, -\mu_c/T, -\psi_c/T, -\mathbf{v}/T\}$, as shown in Fig. 7.3a.[5] In (7.38), only the final chemical reaction term falls into this category. In this case, the volume element need not be in chemical equilibrium, but may be maintained at a higher Planck potential by its chemical composition. This category also includes nuclear and subatomic decay processes, not usually represented in (7.38).

Type II Processes: Those which—although formulated in terms of an infinitesimal volume element—are in fact associated with a physical *flux* which diminishes (or acts conjugate to) a spatial *gradient*. For the *r*th process, this can be written as $\hat{\sigma}_{m,r} = \mathbf{j}_r \cdot \nabla\lambda_r$. The heat, species mass and momentum transport terms in (7.38) all fall into this category. These have two possible physical representations:

(a) If each volume element is considered to be in local spatial equilibrium, as shown in Fig. 7.3a, then no Type II entropy production could occur *within* an element, but only *between* elements. This necessitates analysis of the boundary entropy production terms, which must be integrated over the internal boundaries and/or somehow assigned to each element.

(b) If each volume element need not be in local spatial equilibrium, it can be used to directly represent both the fluxes and gradients, as shown in Fig. 7.3b. Arguably, this gives a more physically defensible representation of a non-equilibrium system—dependent upon the continuum assumption—and for this reason is almost universally adopted throughout fluid mechanics and heat transfer analysis (e.g. in differential derivations of the continuity, Navier–Stokes and energy equations). However, it contradicts the assumption of local equilibrium, creating a philosophical difficulty in the use of intensive variables which, strictly, are defined only at equilibrium [69]. Instead, in this representation, both a value and gradient in each intensive variable are assigned to each point within the infinitesimal element.

Representations II(a) and II(b) involve fundamentally different idealisations of physical transport processes. Their analysis requires different mathematical tools, respectively a hybrid difference-differential calculus and the usual differential calculus.

[5] Some authors unite the variables conjugate to the species flux $\mathbf{j}_c$ into a local electrochemical or gravichemical potential divided by temperature, $-\mu_c^{\mathrm{g}}/T$ [73].

To tease out the distinction between Type II(a)–(b) representations, consider an individual boundary between two infinitesimal elements, as shown in Fig. 7.3a, b. For Type II(b) elements, there is no discontinuity in the rth intensive variable at each boundary, and—by continuity—no change in the rth flux, causing no (or an infinitesimal) change in each non-fluid entropy flux. The fluid-borne entropy flux $\rho s\mathbf{v}$ is similarly unaffected. In consequence, no (or an infinitesimally small) entropy production occurs at the boundary. Type II(a) elements, in contrast, exhibit a step change $\Delta\lambda_r = \lambda_r^+ - \lambda_r^-$ in each spatial intensive variable across the boundary, giving the *net entropy production per unit area* $\left(\mathrm{J\ K^{-1}m^{-2}s^{-1}}\right)$ due to thermodynamic processes at the boundary:

$$\breve{\dot{\sigma}}_m = \left[\Delta(\rho s)\mathbf{v} + \sum_r \mathbf{j}_r \Delta\lambda_r\right] \cdot \mathbf{m} = [\Delta(\rho s)\mathbf{v} + \Delta\mathbf{j}_{S,m}] \cdot \mathbf{m} = \Delta\mathbf{J}_{S,m} \cdot \mathbf{m} \geq 0 \quad (7.39)$$

where $\mathbf{j}_r$ is the flux of the rth conserved quantity, $\mathbf{m}$ is the unit normal to the boundary, $\Delta\mathbf{j}_{S,m}$ is the net non-fluid, non-radiative entropy flux and $\Delta\mathbf{J}_{S,m}$ also includes the net fluid-borne entropy flux (all positive in the direction $\Delta\lambda_r > 0$). In (7.39), it is assumed the fluid-borne entropy flux undergoes a step change at the boundary (e.g. due to a discontinuity $\Delta\rho$ or Δs caused by a step change in $1/T$ or μ_c/T). No step changes are considered in $\mathbf{v}$ or $\mathbf{j}_r$, being fluxes of conserved quantities. The non-radiative entropy production along a boundary Γ is then $\dot{\sigma}_{\Gamma,m} = \iint_\Gamma \breve{\dot{\sigma}}_m dA$.

Often it is desirable to account separately for each side of the boundary, leading to the *absolute* or *half-boundary entropy production per unit area* due to outward flow from a specified face of a volume element:

$$\breve{\dot{\omega}}_m = \left[\rho s\mathbf{v} + \sum_r \mathbf{j}_r \lambda_r\right] \cdot \mathbf{n} = [\rho s\mathbf{v} + \mathbf{j}_{S,m}] \cdot \mathbf{n} = \mathbf{J}_{S,m} \cdot \mathbf{n} \gtrless 0 \quad (7.40)$$

where $\mathbf{n}$ is the outward unit normal. As expected, this depends on the material entropy flux $\mathbf{J}_{S,m}$ at the boundary. From (7.39), $\breve{\dot{\sigma}}_m = \breve{\dot{\omega}}_m^+ - \breve{\dot{\omega}}_m^-$. The total entropy production along Γ is thus given by the two-sided surface integral $\dot{\sigma}_{\Gamma,m} = \iint_{\Gamma^+} \breve{\dot{\omega}}_m^+ dA - \iint_{\Gamma^-} \breve{\dot{\omega}}_m^- dA = \oiint_\Gamma \breve{\dot{\omega}}_m dA = \oiint_\Gamma \mathbf{J}_{S,m} \cdot \mathbf{n} dA$. Applying Gauss' divergence theorem to the surface Γ enclosing the "internal volume" Γ^o, we obtain the interesting result that $\dot{\sigma}_{\Gamma,m} = \iiint_{\Gamma^o} \nabla \cdot \mathbf{J}_{S,m} dV \geq 0$, even though $\iiint_{\Gamma^o} dV = 0$.

From the second law (7.25), each net boundary entropy production (7.39) is non-negative (over a minimum observation time). In contrast, the half-boundary terms (7.40) can be of arbitrary sign, so long as their difference across each internal boundary is non-negative.[6] As a test of consistency, integration of (7.40) over the external control surface yields the net entropy flow rate $\oiint_{CS} \mathbf{J}_{S,m} \cdot \mathbf{n} dA$ contained in (7.32).

Equations (7.39)–(7.40) are used in later sections. They cannot, however, be reconciled in a straightforward manner to the differential equation in (7.34), which corresponds strictly to the Type II(b) or continuum representation.

Effect of Radiation: An important category of processes, omitted from the standard analysis (7.36)–(7.38)—and indeed from most references on non-equilibrium thermodynamics—is the entropy production associated with electromagnetic radiation. Its major principles were however enunciated by Planck [60, 71] over a century ago, and further developed over the past century (e.g. [78–86]). However, there still remains widespread confusion in its calculation, over choices of symbols and preferred parameters, and even in the most appropriate theoretical approach.[7] Many renowned texts on radiation omit the topic entirely (e.g. [87]).

Firstly, the energy of unpolarised electromagnetic radiation per unit frequency travelling through an infinitesimal area (of unit normal $\mathbf{m}$) and infinitesimal solid angle per unit time is represented by its *specific energy intensity* or *energy radiance* I_ν (SI units: W m^{-2}s sr^{-1}). This is a function of the direction $\mathbf{m}$. The *radiative energy flux* or *energy irradiance* $\left(\text{W m}^{-2}\right)$ of radiation striking an infinitesimal area with unit normal $\mathbf{n}$ is then obtained by integration over all incident directions and the spectrum [71]:

$$\mathbf{j}_{E,\nu} = \mathbf{n} \int_0^\infty \iint_{\Omega(m)} I_\nu(\mathbf{m})\, \mathbf{m} \cdot \mathbf{n}\, d\Omega(\mathbf{m})\, d\nu \tag{7.41}$$

where Ω is the solid angle (in steradians) and ν is the frequency. Most authors employ $\mathbf{m} \cdot \mathbf{n} = \cos\theta$ in (7.41), with θ a function of $\mathbf{m}$. Here, (7.41) is integrated over a sphere $\Omega(\mathbf{m}) \in [0, 4\pi]$ to account for travelling radiation from all directions (the *net flux*); for radiation incident on a solid surface, (7.41) is integrated over a hemisphere $\Omega(\mathbf{m}) \in [0, 2\pi]$ (the *absolute flux*). For polarised radiation, the two orthogonal components must be examined separately [84]; an even more general description invokes the two-dimensional complex polarisation tensor, involving conservation of linear and angular momentum as well as energy [88]. Note that (7.41) describes a *reversible* energy flux; this only becomes irreversible in the event of changes in radiance, which necessarily require the interaction of radiation and matter [84].

Similarly, we can consider the *specific entropy intensity* or *entropy radiance* $L_\nu\left(\text{W K}^{-1}\text{m}^{-2}\text{s sr}^{-1}\right)$ of radiation. This is given by [71, 78, 79, 81–85]:

---

[6] In this respect, the half-boundary entropy production terms $\breve{\omega}_m$ are analogous to half-reaction electrode potentials.

[7] For consistency with this chapter, some notational changes are also necessary here.

$$L_\nu(\mathbf{m}) = \frac{2k\nu^2}{c_0^2}\left[\left(\frac{c_0^2 I_\nu(\mathbf{m})}{2\hbar\nu^3}+1\right)\ln\left(\frac{c_0^2 I_\nu(\mathbf{m})}{2\hbar\nu^3}+1\right) - \frac{c_0^2 I_\nu(\mathbf{m})}{2\hbar\nu^3}\ln\frac{c_0^2 I_\nu(\mathbf{m})}{2\hbar\nu^3}\right] \quad (7.42)$$

where k is Boltzmann 's constant, c_0 is the speed of light in a vacuum and $\hbar$ is Planck's constant. Equation (7.42) can be obtained from the Bose–Einstein entropy function, needed to describe electromagnetic radiation [60, 71, 78, 79, 89–91], and is a property of the radiation itself, independent of the entropy produced by its conversion to heat. A different (Fermi–Dirac) relation applies to neutrinos [92]. The *radiative entropy flux* or *entropy irradiance* $(\mathrm{W\,K^{-1}\,m^{-2}})$ is then given by:

$$\mathbf{j}_{S,\nu} = \mathbf{n}\int_0^\infty \iint_{\Omega(m)} L_\nu(\mathbf{m})\,\mathbf{m}\cdot\mathbf{n}\,d\Omega(\mathbf{m})\,d\nu \quad (7.43)$$

For unpolarised radiation emitted from a black-body of temperature T, the specific energy intensity is given by the well-known Planck equation [60, 71]:

$$I_\nu = 2B_\nu = \frac{2\hbar\nu^3}{c_0^2}\frac{1}{\exp(\hbar\nu/kT)-1} \quad (7.44)$$

whereupon (7.43) reduces to $|\mathbf{j}_{S,\nu}| = \frac{4}{3}k_{SB}T^3$, where k_{SB} is the Stefan-Boltzmann constant [81, 82].

We can now construct the local entropy production as the sum of non-radiative (material) and radiative components [81–86]:

$$\hat{\dot{\sigma}} = \hat{\dot{\sigma}}_m + \hat{\dot{\sigma}}_\nu \quad (7.45)$$

From (7.34), applicable equally to either component:

$$\hat{\dot{\sigma}}_m = \frac{\partial}{\partial t}\rho s + \nabla\cdot\mathbf{J}_{S,m} = \frac{\partial}{\partial t}\rho s + \nabla\cdot\mathbf{j}_{S,m} + \nabla\cdot(\rho s\mathbf{v}) \quad (7.46)$$

$$\hat{\dot{\sigma}}_\nu = \frac{\partial}{\partial t}\hat{S}_\nu + \nabla\cdot\mathbf{j}_{S,\nu} \quad (7.47)$$

where $\hat{S}_\nu$ is the entropy per volume due to radiation. Note that only the radiative entropy flux $\mathbf{j}_{S,\nu}$ appears in (7.47); the Clausius heating term $\mathbf{j}_{E,\nu}/T$ due to the radiative energy flux must be incorporated into the thermodynamic entropy flux in (7.46) [81, 82]. Putting these together, the total local entropy production due to material processes and radiation is:

$$\boxed{\hat{\dot{\sigma}} = \frac{\partial}{\partial t}\rho s + \frac{\partial}{\partial t}\hat{S}_\nu + \nabla\cdot(\rho s\mathbf{v}) + \nabla\cdot\mathbf{j}_{S,m} + \nabla\cdot\mathbf{j}_{S,\nu}} \quad (7.48)$$

To reduce (7.48), several approaches have been taken in the literature. Essex [81, 82] applies a volumetric form of the Gibbs equation (7.23) and total energy conservation, for flows only of heat, radiation and chemical constituents, to give:

$$\hat{\dot{\sigma}} = \left\{ \mathbf{j}_Q \cdot \nabla\left(\frac{1}{T}\right) - \sum_c \mathbf{j}_c \cdot \nabla\left(\frac{\mu_c}{T}\right) \right\} - \frac{1}{T}\frac{\partial}{\partial t}\hat{U}_\nu + \frac{\partial}{\partial t}\hat{S}_\nu - \frac{\nabla \cdot \mathbf{j}_{E,\nu}}{T} + \nabla \cdot \mathbf{j}_{S,\nu} \tag{7.49}$$

where $\hat{U}_\nu$ is the energy per volume of radiation, and the braces enclose the material component. Essex [83] extended this to fluid flows with viscous dissipation. Alternatively, Callies and Herbert [84] and Goody and Abdou [86] adopt a Gibbs-like equation for radiation:

$$d\hat{S}_\nu = \frac{1}{T_\nu} d\hat{U}_\nu \tag{7.50}$$

where T_ν is a radiative temperature, defined based on (7.44) as the temperature of matter in equilibrium with radiation of frequency ν. Non-black-body radiation can thus exhibit different radiative temperatures at different wavelengths. For heat and radiative transport only, this leads to [80, 82, 84, 86]:

$$\hat{\dot{\sigma}} = \mathbf{j}_Q \cdot \nabla\left(\frac{1}{T}\right) + \int_0^\infty \iint_{\Omega(\mathbf{m})} \left(\frac{1}{c_0}\frac{\partial I_\nu}{\partial t} + \mathbf{m} \cdot \nabla I_\nu\right) \left(\frac{1}{T_\nu(\mathbf{m})} - \frac{1}{T}\right) d\Omega(\mathbf{m})\, d\nu \tag{7.51}$$

Kröll [80] and Callies and Herbert [84] argue that the integral in (7.51) provides a bilinear formulation of the radiative entropy production, with the first term in brackets (the source function) behaving as an extensive variable. Essex [81, 82], however, disputes this view, since the bilinearity applies to each wavelength and direction. In any case, further corrections are needed in the event of scattering.

As pointed out by Essex [82], integration of the local radiative entropy production (7.48) over a control volume is not straightforward, due to the emission and absorption of radiation by non-adjacent volume elements. This creates direct, non-local connections between every element dV, creating a very different control volume to those usually examined in fluid mechanics. This gives the entropy production term:

$$\dot{\sigma}_\nu^{heat} = \frac{1}{2} \iiint_{CV} \iiint_{CV} h(\mathbf{x}_1, \mathbf{x}_2) \left(\frac{1}{T(\mathbf{x}_2)} - \frac{1}{T(\mathbf{x}_1)}\right) dV dV \tag{7.52}$$

where $h(\mathbf{x}_1, \mathbf{x}_2)$ is the net rate at which heat from position vector $\mathbf{x}_1$ is delivered to $\mathbf{x}_2$ via radiation. Allowing for the loss of energy and entropy radiation from the control volume then gives:

$$\dot{\sigma}_\nu = \frac{1}{2} \iiint_{CV} \iiint_{CV} h(\mathbf{x}_1, \mathbf{x}_2) \left(\frac{1}{T(\mathbf{x}_2)} - \frac{1}{T(\mathbf{x}_1)}\right) dV dV - \iiint_{CV} \frac{f_{CS}(\mathbf{x})}{T(\mathbf{x})} dV + \iint_{CS} \mathbf{j}_{S,\nu} \cdot \mathbf{n} dA \tag{7.53}$$

where $f_{CS}(\mathbf{x})$ is the component of radiative energy from position $\mathbf{x}$ which escapes through the control surface. For a control volume which completely encloses a planet, all terms in (7.53) and all material terms vanish except the entropy radiation, giving $\dot{\sigma} = \iint_{CS} \mathbf{j}_{S,v} \cdot \mathbf{n} dA$ [81, 84]. On these grounds, Essex [81] argues against the MaxEP hypothesis of Paltridge [10], on the grounds that the dominant, radiative entropy production term is missing.

A rather different approach for radiative transfer, involving a minimum entropy production closure of the radiative energy flux (7.41) and higher-order moments, is outlined in [93]. Further treatments of entropy production due to radiative absorption, scattering and other interactions lie beyond the scope of this chapter, and are discussed in the above-cited works.

Compartmentalisation: For many applications, it is desirable to subdivide a control volume into K contiguous compartments. From $\dot{\sigma} = \iiint_{CV} \hat{\dot{\sigma}} dV$, it might be assumed that the global entropy production is simply the sum of that in each compartment. However, this depends on the representation used. For compartments composed of Type II(b) elements, with no intensive variable discontinuities at their boundaries, this assumption is correct. If, however, the compartments are composed of Type II(a) elements, it is also necessary to account for the entropy production due to flows *between* compartments. In consequence, for purely material flows:

$$\begin{aligned} \dot{\sigma}_m &= \sum_{\alpha=1}^{K} \dot{\sigma}_m^{\alpha} + \sum_{\alpha=2}^{K} \sum_{\beta=1}^{\alpha-1} \iint_{CS_{\alpha\beta}} \breve{\dot{\sigma}}_m^{\alpha\beta} \, dA \\ &= \sum_{\alpha=1}^{K} \dot{\sigma}_m^{\alpha} + \sum_{\alpha=2}^{K} \sum_{\beta=1}^{\alpha-1} \iint_{CS_{\alpha\beta}} \left[\Delta(\rho s)^{\alpha\beta} \mathbf{v}^{\alpha\beta} + \sum_r \mathbf{j}_r^{\alpha\beta} \, \Delta\lambda_r^{\alpha\beta} \right] \cdot \mathbf{m} dA \end{aligned} \tag{7.54}$$

where $\dot{\sigma}_m^{\alpha}$ is the material entropy production in the αth compartment, while $\breve{\dot{\sigma}}_m^{\alpha\beta}$ is the material entropy production per area on the control surface $CS_{\alpha\beta}$ between the αth and βth compartments (counted only once and for $\alpha \neq \beta$). In terms of bulk flow rates:

$$\dot{\sigma}_m = \sum_{\alpha=1}^{K} \dot{\sigma}_m^{\alpha} + \sum_{\alpha=2}^{K} \sum_{\beta=1}^{\alpha-1} \left[\Delta\mathcal{F}_{S,f}^{\alpha\beta} + \Delta\mathcal{F}_{S,nf,m}^{\alpha\beta} \right] = \sum_{\alpha=1}^{K} \dot{\sigma}_m^{\alpha} + \sum_{\alpha=2}^{K} \sum_{\beta=1}^{\alpha-1} \Delta\mathcal{F}_{S,m}^{\alpha\beta} \tag{7.55}$$

where $\Delta\mathcal{F}_{S,f}^{\alpha\beta}$, $\Delta\mathcal{F}_{S,nf,m}^{\alpha\beta}$ and $\Delta\mathcal{F}_{S,m}^{\alpha\beta}$ respectively designate the bulk net fluid-borne, non-fluid (non-radiative) and total thermodynamic entropy flow rates normal to the $\alpha\beta$ control surface. If radiative transfer can also take place, (7.54)–(7.55) must be augmented by the three terms in (7.53), with attention to boundary transitions. Relations (7.54)–(7.55) do not require steady state; by definition (7.29), each measurable entropy production term is independently non-negative and therefore additive.

Steady State: Since most entropy-producing systems involve fluctuating conditions, a strict steady state (7.8) or (7.9) is not meaningful. We thus consider the mean steady state $\partial\langle S\rangle_{CV}\partial t = 0$, for which the bulk balance (7.27) gives, in general:

$$\boxed{\langle\dot{\sigma}\rangle = \sum_{\kappa\in CS}[\langle\mathscr{F}^{\kappa}_{S,f}\rangle + \langle\mathscr{F}^{\kappa}_{S,nf}\rangle] = \sum_{\kappa\in CS}\langle\mathscr{F}^{\kappa}_{S,tot}\rangle} \tag{7.56}$$

where $\langle\mathscr{F}^{\kappa}_{S,f}\rangle$, $\langle\mathscr{F}^{\kappa}_{S,nf}\rangle$ and $\langle\mathscr{F}^{\kappa}_{S,tot}\rangle$ are respectively the bulk mean fluid-borne, non-fluid and total entropy flow rates through portion κ of the control surface. Similarly, (7.32)–(7.33) and (7.34) (hence (7.48)) give, respectively:

$$\frac{\partial\langle S\rangle_{CV}}{\partial t} = 0 \Rightarrow \langle\dot{\sigma}\rangle = \oint\!\!\!\oint_{CS}\langle\mathbf{J}_S\rangle\cdot\mathbf{n}dA = \iiint_{CV}\nabla\cdot\langle\mathbf{J}_S\rangle dV \tag{7.57}$$

$$\frac{\partial}{\partial t}\langle\rho s\rangle = 0 \Rightarrow \langle\hat{\dot{\sigma}}\rangle = \nabla\cdot\langle\mathbf{J}_s\rangle = \nabla\cdot\langle\rho s\mathbf{v}\rangle + \nabla\cdot\langle\mathbf{j}_S\rangle \tag{7.58}$$

From (7.56)–(7.58), the mean steady state is quite special, since under this condition, all of the entropy production is exported from the control volume. This restricts the total mean entropy flow terms in (7.56)–(7.58) to be nonnegative. Accordingly, at mean steady state, the total mean entropy production can be calculated either by integration of the mean of (7.38) over the control volume, or more directly from the sum (7.56) or integral (7.57) of mean entropy flows through the control surface.

We therefore see that (7.56)–(7.57) express an *internal-external entropy balance*: at mean steady state, the *total* mean entropy produced within a control volume will exactly balance the *total* mean entropy flow out of its external boundaries. Often this is assumed without proof, but it requires the mean steady state, and applies only to the total quantities. In the presence of radiation, the radiative transport terms must be included within these totals.

7.3.4 Reynolds-Averaged Entropy Production and Closure Problem

We now raise an objection to one feature of previous studies of the MaxEP principle or hypothesis, as applied to planetary climate and other fluid flow systems [10, 11, 13–16]. This objection applies only to the material (non-radiative) component of time-varying, stationary flows, amenable to the Reynolds decomposition and averaging method [46, 51, 57, 58]. Although not stated explicitly, the vast majority of such studies do not actually use the mean steady-state entropy production $\langle \hat{\sigma}_m \rangle = \sum_\ell \langle \mathbf{j}_\ell \cdot \mathbf{F}_\ell \rangle$ (7.58) or its global form (7.56)–(7.57). Instead, they invoke a different quantity: the *steady-state entropy production in the mean*, $\lfloor \hat{\sigma}_m \rfloor = \sum_\ell \langle \mathbf{j}_\ell \rangle \cdot \langle \mathbf{F}_\ell \rangle$, based on products of mean fluxes or rates and their conjugate mean gradients or forces. These two quantities are not the same. By Reynolds decomposition of each independent quantity $a = \langle a \rangle + a'$, where $a'(\mathbf{x}, t)$ is the time-varying component, subject to the usual averaging rules,[8] the difference is:

$$\begin{aligned} (\!| \hat{\sigma}_m |\!) &= \langle \hat{\sigma}_m \rangle - \lfloor \hat{\sigma}_m \rfloor = \sum\nolimits_\ell \langle \mathbf{j}'_l \cdot \mathbf{F}'_l \rangle \\ &= \left\langle \mathbf{j}'_Q \cdot \nabla \left(\frac{1}{T} \right)' \right\rangle - \sum_c \left\langle \mathbf{j}'_c \cdot \nabla \left(\frac{\mu_c}{T} \right)' \right\rangle - \sum_c \left\langle \mathbf{j}'_c \cdot \left(\frac{M_c \nabla \psi_c}{T} \right)' \right\rangle \\ &\quad - \left\langle \tau' : \left(\frac{\nabla \mathbf{v}}{T} \right)' \right\rangle - \sum_d \left\langle \hat{\dot{\xi}}'_d \Delta \left(\frac{\tilde{G}_d}{T} \right)' \right\rangle \geq 0 \end{aligned} \tag{7.59}$$

Usually the flux and rate terms in (7.59) are linearised using Onsager coefficients as functions of the forces, giving a sum of quadratic fluctuation terms (see [94, 95]). Depending on its cause, the body force may be strictly steady and so disappear from (7.59). All other terms, however, consist of nonzero nonlinear products, except under strict steady-state conditions.

In dissipative systems far from equilibrium, the mean fluctuating entropy production $(\!| \hat{\sigma}_m |\!)$ (7.59) can be considerably larger—in many cases by orders of magnitude—than the entropy production in the mean $\lfloor \hat{\sigma}_m \rfloor$ [46, 51, 57, 58]. It is therefore difficult, a priori, to see why the latter should constitute the objective function for a variational principle. As shown in Sect. 7.4, however, precisely this function emerges from a judicious MaxEnt analysis of a non-equilibrium system at steady state.

We now incorporate fluctuating radiation with mean entropy production $\langle \hat{\sigma}_\nu \rangle$ and mean net entropy flux $\langle \mathbf{j}_{S,\nu} \rangle$. Writing $\lfloor \mathbf{j}_{S,m} \rfloor = \sum_\ell \langle \mathbf{j}_\ell \rangle \langle \lambda_\ell \rangle$ for the material entropy flux in the mean and $(\!| \mathbf{j}_{S,m} |\!) = \sum_\ell \langle \mathbf{j}'_l \lambda'_\ell \rangle$ for its mean fluctuation, Reynolds averaging of the local entropy balance (7.48) yields:

$$\boxed{\langle \hat{\sigma} \rangle = \lfloor \hat{\sigma}_m \rfloor + (\!| \hat{\sigma}_m |\!) + \langle \hat{\sigma}_\nu \rangle = \nabla \cdot \{ \lfloor \rho s \mathbf{v} \rfloor + (\!| \varsigma_\mathbf{v} |\!) + \lfloor \mathbf{j}_{S,m} \rfloor + (\!| \mathbf{j}_{S,m} |\!) + \langle \mathbf{j}_{S,\nu} \rangle \}} \tag{7.60}$$

with $\lfloor \rho s \mathbf{v} \rfloor = \langle \rho \rangle \langle s \rangle \langle \mathbf{v} \rangle$ and $(\!| \varsigma_\mathbf{v} |\!) = \langle \rho' s' \rangle \langle \mathbf{v} \rangle + \langle s' \mathbf{v}' \rangle \langle \rho \rangle + \langle \rho' \mathbf{v}' \rangle \langle s \rangle + \langle \rho' s' \mathbf{v}' \rangle$. On integration and application of Gauss' theorem:

---

[8] Typical Reynolds averaging rules for irreducible parameters a and b are: $\langle 1 \rangle = 1$, $\langle \langle a \rangle \rangle = \langle a \rangle$, $\langle a + b \rangle = \langle a \rangle + \langle b \rangle$, $\langle a \langle b \rangle \rangle = \langle a \rangle \langle b \rangle$, $\langle a' \rangle = 0$, $\langle \partial a / \partial x \rangle = \partial \langle a \rangle / \partial x$ and $\langle \int a dx \rangle = \int \langle a \rangle dx$ [46, 51, 57, 58].

$$\langle \dot{\sigma} \rangle = \iiint_{CV} \{ \lfloor \hat{\dot{\sigma}}_m \rfloor + (\!| \hat{\dot{\sigma}}_m |\!) + \langle \hat{\dot{\sigma}}_v \rangle \} dV$$
$$= \oiint_{CS} \{ \lfloor \rho s \mathbf{v} \rfloor + \lfloor \mathbf{j}_{S,m} \rfloor \} \cdot \mathbf{n} dA + \oiint_{CS} \{ (\!| \varsigma_v |\!) + (\!| \mathbf{j}_{S,m} |\!) \} \cdot \mathbf{n} dA + \oiint_{CS} \langle \mathbf{j}_{S,v} \rangle \cdot \mathbf{n} dA \tag{7.61}$$

or, in macroscopic terms:

$$\boxed{\langle \dot{\sigma} \rangle = \lfloor \dot{\sigma}_m \rfloor + (\!| \dot{\sigma}_m |\!) + \langle \dot{\sigma}_v \rangle = \sum_{\kappa \in CS} \left\{ \left\lfloor \mathscr{F}^{\kappa}_{S,m} \right\rfloor + \left(\!\left| \mathscr{F}^{\kappa}_{S,m} \right|\!\right) + \langle \mathscr{F}^{\kappa}_{S,v} \rangle \right\}} \tag{7.62}$$

The non-vanishing mean fluctuation terms of the material flows in (7.60)–(7.62) create many difficulties. Firstly, there is no guarantee—even at steady state—that the material entropy production in the mean $\lfloor \dot{\sigma}_m \rfloor$ will be in balance with the net outward material entropy flow in the mean $\sum_{\kappa \in CS} \left\lfloor \mathscr{F}^{\kappa}_{S,m} \right\rfloor$. In other words, it is possible that part of the mean fluctuating component of the material entropy production $(\!| \dot{\sigma}_m |\!)$ is converted into outward entropy flows in the mean $\left\lfloor \mathscr{F}^{\kappa}_{S,m} \right\rfloor$, or into the mean radiative flux $\langle \mathscr{F}^{\kappa}_{S,v} \rangle$. Alternatively, some of the material entropy production in the mean $\lfloor \dot{\sigma}_m \rfloor$ could be converted into mean fluctuating entropy flows $\left(\!\left| \mathscr{F}^{\kappa}_{S,m} \right|\!\right)$ or carried by radiation. It is therefore not possible to claim, without further proof, that the extremum calculated using one of $\lfloor \dot{\sigma}_m \rfloor$ or $\sum_{\kappa \in CS} \left\lfloor \mathscr{F}^{\kappa}_{S,m} \right\rfloor$, or one such term plus its corresponding radiative term, is equivalent to the extremum based on the other. Secondly, it is not possible, even in principle, to calculate the fluctuation terms from the mean quantities, since they contain additional unknown (and correlated) parameters, unless some other theoretical principles or constitutive relations can be invoked.

These features of fluctuating, dissipative flow systems are well-known in fluid mechanics, but are here generalised to all non-equilibrium systems with fluid and non-fluid flows. They can collectively be referred to as the *entropy production closure problem*. This problem affects the vast majority of previous studies on entropy production extremum principles, in which the distinction between in-the-mean and total mean components is not taken explicitly into account.

7.4 MaxEnt Analysis of Flow Systems

We now close our discussion of control volume analysis and entropy balance by a direct MaxEnt analysis of a flow system [39–41]. This provides a fundamental framework for the analysis of non-equilibrium systems—indeed, as fundamental as thermodynamics itself—yet underpinned by the same generic foundation provided by Jaynes' method. The analysis can be applied at any scale, integral or differential [41]; here we only examine the local scale, in the absence of radiation.

Consider an infinitesimal volume element within a control volume, as shown in Fig. 7.1b, using the Type II(b) continuum representation. Such a fluid element experiences instantaneous values of various fluxes and rates $\mathbf{j}_{\ell,\mathrm{i}} \in \{\mathbf{j}_Q, \mathbf{j}_c, \tau, \hat{\dot{\xi}}_d\}$. At the mean steady state, these are constrained by their mean values $\langle \mathbf{j}_\ell \rangle \in \{\langle \mathbf{j}_Q \rangle, \langle \mathbf{j}_c \rangle, \langle \tau \rangle, \langle \hat{\dot{\xi}}_d \rangle\}$. We therefore adopt the multivariate relative entropy $\mathfrak{H}_{st} = -\sum_{\mathrm{i}} p_{\mathrm{i}} \ln(p_{\mathrm{i}}/q_{\mathrm{i}})$—here termed the *flux entropy* [39]—as a measure of the variability or uncertainty in the allocation of fluxes and rates to possible instantaneous values. Combining the entropy and constraints gives the Lagrangian:

$$L_{st} = -\sum_{\mathrm{i}} p_{\mathrm{i}} \ln \frac{p_{\mathrm{i}}}{q_{\mathrm{i}}} - \zeta_0 \left(\sum_{\mathrm{i}} p_{\mathrm{i}} - 1 \right) - \sum_{\ell} \zeta_\ell \cdot \left(\sum_{\mathrm{i}} p_{\mathrm{i}} \mathbf{j}_{\ell,\mathrm{i}} - \langle \mathbf{j}_\ell \rangle \right) \tag{7.63}$$

where ζ_0 and ζ_ℓ are Lagrangian multipliers for normalisation and the ℓth constraint. Maximisation yields the most probable realization and maximum flux entropy:

$$p_{\mathrm{i}}^* = \frac{q_{\mathrm{i}}}{Z_{st}} \exp\left(-\sum_{\ell} \zeta_\ell \cdot \mathbf{j}_{\ell,\mathrm{i}} \right), \quad \text{with} \quad Z_{st} = \sum_{\mathrm{i}} q_{\mathrm{i}} \exp\left(-\sum_{\ell} \zeta_\ell \cdot \mathbf{j}_{\ell,\mathrm{i}} \right) \tag{7.64}$$

$$\mathfrak{H}_{st}^* = \ln Z_{st} + \sum_{\ell} \zeta_\ell \cdot \langle \mathbf{j}_\ell \rangle = -\Phi_{st} + \sum_{\ell} \zeta_\ell \cdot \langle \mathbf{j}_\ell \rangle \tag{7.65}$$

where Z_{st} is the flux partition function and $\Phi_{st} = -\zeta_0$ can be interpreted as a local *flux potential* for non-equilbrium systems, analogous to the Planck potential in equilibrium thermodynamics. Comparing (7.65) to the local material entropy production (7.38), we recognise the multipliers as proportional to the mean gradients or forces:

$$\zeta_\ell = -\frac{\langle \mathbf{F}_\ell \rangle}{\mathcal{K}} \in \frac{1}{\mathcal{K}} \left\{ -\left\langle \nabla \frac{1}{T} \right\rangle, \left\langle \nabla \frac{\mu_c}{T} + \frac{M_c \nabla \psi_c}{T} \right\rangle, \left\langle \frac{\nabla \mathbf{v}^{\mathrm{T}}}{T} \right\rangle, \left\langle \Delta \frac{\tilde{G}_d}{T} \right\rangle \right\} \tag{7.66}$$

where $\mathcal{K}$ is a positive constant $\left(\mathrm{J\,K^{-1}\,m^{-3}\,s^{-1}}\right)$. Equations (7.64)–(7.65) then give:

$$p_{\mathrm{i}}^* = \frac{q_{\mathrm{i}}}{Z_{st}} \exp \frac{\hat{\dot{\sigma}}_{m,\mathrm{i}}}{\mathcal{K}} \tag{7.67}$$

$$\mathfrak{H}_{st}^* = -\Phi_{st} - \frac{\lfloor \hat{\dot{\sigma}}_m \rfloor}{\mathcal{K}} \tag{7.68}$$

where $\hat{\dot{\sigma}}_{m,\mathrm{i}} = \sum_\ell \langle \mathbf{F}_\ell \rangle \cdot \mathbf{j}_{\ell,\mathrm{i}}$ is the local material entropy production for the **i**th category or state, based on mean gradients or forces. We therefore obtain a Gibbs-like relation (7.68) for a steady-state flow system, analogous to (7.23) for equilibrium systems, which contains the local material entropy production in the mean

$\lfloor \hat{\dot{\sigma}}_m \rfloor$. Further analyses, analogous to those in Sect.3.1, provide a set of derivative relations and Legendre duality between $\mathfrak{H}^*_{st}$ and Φ_{st} [39–41].

Just as in equilibrium thermodynamics (see Sect. 7.3.2), we can interpret the potential Φ_{st} as the state function which is *minimised* to give the most probable state of a "universe", consisting of the flow system (control volume) and its controlling environment. Rewriting (7.68) using generalised heat and work concepts [31]:

$$\boxed{d\Phi_{st} = -d\mathfrak{H}^*_{st} - \frac{d\lfloor \hat{\dot{\sigma}}_m \rfloor}{\mathcal{K}}} \tag{7.69}$$

in which each quantity Φ_{st}, $\mathfrak{H}^*_{st}$ and $\lfloor \hat{\dot{\sigma}}_m \rfloor$ is a state function, we again obtain the step change $\Delta\Phi_{st} = -\Delta\mathfrak{H}^*_{st} - \Delta\lfloor \hat{\dot{\sigma}}_m \rfloor/\mathcal{K}$, given by integration $\int_{C_{st} \in \mathcal{C}_{st}} \Delta\phi_{st}$ over some path C_{st} from a set of allowable paths $\mathcal{C}_{st}$. We again see that minimisation of Φ_{st} to give $\Delta\Phi_{st} < 0$ can occur in three ways:

1. By a coupled increase in both $\mathfrak{H}^*_{st}$ and $\lfloor \hat{\dot{\sigma}}_m \rfloor$ along C_{st}, whence $\Delta\mathfrak{H}^*_{st} > 0$ and $\Delta\lfloor \hat{\dot{\sigma}}_m \rfloor > 0$;
2. By a coupled increase in $\mathfrak{H}^*_{st}$ and decrease in $\lfloor \hat{\dot{\sigma}}_m \rfloor$ along C_{st}, such that $\Delta\mathfrak{H}^*_{st} > |\Delta\lfloor \hat{\dot{\sigma}}_m \rfloor/\mathcal{K}| > 0$; or
3. By a coupled decrease in $\mathfrak{H}^*_{st}$ and increase in $\lfloor \hat{\dot{\sigma}}_m \rfloor$ along C_{st}, such that $\Delta\lfloor \hat{\dot{\sigma}}_m \rfloor/\mathcal{K} > |\Delta\mathfrak{H}^*_{st}| > 0$.

The first and third scenarios can be interpreted as a constrained maximisation of the entropy production (MaxEP) in the mean, over the set of paths $\mathcal{C}_{st}$. In contrast, the second scenario can be viewed as a constrained minimisation of the entropy production (MinEP) in the mean, over $\mathcal{C}_{st}$. Such interpretations do not, however, represent the whole picture, since they fail to account for changes in the flux entropy $\mathfrak{H}^*_{st}$, which can also be interpreted as being maximised in scenarios 1 and 2 and minimised in scenario 3. For maximum generality, the three scenarios can be united into a *minimum flux potential* principle which controls the state of an infinitesimal flow system.

Further treatments of this analysis are available elsewhere [39–41, 96, 97]. An integral formulation can also be developed, applicable to an entire control volume at mean steady state [41]. The connection between global and local formulations—especially a formulation which includes radiation (7.48) or which takes account of the entropy production closure problem (Sect. 7.3.4)—remains unresolved and requires further research.

To summarise, the foregoing MaxEnt analysis indicates that there is no universal MaxEP or MinEP principle applicable to non-equilibrium flow systems. Instead, such "principles" emerge—in the mean—as subsidiary effects under particular conditions. This conclusion is supported by convincing experimental evidence, at least at the integral scale of analysis. This includes inversion of the Paltridge MaxEP principle for fluid flow in pipes, subject either to constraints on the flow rates or the conjugate pressure gradients [27–30]. Analogous extremum inversions are also observed or suggested by theoretical analyses of plasmas [98, 99], turbulent shear flows [100], Rayleigh-Bénard convection [101], heat or momentum transfer with advection [102] and flows around particles [103]. According to the present analysis, such phenomena will be governed by a more general principle involving minimisation of some quantity, related to the flux potential Φ_{st}; the ongoing challenge is to enlarge the underlying theoretical framework.

7.5 Conclusions

This chapter explores the foundations of the entropy and entropy production concepts, using the engineering tool of "control volume analysis" for the analysis of fluid flow systems. Firstly, the principles of control volume analysis are enunciated and applied to flows of conserved quantities (e.g. mass, momentum, energy) through a control volume, giving integral (Reynolds transport theorem) and differential forms of the conservation equations. Strict (instantaneous) and mean definitions of the steady state are provided, based on a stationary first moment or "Reynolds average". The generic entropy concept $\mathfrak{H}$—and the purpose of the maximum entropy (MaxEnt) principle—are established by combinatorial arguments (the Boltzmann principle). An entropic analysis of an equilibrium thermodynamic system is then conducted, giving the thermodynamic entropy S. Control volume analyses of a flow system then gives the "entropy production" concept for simple, integral and infinitesimal flow systems. Some technical features of such systems are then examined, including discrete and continuum representations of volume elements, the effect of radiation, and the analysis of systems subdivided into compartments. A Reynolds decomposition of the entropy production equation then reveals an "entropy production closure problem" in fluctuating dissipative systems: even at steady state, the entropy production based on mean flow rates and gradients is not necessarily in balance with the outward entropy fluxes based on mean quantities. Finally, the direct application of Jaynes' MaxEnt method yields a theoretical framework with which to predict the steady state of a flow system. This is cast in terms of a "minimum flux potential" principle, which reduces, in different circumstances, to maximum or minimum entropy production (MaxEP or MinEP) principles based on mean flows and gradients.

Further, substantial research is required on many of the formulations presented in this chapter, especially on the newly disclosed entropy production closure problem (Sect. 7.3.4) and on the MaxEnt analysis of steady-state flow systems (Sect. 7.4). Within the MaxEnt formulation, the effects of local to global scaling (see Sect. 7.2 and [39–41, 96]) and compartmentalisation (Sect. 7.3.3); of time versus ensemble averaging and associated ergodic and transient effects; of non-local interactions by electromagnetic, neutrino or other radiation Sect. 7.3.3); and of the closure problem Sect. 7.3.4), remain unresolved. It is hoped that this chapter inspires others to attain a deeper understanding and higher technical rigour in the calculation and extremisation of the entropy production in flow systems of all types.

Acknowledgments The authors acknowledge funding from the Chair of Excellence Closed-loop control of turbulent shear flows using reduced-order models (TUCOROM) of the French Agence Nationale de la Recherche (ANR), hosted by Institute PPrime, Poitiers, France; and a Rector Funded Visiting Fellowship and travel funding from UNSW. We appreciate valuable stimulating discussions at the MaxEP workshop, Canberra, Sept. 2011 and at MaxEnt 2012, Garching, Germany, July 2012, and with Filip Meysman, Bjarne Andresen and Markus Abel. The three reviewers are sincerely thanked for their comments.

References

1. Helmholtz, H., Zur Theorie der stationäiren Ströme in reibenden Flüssigkeiten, Wiss. Abh., **1**, 223–230 (1868)
2. Rayleigh, L.: On the motion of a viscous fluid. Phil. Mag. **26**, 776–786 (1913)
3. Malkus, W.V.R.: Outline of a theory of turbulent shear flow. J. Fluid Mech. **1**, 521–539 (1956)
4. Busse, F.H.: Bounds for turbulent shear flow. J. Fluid Mech. **41**(1), 219–240 (1970)
5. Kerswell, R.R.: Upper bounds on general dissipation functionals in turbulent shear flows: Revisiting the 'efficiency' functional. J. Fluid Mech. **461**, 239–275 (2002)
6. Onsager, L.: Reciprocal relations in irreversible processes I. Phys. Rev. **37**, 405–426 (1931)
7. Onsager, L.: Reciprocal relations in irreversible processes II. Phys. Rev. **38**, 2265–2279 (1931)
8. Prigogine, I.: Introduction to Thermodynamics of Irreversible Processes, 3rd edn. Interscience Publications, New York (1967)
9. Kondepudi, D., Prigogine, I.: Modern Thermodynamics: From Heat Engines to Dissipative Structures. Wiley, Chichester (1998)
10. Paltridge, G.W.: Global dynamics and climate—a system of minimum entropy exchange. Quart. J. Royal Meteorol. Soc. **101**, 475–484 (1975)
11. Paltridge, G.W.: The steady–state format of global climate. Quart. J. Royal Meteorol. Soc. **104**, 927–945 (1978)
12. Ziegler, H.: An Introduction to Thermomechanics. North–Holland Publ Co, New York (1977)
13. Ozawa, H., Ohmura, A., Lorenz, R.D., Pujol, T.: The second law of thermodynamics and the global climate system: A review of the maximum entropy production principle. Rev. Geophys. **41**(4), 1–24 (2003)
14. Kleidon, A., Lorenz, R.D. (eds.): Non-equilibrium Thermodynamics and the Production of Entropy: Life, Earth and Beyond. Springer, Heidelberg (2005)
15. Martyushev, L.M., Seleznev, V.D.: Maximum entropy production principle in physics, chemistry and biology. Phys. Rep. **426**, 1–45 (2006)
16. Bruers, S., Classification and discussion of macroscopic entropy production principles, *arXiv:cond–mat/0604482v3* (2007)
17. Bejan, A.: Entropy Generation Minimization. CRC Press, Boca Raton (1996)
18. Salamon, P., Andresen, B., Gait, P.D., Berry, R.S., The significance of Weinhold's length, J. Chem. Phys., **73**, 1001–1002 (1980), erratum 73, 5407 (1980)
19. Salamon, P., Berry, R.S.: Thermodynamic length and dissipated availability. Phys. Rev. Lett. **51**, 1127–1130 (1983)

20. Nulton, J., Salamon, P., Andresen, B., Anmin, Q.: Quasistatic processes as step equilibrations. J. Chem. Phys. **83**, 334–338 (1985)
21. Niven, R.K., Andresen, B., Jaynes' maximum entropy principle, Riemannian metrics and generalised least action bound. In: Dewar, R.L., Detering, F. (eds.) Complex Physical, Biophysical and Econophysical Systems, World Scientific Lecture Notes in Complex Systems, World Scientific, vol. 9, pp. 283–318. Hackensack, NJ, (2009)
22. Jeans, J., The Mathematical Theory of Electricity and Magnetism, 5th edn., Cambridge U.P. (1925)
23. Landauer, R., Stability and entropy production in electrical circuits. J. Stat. Phys. **13**, 1–16 (1975)
24. Jaynes, E.T.: The minimum entropy production principle. Ann. Rev. Phys. Chem. **31**, 579–601 (1980)
25. Županović, P., Juretić, D., Botrić, S.: Kirchhoff's loop law and the maximum entropy production principle. Phys. Rev. E **70**, 056108 (2004)
26. Christen, T.: Application of the maximum entropy production principle to electrical systems. J. Phys. D Appl. Phys. **39**, 4497–4503 (2006)
27. Thomas, T.Y.: Qualitative analysis of the flow of fluids in pipes. Am. J. Math. **64**(1), 754–767 (1942)
28. Paulus, D.M., Gaggioli, R.A.: Some observations of entropy extrema in fluid flow. Energy **29**, 2487–2500 (2004)
29. Martyushev, L.M.: Some interesting consequences of the maximum entropy production principle. J. Exper. Theor. Phys. **104**, 651–654 (2007)
30. Niven, R.K.: Simultaneous extrema in the entropy production for steady–state fluid flow in parallel pipes. J. Non-Equil. Thermodyn. **35**, 347–378 (2010)
31. Jaynes, E.T.: Information theory and statistical mechanics. Phys. Rev. **106**, 620–630 (1957)
32. Jaynes, E.T.: Information theory and statistical mechanics. In: Ford, K.W. (ed), Brandeis University Summer Institute, Lectures in Theoretical Physics, Statistical Physics, Benjamin–Cummings Publ. Co., vol. 3, pp. 181–218. (1963)
33. Jaynes, E.T. (Bretthorst, G.L. ed.): Probability Theory: The Logic of Science, Cambridge U.P., Cambridge (2003)
34. Tribus, M.: Information theory as the basis for thermostatics and thermodynamics. J. Appl. Mech. Trans. ASME **28**, 1–8 (1961)
35. Tribus, M.: Thermostatics and Thermodynamics. D. Van Nostrand Co. Inc., Princeton (1961)
36. Kapur, J.N., Kesevan, H.K.: Entropy Optimization Principles with Applications. Academic Press, Inc., Boston (1992)
37. Dewar, R.C.: Information theory explanation of the fluctuation theorem, maximum entropy production and self–organized criticality in non–equilibrium stationary states. J. Phys. A: Math. Gen. **36**, 631–641 (2003)
38. Dewar, R.C.: Maximum entropy production and the fluctuation theorem. J. Phys. A: Math. Gen. **38**, L371–L381 (2005)
39. Niven, R.K.: Steady state of a dissipative flow–controlled system and the maximum entropy production principle. Phys. Rev. E **80**, 021113 (2009)
40. Niven, R.K.: Minimisation of a free–energy–like potential for non–equilibrium systems at steady state. Phil. Trans. B **365**, 1323–1331 (2010)
41. Niven, R.K.: Maximum entropy analysis of steady–state flow systems (and extremum entropy production principles). In: Goyal, P., Giffins, A., Knuth, K.H., Vrscay, E. (eds.), AIP Conference Proceedings, vol. 1443, pp. 270–281. Melville, New York, (2012)
42. Prager, W.: Introduction to Mechanics of Continua. Ginn & Co., Boston (1961)
43. Aris, R.: Vectors, Tensors, and the Basic Equations of Fluid Mechanics. Prentice-Hall, Englewood Cliffs (1962)
44. Tai, C.-T.: Generalized Vector and Dyadic Analysis. IEEE, New York (1992)
45. Street, R.L., Watters, G.Z., Vennard, J.K.: Elementary Fluid Mechanics, 7th edn. John Wiley, New York (1996)

46. Spurk, J.H.: Fluid Mechanics. Springer, Berlin (1997)
47. Munson, B.R., Young, D.F., Okiishi, T.H., Huebsch, W.W.: Fundamentals of Fluid Mechanics, 6th SI ed., John Wiley, New York (2010)
48. Lanczos, C.: The Variational Principles of Mechanics, 4th edn. Dover Publications, New York (1970)
49. Harris, S.: An Introduction to the Theory of the Boltzmann Equation. Dover Publications, New York (1971)
50. Risken, H.: The Fokker-Planck Equation: Methods of Solution and Applications. Springer, Berlin (1984)
51. Durst, F.: Fluid Mechanics. Springer, Berlin (2008)
52. de Groot, S.R., Mazur, P.: Non-Equilibrium Thermodynamics. Dover Publications, New York (1984)
53. Bird, R.B., Stewart, W.E., Lightfoot, E.N.: Transport Phenomena, 2nd edn. Wiley, New York (2006)
54. Kuiken, G.D.C.: Thermodynamics of Irreversible Processes. Wiley, Chichester (1994)
55. Demirel, Y.: Nonequilibrium Thermodynamics. Elsevier, New York (2002)
56. Kreyszig, E.: Advanced Engineering Mathematics, 7th edn. Wiley, New York (1993)
57. Schlichting, H., Gersten, K.: Boundary Layer Theory, 8th edn. Springer, New York (2001)
58. White, F.M.: Viscous Fluid Flow, 3rd int. ed., McGraw-Hill, New York (2006)
59. Boltzmann, L.: Über die Beziehung zwischen dem zweiten Hauptsatze dewr mechanischen Wärmetheorie und der Wahrscheinlichkeitsrechnung, respective den Sätzen über das Wärmegleichgewicht, Wien. Ber. **76**, 373–435 (1877); English transl.: Le Roux, J. http://www.essi.fr/~leroux/ (2002)
60. Planck, M.: Über das Gesetz der Energieverteilung im Normalspektrum, Annalen der Physik **4**, 553–563 (1901)
61. Sanov, I. N.: On the probability of large deviations of random variables. Mat. Sbornik **42**, 11–44 (1957) (Russian)
62. Kullback, S., Leibler, R.A.: On information and sufficiency. Annals Math. Stat. **22**, 79–86 (1951)
63. Shannon, C.E.: A mathematical theory of communication. Bell Sys. Tech. J. **27**(379–423), 623–659 (1948)
64. Dewar, R.C., Maritan, A.: A theoretical basis for maximum entropy production. In: Dewar, R.C., Lineweaver, C., Niven, R.K., Regenauer–Lieb, K., (eds.) Beyond the Second Law: Entropy Production and Non-Equilibrium Systems, Springer, Heidelberg (2013)
65. Gibbs, J.W.: Elementary Principles of Statistical Mechanics. Dover Publications, New York (1902)
66. Schrödinger, E.: Statistical Thermodynamics. Cambridge U.P., Cambridge (1952)
67. Hill, T.L.: Statistical Mechanics. McGraw-Hill, New York (1956)
68. Atkins, P.W.: Physical Chemistry, 2nd edn. Oxford University Press, Oxford (1982)
69. Callen, H.B.: Thermodynamics and an Introduction to Thermostatistics, 2nd edn. John Wiley, New York (1985)
70. Niven, R.K.: Maximum–entropy weighting of multiple Earth climate models. Clim. Dyn. **39**(3), 755–765 (2012)
71. Planck, M.: The Theory of Heat Radiation, 2nd edn. Engl. transl., Dover Publications, New York (1914)
72. Planck, M.: Treatise on Thermodynamics, 3rd edn. Engl. transl., Dover Publications, New York (1945)
73. Guggenheim, E.A.: Thermodynamics: An Advanced Treatment for Chemists and Physicists. North–Holland Publ. Co, Amsterdam (1967)
74. Gibbs, J.W.: On the equilibrium of heterogeneous substances, Trans. Connecticut Acad. **3**, 108–248 (1875–1876); **3**, 343–524 (1877–1878)
75. Moran, M.J., Shapiro, H.N.: Fundamentals of Engineering Thermodynamics, 5th edn. John Wiley, New York (2006)

76. Jaumann, G:. Geschlossenes System physikalischer und chemischer Differentialgesetze, Sitzungsberichte Akademie der Wisenschaften Wien, Mathematisch–Naturwissenschaftliche Klasse, Abt. 2a, **120**, 385–530 (1911)
77. Kjelstrup, S., Bedeaux, D., Johannessen, E., Gross, J.: Non-Equilibrium Thermodynamics for Engineers. World Scientific, New Jersey (2010)
78. Rosen, P.: Entropy of radiation. Phys. Rev. **96**(3), 555 (1954)
79. Ore, A.: Entropy of radiation. Phys. Rev. **98**(4), 887–888 (1955)
80. Kröll, W.: Properties of the entropy production due to radiative transfer. J. Quant. Spectrosc. Radiat. Transfer **7**(5), 715–723 (1967)
81. Essex, C.: Radiation and the irreversible thermodynamics of climate. J. Atm. Sci. **41**(12), 1985–1991 (1984)
82. Essex, C.: Radiation and the violation of bilinearity in the thermodynamics of irreversible processes. Planet. Space Sci. **32**(8), 1035–1043 (1984)
83. Essex, C.: Global thermodynamics, the Clausius inequality and entropy radiation. Geophys. Astrophys. Fluid Dyn. **38**, 1–13 (1987)
84. Callies, U., Herbert, F.: Radiative processes and non–equilibrium thermodynamics. J. Appl. Math. Phys. (ZAMP) **39**, 242–266 (1988)
85. Pelkowski, J.: Towards an accurate estimate of the entropy production due to radiative processes: Results with a gray atmosphere model. Meteorol. Atmos. Phys. **53**, 1–17 (1994)
86. Goody, R.M., Abdou, W.: Reversible and irreversible sources of radiation entropy. Quart. J. Royal Meteorol. Soc. **122**, 483–494 (1996)
87. Goody, R.M., Yung, Y.L.: Atmospheric Radiation: Theoretical Basis, 2nd edn. Oxford University Press, New York (1989)
88. Landau, L.D., Lifshitz, E.M.: A Shorter Course of Theoretical Physics, vol. 1, Mechanics and Electrodynamics, Permagon Press, Oxford (1972)
89. Niven, R.K.: Exact Maxwell-Boltzmann, Bose-Einstein and Fermi-Dirac statistics. Phys. Let. A **342**(4), 286–293 (2005)
90. Niven, R.K.: Cost of s–fold decisions in exact Maxwell-Boltzmann. Bose-Einstein and Fermi-Dirac statistics, Physica A **365**(1), 142–149 (2006)
91. Niven, R.K.: Combinatorial entropies and statistics. Eur. Phys. J. B **70**, 49–63 (2009)
92. Essex, C., Kennedy, D.C.: Minimum entropy production of neutrino radiation in the steady state. J. Stat. Phys. **94**(1/2), 253–267 (1999)
93. Christen, T., Kassubek, F.: Entropy production–based closure of the moment equations for radiative transfer. In: Dewar, R.C., Lineweaver, C., Niven, R.K., Regenauer–Lieb, K., (eds.) Beyond the Second Law: Entropy Production and Non-Equilibrium Systems, Springer, Heidelberg (2013)
94. Kock, F., Herwig, H.: Local entropy production in turbulent shear flows: A high–Reynolds number model with wall functions. Int. J. Heat Mass Transf. **47**, 2205–2215 (2004)
95. Naterer, G.F., Camberos, J.A.: Entropy–Based Design and Analysis of Fluids Engineering Systems. CRC Press, Boca Raton (2008)
96. Noack, B.R., Niven, R.K.: Maximum–entropy closure for a Galerkin system of incompressible shear flow. J. Fluid Mech. **700**, 187–213 (2012)
97. Noack, B.R., Niven, R.K.: A hierarchy of maximum entropy closures for Galerkin systems of incompressible. Comput. Math. Appl. **65**, 1558–1574 (2013)
98. Kawazura, Y., Yoshida, Z.: Entropy production rate in a flux–driven self–organizing system. Phys. Rev. E **82**, 066403 (2010)
99. Kawazura, Y., Yoshida, Z.: Comparison of entropy production rates in two different types of self–organized flows: Bénard convection and zonal flow. Phys. Plasmas **19**, 012305 (2012)
100. Ozawa, H., Shikokawa, S., Sakuma, H.: Thermodynamics of fluid turbulence: A unified approach to the maximum transport properties. Phys. Rev. E **64**, 026303 (2001)
101. Weaver, I., Dyke, J.G., Oliver, K.: Can the principle of maximum entropy production be used to predict the steady states of a Rayleigh–Bénard convective system? In: Dewar, R.C., Lineweaver, C., Niven, R.K., Regenauer–Lieb, K. (eds.) Beyond the Second Law: Entropy Production and Non-Equilibrium Systems, Springer, Heidelberg (2013)

102. Ozawa, H., Shimokawa, S.: The time evolution of entropy production in nonlinear dynamic systems. In: Dewar, R.C., Lineweaver, C., Niven, R.K., Regenauer–Lieb, K. (eds.) Beyond the Second Law: Entropy Production and Non-Equilibrium Systems, Springer, Heidelberg (2013)
103. Vaidya, A.: Maximum entropy production and stable configurations in fluid–solid interactions. In: Dewar, R.C., Lineweaver, C., Niven, R.K., Regenauer–Lieb, K. (eds.) Beyond the Second Law: Entropy Production and Non-Equilibrium Systems, Springer, Heidelberg (2013)

Chapter 8
Earth System Dynamics Beyond the Second Law: Maximum Power Limits, Dissipative Structures, and Planetary Interactions

Axel Kleidon, Erwin Zehe, Uwe Ehret and Ulrike Scherer

Abstract Planet Earth is a thermodynamic system far from equilibrium and its functioning—obviously—obeys the second law of thermodynamics, at the detailed level of processes, but also at the planetary scale of the whole system. Here, we describe the dynamics of the Earth system as the consequence of sequences of energy conversions that are constrained by thermodynamics. We first describe the well-established Carnot limit and show how it results in a maximum power limit when interactions with the boundary conditions are being allowed for. To understand how the dynamics within a system can achieve this limit, we then explore with a simple model how different configurations of flow structures are associated with different intensities of dissipation. When the generation of power and these different configuration of flow structures are combined, one can associate the dynamics towards the maximum power limit with a fast, positive and a slow, negative feedback that compensate each other at the maximum power state. We close with a discussion of the importance of a planetary, thermodynamic view of the whole Earth system, in which thermodynamics limits the intensity of the dynamics, interactions strongly shape these limits, and the spatial organization of flow represents the means to reach these limits.

A. Kleidon (✉)
Max-Planck-Institute for Biogeochemistry, Hans-Knöll-Strasse 10, 07745 Jena, Germany
e-mail: akleidon@bgc-jena.mpg.de

E. Zehe · U. Ehret · U. Scherer
Institute of Water Resources and River Basin Management Karlsruhe Institute of Technology KIT, Kaiserstrasse 12, 76129 Karlsruhe, Germany
e-mail: erwin.zehe@kit.edu

U. Ehret
e-mail: uwe.ehret@kit.edu

U. Scherer
e-mail: ulrike.scherer@kit.edu

R. C. Dewar et al. (eds.), *Beyond the Second Law*,
Understanding Complex Systems, DOI: 10.1007/978-3-642-40154-1_8,

8.1 Introduction

Heat flows from warm to cold, water flows from the mountain top to the valley floor and wood burns into ashes. The reverse direction for these processes does not quite make sense. Heat does not spontaneously flow from cold to warm, water does not flow uphill, and wood does not emerge from the ashes. These directions reflect the fundamental direction imposed by the second law of thermodynamics. Yet, at the same time, this implication is almost trivial, in the sense that nobody would seriously question these directions as this is what we observe in nature over and over again. In this respect, the second law does not seem to contain much value to learn more about these processes because this knowledge is already established in the many mathematical descriptions we use to describe Earth system processes. So the basic question is whether there is more to learn from the second law of thermodynamics beyond these general trivialities?

The proposed principle of Maximum Entropy Production (MaxEP) seems to suggest that there is more to learn. The MaxEP principle states that processes do not merely follow the second law, but proceed at a maximum rate at which the rate of entropy production is maximized. This would essentially mean that the dynamics of isolated systems do not merely evolve towards a state of thermodynamic equilibrium, but that they would do so at the fastest possible rate. At the core of the thermodynamic interpretation of MaxEP is a trade-off, by which a greater flux is associated with a more depleted gradient. Since entropy production is expressed by the product of flux and gradient, the trade-off between flux and gradient results in a state of maximum entropy production at intermediate values for the flux and gradient. Over the last 15 years, renewed attention has been given to this principle, in terms of its theoretical basis [1–5] as well as its application to Earth and environmental processes [6–8]. While there are some indications for support, e.g. regarding heat transport by planetary atmospheres in simple climate models [9, 10] as well as general circulation models [11, 12], there are also quite a number of issues that still need to be resolved [13–15]. For instance, why should environmental systems "care" about entropy production, rather than more traditional quantities such as mechanical forces or mass fluxes? How would systems know that they "need" to maximize entropy production? With the breadth of competing processes shaping the Earth system, how do we know which entropy production should be maximized? And what new insights can be provided by MaxEP or similar maximization principles that we cannot get without these principles?

In this chapter we provide a brief overview of how these shortcomings of the proposed MaxEP principle can be overcome by shifting the focus onto the maximization of power, i.e. work through time, within an Earth system context. This maximization of power yields states that are nearly indistinguishable from equivalent MaxEP states, but it provides a clearer basis to understand which aspect is maximized within a system and to understand how this maximization is achieved. To do so, we first derive the Carnot limit from the laws of thermodynamics and then relate it to the maximum power limit in the next section. Then we illustrate how maximization can be achieved by the organization of flow and how it relates to basic feedbacks that shape the evolutionary dynamics. This is then applied to the Earth system at large to explain how the dynamics and couplings of the planet essentially reflect the acceleration of the second law at the planetary scale. We close with a brief summary and conclusions.

8.2 Maximum Power Limits

Thermodynamics informs us about the limits of how much work can be derived from a heating gradient. The best known limit is the Carnot limit, which represents the best case for extracting work from a heating gradient that satisfies the first and second law of thermodynamics. For its derivation, we consider a system shown in Fig. 8.1a as a dashed box labeled "heat engine" that is situated between a hot reservoir with temperature T_h and a cold reservoir with temperature T_c. Applied to this setting, the first law in a steady state in which the internal energy does not change in time is represented by the balance of the heating by the heat flux J_{in}, the cooling by J_{out}, and the mechanical work done through time (or power), P_{ex}:

$$0 = J_{in} - J_{out} - P_{ex} \tag{8.1}$$

To identify the constraints imposed by the second law, we need to consider the entropy balance of the system . When we consider this balance in a steady state in which the entropy of the system does not change in time, this balance is represented by the entropy production due to irreversible processes within the system, σ, the entropy import by heating, J_{in}/T_h, and the entropy export by cooling, J_{out}/T_c:

$$0 = \sigma + \frac{J_{in}}{T_h} - \frac{J_{out}}{T_c} \tag{8.2}$$

The second law requires that $\sigma \geq 0$. With this requirement, we can combine Eqs. (8.1) and (8.2) and solve for P_{ex}:

$$P_{ex} \leq J_{in} \frac{T_h - T_c}{T_h} \tag{8.3}$$

The best case is given when the power equals the right hand side of this equation and this is known as the Carnot limit. It expresses the maximum rate by which heat can be converted into mechanical work that is permitted by the first and second law. Greater values of $P_{ex} = J_{in} - J_{out}$ would require $\sigma < 0$ in Eq. (8.2), which would violate the second law. Such conditions are shown in the upper right in Fig. 8.1a by the shaded area.

In this derivation of P_{ex}, it is assumed that no entropy is associated with P_{ex}, so that all of P_{ex} is associated with performing work. In the Earth system, such work is needed, for instance, to generate kinetic energy associated with motion, or to lift

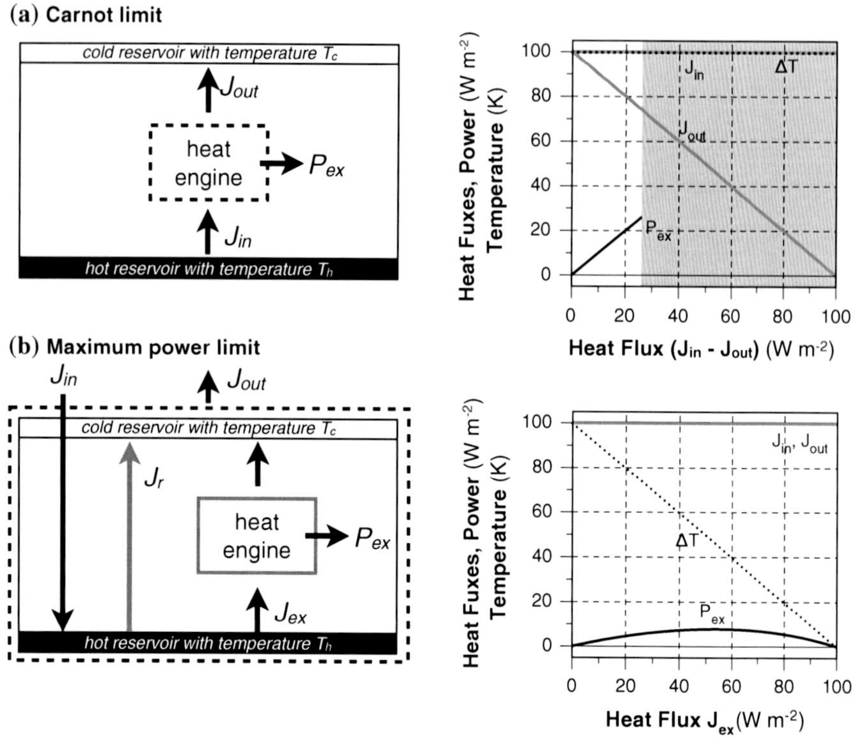

Fig. 8.1 Systems used in the text to describe **a** the Carnot limit and **b** the maximum power limit. The *dashed line* in the figures on the *left* show the delineation of the system boundary. The *right panels* show the sensitivity of the heat fluxes J_{in} and J_{out}, the temperature gradient $\Delta T = T_h - T_c$ and the extracted power P_{ex} to the heat flux utilized by the engine. The area *shaded grey* in the *upper right* plot shows conditions that is not permitted by the second law as it would require negative entropy production within the system, so that the Carnot limit of maximum power in the upper system is located at the edge of the *shaded* area. In the system shown in **b**, the maximum power limit results from the trade-off between a greater heat flux J_{ex} and the reduced temperature difference ΔT. After [17]

material against gravity and generate potential energy. We will refer to the generated form of energy as "free energy" here in a general sense because it is equivalent to the capacity of a system to perform work. The free energy is associated with a gradient of a different variable. For instance, when work is performed to generate motion, the form of free energy is kinetic energy, and the associated gradient is in the associated momentum. Hence, the Carnot limit can be seen as the maximum rate by which a heating gradient can be converted into a gradient of another variable.

The derivation of the Carnot limit makes two, important assumptions: (1) the two heat reservoirs that drive the heat engine remain at fixed temperatures and are not affected by the generation of work within the system; and (2) no irreversible process takes place within the system. These assumptions cannot be made for many Earth system processes. Atmospheric convection, for instance, is driven by the heating associated with the absorption of solar radiation at the surface and the cooling aloft through the emission of terrestrial radiation. The convective motion transports about 99 W m^{-2} from the surface to the atmosphere in the global mean [16], which is more than half of the surface solar radiative heating of 160 W m^{-2}. Hence, the surface temperature is not a fixed boundary condition that drives the convective heat engine, but it is strongly affected by the intensity of convective cooling. In addition to this convective cooling, the surface is cooled by the net emission of terrestrial radiation of about 61 W m^{-2}, which is associated with irreversible radiative transfer that produces considerable entropy. Hence, the assumption that no entropy is produced is not fulfilled either. Similar arguments can be made for the large-scale poleward transport of heat by the climate system as well as for other processes, e.g. mantle convection in the Earth's interior. It would thus seem that the assumptions being made to derive the Carnot limit would not apply to quite a range of Earth system processes while the laws of thermodynamics naturally apply and limit the rate at which these processes can perform work.

We can nevertheless derive a maximum power limit for a slightly altered setup as shown in Fig. 8.1b that is more representative of Earth systems [17]. The two differences to the typical Carnot limit are that (1) the heat balances for T_h and T_c are part of the system and can therefore react to the rate at which work is performed within the system, and that (2) there is an additional process (radiative transfer, J_r, in Fig. 8.1b) that depletes the temperature gradient and produces entropy within the system. In this setup, we can use the steady-state surface energy balance (i.e. $dT_h/dt = 0$) as a constraint to express the temperature gradient $T_h - T_c$ as a function of the surface solar radiative heating, J_{in}, and the convective heat flux J_{ex} utilized by the convective heat engine:

$$0 = J_{in} - k_r(T_h - T_c) - J_{ex} \tag{8.4}$$

For simplicity, the net radiative exchange between the reservoirs is represented in a linearized way by $J_r = k_r(T_h - T_c)$, which is derived from the linearization of the Stefan-Boltzmann law.

To derive the maximum in power that can be derived from the heating difference $T_h - T_c$, we apply the Carnot limit to J_{ex}, use Eq. (8.4) to express $T_h - T_c$ in terms of J_{in} and J_{ex}, and get an expression of power P_{ex} that depends quadratically on J_{ex}:

$$P_{ex} = J_{ex}\frac{T_h - T_c}{T_h} = J_{ex}\frac{(J_{in} - J_{ex})}{k_r T_h} \tag{8.5}$$

When we neglect the dependence of T_h on J_{ex} in the denominator, this expression achieves a maximum value $P_{\max}$ for a convective heat flux $J_{ex} = J_{in}/2$ of

$$P_{\max} = \frac{J_{in}^2}{4k_r T_h} = \frac{1}{4} J_{in} \frac{T_{h,0} - T_{c,0}}{T_h} \tag{8.6}$$

where $T_{h,0}$ and $T_{c,0}$ are the temperatures for $J_{ex} = 0$.

The maximum power limit expressed in Eq. (8.6) looks like a Carnot limit [cf. Eq. (8.3), in particular when approximating $T_h \approx T_{h,0}$] with fixed, radiative equilibrium temperatures, except that it is reduced by a factor of 4. This reduction can directly be seen when comparing the maximum values of P_{ex} in Fig. 8.1a and b. The lower limit results from the reduction of the temperature gradient to half its maximum value and from the "competing" process of radiative transfer that consumes some of the heat flux J_{in}. This expression is essentially identical to the maximum power limit that is well known in electrical engineering and, when applied to typical atmospheric conditions, yields maximum estimates of power involved in atmospheric motion that are consistent with observations [17].

The maximum power limit is nearly identical to the Maximum Entropy Production (MaxEP) state reported earlier in atmospheric applications [9, 18]. This can be seen when considering the entropy budget of the system. In steady state, this budget is given by the import of entropy associated with the absorption of J_{in} at a temperature T_h, and the export of entropy associated with the emission of $J_{out} = J_{in}$ at a temperature T_c. The entropy production within the system due to radiative exchange, σ_r, and due to the convective heat flux, σ_{ex}, is balanced by the net entropy export by the system, so that

$$0 = \sigma_r + \sigma_{ex} + \frac{J_{in}}{T_h} - \frac{J_{in}}{T_c} \tag{8.7}$$

Noting that $J_{in} = J_r + J_{ex}$ and $\sigma_r = J_r(1/T_c - 1/T_h)$, we obtain

$$\sigma_{ex} = J_{ex}\left(\frac{1}{T_c} - \frac{1}{T_h}\right) = J_{ex}\frac{T_h - T_c}{T_h T_c} = \frac{P_{ex}}{T_c} \tag{8.8}$$

where we used the expression of P_{ex} from Eq. (8.5). In steady state, power equals dissipation, $P_{ex} = D$, in this closed system, so that the maximum power limit is equivalent to the maximization of σ_{ex} if all of the mechanical energy generated by P_{ex} is dissipated at the temperature T_c of the cold reservoir. This is typically not the case for the Earth's atmosphere. About half of the generated kinetic energy is either dissipated near the heated surface or is transferred into the ocean. Hence, the maximum power limit should be slightly below the MaxEP state. Nevertheless, the difference is hardly distinguishable using realistic numbers, so that the examples that provide support for the MaxEP principle concerning atmospheric heat transport [9–12], can equally be interpreted as an indication that the atmospheric circulation operates very close to the maximum power limit.

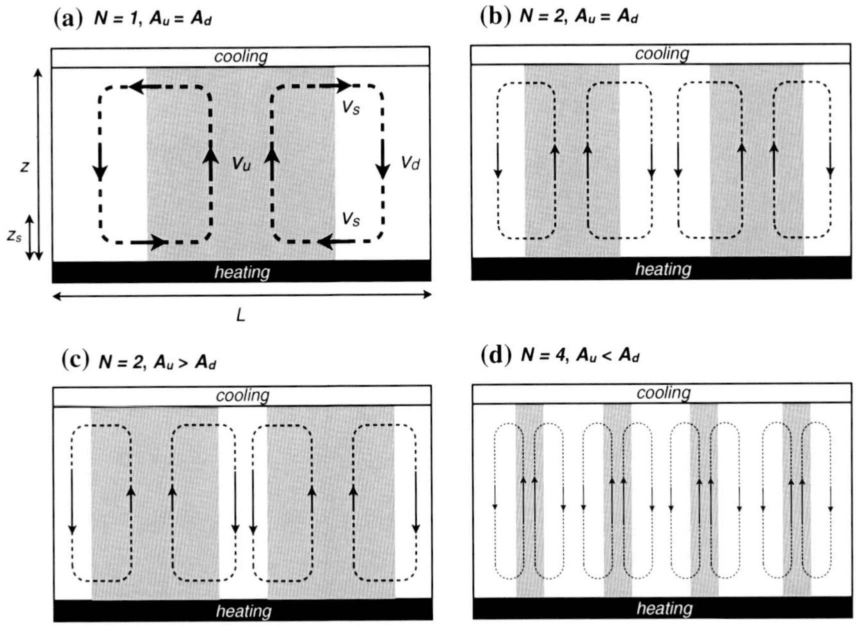

Fig. 8.2 Four examples of different flow structures that differ in the number of convection cells N and the areas of updraft A_u (*shaded grey*) in relation to the area of downdraft A_d of each cell. These different arrangements result in different intensities of frictional dissipation

8.3 Maximization Through Structure

The maximum power limit given by Eq. (8.6) establishes the upper limit to how much dynamics can be generated within a system, but it does not tell us how this maximization would be achieved, neither in terms of the evolutionary dynamics nor which aspect of motion would allow for the needed flexibility to achieve maximization. To explore the latter aspect (the former is dealt with in the next section), we need to look at how the flow is organized in space and time and how this organization affects the ability of the flow to generate and dissipate kinetic energy. This is done in the following using simple, conceptual considerations with some quantitative illustrations.

When motion is generated at a certain rate, the resulting flow can take various forms, as illustrated in Fig. 8.2. For instance, the flow can be accomplished by few, or many, convection cells N , and it can be associated with different areas over which updrafts take place. The physical balances that constrain these flow structures are the conservation of energy, mass and momentum, which apply to the local scale, but also at the global, system-level scale. At the local scale, momentum conservation leads to the well-known Navier–Stokes equation of fluid dynamics. What we aim for here is a system-level description of the dynamics that does not require the information from the local scale, but only considers the balances at the global, aggregated level. The maximum power limit applied to the system level would then act as a constraint to the overall dynamics within the system. To understand how different flow configurations affect the system-level properties, particularly dissipation, we consider a simple, aggregated description of the system shown in Fig. 8.2 in the following.

The system depicted in Fig. 8.2 is a closed system that only exchanges heat with its environment. The dynamics of the total kinetic energy, A_{ke}, within the system then include only the generation and dissipation of kinetic energy, but does not include its exchange associated with mass exchange across the system boundary:

$$\frac{dA_{ke}}{dt} = P_{ex} - D \tag{8.9}$$

where P_{ex} is the generation rate of kinetic energy (being equivalent to the power of the heat engine from the previous section), and D is the rate of frictional dissipation. The main point about structure and maximization made here is that the many different ways in which the flow takes place (as, e.g., those shown in Fig. 8.2) are associated with different intensities of D, so that for the same P_{ex}, different values of A_{ke} can be achieved. Since a higher value of A_{ke} transports more heat, i.e. results in a greater value of J_{ex}, rearrangements in the flow can then form the basis for a positive feedback by which a higher value of A_{ke} results in a greater value of P_{ex} (which is explored further in the next section).

To demonstrate the different intensities of D associated with different flow patterns, we consider an area of size $A_{tot} = L^2$, where L is the horizontal dimension, a height of convection z and a boundary layer height of z_s. The updraft in each cell is assumed to take place with a uniform updraft velocity v_u through a horizontal, circular cross-sectional area of one updraft cell, $A_u = \pi r_u^2$, the downdraft takes place with a uniform downdraft velocity v_d and a cross-sectional area of $A_d = A_{tot}/N - A_u$, and a velocity v_s near the surface through a vertical, cylindrical cross-section at the bottom of the updraft cells of $A_s = 2\pi r_u z_s$.

Continuity requires that the mass fluxes J_m within a convection cell balance, i.e. that the mass lifted in the updraft is balanced by the mass transported by the downdraft and along the surface:

$$J_m = \rho A_u v_u = \rho A_d v_d = \rho A_s v_s \tag{8.10}$$

where we assume the same air density ρ for simplicity. This requirement yields the following expressions for the three velocities:

$$v_u = \frac{J_m}{\rho A_u} \qquad v_d = \frac{J_m}{\rho (A_{tot}/N - A_u)} \qquad v_s = \frac{J_m}{2\sqrt{\pi}\rho\sqrt{A_u} z_s} \tag{8.11}$$

noting that $r_u = \sqrt{A_u/\pi}$.

The total frictional dissipation D results from the friction within the fluid between the updrafts and downdrafts, D_a, from the contact with the surface, D_s, and at the upper boundary at the top of the convection cell D_u, which we will assume to be equal to D_s for simplicity in the following (although friction at a solid surface may not be equal to internal friction within air):

$$D = D_a + 2D_s \tag{8.12}$$

Frictional dissipation within the fluid, D_a, is given by the viscosity of the fluid, μ, the velocity gradient between the cells, which is approximated by the difference in updraft- and downdraft velocities $(v_u - v_d)$ divided by the mean distance between the up- and downdraft within a cell, $L/2N$, through the vertical, cylindrical surface area of the updraft cell, $2\pi r_u(z - 2z_s)$:

$$\begin{aligned} D_a &= \mu \left(\frac{\partial v}{\partial x}\right)^2 A = \mu \left(\frac{v_u - v_d}{L/2N}\right)^2 (2\pi r_u)\,(z - 2z_s)N \\ &= \frac{\mu}{\rho^2}\gamma_a J_{m,\,tot}^2 \end{aligned} \tag{8.13}$$

where the geometric factor γ_a is given by

$$\gamma_a = 8\sqrt{\pi}\frac{A_{tot}}{A_u^{3/2}(A_{tot} - NA_u)^2}(z - 2z_s)N \tag{8.14}$$

In other words, the overall frictional dissipation within the fluid depends on material properties (viscosity μ and density ρ), the total mass flux $J_{m,\,tot} = NJ_m$ that is associated with the kinetic energy of convective motion, but also to some extent on a purely geometric factor, γ_a, that is associated with the organization of the mass flux in terms of the number of convection cells N as well as the cross section of the updraft A_u.

Frictional dissipation at the surface, D_s, is expressed similarly in terms of a velocity gradient, v_s/z_s, and the surface area, A_{tot}:

$$\begin{aligned} D_s &= \mu \left(\frac{v_s}{z_s}\right)^2 A_{tot} \\ &= \frac{\mu}{\rho^2}\gamma_s J_{m,\,tot}^2 \end{aligned} \tag{8.15}$$

where the geometric factor γ_s is given by

$$\gamma_s = \frac{1}{4\pi}\frac{A_{tot}}{A_u}\frac{1}{z_s^2}\frac{1}{N^2} \tag{8.16}$$

This expression is similar to Eq. (8.13) above in that it also depends on purely material properties, the total mass flux, as well as a geometric factor γ_s.

The total frictional dissipation D can then be expressed as:

$$D = \frac{\mu}{\rho^2}(\gamma_a + 2\gamma_s)J_{m,\,tot}^2 \propto \gamma A_{ke} \tag{8.17}$$

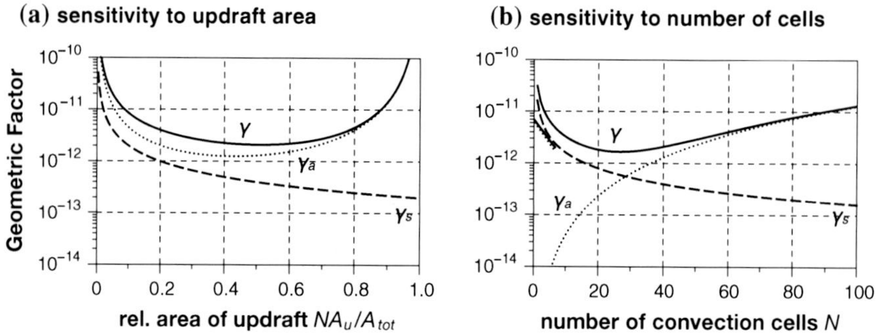

Fig. 8.3 Sensitivity of the geometric factors of flow organization related to interior friction within the fluid, γ_a, friction with the surface, γ_s, and total, $\gamma = \gamma_a + 2\gamma_s$, **a** to the total updraft area, NA_u/A_{tot} and **b** the number of convection cells, N. For both plots, the values $A_{tot} = 10^6$ m^2, $z = 1{,}000$ m, and $z_s = 100$ m are used. For the *left* plot a value of $N = 40$ was used, for the *right* plot a value of $NA_u/A_{tot} = 0.5$

which, by using the continuity requirement (Eq. 8.10), could be formulated in terms of the square of a velocity (e.g. v_u) to yield a typical parameterization for frictional dissipation, or in terms of the kinetic energy A_{ke} within the system.

Figure 8.3 shows the extent to which the total dissipation, D, depends on the flow configuration, as characterized by the geometric factors. As can be seen by the sensitivities, the geometric factors vary by an order of magnitude or more by the variation of A_u and N. Both sensitivities of the total geometric factor, $\gamma = \gamma_a + 2\gamma_s$, exhibit a characteristic minimum at which frictional dissipation is reduced merely by rearrangement of the flow.

As a consequence of this sensitivity of D to the arrangement of the flow, different values of the kinetic energy A_{ke} of the system can be achieved for the same generation rate P_{ex}. The critical link between this flexibility in A_{ke} and maximum power P_{ex} is that the amount of kinetic energy A_{ke} reflects the speed of motion within the convection cell, which in turn is related to the convective heat flux J_{ex} by

$$J_{ex} = c_p(T_h - T_c)J_{m,tot} \propto \Delta T \sqrt{A_{ke}} \tag{8.18}$$

In other words, a rearrangement in the flow can lower its frictional dissipation D (through the effect on γ in Eq. 8.17), enhance the flow velocity, transport more mass and heat (cf. Eq. 8.18), and thereby generate more power P_{ex} to drive the flow (cf. Eq. 8.5). This latter enhancement of P_{ex} through structured flow results from the concentration of the driving gradient at the boundary of the system (see also [19]). This effect can be seen by reformulating the expression of maximum power P_{ex} (Eq. 8.6) in terms of the temperature gradient by using the energy balance (Eq. 8.4):

$$P_{max} = \frac{k_r}{T_h}(T_h - T_c)^2 \tag{8.19}$$

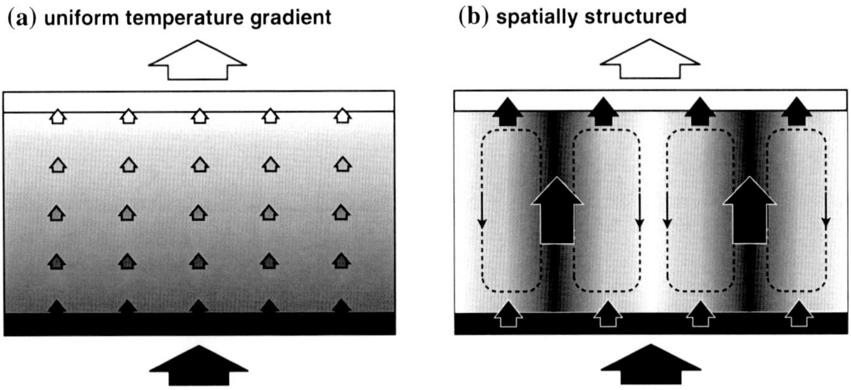

Fig. 8.4 Schematic illustration of the effect of structured flow on the distribution of temperature gradients. **a** In the absence of motion, the temperature gradient is uniformly distributed across the system between the heating source below, and the cooling source aloft. **b** When structured flow takes place, temperature gradients are confined to small regions at the interface to the heating and cooling source and are able to enhance the generation rate of motion

In other words, P_{max} depends quadratically on the temperature gradient. The implication of this is that for the same rate of heating, J_{in}, a uniform distribution of the temperature gradient yields less power than a non-uniform distribution of the gradient at the system boundary. This effect is qualitatively illustrated in Fig. 8.4. With stronger motion, more cooled air is advected by the convection cell to the heated surface, thereby concentrating the temperature gradient to the area near the surface. The key insight here is that a non-uniform distribution of the driving temperature gradient is intimately linked with the development of structured flow and that this affects the ability of the system to derive power from the temperature gradient.

Of course, this simple example makes several assumptions, such as constant density and a simple geometry, and treats convection in a highly simplistic way. It nevertheless substantiates the point that the formation of specific flow structures such as convection cells affect the *intensity* by which kinetic energy is dissipated and thereby constitute "degrees of freedom" that allow the fluid to adjust to a state of maximum power.

8.4 Dynamics and Feedbacks Associated with Maximization Through Structure

We now ask why the evolution and the dynamics of a system would inevitably evolve to a maximum power state. The following discussion on feedbacks show rather general mechanisms that, in principle, should be transferrable to very different structures as well (for instance water flow in river basin networks, [20]).

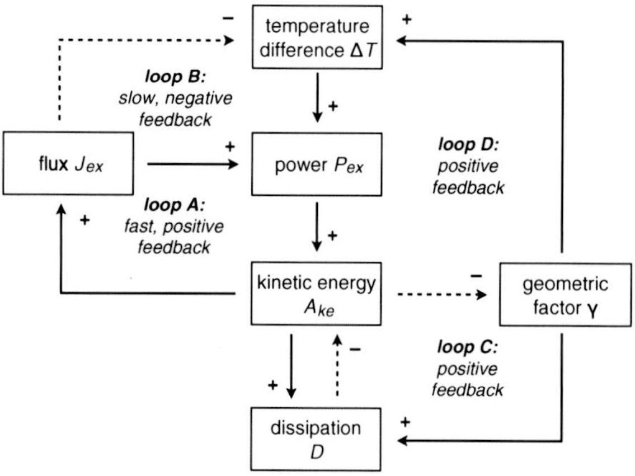

Fig. 8.5 A feedback diagram to illustrate how the dynamics of kinetic energy generation and dissipation relate to the maximization of power and structure formation. *Solid lines* with "+" indicate positive influences (e.g., a larger temperature difference results in a greater power, i.e. the derivative $\partial P_{ex}/\partial \Delta T > 0$). *Dashed lines* with "−" show negative influences (e.g., an enhanced heat flux reduces the driving gradient, i.e. $\partial \Delta T/\partial J_{ex} < 0$). Four feedback loops (A, B, C, D) are shown: Feedbacks A and B on the *left* relate to the maximum power limit, and the feedbacks C and D on the *right* relate to how structured flow can achieve this limit. After [20]

We illustrate these feedbacks in the following specifically with the example of convective motion given above, bringing together and summarizing the previous sections. We describe these feedbacks similar to the feedback analysis that is common in climatology [21].

Imagine if the system shown in Fig. 8.1b is initially at rest, i.e. in a state of no convective heat flux ($J_{ex} = 0$) and no kinetic energy ($A_{ke} = 0$). This state satisfies the energy-, mass-, and momentum balances of the system. We now need to understand why such a state, when perturbed, would evolve towards a dynamic state with $A_{ke} > 0$, and why this evolution would "stop" at a maximum power state. In the context of feedbacks, we need to identify a positive feedback that amplifies the growth of the initial perturbation, and a negative feedback that stops the growth at the maximum power state. Before we identify these feedbacks, let us first go through the relationships between the different variables of the system as shown in Fig. 8.5 and relate them to the above equations.

The source for the dynamics is the temperature difference ($\Delta T = T_h - T_c$) in the system that develops due to the uneven heating and cooling of the system due to J_{in} and J_{out}. This gradient generates buoyancy and motion, and enters directly the expression for the generation rate of kinetic energy (P_{ex}, Eq. 8.5). This power (P_{ex}) generates kinetic energy (A_{ke}, Eq. 8.9), which is then subsequently dissipated (D). The rate of dissipation (D) depends on the kinetic energy (A_{ke}) as well as the spatial organization of the flow, characterized by the geometric factor γ (Eq. 8.17). Motion in the system results in the convective heat flux (J_{ex}), so that this heat flux depends on the kinetic energy A_{ke} and the temperature difference ΔT (Eq. 8.18).

The following feedbacks are established because the temperature difference (ΔT) as well as the power (P_{ex}) depend on the convective heat flux J_{ex} (Eqs. 8.4 and 8.5).

We now look at the consequences of a perturbation in the temperature difference ($d\Delta T$) on the generation of kinetic energy (dP_{ex}). The generation rate (P_{ex}) depends on the convective heat flux (J_{ex}) and on the temperature difference (ΔT), so that, effectively, $P_{ex} = P_{ex}(J_{ex}(A_{ke}(P_{ex}, D)), \Delta T_{ex})$. Hence, the overall change in dP_{ex} depends on the direct effect of ΔT on P_{ex} (i.e. $\partial P_{ex}/\partial \Delta T$), and indirect effects due to the various interdependencies, which are described by a product of partial derivatives

$$\frac{dP_{ex}}{d\Delta T} = \frac{\partial P_{ex}}{\partial J_{ex}}\frac{\partial J_{ex}}{\partial A_{ke}}\frac{\partial A_{ke}}{\partial P_{ex}}\frac{\partial P_{ex}}{\partial \Delta T} + \frac{\partial P_{ex}}{\partial \Delta T}\frac{\partial \Delta T}{\partial J_{ex}}\frac{\partial J_{ex}}{\partial A_{ke}}\frac{\partial A_{ke}}{\partial P_{ex}}\frac{\partial P_{ex}}{\partial \Delta T} \tag{8.20}$$

The first term on the right hand side represents a positive feedback (feedback A in Fig. 8.5). An increase in the generation rate results in an increase in kinetic energy ($\partial A_{ke}/\partial P_{ex} > 0$), which causes an increase in the heat flux ($\partial J_{ex}/\partial A_{ke} > 0$) which in turn results in greater power ($\partial P_{ex}/\partial J_{ex} > 0$). Since all derivatives are positive, the initial change is amplified and this constitutes a positive feedback. The second term on the right hand side represents a negative feedback (feedback B). The last three derivatives are the same as in the first term and describe the increase of the heat flux J_{ex} due to the initial change in ΔT. The greater heat flux also results in a decrease in the temperature difference ($\partial \Delta T/\partial J_{ex} < 0$), and a decrease in temperature difference decreases the power ($\partial P_{ex}/\partial \Delta T > 0$). Hence, the product of these derivatives is negative, so that these effects constitute a negative feedback. Since temperature changes involve changes in thermal inertia, this feedback is likely to act more slowly than feedback A.

With increasing values of kinetic energy (A_{ke}) in the system, the derivatives change their values, and so do the strengths of the two feedbacks. The deciding difference in these feedbacks relates to the terms ($\partial P_{ex}/\partial J_{ex}$) and $(\partial P_{ex}/\partial \Delta T)(\partial \Delta T/\partial J_{ex})$, while the other terms could be factored out in the above Eq. (8.20). Because $(\partial P_{ex}/\partial J_{ex}) = \Delta T/T_h$, $(\partial P_{ex}/\partial \Delta T) = J_{ex}/T_h$, and $(\partial \Delta T/\partial J_{ex}) = -1/k_r$, the sum of these terms $(\partial P_{ex}/\partial J_{ex}) + (\partial P_{ex}/\partial \Delta T)(\partial \Delta T/\partial J_{ex})$ decreases with an increasing values of J_{ex}, and cancel each other exactly at the maximum power state, when $\Delta T/T_h - J_{ex}/(k_r T_h) = 0$, or $J_{ex} = k_r \Delta T = J_{in}/2$. In other words, at the maximum power state, the feedbacks A and B operate with same strength, but with opposite signs, so that the maximum power state should be the state that is dynamically the most stable. Noting that power equals dissipation in steady state, this line of reasoning is consistent with the dynamic stability analysis of Malkus [22], with the derivation by Dewar and Maritan (Chap. 3), and with the reasoning behind the MaxEP state by Ozawa et al. [6].

The spatial organization of the flow affects the two feedbacks described above. In steady state, we have $P_{ex} = D \propto \gamma A_{ke}$, so that $A_{ke} \propto P_{ex}/\gamma$. Hence, the derivative $\partial A_{ke}/\partial P_{ex} \propto \gamma^{-1}$ depends on the spatial arrangement of the flow. To relate structure formation to the maximization of power, we note that changes in ΔT can also result from the internal dynamics and, specifically, the spatial organization as shown in Fig. 8.4. The feedbacks that are related to structure formation are shown in Fig. 8.5 in terms of feedbacks C and D. Feedback C characterizes the reduction in frictional dissipation due to the development of structured flow that is captured by the geometric factor γ. In other words, when an increase in kinetic energy results in a change in spatial organization and a reduction in the geometric factor ($\partial\gamma/\partial A_{ke} < 0$), this would reduce dissipation ($\partial D/\partial\gamma > 0$), resulting in an increase in kinetic energy ($\partial A_{ke}/\partial D < 0$). Overall, this feedback constitutes a positive feedback related to the reduction of internal dissipation due to spatial reorganization of the flow. The implication of this feedback is that for a given generation rate P_{ex}, a reduction in γ would enhance A_{ke}, J_{ex}, and thus P_{ex}. Hence, those perturbations in the spatial organization of the flow that enhance power would continue to grow and play an important part of feedback A. The confinement of temperature gradients to the system boundary that was qualitatively discussed above constitutes a further feedback (feedback D, see also Schneider and Kay [19] for relevant discussion on temperature profiles in convective cells), in which a change in spatial organization would affect the temperature difference (ΔT), power (P_{ex}), kinetic energy (A_{ke}), which could then feed back to the value of the geometric factor. While we did not provide mathematical relationships to express this feedback in detail, these effects would be reflected in the partial derivatives of $\partial A_{ke}/\partial P_{ex}$ and $\partial\Delta T/\partial J_{ex}$, thereby affecting feedback B.

To sum up, this discussion on dynamics and feedbacks suggests that a state of maximum power would naturally emerge from the dynamics within a system. First, a fast, positive feedback enhances free energy generation within the system through the formation of structured flow. This positive feedback is eventually balanced by the development of a negative feedback associated with the depletion of the driving gradient through the enhanced heat transport, so that the dynamics should be maintained in a steady state near the maximum power limit.

8.5 Implications of Maximum Power for Planetary Interactions

When we apply maximum power limits to the Earth system, we need to recognize that essentially all forms of free energy originate directly or indirectly from the planetary drivers: solar radiation and the cooling of the Earth's interior. These maintain the ultimate driving gradients from which free energy is generated, which is then either dissipated directly, or converted into other forms of free energy and dissipated subsequently. For instance, heating gradients generated by differences in the absorption of solar radiation result in the generation of kinetic energy and associated momentum gradients. These gradients are either dissipated by friction, or used to dehumidify the atmosphere and lift water vapor to the height at which it condenses. Subsequently, surface evaporation dissipates the gradient in specific humidity and falling raindrops dissipate the potential energy. Hence, the dynamics

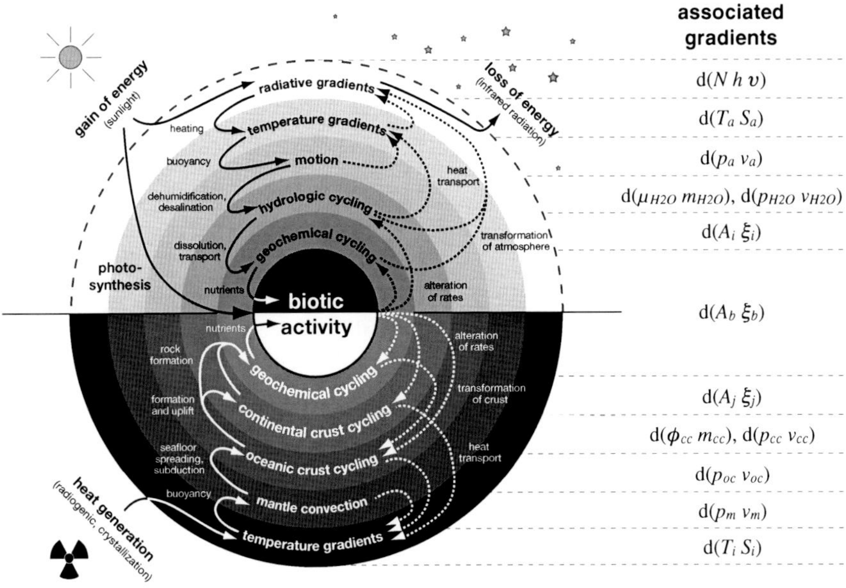

Fig. 8.6 Schematic diagram of the planetary hierarchy of free energy generation, transfer and dissipation to different forms (*solid lines*) and associated effects on the driving gradients (*dotted lines*). The different layers are associated with different forms of gradients, free energy and disequilibrium. The associated gradients that express these forms of free energy are shown on the right, with radiant energy expressed by the number of photons N, and the energy per photon $h\nu$ with frequency ν, thermal energy by temperature T and entropy S, kinetic energy by momentum p and velocity v, binding energy by chemical potential μ and mass m, potential energy by geopotential ϕ and mass m, and chemical energy by affinity A and extent of reaction ξ. After [37]

of the Earth system can be viewed as an interconnected cascade of energy conversions, as illustrated in Fig. 8.6.

The second law and the maximum power limit have five important, broader implications for the cascades of energy conversions within the Earth system:

Hierarchy of free energy generation and driving gradients. The generation of different forms of free energy within the Earth system do not take place independently, but the free energy and the associated gradients generated by one process typically form the driving gradient of another process. This connectedness of the free energy generation terms is shown by the solid lines in Fig. 8.6. For instance, the gradient $d(Nh\nu)$ in radiative exchange at the Earth-space boundary causes gradients in radiative heating, $d(TS)$, which is in part converted into the kinetic energy, $d(pv)$, associated with atmospheric motion and gradients in velocity v. Motion in turn is in part dissipated by friction, that is, kinetic energy $d(pv)$ is converted into heat $d(TS)$, but also performs other types of work, e.g. lifting dust and moisture or forming waves and currents in the ocean. These transfer processes generate potential energy, $d(\phi m)$, out of the kinetic energy of motion $d(pv)$. As a consequence, the dynamics of free energy are then not simply formed by generation and dissipation terms, as it is often being done (e.g. in the form of the Lorenz energy cycle in atmospheric dynamics [23, 24], or as represented by the simple representation in Eq. 8.9 above), but include transfer terms to other forms of free energy. These transfer terms of energy play the critical role of the "glue" that connects processes and that in the end result in the highly complex and interacting Earth system. The formulation of these dynamics in terms of driving gradients and resulting forms of free energy provides a clear direction and causality despite the complexity that is involved.

Interactions and feedbacks to driving gradients at higher levels. When free energy is generated from a driving gradient, the driving gradient is inevitably depleted. In the simple example in Sect. 8.2, the generation of motion inevitably results in a convective heat flux that depletes the temperature gradient (as also shown in the feedbacks in Fig. 8.5). Consequently, each conversion along the solid lines shown in Fig. 8.6 is associated with inevitable effects on the driving gradients, as indicated by the dashed lines in Fig. 8.6, and therefore on the whole chain of conversions. For instance, when motion is generated by differential radiative heating, the resulting motion transports heat that accelerates the depletion of the differential radiative heating. When motion lifts vapor to greater heights and colder temperatures, it brings vapor to condensation. This dehumidification of the atmosphere by motion results in the transport of latent heat that reduces the heat available for driving the atmospheric heat engine [25–28]. Hence, each conversion along the solid lines in Fig. 8.6 results in inevitable interactions between processes that affect free energy generation by these processes and, ultimately, the exchange of radiation and entropy with space.

Maximum power limits. The conversions of gradients into different forms of free energy down the hierarchy shown by the solid lines in Fig. 8.6 is restricted by the rate by which the gradient is generated in the layer above. At best, all of the driving gradient can be converted into free energy. For most of the conversions, however, free energy can either be dissipated directly or converted into another form of free energy further down the layers and dissipated subsequently. These two "options" for the fate of the free energy imply maximum power limits akin to the one shown in Sect. 8.2 in which the direct dissipation is associated with radiative transfer (J_r), while the conversion to free energy and its subsequent dissipation is associated with P_{ex} and D, respectively. Consequently, each of the free energy conversions down from the planetary driver involves some direct dissipation, so that less free energy can be generated with each additional conversion.

What this then implies is that because of these limits, abiotic processes cannot generate substantial amounts of chemical free energy ($d(A_i\xi_i)$ and $d(A_j\xi_j)$ in Fig. 8.6) that could transform the chemical composition of the atmosphere. In contrast, photosynthetic life avoids these dissipative losses by generating chemical free energy ($d(A_b\xi_b)$ in Fig. 8.6) directly by exploiting $d(Nh\nu)$ by photochemistry. This insight is consistent with the common attribution of the chemical disequilibrium in the Earth's atmosphere to the presence of abundant life [29, 30]. By formulating biotic activity in such thermodynamic terms, one may also explore maximum power limits in biochemical processes [31] and the biosphere [32, 33].

Maximization by dissipative structures. Maximization of the different rates of free energy conversions can take place by developing structures that are able to reduce the extent of internal friction and dissipation. While the maximum power limit is given by the constraints imposed by the driving gradient, this limit can only be achieved by adjustments of the dynamics associated with the generated form of free energy. This possibility of the flow to adjust its level of frictional dissipation was demonstrated in Sect. 8.3 for convective flow. The widespread presence of similar, reproducible structures, such as convection cells, waves or fractal networks in Earth system processes can be seen as the manifestation of maximization through structure at different spatial and temporal scales. This interpretation of dissipative structures as the means to achieve the maximum power limit provides a new and broader basis to link previous work along similar lines (e.g. Prigogine's "dissipative structures" [34], Bejan's "constructal law" [35] and the assumption of minimum energy dissipation in fractal networks by [36]) to the flexible boundary conditions and interactions within the planetary context.

The Second Law at the planetary scale. Each of the conversions of gradients among the different layers in Fig. 8.6 obey the second law, which is contained in the maximum power limit by the assumption that P_{ex} operates at the Carnot limit. In fact, the dynamics of free energy generation, transfer, and dissipation are such that they enhance gradient depletion and thereby accelerate processes in the direction of the second law when evaluated at the scale of the gradient that is at a higher layer within the hierarchy. In the simple example in Sect. 8.2 this acceleration is reflected in the depleted temperature gradient $T_h - T_c$, with a maximum possible reduction at the maximum power limit to $T_h - T_c = (T_{h,0} - T_{c,0})/2$. In the Earth's atmosphere, this depletion is reflected in the reduced gradient in the net radiative exchange at the top of the atmosphere as a result of large-scale atmospheric heat transport (which, in the context of the system described here, would correspond to spatial differences in $J_{in} - J_{out}$). When we generalize this effect and apply it to the planetary system, this would imply that the overall dynamics of free energy generation and transfer among the different layers in Fig. 8.6 are such that these deplete gradients faster, possibly as fast as possible (cf. "maximization by structure"), so that the whole system should deplete the driving, radiative gradients by as much as possible. This would then imply that the complex dynamics of the Earth system would result in the maximum rate of radiative entropy production to the extent that this is possible by the dynamics associated with free energy transformations. This latter restriction is important: The heat transport by convection, for instance, could not completely level out the temperature gradient (i.e. $T_h - T_c = 0$ cannot be achieved by the heat transported by motion in steady state) but is restricted to states below or at the maximum power limit in steady state (as discussed above in the context of feedbacks). This perspective of the hierarchy shown in Fig. 8.6 as the implementation of a "planetary accelerator" of the second law provides a powerful general direction to the complex and seemingly arbitrary conversions and interactions within the Earth system.

8.6 Summary and Conclusions

In this chapter, we described how the second law of thermodynamics sets the direction and constraints for the dynamics of the whole Earth system, but also how the dynamics act to accelerate the second law towards a state of thermodynamic equilibrium. The sequence of generation, dissipation, and transfer of free energy to different forms acts to accelerate the progress into the direction imposed by the second law. At the same time, the second law imposes a fundamental constraint on the strength of this sequence by setting the maximum power limit. This limit can be achieved by the internal dynamics of the system through adjustments of the flow into structures, such as convection cells. The development of such "dissipative structures" reduces internal dissipation, so that for the same generation rate, more free energy can be maintained within the system. This results in a positive feedback that enhances free energy generation and structure formation up to the maximum power limit. At this limit, the negative feedback resulting from the accelerated depletion of the driving gradient compensates the positive feedback, resulting in dynamics that should be maintained near a steady state of maximum power.

When this perspective is applied to the dynamics of the Earth system as a whole, this results in a hierarchy of free energy generation and transfer, where one form of generated free energy constitutes the driving gradient for the generation of another form of free energy. Overall, such a planetary hierarchy of free energy conversions should represent a "planetary accelerator" towards a state of thermodynamic equilibrium and, when maintained at maximum power, reflect the means to deplete the planetary driving gradients as fast as possible. Since the Earth exchanges mostly radiation of different entropy with space, this would constitute the means to overall produce radiative entropy at the maximum possible rate by these dynamics that involve the conversions and dissipation of the various forms of free energy.

The thermodynamic limits in this chapter were formulated in terms of maximum power limits rather than in terms of the proposed principle of Maximum Entropy Production (MaxEP). The outcomes of both, maximum power or MaxEP, are essentially indistinguishable in terms of the associated temperature gradients and heat fluxes when applied to e.g. a convective system. The maximum power limit has the advantage that it specifically describes the driving gradient and the dynamical processes involved, which should facilitate the application of this limit to Earth system processes. In comparison, the use of MaxEP is often ambiguous because it is not clear which entropy production is to be maximized and why the dynamics would be such that they result in maximization of entropy production compared to other aspects that are more directly involved in the dynamics (such as forces, energy or power). In this sense, the shift in focus to maximum power should not be seen as a contradiction to previous work on MaxEP, but rather as a continuation and sharpening of the application of thermodynamic limits to Earth system processes. It is quite likely that the maximization of power by systems can be derived as a form of entropy production maximization that is constrained by more than the energy- and mass balance, e.g. by the momentum balance [22] (Dewar and Maritan, Chap. 3). No matter whether power or entropy production is maximized, the key aspect in the maximization is that the boundary conditions are not fixed, but react to the dynamics within the system and accelerate the depletion of the driving gradient. Hence, the maximization reflects the central role of interactions between the system dynamics and the boundary conditions. The shift in emphasis from MaxEP to maximum power led to the insight that systems are able to adjust to maximum power states through the development of structured flow that reduces frictional dissipation within the system.

This perspective needs to be developed further in the future, as it allows us to become more specific regarding the conditions under which the maximum power state is achievable. For instance, the minimum dissipation solution in the example presented in Sect. 8.3 depends on the total size of the system (i.e. A_{tot}), while the value of N is constrained to integer values $N \geq 1$. Even in this simple example one can envision situations where the system is too small to be flexible enough to minimize internal dissipation and therefore being unable to evolve to the maximum power state. In such a case, the dynamics are too constrained, or, formulated differently, the degrees of freedom within the system are too low to achieve the maximum power state.

Overall, the progression presented here from a relatively simple MaxEP view of the dynamics of Earth system processes to "maximization of power through structure" within the context of the whole Earth system should provide a much more specific basis to demonstrate the relevance of thermodynamic limits to the structure and functioning of the planetary dynamics of the Earth system.

Acknowledgments This research contributes to the Helmholtz Alliance "Planetary Evolution and Life". The authors thank Roderick Dewar and two anonymous reviewers for their constructive comments.

References

1. Dewar, R.C.: Maximum Entropy Production and non-equilibrium statistical mechanics. In: Kleidon, A., Lorenz, R. D. (eds.) Non-Equilibrium Thermodynamics and the Production of Entropy: Life, Earth, and Beyond, pp. 41–56, Springer, Heidelberg (2005)
2. Dewar, R.C.: J. Phys. A **38**, L371 (2005). doi:10.1088/0305-4470/38/21/L01
3. Niven, R.K.: Phys. Rev. E **80**, 021113 (2009)
4. Dewar, R.C.: Entropy **11**(4), 931 (2010)
5. Niven, R.K.: Phil. Trans. R. Soc. B **365**, 1323 (2010)
6. Ozawa, H., Ohmura, A., Lorenz, R.D., Pujol, T.: Rev. Geophys. **41**, 1018 (2003)

7. Kleidon, A., Lorenz, R.D. (eds.): Non-Equilibrium Thermodynamics and the Production of Entropy: Life, Earth, and Beyond. Springer, Heidelberg (2005)
8. Kleidon, A. Malhi, Y. Cox, P.M.: Phil. Trans. R. Soc. B **365**, 1297 (2010)
9. Lorenz, R.D. Lunine, J.I., Withers, P.G. McKay, C.P.: Geophys. Res. Lett. **28**, 415 (2001)
10. Lorenz, R.D., Mckay, C.P.: Icarus **165**(2), 407 (2003)
11. Kleidon, A., Fraedrich, K., Kunz, T., Lunkeit, F.: Geophys. Res. Lett. **30**, 2223 (2003). doi:10.1029/2003GL018363
12. Kleidon, A., Fraedrich, K., Kirk, E., Lunkeit, F.: Geophys. Res. Lett. **33**, L06706 (2006). doi:10.1029/2005GL025373
13. Goody, R.: J. Atmos. Sci. **64**, 2735 (2007)
14. Volk, T.: Clim. Ch. **85**, 251 (2007)
15. Volk, T., Pauluis, O.: Phil. Trans. R. Soc. B **365**, 1317 (2010)
16. Kiehl, J.T., Trenberth, K.E.: Bull. Amer. Meteorol. Soc. **78**, 197 (1997)
17. Kleidon, A.: Phil. Trans. R. Soc. A **370**, 1012 (2012)
18. Kleidon, A.: Clim. Ch. **66**, 271 (2004)
19. Schneider, E.D., Kay, J.J.: Math. Comput. Modeling **19**, 25 (1994)
20. Kleidon, A., Zehe, E., Ehret, U., Scherer, U.: Hydrol. Earth Syst. Sci. **17**, 225 (2013)
21. Hansen, J., Lacis, A., Rind, D., Russell, G., Stone, P., Fung, I., Ruedy, R., Lerner, J.: Climate processes and climate sensitivity. Geophys. Monogr. **29** (1984), (American Geophysical Union)
22. Malkus, W.V.R.: Phys. Fluids **8**, 1582 (1996)
23. Lorenz, E.N.: Tellus **7**, 157 (1955)
24. Lorenz, E.N.: Dynamics of Climate. In: Pfeffer, R.C. (ed.) pp. 86–92. Pergamon Press, Oxford (1960)
25. Pauluis, O., Held, I.M.: J. Atmos. Sci. **59**, 126 (2002)
26. Pauluis, O., Held, I.M.: J. Atmos. Sci. **59**, 140 (2002)
27. Pauluis, O.: Water vapor and entropy production in the Earth's atmosphere. In: Kleidon, A., Lorenz, R. D. (eds.) Non-Equilibrium Thermodynamics and the Production of Entropy: Life, Earth, and Beyond, pp. 173–190, Springer, Heidelberg (2005)
28. Kleidon, A., Renner, M.: Hydrol. Earth Syst. Sci. 17, 2873–2892 (2013), doi:10.5194/hess-17-2873-2013
29. Lovelock, J.E.: Nature **207**, 568 (1965)
30. Lovelock, J.E.: Proc. Roy. Soc. Lond. B **189**, 167 (1975)
31. Juretic, D., Zupanovic, P.: The free-energy transduction and entropy production in initial photosynthetic reactions. In: Kleidon, A., Lorenz, R. D. (eds.) Non-Equilibrium Thermodynamics and the Production of Entropy: Life, Earth, and Beyond, pp. 161–172, Springer, Heidelberg (2005)
32. Lotka, A.J.: Proc. Natl. Acad. Sci. U.S.A. **8**, 147 (1922)
33. Lotka, A.J.: Proc. Natl. Acad. Sci. U.S.A. **8**, 151 (1922)
34. Prigogine, I.: Science **201**, 777 (1978)
35. Bejan, A., Lorente, S.: Phil. Trans. R. Soc. B **365**, 1335 (2010)
36. West, G.B., Brown, J.H., Enquist, B.J.: Science **276**, 122 (1997)
37. Kleidon, A.: Phys. Life Rev. **7**, 424 (2010)

Part III
Applications to Non-equilibrium Systems

Chapter 9
Predictive Use of the Maximum Entropy Production Principle for Past and Present Climates

Corentin Herbert and Didier Paillard

Abstract In this chapter, we show how the MaxEP hypothesis may be used to build simple climate models without representing explicitly the energy transport by the atmosphere. The purpose is twofold. First, we assess the performance of the MaxEP hypothesis by comparing a simple model with minimal input data to a complex, state-of-the-art General Circulation Model. Next, we show how to improve the realism of MaxEP climate models by including climate feedbacks, focusing on the case of the water-vapour feedback. We also discuss the dependence of the entropy production rate and predicted surface temperature on the resolution of the model.

9.1 Introduction

Although it is not straightforward to define what climate is precisely, one may suggest that what we call *the climate system* is made up of the atmosphere, the oceans, the cryosphere, the biosphere and the lithosphere [1]. The different components interact in various ways, and their relative importance depends on the question asked. For instance in numerical weather prediction, taking place on a timescale of a few days, the main dynamical component is the atmosphere and all the other components may be regarded as prescribed. On the contrary, the evolution of climate on very long timescales (of the order of tenths or hundreds million year) is essentially determined by the exchanges of carbon between the land, the oceans and the atmosphere.

C. Herbert (✉)
National Center for Atmospheric Research, P.O. Box 3000 Boulder, CO 80307, USA
e-mail: cherbert@ucar.edu

D. Paillard
Laboratoire des Sciences du Climat et de l'Environnement, IPSL, CEA-CNRS-UVSQ, UMR 8212, 91191 Gif-sur-Yvette, France
e-mail: didier.paillard@lsce.ipsl.fr

R. C. Dewar et al. (eds.), *Beyond the Second Law*,
Understanding Complex Systems, DOI: 10.1007/978-3-642-40154-1_9,

The distribution of surface temperature is of primary interest. It depends on a large number of factors, such as the composition of the atmosphere (upon which the radiative energy exchanges depend), the circulation of the atmosphere and oceans, the ocean salinity, the presence of ice-sheets, the type of terrestrial vegetation cover,... State-of-the-art climate models, usually referred to as *General Circulation Models* (GCMs), now include many of the above factors (the term *Earth System Models* is starting to emerge).

However, not all this complexity is necessary to obtain a rough estimate of the temperature of a planetary atmosphere: perhaps the simplest approach is to balance the incoming solar radiation with the outgoing planetary radiation. Again this can be done at various levels of accuracy, depending on the knowledge we have of the concentration of the radiatively active constituents of the atmosphere (e.g. water-vapour and carbon dioxide). Imposing a local radiative equilibrium is in fact misleading: latitudinal and vertical differential heating trigger atmospheric motions, which carry heat to mitigate the temperature gradients that would exist at radiative equilibrium. The resulting energy transport term can be parametrized (for instance as a diffusion process with empirical diffusivity) as a function of the temperature distribution, so that we can solve the model without resolving explicitly the motions of the atmosphere. Such models, consisting of a radiative model and a parameterization of the energy transport by the atmosphere are called *Energy Balance Models* (EBMs). Alternatively, one may solve the fluid dynamics problem and compute explicitly the velocity field: this is what GCMs do. The hierarchy of climate models, ranging from simple EBMs to complex GCMs, also comprises the so-called *intermediate complexity models* (EMIC), which offer a variety of simplified representations of the atmospheric and oceanic circulation and other phenomena [2]. The main interest of EMICs is their relatively low computational cost, compared to GCMs, which make them particularly suitable for the study of palaeoclimates. Indeed, the timescales involved in such problems reduce the role of GCMs to simulating snapshots. Both GCMs and EMICs require a certain amount of *parameter tuning*. This is sometimes a problem when studying past climates for which little data is available on which to base adjustment procedures, and even more so for other planetary climates, where many features differ tremendously from the terrestrial conditions on which the empirical parameterizations were tested.

Nevertheless, the laws of physics remain the same when going back into time or out into the cosmos. The three branches of physics which play a fundamental part in setting the climate of a planet are radiation physics [3], fluid dynamics [4, 5] and thermodynamics [6]. One fundamental principle which is always present, even in simple models like EBMs, is the first law of thermodynamics, because it describes the exchanges of energy in a system. To energy exchanges are associated equilibrium temperature distributions. On the other hand, even in the most sophisticated climate models to date, the second law of thermodynamics, which also describes the exchanges of energy in a system but in a qualitative rather than quantitative way, is not taken into account. When subgrid-scale parameterizations are involved, classical models may even violate the second law of thermodynamics [7]. It has also been suggested that spurious sources of entropy production could lead to a global cold bias in climate models [8]. Henceforth, a number of diagnostic tools emerged to study the thermodynamic properties of climate models [9] (see also Chap. 10). Besides, postulating that the system chooses the steady-state with maximum entropy production given certain constraints leads to a variational problem which has proved very efficient for predictive use. This is the so-called *Maximum Entropy Production principle* [10–12]. We shall not discuss here the theoretical foundations (or lack thereof) of this hypothesis (see [13–16]), but only its consequences for climate modelling. Hitherto, mainly two approaches have been developed. One point of view is that the MaxEP principle can be useful to select the value of adjustable parameters in empirical parameterizations from existing models, in an *objective* way [17–21]. In the second approach, the purpose is to build simple climate models based on the MaxEP hypothesis for describing unresolved processes. We shall present the latter approach in this chapter. After briefly reviewing earlier attempts (Sect. 9.2) we build a MaxEP climate model devoid of *ad hoc* assumptions and we show how to include feedbacks like the water-vapour feedback (Sect. 9.3). The model is then tested for pre-industrial and Last Glacial Maximum conditions (Sect. 9.4).

9.2 The Paltridge Model

A typical one-dimensional EBM consists of a certain number of *boxes*, representing latitudinal zones, characterized by a single temperature. Each box receives energy from the outside in the form of solar radiation, and radiates back to space in the longwave domain. The difference of these two terms, which is usually called the *radiative budget* of the box, does not necessarily vanish: there are also energy exchanges with the neighbouring boxes due to atmospheric (and oceanic) transport of heat. Hence, for box i, the total energy budget reads

$$c_{pi}\frac{dT_i}{dt} = R_i + \gamma_i, \tag{9.1}$$

where c_{pi}, T_i, R_i and γ_i denotes respectively the heat capacity, temperature, radiative budget and atmospheric (or oceanic) convergence for box i. A *radiative scheme* provides R_i as a function of T_i: e.g. $R_i = \xi_i S - \varepsilon_i \sigma T_i^4$ where S is the solar constant, ξ_i represents the projection of the surface of the latitude belt onto the sphere centered on the sun, σ is the Stefan-Boltzmann constant and ε_i the emissivity of the surface. In such a radiative scheme, the greenhouse effect is not taken into account. In contrast, there is no simple expression for γ_i which can be justified from first principles. A standard *parameterization* in this context is to assume a diffusion-like term, but there is no justification for this hypothesis and the diffusion coefficient has to be chosen empirically.

Paltridge [22] suggested a model, with a more elaborate radiative scheme—involving in particular a cloud cover variable θ_i in each box—than our above example, in which γ_i is not empirically parameterized as a function of the temperatures T_i, but instead satisfies a maximum entropy production principle. He postulates that the steady-state temperature distribution T_i is such that the material entropy production rate $\sigma = \sum_i \frac{\gamma_i}{T_i}$ is maximum, subject to the global steady-state constraint $\sum_i \gamma_i = 0$. At steady state, $\gamma_i = -R_i$ and σ is a function of the temperatures T_i only. At steady-state, the distributions of temperature, cloud cover, atmospheric and oceanic meridional fluxes obtained are in striking accordance with observations. In spite of this apparent success, some major criticism remain. First of all, the planetary rotation rate is believed to be a major driver of the latitudinal distribution of temperatures, but it does not appear at all in Paltridge's model. Besides, it is clear that the principle does not hold in the case of a planet without atmosphere (see Chap. 11). One may thus wonder if it is not pure coincidence that it seems to apply to the Earth's atmosphere [23]. Last but not least, there is no theoretical justification for the principle of maximum entropy production.

The thread was taken up in a series of papers [24–27], verifying Paltridge's results in different variants of the original model, but the fundamental objections mentioned above remained unanswered. More recently, Lorenz [28] added some support to the idea that the agreement between the model and observations is not a coincidence, by showing that it gives acceptable results for Titan and Mars as well. The question of the independence with respect to the planetary rotation rate was also adressed by Jupp [29] in a MaxEP model with a simple parameterization of atmospherics dynamics. Nevertheless, one fundamental concern remains: the Paltridge model and its variants still contain a large number of parameterizations, *ad hoc* hypothesis and empirical coefficients, for instance in the radiative scheme, in the cloud parameterization or in the treatment of surface heat fluxes (maximum convective hypothesis). Is it possible to get rid of these potential biases to assess the intrinsic value of the MaxEP conjecture in the climate modelling framework? This is the question we address in the next section.

9.3 A Simple MEP Model with Water-vapour Feedback

9.3.1 NEF Radiative Scheme

A possible strategy to assess the degree of coincidence in Paltridge's results may be to build a MaxEP model devoid of any *ad-hoc* parameter and assumptions. To that end, we suggest a new radiative scheme based on the Net Exchange Formulation (NEF), which only involves physical quantities (values of which are known a priori). Following [30], we introduce a two dimensional model with two layers: for each grid point characterized by a latitude and a longitude, there is a surface layer with a temperature T_g and an atmospheric layer with a temperature T_a.

Each layer absorbs an amount of solar radiation (Ψ_{gs}^{SW} for the surface layer and Ψ_{as}^{SW} for the atmosphere) given by (Fig. 9.1):

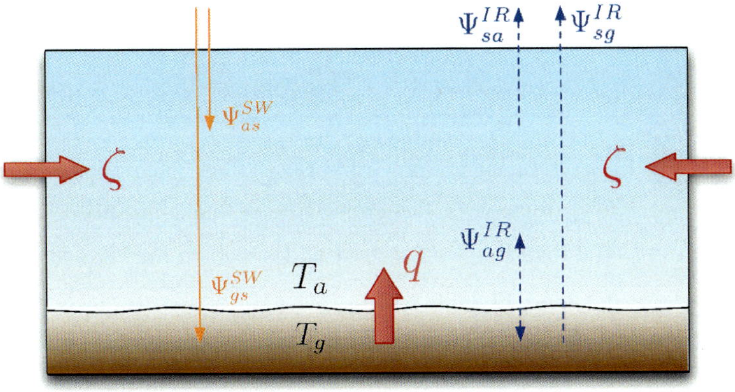

Fig. 9.1 One grid cell of a two-layer MEP model. The surface layer has temperature T_g and exchanges heat q (*thick solid red arrow*) with the overlying atmospheric layer of temperature T_a. Both layers absorb solar radiation (*thin solid yellow arrows*) and emit and absorb longwave radiation (*thin dashed blue arrows*). The atmospheric layer exchanges energy with the surrounding cells: the convergence of the atmospheric heat flux is ζ (*thick solid red arrow*)

$$\Psi_{gs}^{SW} = (\bar{s}(\alpha_g) - s)(1 - \alpha_g)\xi S, \tag{9.2}$$

$$\Psi_{as}^{SW} = (s + \alpha_g s^*)\xi S, \tag{9.3}$$

where S is the solar constant, ξ the projection of the cell area onto the sphere, α_g the surface albedo, and the coefficients s, s^* and $\bar{s}$ are adapted from the classical Lacis and Hansen scheme [31]:

$$\bar{s}(\alpha_g) = 0.353 + \frac{0.647 - \bar{R}_r(\xi) - A_{oz}(Mu_{O_3})}{1 - \bar{\bar{R}}_r^*\alpha_g}, \tag{9.4}$$

$$s = A_{wv}(M\tilde{u}), \tag{9.5}$$

$$s^* = A_{wv}\left(\left(M + \frac{5}{3}\right)\tilde{u}\right) - A_{wv}(M\tilde{u}). \tag{9.6}$$

Here $u_{O_3}, \tilde{u}$ represent respectively the vertically integrated ozone and water vapour density (including pressure scaling [32]), M accounts for the slant path of solar rays, $\bar{R}_r(\xi)$ and $\bar{\bar{R}}_r^*$ account for Rayleigh scattering in the atmosphere, and A_{oz}, A_{wv} are absorption functions for ozone and water vapour. See [30–32] for details.

The long-wave radiative exchanges can be written in a simple form using the Net Exchange Formulation [33]. The surface layer and the atmosphere exchange a net amount of energy Ψ_{ag}^{IR} through infrared radiation, while the surface and the atmosphere radiate respectively Ψ_{sg}^{IR} and Ψ_{sa}^{IR} to space (see [30] for a derivation):

$$\Psi_{ag}^{IR} = t(T_g)\sigma T_g^4 - t(T_a)\sigma T_a^4, \tag{9.7}$$

$$\Psi_{sa}^{IR} = t(T_a)\sigma T_a^4, \tag{9.8}$$

$$\Psi_{sg}^{IR} = \left(1 - \frac{t(T_g)}{\mu}\right)\sigma T_g^4, \tag{9.9}$$

where μ is the Elsasser factor arising from the angular integration, and $t(T) = \mu\left(1 - \int_0^{+\infty} \frac{B_\nu(T)}{\sigma T^4}\tau_\nu d\nu\right)$ represents the emissivity of the atmosphere (B_ν is the Planck function). The transmission function τ_ν depends on the vertical profiles of absorbing gases, pressure and temperature: $\tau_\nu = \exp\left(-\frac{1}{\mu}\int_0^H k_\nu(z)dz\right)$, where k_ν is the absorption coefficient, and H the total height of the atmosphere. To sum up, the only parameters required by the radiative scheme are the vertically integrated concentrations of water vapour $\tilde{u}$, carbon dioxide u_{CO_2} (they determine k_ν), ozone u_{O_3} and the surface albedo α_g.

The steady-state condition for each box reads, for every grid point:

$$\Psi_{gs}^{SW} + \Psi_{as}^{SW} - \Psi_{sg}^{IR} - \Psi_{sa}^{IR} + \zeta = 0, \tag{9.10}$$

$$\Psi_{gs}^{SW} - \Psi_{ag}^{IR} - \Psi_{sg}^{IR} - q = 0, \tag{9.11}$$

where ζ is the horizontal convergence of atmospheric heat fluxes and q the surface to atmosphere heat flux. The total material entropy production is given by

$$\sigma_M(\{T_{a,ij}, T_{g,ij}\}) = \sum_{i=1}^{N_{lat}}\sum_{j=1}^{N_{lon}}\left(\frac{q_{ij}}{T_{a,ij}} - \frac{q_{ij}}{T_{g,ij}} + \frac{\zeta_{ij}}{T_{a,ij}}\right)A_{ij}, \tag{9.12}$$

where A_{ij} is the area of the grid cell in position (i,j) and q_{ij}, ζ_{ij} are functions of $T_{a,ij}, T_{g,ij}$ given by (9.10), (9.11). We are interested in the fields that maximize σ_M while satisfying the global constraint $\sum_{i,j} A_{ij}\zeta_{ij} = 0$, which can be translated into an unconstrained variational principle using Lagrange multipliers.

9.3.2 Different Versions of the Model

The MaxEP model described in the previous section requires only physical parameters as an input. In a first step, we compute the horizontal distribution of $\tilde{u}$ (vertically integrated water vapour density) and u_{O_3} by linear interpolation of standard atmospheric profiles [34] (depending only on the latitude). To compare with the results of Paltridge, we also assume that the coefficients $t(T)$ in Eq. 9.7 are fixed, with a prescribed reference temperature T_{ref} (dependent on the latitude) also computed from the standard profiles [34] (version v0 in Table 9.1). However, the assumption of constant $t(T)$ coefficient is very unrealistic: the shift in the Planck spectrum associated with a variation in temperature of the surface or atmospheric layer has a strong impact on the optical properties of the atmosphere.

Table 9.1 Different versions of the MaxEP model and the resulting global mean surface temperature for pre-industrial (PI) and last glacial maximum (LGM) climates, compared to GCM runs with the IPSL_CM4 model.

Model version	$\tilde{u}$	u_{O_3}	u_{CO_2} (ppmv)	$t(T)$	$\langle T_{PI} \rangle$ (°C)	$\langle T_{LGM} - T_{PI} \rangle$
MaxEP v0	MC	MC	280	$T = T_{ref}$ (MC)	22.9	−1.98
MaxEP v1	MC	MC	280	$T = T_a, T_g$	22.3	−1.84
MaxEP v2	MC	0	280	$T = T_a, T_g$	22.5	−1.84
MaxEP v3	$u^*(T_a)$	0	280	$T = T_a, T_g$	19.9	−2.9
IPSL	–	–	280	–	15.7	−2.53

"*MC*" stands for the integrated standard McClatchey profiles, and the angular brackets mean global average. See Sect. 9.3.2 for the definition of the different versions and Sect. 9.4 for the discussion of the results

In version v1, we retain the dependence of the emissivity of the atmosphere on surface and atmospheric temperatures. Besides, fixing the profiles of water-vapour and ozone is also a restrictive hypothesis, especially in view of potential applications to different climates for which standard profiles are not well known. As far as ozone is concerned, we can simply examine a version of the model in which we completely ignore ozone (version v2). For water-vapour, the situation is slightly more complicated: the atmospheric temperature is linked via the Clausius-Clapeyron relation to the water vapour content, which itself feeds back onto the temperature via the greenhouse effect. Yet, in the previous versions (v0–v2) of the MaxEP model, we kept fixed the absolute amount of water vapour in the atmosphere, independently of the temperature. In version v3, we fix the relative humidity $RH = P_{H_2O}/P_{sat}(T)$. The vertically integrated density of water vapour is related to the relative humidity, temperature and pressure profiles through:

$$u^*_{H_2O} = \frac{1}{g}\frac{M_{H_2O}}{M_{air}} \int_0^{P_s} RH \times P_{sat}(T) \frac{dp}{p}, \tag{9.13}$$

where M_{H_2O}, M_{air} are the molar masses of water and air, g is the gravity and P_s the surface pressure. In our model with one atmospheric layer, we may assume that the relative humidity is uniformly distributed in each atmospheric cell, with a vertical extent equal to the scale height for water vapour. Relation (9.13) then becomes $u^*_{H_2O} \approx M_{H_2O}/(gM_{air}) \times RH \times P_{sat}(T)$ (version 3). The different versions are summarized in Table 9.1. The purpose of comparing these different versions of the model is at the same time to test the impact of reducing the quantity of input parameters (no T_{ref}, no u_{O_3}) and to improve the realism (Planck spectrum, water-vapour feedback).

9.3.3 Water-vapour Feedback and Multiple Steady States

The physical quantities involved in the climate system are related in many ways, so that a change in one of these quantities can have an influence on another one, feeding back onto the original quantity, either moving it closer (negative feedback) or farther (positive feedback) from its initial value. A classical example of positive feedback is the water-vapour feedback. If the temperature increases locally, the water vapour saturation pressure will increase so that more water (if available) may evaporate in the atmosphere, leading to stronger greenhouse effect and thus further increase of the temperature. Feedbacks of this sort can lead to multiple equilibria, bifurcations and hysteresis phenomena. For a given relative humidity distribution, equilibrium states with radically different temperatures are simultaneously possible [35]. The water-vapour feedback has been shown to play a major part in important climate problems [36], exactly like feedbacks of different natures [37, 38]. Hence, it is essential to be able to represent them correctly in a climate model. In the context of MaxEP models, it was shown in [39] that the ice-albedo feedback gives rise to multiple local maxima in the entropy production rate, corresponding to the multiple equilibria that appear in a traditional EBM (see also Chap. 10). Here, we observe multiple local maxima of the entropy production rate in a certain range of solar constant and relative humidity. One great advantage of MaxEP is the small computational cost of maximizing a function as compared to integrating a complex differential equation. Of course this is no longer true if the function, or the submanifold on which to search for the maximum, becomes too complicated. Already, in the presence of multiple maxima, this difficulty has to be dealt with as the steady-state selected by the maximization algorithm may depend on the initial value. To avoid being trapped in an irrelevant state, several methods may be investigated. First it is possible to further restrict the manifold defined by the constraints to ensure that it contains only one local maximum of the entropy production. In the case of the water-vapour feedback in our two-layer model, solving the radiative balance for the whole column in terms of the atmospheric temperature may lead to several solutions. Selecting systematically one of them before computing the entropy forces the system to remain on the portion of interest in phase space. This is the technique that we use here. Alternatively, introducing the time dimension and assuming that at each time step, the system maximizes instantaneous entropy production with an additional term corresponding to time derivatives, it was suggested in [39] to use *relaxation equations* as a numerical algorithm to compute the final state (see also Chap. 18).

9.4 Results: Present and Last Glacial Maximum Climates

We compared the surface temperature distribution obtained from MaxEP with that obtained from a state-of-the-art GCM, the IPSL_CM4 model. The IPSL model is a coupled atmosphere–ocean model [40] used for the Fourth Assessment Report (AR4) of the Intergovernmental Panel on Climate Change (IPCC) [41]. For pre-industrial climate, the forcings in the IPSL model are: pre-industrial greenhouse gas concentration ($CO_2 = 280$ ppm, $CH_4 = 760$ ppb, $N_2 = 270$ ppb), insolation, coastlines, topography and land-ice extent. The surface albedo is computed from

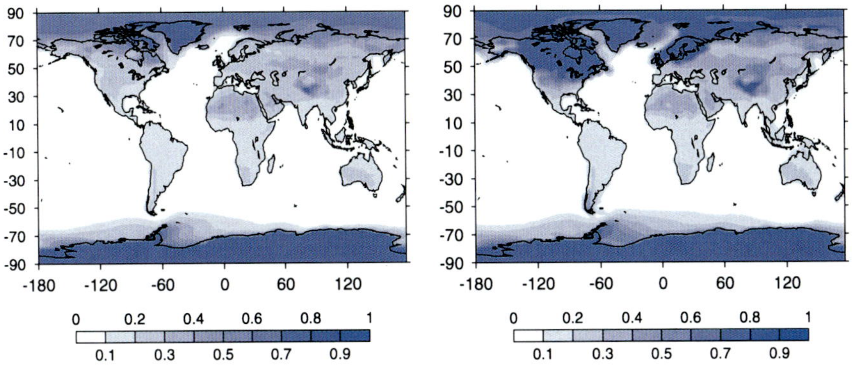

Fig. 9.2 Surface albedo α_g in the IPSL model, for pre-industrial (*left*) and Last Glacial Maximum (*right*) conditions

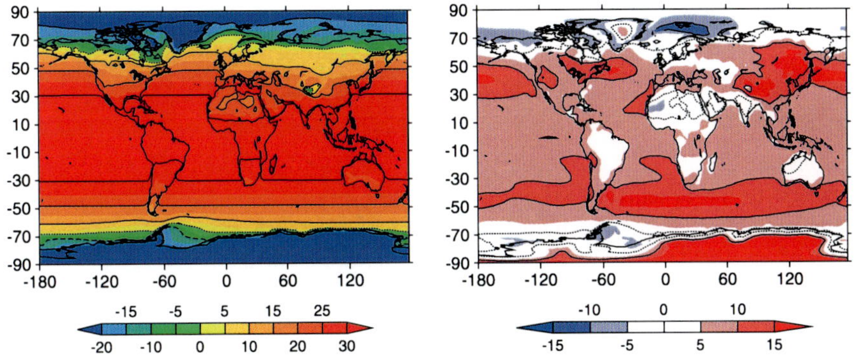

Fig. 9.3 *Left* surface temperature T_g for pre-industrial conditions obtained with the MaxEP model (version v0). *Right* Difference between the surface temperature T_g in the MaxEP model and the IPSL model for pre-industrial conditions. *Contour lines* interval is 10 °C, positive contours are drawn in *solid lines*, negative contours in *dashed lines* and the null contour as a *dotted line*

the IPSL_CM4 pre-industrial simulation and used as a forcing for the MaxEP model (Fig. 9.2, left).

The surface temperature distribution obtained with the MaxEP model is represented in Fig. 9.3 along with the difference between the MaxEP model and the IPSL model. The global mean surface temperature for the MaxEP model is $\langle T_{PI} \rangle = 22.9$ °C. By comparison, $\langle T_{PI} \rangle$ in the IPSL simulation is approximately 7 °C lower (Table 9.1); as Fig. 9.3 reveals, the major part of this difference comes from areas where the cloud cover is important, or elevated areas like the Antarctica. It is shown in [30] that a crude estimation of the effect of clouds and elevation suffices to explain the major part of the difference with the IPSL model. Figure 9.4 shows the meridional energy transport as a function of latitude for both the MaxEP model and the IPSL model for pre-industrial conditions. The agreement is remarkable given the simplicity of the MaxEP model.

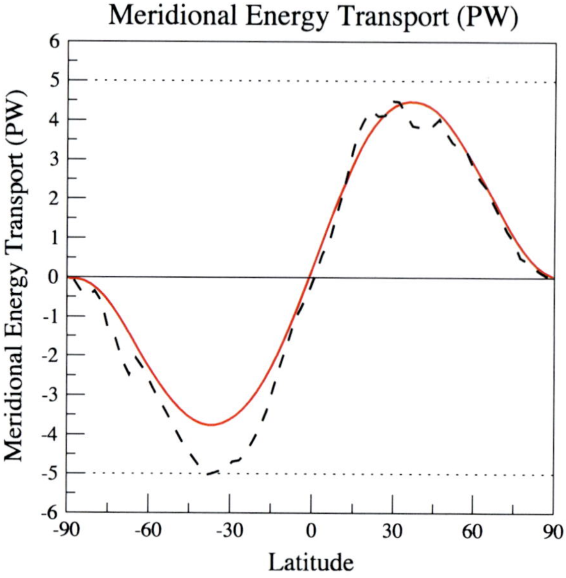

Fig. 9.4 Meridional energy transport as a function of the latitude in the MaxEP model (version v0, *solid red line*) and the IPSL model (*dashed black line*), for pre-industrial conditions

One advantage of the reformulation of the Paltridge model presented here is that due to the absence of *ad-hoc* parameters, it is possible to test the model on climates other than the Pre-Industrial period. For instance, it is possible to change the surface albedo to take into account the variations of ice or vegetation extent. A time period which is well documented and for which simulations with GCMs are available is the *Last Glacial Maximum* (LGM). It corresponds to the time during the last glacial period when the ice-sheets extent was maximum, roughly 21,000 years ago [42]. At that time, large ice-sheets covered North America and Northern Europe, and the global mean temperature was approximately 5 °C lower than present. In the MaxEP model, it is only possible to take into account the effect in surface albedo due to the presence of the ice-sheets at the LGM (Fig. 9.2, right), and not, for instance the associated topography effect. To ensure the comparison with the IPSL model is as direct as possible, we use a simulation where only the albedo effect is taken into account in the GCM. The resulting surface temperature difference between the LGM and the PI is shown in Fig. 9.5 for both models. The global mean difference is ≈ -2 °C in the case of the MaxEP model and ≈ -2.5 °C for the IPSL model. However, in the IPSL model the temperature anomaly spreads over a large area in the Northern Hemisphere, while in the MaxEP model, it concentrates over the area where the ice-sheets are.

Table 9.1 compares the global mean surface temperatures obtained using the different models, for both Pre-Industrial and Last Glacial Maximum conditions. Including the interactive Planck spectrum (version v1 compared to version v0) leads to a slight cooling (0.6 °C) and a smaller albedo sensitivity, while turning off the ozone (version v2 compared to version v1) yields a very small warming (0.2 °C) and does not change the sensitivity. Figure 9.6 shows the dependence of

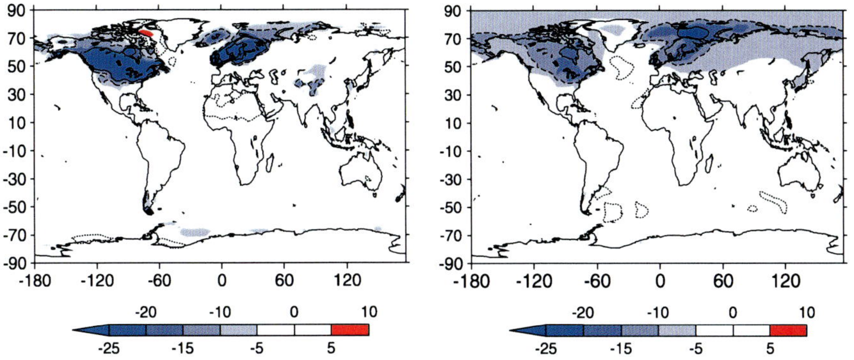

Fig. 9.5 Surface temperature difference between the Last Glacial Maximum and the pre-industrial, in the MaxEP model (*left*, version v0) and in the IPSL model (*right*). Contour lines space is 10 °C, positive contours are drawn in solid lines, negative contours in *dashed lines* and the null contour as a *dotted line*

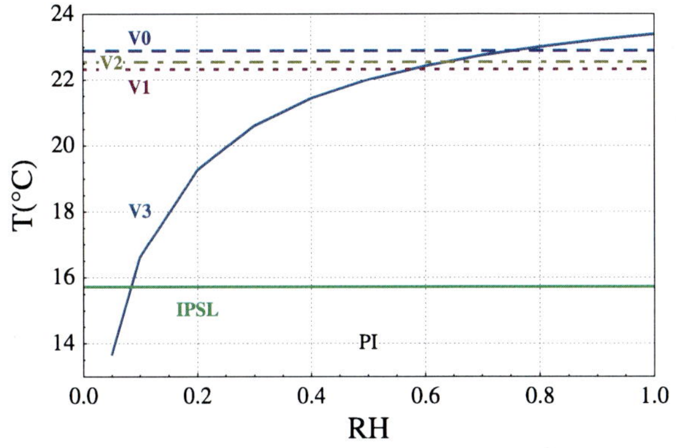

Fig. 9.6 *Solid blue curve* Global mean surface temperature T_g as a function of relative humidity (with a homogeneous distribution). The horizontal lines indicate the temperature obtained by fixing the absolute humidity in the MaxEP model, versions v0 (*dashed blue*), v1 (*dotted red*) and v2 (*dashed-dotted yellow*), and for the IPSL model (*green solid line*)

the global mean surface temperature on relative humidity. For simplicity, a horizontally homogeneous relative humidity distribution is used. The global mean surface temperature spans a wide interval, between approximately 14 and 24 °C. In particular, it encompasses the global mean surface temperature obtained with other versions of the MaxEP model and with the IPSL model.

The latitudinal dependence of surface temperature distributions obtained from the different models[1] is shown in Fig. 9.7, for both pre-industrial and LGM conditions. When the water vapour feedback is active (version v3), the surface

[1] The uniform relative humidity in version 3 is chosen as the mean relative humidity in the MaxEP v0 case.

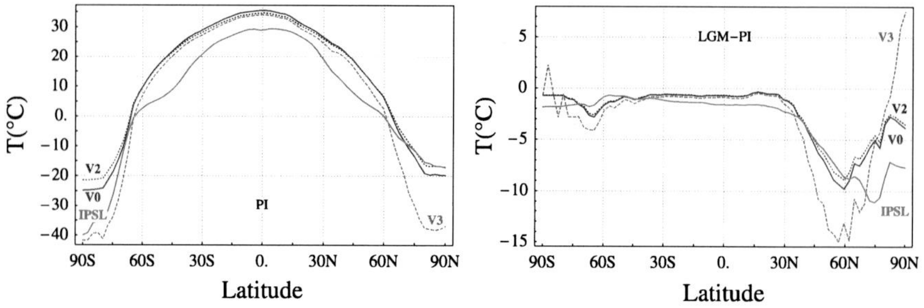

Fig. 9.7 *Left* Surface temperature T_g for pre-industrial conditions, for the different versions of the MaxEP model: version v0 (*solid blue*), v2 (*dotted red*), v3 (*dashed yellow*) and for the IPSL model (*solid green*). *Right* Surface temperature difference between the Last Glacial Maximum and pre-industrial

temperature is much lower in the polar regions than with other versions of the MaxEP model. For the same reason, the response to the albedo change at the LGM is also stronger (Fig. 9.7, right). Globally, the temperature response is approximately 1 °C stronger than in the absence of the water vapour feedback (Table 9.1).

9.5 The Importance of Spatial Resolution

In the MaxEP procedure, it is traditionally argued that maximizing the entropy production constitutes a way to represent the effect of small, unresolved scales, on the large, resolved scales. In the case of meridional heat transport in (dry) planetary atmospheres, the energy is carried partly by the mean flow and partly by turbulent fluctuations. Nevertheless, even a model accounting for no dynamics at all like the MaxEP model shown here presents reasonable transport curves. For the sake of the comparison with the IPSL model, we started with an identical resolution for the GCM and the MaxEP model ($N_{lat} = 72$ and $N_{lon} = 96$, corresponding to a $3.7 \times 2.5°$ grid). In the MaxEP model, the resolution is somewhat arbitrary as the computational cost is negligible. In the light of the interpretation of MaxEP as a parameterization of small-scale processes, one may naturally ask how the results of the MaxEP model depend on the resolution.

Figure 9.8 shows the curves of total material entropy production and globally averaged surface temperature obtained with the MaxEP v0 model with different resolutions. We keep a constant aspect ratio $N_{lat}/N_{lon} = 3/4$ and vary the total number of boxes. Both curves are monotonically increasing with resolution. Although there is no explicit representation of the dynamics here, the dependence on resolution is very similar to the findings of [19] for a GCM. In particular, it shows that the results of the MaxEP model converge when the resolution increase.

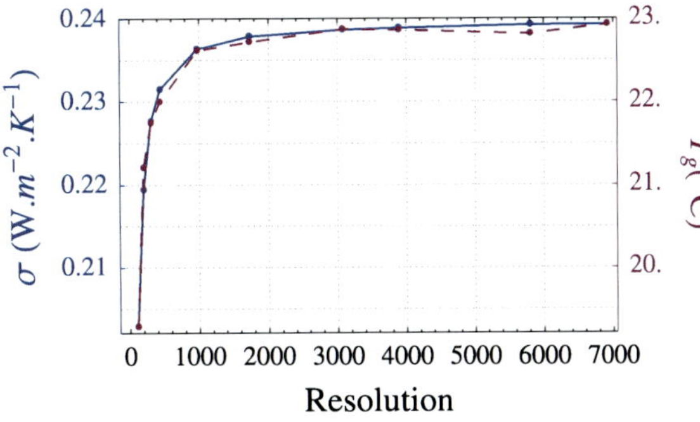

Fig. 9.8 Total material entropy production σ (*solid blue*) and global mean surface temperature $\langle T_g \rangle$ (*dashed red*) as functions of the resolution (number of cells). The aspect ratio is maintained equal to 3/4

9.6 Future Challenges for MaxEP Climate Modelling

In this chapter, we have presented a detailed account of how the MaxEP conjecture can be applied to climate modelling. We have shown how a MaxEP model without *ad-hoc* hypotheses could be built and we have compared its performances in simulating both the pre-industrial and Last Glacial Maximum climates with a coupled atmosphere–ocean GCM. The results appear to be robust with respect to minor modifications (versions v0–v2) of the model. To go beyond these results, we argue that it is necessary to account for some feedbacks, and show how to treat them in the MaxEP framework. We stress the importance of the water vapour feedback (version v3) on the surface temperature. Going further would now require the ability to include a water-cycle model in our MaxEP model. From there one may hope to be able to represent clouds in a more robust way than in the original Paltridge model. To become a realistic climate model, the MaxEP model would still require important features, like a seasonal cycle (see [43]), a representation of atmospheric dynamics, a more accurate description of the vertical structure, etc., but there are reasons to believe that this would not be completely out of reach. This key challenge would have to be taken up without sacrificing the original strengths of the MaxEP model (absence of empirical parameterizations and *ad-hoc* coefficients, rapidity, conceptual simplicity). Another major point which would deserve clarification is the theoretical basis of the MaxEP principle (see Chap. 3). In particular, it would be desirable to establish which entropy production should be maximized: Is it always the material entropy production? (See for instance [21]).

If this program could be achieved, the climate modelling community would acquire a valuable new tool, in addition to the existing hierarchy of models, to improve our understanding of past, present and future climates, on Earth and beyond.

References

1. Peixoto, J.P., Oort, A.H.: Physics of Climate. Springer, New-York (1992)
2. McGuffie, K., Henderson-Sellers, A.: A Climate Modelling primer. John Wiley (2005)
3. Goody, R., Yung, Y.: Atmospheric Radiation: Theoretical Basis. Oxford University Press, Oxford (1995)
4. Holton, J.: An Introduction to Dynamic Meteorology. Academic Press, New York (2004)
5. Pedlosky, J.: Geophysical Fluid Dynamics. Springer, Berlin (1987)
6. Ambaum, M.H.P.: Thermal Physics of the Atmosphere. Wiley, Chichester (2010)
7. Holloway, G.: From classical to statistical ocean dynamics. Surv. Geophys. **25**, 203–219 (2004)
8. Johnson, D.: "General coldness of climate models" and the second law: implications for modeling the earth system. J. Climate **10**, 2826 (1997)
9. Lucarini, V.: Thermodynamic efficiency and entropy production in the climate system. Phys. Rev. E 80, 021118 (2009)
10. Kleidon, A., Lorenz, R. (eds.): Non-equilibrium Thermodynamics and the Production of Entropy: Life, Earth, and Beyond. Springer, Berlin (2005)
11. Martyushev, L., Seleznev, V.: Maximum entropy production principle in physics, chemistry and biology. Phys. Rep. **426**, 1–45 (2006)
12. Ozawa, H., Ohmura, A., Lorenz, R., Pujol, T.: The second law of thermodynamics and the global climate system: a review of the maximum entropy production principle. Rev. Geophys. **41**, 1018 (2003)
13. Bruers, S.: A discussion on maximum entropy production and information theory. J. Phys. A **40**, 7441–7450 (2007)
14. Dewar, R.: Information theory explanation of the fluctuation theorem, maximum entropy production and self-organized criticality in non-equilibrium stationary states. J. Phys. A **36**, 631–641 (2003)
15. Dewar, R.: Maximum entropy production and non-equilibrium statistical mechanics. In: Kleidon, A., Lorenz, R. (eds.) Non-equilibrium Thermodynamics and the Production of Entropy: Life, Earth, and Beyond. Springer, Heidelberg (2004)
16. Grinstein, G., Linsker, R.: Comments on a derivation and application of the 'maximum entropy production' principle. J. Phys. A **40**, 9717–9720 (2007)
17. Ito, T., Kleidon, A.: Entropy production of atmospheric heat transport. In: Kleidon, A., Lorenz, R. (eds.) Non-equilibrium Thermodynamics and the Production of Entropy: Life, Earth, and Beyond. Springer, Heidelberg (2004)
18. Kleidon, A., Fraedrich, K., Kirk, E., Lunkeit, F.: Maximum entropy production and the strength of boundary layer exchange in an atmospheric general circulation model. Geophys. Res. Lett. **33**, 1627–1643 (2006)
19. Kleidon, A., Fraedrich, K., Kunz, T., Lunkeit, F.: The atmospheric circulation and states of maximum entropy production. Geophys. Res. Lett. **30**, 2223 (2003)
20. Kunz, T., Fraedrich, K., Kirk, E.: Optimisation of simplified GCMs using circulation indices and maximum entropy production. Clim. Dyn. **30**, 803–813 (2008)
21. Pascale, S., Gregory, J.M., Ambaum, M.H.P., Tailleux, R.: A parametric sensitivity study of entropy production and kinetic energy dissipation using the FAMOUS AOGCM. Clim. Dyn. **38**, 1211–1227 (2012)
22. Paltridge, G.: Global dynamics and climate-a system of minimum entropy exchange. Q. J. R. Meteorol. Soc. **101**, 475–484 (1975)
23. Rodgers, C.: Comments on Paltridge's "minimum entropy exchange" principle. Q. J. R. Meteorol. Soc. **102**, 455–457 (1976)
24. Gerard, J., Delcourt, D., Francois, L.: The maximum entropy production principle in climate models: application to the faint young sun paradox. Q. J. R. Meteorol. Soc. **116**, 1123–1132 (1990)

25. Grassl, H.: The climate at maximum entropy production by meridional atmospheric and oceanic heat fluxes. Q. J. R. Meteorol. Soc. **107**, 153–166 (1981)
26. Paltridge, G.: The steady-state format of global climate. Q. J. R. Meteorol. Soc. **104**, 927–945 (1978)
27. Wyant, P., Mongroo, A., Hameed, S.: Determination of the heat-transport coefficient in energy-balance climate models by extremization of entropy production. J. Atmos. Sci. **45**, 189–193 (1988)
28. Lorenz, R., Lunine, J., Withers, P., McKay, C.: Titan, Mars and Earth: Entropy production by latitudinal heat transport. Geophys. Res. Lett. **28**, 415–418 (2001)
29. Jupp, T.E., Cox, P.: MEP and planetary climates: insights from a two-box climate model containing atmospheric dynamics. Phil. Trans. R. Soc. B **365**, 1355–1365 (2010)
30. Herbert, C., Paillard, D., Kageyama, M., Dubrulle, B.: Present and Last Glacial Maximum climates as states of maximum entropy production. Q. J. R. Meteorol. Soc. **137**, 1059–1069 (2011)
31. Lacis, A., Hansen, J.: A parameterization for the absorption of solar radiation in the earth's atmosphere. J. Atmos. Sci. **31**, 118–133 (1974)
32. Stephens, G.: The parameterization of radiation for numerical weather prediction and climate models. Mon. Wea. Rev. **112**, 826–867 (1984)
33. Dufresne, J.L., Fournier, R., Hourdin, C., Hourdin, F.: Net Exchange Reformulation of Radiative Transfer in the CO_2 15 μm Band on Mars. J. Atmos. Sci. **62**, 3303–3319 (2005)
34. McClatchey, R., Selby, J., Volz, F., Fenn, R., Garing, J.: Optical properties of the atmosphere. Air Force Camb. Res., Lab (1972)
35. Renno, N.: Multiple equilibria in radiative-convective atmospheres. Tellus A **49**, 423–438 (1997)
36. Pierrehumbert, R.: The hydrologic cycle in deep-time climate problems. Nature **419**, 191 (2002)
37. Lenton, T., Held, H., Kriegler, E., Hall, J.W., Lucht, W., Rahmstorf, S., Schellnhuber, H.: Tipping elements in the earth's climate system. Proc. Natl. Aca. Sci. U.S.A. **105**, 1786–1793 (2008)
38. Roe, G., Baker, M.: Why is climate sensitivity so unpredictable? Science **318**, 629 (2007)
39. Herbert, C., Paillard, D., Dubrulle, B.: Entropy production and multiple equilibria: the case of the ice-albedo feedback. Earth Syst. Dynam. **2**, 13–23 (2011)
40. Marti, O., Braconnot, P., Dufresne, J.L., Bellier, J., Benshila, R., Bony, S., Brockmann, P., Cadule, P., Caubel, A., Codron, F., de Noblet-Decoudre, N., Denvil, S., Fairhead, L., Fichefet, T., Foujols, M.A., Friedlingstein, P., Goosse, H., Grandpeix, J.Y., Guilyardi, E., Hourdin, F., Idelkadi, A., Kageyama, M., Krinner, G., L′evy, C., Madec, G., Mignot, J., Musat, I., Swingedouw, D., Talandier, C.: Key features of the IPSL ocean atmosphere model and its sensitivity to atmospheric resolution. Clim. Dyn. **34**, 1–26 (2010)
41. IPCC: Climate Change 2007: The Physical Science Basis. Contribution of Working Group I to the Fourth Assessment Report of the Intergovernmental Panel on Climate Change. Cambridge University Press, Cambridge, United Kingdom and New York, NY, USA (2007)
42. Crowley, T.J., North, G.R.: Paleoclimatology. Oxford University Press, Oxford (1996)
43. Paillard, D., Herbert, C.: Maximum entropy production and time varying problems: the seasonal cycle in a conceptual climate model. Entropy **15**, 2846–2860 (2013)

Chapter 10
Thermodynamic Insights into Transitions Between Climate States Under Changes in Solar and Greenhouse Forcing

Robert Boschi, Valerio Lucarini and Salvatore Pascale

Abstract A detailed thermodynamic, sensitivity analysis of the steady state climate system is performed with respect to the solar constant S^* and the carbon dioxide concentration of the atmosphere, [CO_2]. Using PlaSim, an Earth-like general circulation model of intermediate complexity, S^* is modulated between 1,160 and 1,510 Wm^{-2} for values of [CO_2] ranging from 90 to 2,880 ppm. It is observed that in a wide parameter range, which includes the present climate conditions, the climate is multistable, i.e. there are two coexisting attractors, one characterised by warm, moist climates (*W*) and the other by a completely frozen sea surface (Snowball Earth, *SB*). For both sets of states, empirical relationships for surface temperature, material entropy production, meridional energy transport, Carnot efficiency and dissipation of kinetic energy are constructed in the parametric plane ([CO_2], S^*). Linear relationships are found for the two transition lines ($W \rightarrow SB$ and $SB \rightarrow W$) in ([CO_2], S^*) between S^* and the logarithm of [CO_2]. The dynamical and thermodynamical properties of *W* and *SB* are completely different. *W* states are dominated by the hydrological cycle and latent heat is prominent in the material entropy production. The *SB* states are mainly dry climates where heat transport is realized through sensible heat fluxes and entropy mostly generated by dissipation of kinetic energy. It is also shown that the Carnot-like efficiency regularly increases towards each transition between *W* and *SB* and that each transition is associated with a large decrease of the Carnot efficiency indicating a restabilisation of the system. Furthermore, it has been found that in *SB* states, changes in the vertical temperature structure are responsible for the observed changes in the meridional transport.

R. Boschi (✉) · V. Lucarini · S. Pascale
Meteorologisches Institut, Klima Campus, University of Hamburg, Hamburg, Germany
e-mail: robert.boschi@zmaw.de

V. Lucarini
Department of Mathematics and Statistics, University of Reading, Reading, UK

R. C. Dewar et al. (eds.), *Beyond the Second Law*,
Understanding Complex Systems, DOI: 10.1007/978-3-642-40154-1_10,

10.1 Introduction

Probably the most notable examples of climate change events occurred during the Neoproterozoic (period spanning from 1,000 to 540 million years ago), when the Earth is believed to have suffered two of its most severe periods of glaciation [16] and entered into what is often referred to as a snowball Earth (*SB*) climate state. The SB climate is characterized by an almost completely ice covered planet with global temperatures well below 0 °C and an extremely dry atmosphere. This period coincided with large carbon dioxide fluctuations, while the solar constant (about 1,365 Wm^{-2} in present conditions) is believed to have been 94 % of current levels, rising to 95 % by the end of the Neoproterozoic [15, 49].

The two main factors effecting the concentration of atmospheric CO_2 are biotic activities and volcanism. Volcanic eruptions bring about very sudden and dramatic increases in CO_2 concentration ([CO_2]). This is a short-lived process and provides a means by which to exit a *SB* climate state by increasing the opacity of the atmosphere and enhancing the greenhouse effect. By contrast, biospheric effects tend to occur more gradually as biotic activity and atmospheric composition are coupled so that large fluctuations in the carbon pools occur over relatively long time scales.

The effect of *SB* events on the biosphere is believed to have been disastrous. Carbon-isotope ratios characteristic of Earth's mantle [19, 23] rather than of life processes, were recorded immediately below and above the glacial deposits, implying that oceanic photosynthesis was effectively non-existent during *SB* events. The result of this and anoxic conditions beneath the ice should have lead to the disappearance of most forms of life except bacteria. The final disappearance of snowball conditions since the Neoproterozoic may have been the main contributing factor in the development of complex multi-cellular life that began around 565 million years ago.

Based on the evidence supported by [18, 19], it is therefore expected that the Earth is potentially capable of supporting multiple steady states for the same values of some parameters such as the solar constant and [CO_2], which directly affect the radiative forcing. It is important therefore to explore this hypothesis, due to the relevance for the history of our planet but also to help understand other planets capabilities for supporting life.

Initial research using simple 0-D models [2, 51], 1-D models [3, 13] as well as more recent analyses performed using complex 3-D general circulation models [39, 49, 56], provide support for the existence of such bistability. The $SB \rightarrow W$ and $W \rightarrow SB$ transitions tend to occur in an abrupt rather than a smooth transition. The main mechanism triggering such abrupt transitions is the positive ice-albedo feedback [2, 54]. Such a feedback is associated with the fact that as temperatures increase, the extent of snow and ice cover decreases thus reducing the albedo and therefore increasing the amount of sunlight absorbed by the Earth system. Conversely, a negative fluctuation in the temperature leads to an increase in the albedo therefore reinforcing the cooling.

The presence of such catastrophic climate shifts [1] suggest the existence of a global bifurcation in the climate system for certain combinations of its descriptive parameters [8]. The loss of stability realized in the $W \rightarrow SB$ and $SB \rightarrow W$ transitions is related to the catastrophic disappearance of one of the two attractors describing the two possible climatic states, as a result of a set of complicated bifurcations.

Starting from present conditions, the most obvious physical parameters to modulate in order to bring about the transition to the *SB* state is the solar constant [37, 38]. Even if other model experiments [56] show that a decrease in CO_2 alone can bring about transition to the *SB* state, though this requires a reduction of more than 80 %, compared to a decrease of less than 10 % for the solar luminosity. The Neoproterozoic however highlights the importance of considering changes in the levels of $[CO_2]$ as a mechanism for the transitions to and from the *SB* climate state, and therefore the dramatic impact it can have on the overall state of the climate system. It is therefore interesting to alter both the solar luminosity and the atmospheric opacity as these are two important parameters affecting the overall properties of the system. If one wants to explore extensively the parametric space of climate steady states, it is therefore necessary to consider a wide range of values for both of these parameters. The originally of the work present in the following sections, comes from exploring both the solar and greenhouse forcings together but also from our analysis of the transitions between both states of the climate system from the point of view of the Carnot-like efficiency.

When using a complex climate model to study transitions, it is important to choose the correct physical observables to provide information about the global properties of the system. The temperature, which is the variable traditionally investigated in climate sciences, gives an overall view of the state of the system but does not give immediate information about the processes occurring within it. In addition to temperature, it is therefore important to consider diagnostic quantities that provide information on the behaviour of processes occurring within the system. Since the climate system is in a non-equilibrium state [5, 25] the best way is to approach this problem from a non-equilibrium thermodynamics point of view. This means introducing diagnostic tools which complement the more traditional diagnostics based on classical climatological fields as temperature, precipitation and winds. It will be discovered at the end of this study that the temperature is a good physical observable for the climate as it is intimately interconnected with the thermodynamical quantities.

Recently a great deal of work has been carried out in studying climate irreversibility and entropy production in the climate system associated with dissipative processes [9, 11, 14, 24, 42–46]. In particular, a recent study by [37] focused on investigating the multiple steady states of *SB* and *W* and the $SB \rightarrow W$ and $W \rightarrow SB$ transitions in terms of the thermodynamic properties of the system. The $[CO_2]$ was fixed at present levels while the solar constant was varied. Initially, S^* was decreased from present levels until transition to the *SB* state (see Fig. 10.1a and b). The solar constant was again increased moving right along the blue line until the transition back up to the red line. The area marked by the solid and dashed lines

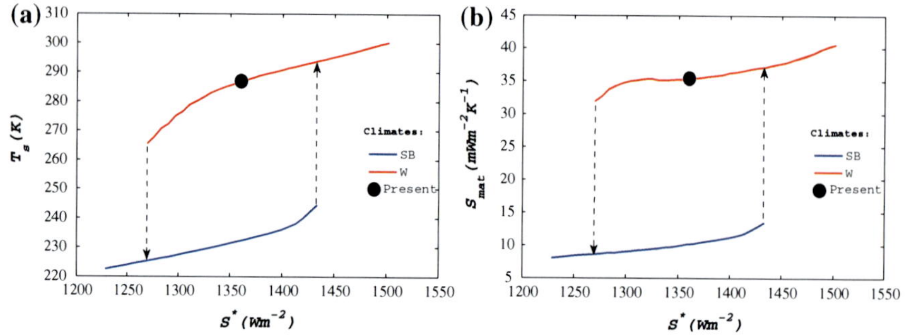

Fig. 10.1 (**a**) Surface temperature and (**b**) material entropy production for steady states obtained for different values of the solar constant S^* while maintaining [CO_2] at 360 ppm. The present climate is marked with a *black circle*, the warm (*W*) states in *red* and the snowball (*SB*) states in *blue*. From [38]. *W* states are warm moist climates similar to our own while the *SB* states are climates very dry atmospheres, with an Earth's surface almost completely covered in ice

indicates the hysteresis of the climate and was found in terms of the surface temperature, T_S and material entropy production, $\dot{S}_{mat}$. As shown in Fig. 10.1a and b, $\dot{S}_{mat}$ is an even clearer indicator of the $W \rightarrow SB$ and $SB \rightarrow W$ transitions than temperature, because while temperature difference between two coexisting states is of the order of 50 K (or 20 %), in the case of the material entropy production the *W* state features values larger by a factor of 4 than the corresponding *SB* state.

The work presented in this chapter builds on and unifies the work done by [37] with the analysis performed in [38], where [CO_2] variations alone are considered, in order to obtain a more complete picture of how radiative and dynamical processes are coupled in a vast range of climates. Specifically, a detailed thermodynamic, parametric sensitivity study of the steady state climate system is performed with respect to S^* and [CO_2]. Using PlaSim, a general circulation model of intermediate complexity [10], we study the climate states realised when the solar constant is modulated between 1,160 and 1,510 Wm^{-2} and the values of [CO_2] are varied between 90 and 2,880 ppm. Our aim here is to produce a simulation-based reconstruction of the global structural properties of the climatic attractors. For both *W* and *SB* states we compute surface temperature, material entropy production, meridional energy transport, Carnot efficiency [22, 35] and dissipation of kinetic energy and propose empirical relationships in the parametric plane ([CO_2], S^*). We will look for an empirical relation for the two transition lines ($W \rightarrow SB$ and $SB \rightarrow W$) in the parametric plane between S^* and the natural logarithm of [CO_2] which marks the boundaries of the hysteresis in the climate system. The aforementioned quantities will be used to explain changes in large-scale climate behaviour and the effect of climate change on features such as stratification and baroclinicity in order to understand changes in the meridional heat transport across the parameter range. It will also be shown that the Carnot-like efficiency has a key role in defining the stability of the system, which is related to abrupt climatic shifts. Note, our work differs from most other work done on the snowball state, in

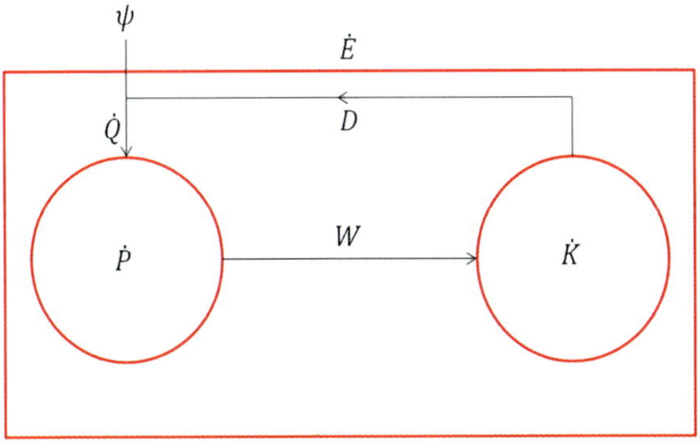

Fig. 10.2 Schematic figure showing the two main classes of energy: Potential (*P*) and Kinetic (*K*); and the transfer processes (ψ, W, D) which exchange energy between them. $\dot{E}$ is the total energy within the system and $\dot{Q}(=\psi + D)$, is the total heating rate due to dissipation (D) and the convergence of heat fluxes (ψ). W is the work done by the system. ψ includes the external radiative input to the climate system from the sun, therefore $\dot{E} = \psi$

the sense that we explore the snowball state using the Earth's present land configuration as opposed how it was during the Neoproterozoic era.

The chapter is structured in the following way: in Sect. 10.2 we will describe details of the non-equilibrium thermodynamics of the climate and the diagnostic tools used. Section 10.3 will describe the PlaSim climate model and the steps taken in the simulation procedure, the results of which are discussed in Sects. 10.4 and 10.5.

10.2 Non-equilibrium Thermodynamics of the Climate

In this section we recapitulate some thermodynamic properties of the climate system and introduce the notation used throughout this chapter. We follow what has been previously presented in [35]. If the Earth's global climate system (surface and atmosphere) is encompassed by a domain Ω, the total energy budget is given by $E(\Omega) = P(\Omega) + K(\Omega)$, where K represents the total kinetic energy and P is the moist static potential energy, in which we adopt the usual Lorenz approach of combining the contributions from the thermal (including latent heat) and potential energy [22, 46]. The time derivative of K and P can be found to be $\dot{K} = -D + W$ and $\dot{P} = \psi + D - W$, where D is the dissipation and therefore always positive, W is the instantaneous work done by the system and ψ which is the heating due to convergence of turbulent heat fluxes and radiative heat, such that $\dot{E} = \psi$ (see Fig. 10.2). The dependence on Ω has been dropped for convenience. The total heating rate can therefore be written as $\dot{Q} = \psi + D$. Considering the climate as a non-equilibrium steady state system (NESS, see [12]), over long time scales $\bar{\dot{E}} = \bar{\dot{P}} = \bar{\dot{K}} = 0$ (the bar indicates averaging over long time periods).

Let us define $\dot{q}$ as the local heating rate so that $\dot{q} = \rho\left(\epsilon^2 - \vec{\nabla}.\mathbf{H}\right)$ [37] where $\epsilon^2 > 0$ is the local rate of heating due to viscous dissipation of kinetic energy and $\mathbf{H}$ is given by the sum of turbulent heat fluxes plus radiative energy fluxes, such that over the whole domain, $D = \int_\Omega dV\rho\varepsilon^2$ and $\psi = \int_\Omega dV\rho\vec{\nabla}.\mathbf{H}$, respectively. Dividing Ω into two subdomains: Ω^+, where $\dot{q} = \dot{q}^+ > 0$, and Ω^- in which $\dot{q} = \dot{q}^- < 0$, we find that $\dot{q}^+$ plus $\dot{q}^-$ integrated over Ω equals the derivative of total heating ($\dot{\mathrm{Q}}$) due to dissipation (D) and the convergence of heat fluxes ($\dot{\psi}$) :

$$\psi + D = \dot{P} + W = \int_{\Omega^+} dV\rho\dot{q}^+ + \int_{\Omega^-} dV\rho\dot{q}^- = \dot{\mathrm{Q}}^+ + \dot{\mathrm{Q}}^- = \dot{\mathrm{Q}}, \tag{10.1}$$

where the quantities $\dot{\mathrm{Q}}^+$ and $\dot{\mathrm{Q}}^-$ are positive and negative at all times, respectively. Since dissipation is positive definite, $-\bar{\dot{K}} + \bar{W} = \bar{D} = \bar{\dot{P}} + \bar{W} = \bar{W} = \overline{\dot{\mathrm{Q}}^+} + \overline{\dot{\mathrm{Q}}^-} > 0$.

On spatial scales far smaller than Ω itself, it is practical to assume local equilibrium (local thermodynamic equilibrium hypothesis, [5]) so that locally $\dot{q} = \dot{s}T$ with $\dot{s}$ the time derivative of the entropy density. The total rate of change of the entropy of the system is:

$$\dot{S} = \int_{\Omega^+} dV\rho\frac{\dot{q}^+}{T} + \int_{\Omega^-} dV\rho\frac{\dot{q}^-}{T} = \int_{\Omega^+} dV\rho\dot{s}^+ + \int_{\Omega^-} dV\rho\dot{s}^- = \dot{\Sigma}^+ + \dot{\Sigma}^-, \tag{10.2}$$

where $\dot{\Sigma}^+ > 0$ *and* $\dot{\Sigma}^- < 0$. Using Eq. (10.2) and assuming that the Earth system is in a steady state, over a long time average, $\overline{\dot{\Sigma}^+} = -\overline{\dot{\Sigma}^-}$ as $\bar{\dot{S}} = 0$. Therefore, $2\overline{\dot{\Sigma}^+} = \overline{\int_\Omega dV\rho|\dot{s}|}$, so that $\overline{\dot{\Sigma}^+}$ measures the absolute value of the entropy fluctuations throughout the domain.

When integrating over the whole domain and considering long time averages, we have the following equivalent expressions for the thermodynamic quantities: $\overline{\dot{\mathrm{Q}}^+} = \overline{\dot{\Sigma}^+}\Theta^+$ and $\overline{\dot{\mathrm{Q}}^-} = \overline{\dot{\Sigma}^-}\Theta^-$, where Θ^+ and Θ^- are the time and space averaged temperatures of the Ω^+ and Ω^- domains respectively. These expressions are only valid if correlations between T and $\dot{s}^+$ or $\dot{s}^-$ are ignored. Since $\left|\overline{\dot{\Sigma}^+}\right| = \left|\overline{\dot{\Sigma}^-}\right|$ and $\left|\overline{\dot{\mathrm{Q}}^+}\right| > \left|\overline{\dot{\mathrm{Q}}^-}\right|$, it can be shown that $\Theta^+ > \Theta^-$, i.e. absorption typically occurs at higher temperature than release of heat [21, 22, 46]. The Earth's climate system can be considered like a Carnot heat engine. For conceptual purposes, in its simplest form, we can think of the atmosphere as being a fluid which exchanges heat between two thermal reservoirs: from a region of net warming mostly in Tropics and extra Tropics, to a region of net cooling in the extra-tropics and the poles. The Carnot–like efficiency of the system can therefore be defined as:

$$\eta = \frac{\int_{\Omega^+} dV\rho\dot{q}^+ + \int_{\Omega^-} dV\rho\dot{q}^-}{\int_{\Omega^+} dV\rho\dot{q}^+} = \frac{\overline{\dot{\mathrm{Q}}^+} + \overline{\dot{\mathrm{Q}}^-}}{\overline{\dot{\mathrm{Q}}^+}} = \frac{\bar{W}}{\overline{\dot{\mathrm{Q}}^+}}, \tag{10.3}$$

where, the work done by the Carnot engine of the climate system is found to be, $\bar{W} = \eta\overline{\dot{Q}^+}$.

As shown in [31, 32]—and clarified in [22]—the long term average of the work performed by the system is equal to the long-term average of the generation of available potential energy, as typical of forced-dissipative steady states. The Earth exists in a steady state maintained far from equilibrium by net radiative heating at the equator and net cooling at the poles which results in its observed vertical and meridional heat transports. This gives rise to on-going irreversible processes, including phase transitions in H_2O and frictional dissipation, which are characterized by a positive entropy production. The entropy production due to the irreversibility of the processes occurring within the climatic fluid is called the material entropy production, $\dot{S}_{mat}$ and can be written in general terms as:

$$\overline{\dot{S}_{mat}} = \overline{\int_\Omega \frac{\varepsilon^2}{T} dV} + \overline{\int_\Omega \vec{F}_{SH} \cdot \vec{\nabla}\frac{1}{T} dV} + \overline{\int_\Omega \vec{F}_{LH} \cdot \vec{\nabla}\frac{1}{T} dV}, \tag{10.4}$$

where the first, second, and third terms on the RHS are related to the dissipation of kinetic energy, and to the transport of sensible and latent heat respectively. We now wish to link the terms of the entropy budget in Eq. (10.2) with those of the entropy production in Eq. (10.4).

The second law of thermodynamics states that the entropy variation of a system at temperature, T receiving an amount of heat δq is larger than or of equal to $\delta q/T$ [29]. In this case:

$$\begin{aligned}\overline{\dot{S}_{mat}(\Omega)} \geq \overline{\dot{S}_{min}(\Omega)} &= \overline{\left(\frac{\int_\Omega dV\rho\dot{q}}{\int_\Omega dV\rho T}\right)} \\ &= \overline{\left(\frac{\dot{Q}^+ + \dot{Q}^-}{\Theta}\right)} \approx \frac{\overline{\dot{Q}^+} + \overline{\dot{Q}^-}}{\overline{\langle\Theta\rangle}} \approx \frac{\overline{\dot{Q}^+} + \overline{\dot{Q}^-}}{(\Theta^+ + \Theta^-)/2} = \frac{\bar{W}}{(\Theta^+ + \Theta^-)/2},\end{aligned} \tag{10.5}$$

where $\overline{\dot{S}_{mat}(\Omega)}$ is the long-term average of the material entropy production, $\overline{\dot{S}_{min}(\Omega)}$ is its lower bound, i.e. the minimal value of the entropy production compatible with the presence of a Lorenz energy cycle with average intensity $\overline{\dot{W}}$ and $\langle\Theta\rangle$ is the density averaged temperature of the system. The approximation holds as long as we can neglect the impact of the cross-correlation between the total net heat balance and the average temperature and we can assume that $\langle\Theta\rangle$ can be approximated by the mean of the two Carnot temperatures Θ^+ and Θ^-.

This is because on Earth, the regions of net atmospheric warming are well approximated to be centered around the Equator between about [30°S, 30°N] i.e. the Tropics, with all other regions associated with net cooling. Therefore, the total mass in each of these regions is comparable. The boundaries between these two regions, the tropical/extratropical transition, are located at the peak of the meridional energy transport, which [55], has shown is constrained to be close to what is observed today, for a vast array of climates. The dominating factor which sets the latitudinal extent of the region of net heating is the incident angle of solar radiation [6]. This implies that we can assume the regions of net heating and cooling are comparable in all climate scenarios considered here, assuming the atmosphere remains relatively transparent to SW radiation.

Note, in the real atmosphere, the vertical structure of the zonally averaged heating pattern is not homogeneous, in large part, due to the contribution of latent heat. In particular, we observe regions of net heating in the lowest third of the troposphere extending towards the poles by up to 60° latitude, while in the top two-thirds, net heating is mostly seen out to about 15° away from the equator [38]. The masses of the two portion of the atmosphere, characterized by net heating and net cooling, respectively, are almost the same. We can therefore explicitly write $\overline{\dot{S}_{min}(\Omega)}$ as:

$$\begin{aligned} \overline{\dot{S}_{min}(\Omega)} &\approx \frac{\bar{\dot{\mathrm{W}}}}{(\Theta^+ + \Theta^-)/2} = \frac{\eta\overline{\dot{\Phi}^+}}{(\Theta^+ + \Theta^-)/2} \\ &= \eta\frac{\Theta^+}{(\Theta^+ + \Theta^-)/2}\overline{\dot{\Sigma}^+} = \eta\frac{1}{1-\eta/2}\overline{\dot{\Sigma}^+} \approx \eta\overline{\dot{\Sigma}^+}, \end{aligned} \tag{10.6}$$

where the last approximation holds as long as $\eta \ll 1$, which applies in the case of the climate system. Therefore, η sets also the proportionality factor relating the lower bound to the entropy production of the system $\overline{\dot{S}_{min}(\Omega)}$—due to macroscopically irreversible processes—to the absolute value of the entropy fluctuations inside the system due to macroscopically reversible heating or cooling processes. Note that if the system is isothermal and at equilibrium the internal entropy production is zero, since $\eta \to 0$. The lower bound to the material entropy production corresponds to the contribution coming from the dissipation of kinetic energy through viscous processes. Therefore, the average material entropy production can be expressed as $\overline{\dot{S}_{mat}} = \overline{\dot{S}_{min}} + \overline{\dot{S}_{exc}}$, where $\overline{\dot{S}_{exc}}$ is the excess of entropy production with respect to the minimum, which results from the heat transport down the temperature gradient [35]. We can define:

$$\alpha \approx \frac{\overline{\dot{S}_{exc}}}{\overline{\dot{S}_{min}}} \approx \frac{\int_\Omega dV\overline{\mathbf{H}.\vec{\nabla}\left(\frac{1}{T}\right)}}{\frac{\bar{\dot{\mathrm{W}}}}{\langle\Theta\rangle}} \geq 0, \tag{10.7}$$

as a parameter of the irreversibility of the system, which is zero if all the production of entropy is due to the—unavoidable—viscous dissipation of the mechanical energy. As $\overline{\dot{S}_{mat}} \approx \eta\overline{\dot{\Sigma}^+}(1+\alpha)$, we have that the entropy production is maximized if we have a joint optimization of heat transport and the production of mechanical work. Note that, if heat transport down the temperature gradient is very strong, the efficiency η is small because the difference between the temperatures of the warm and cold reservoirs is greatly reduced (the system is almost isothermal), whereas, if the transport is very weak, the factor α is small.

10.3 Simulation Procedure

The dynamical and thermodynamical properties of the Earth's climate are studied here using PlaSim [11], a climate model of intermediate complexity, freely available at http://www.mi.uni-hamburg.de/plasim. Its dynamical core is formulated using the primitive equations for vorticity, divergence, temperature and the logarithm of surface pressure, solved using the spectral transform method [7, 41], where the prognostic variables are represented as the sum of a series of complex exponential functions. Unresolved processes for long [50] and short [28] wave radiation, shallow, moist [26, 27] and dry convection, cloud formation [52–54] and large scale precipitation, latent and sensible heat boundary layer fluxes, horizontal and vertical diffusion [30, 33, 34] are parameterized.

The model is coupled to a 50 m deep mixed layer ocean which contains a thermodynamic sea-ice model. The advantage of using a slab ocean as opposed to a full ocean is that it allows for the climate system to reach a stead state in less than 35 years after a change in e.g. the solar constant. With full ocean coupling, the integration time of the model and the time needed to reach a steady state would be an order of magnitude larger [56].

We wish to emphasize that whereas most state-of-the-art general circulation models feature considerable energy imbalances, as highlighted by [36], the energy bias is of the order of 0.5 Wm^{-2}, almost an order of magnitude less than in most GCMs and an entropy diagnostic is available [11], thus making it well suited for this work.

The model is run at T21 resolution (approximately 5.6° × 5.6°) with 10 vertical levels. Modulating S^* with respect to [CO_2] of 90, 180, 270, 360, 540, 720, 1,080, 1,440, 2,160 and 2,880 ppm, we are able to reconstruct the *SB* and *W* climate states. The procedure occurs as follows for each of the considered values of [CO_2]:

1. the model is initially run to a *W* steady state for 100 years with S^* equal to 1,415 Wm^{-2};
2. S^* is decreased by a small amount for each value of [CO_2] and the model run is continued until a steady state is reached;
3. step 2 is repeated until S^* is reduced to 1,165 Wm^{-2}; the point of $W \rightarrow SB$ transition is noted down;
4. the reverse operation is then performed with S^* increased by intervals of 15 Wm^{-2}, from 1,165 Wm^{-2} up to a value of 1,510 Wm^{-2}, each time allowing the system to reach a steady state; the point of $SB \rightarrow W$ transition is noted down.

Further to this, we identify the position of the transition to a higher resolution than the rest of the parameter range in the direction of S^*. For values of S^* within 10 Wm^{-2} before the transition, S^* is decreased in intervals of 1 Wm^{-2}, each time permitting 50 years for the system to reach a steady state, until after the transition is observed.

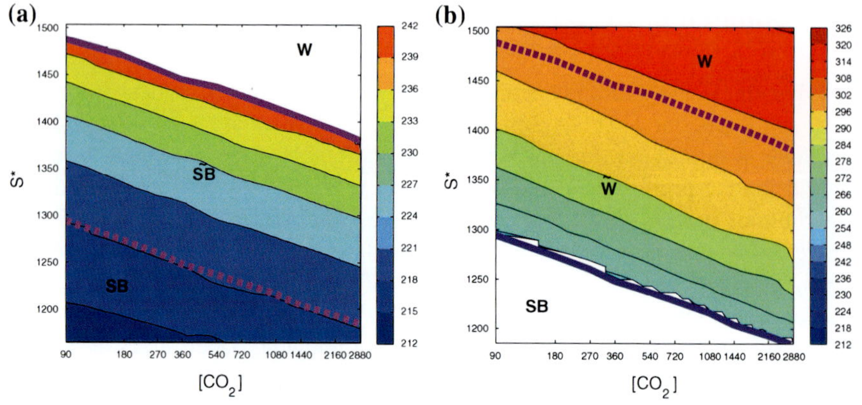

Fig. 10.3 Contour plot of surface temperature (*K*) as a function of S^* and $[CO_2]$. The *lower SB* (**a**) and *upper W* (**b**) manifolds are shown. The transition *SB* → *W* and *W* → *SB* are shown by the *upper* and *lower purple lines* respectively. The *solid purple line* indicating the active transition

10.4 Results: Hysteresis, Bistability and Regime Boundaries in a Parametric Space

10.4.1 Temperature and Entropy Production

Initially, focus is put on analysing the parametric plane ($[CO_2]$, S^*), referred to as the CS space, which in the following shall be considered in terms of global mean surface temperature, T_s. The transition zones between the main climate states are clearly defined from the dependence of surface temperature on $[CO_2]$ and S^*. Note, the qualitative properties of the climate system in the CS space, namely the presence of bistability between the *SB/W* states, can be reconstructed from any observable of the climate state, but it is most instructive to select first the surface temperature because it is also experimentally most relevant. The change in surface temperature through the CS space is illustrated in Fig. 10.3a and b.

We identify two main climatic regimes, observed as two distinct manifolds ($[CO_2]$, S^*, T_s) and characterized by a sharp change in the profile of T_s when jumping from one manifold to another. We refer to these as the upper and lower manifolds, representative of the *W* and *SB* regimes respectively. As would be expected, there is a monotonic increase of temperature with increasing $[CO_2]$ or S^* on both manifolds [47, 48, 56]. The temperature range on the *SB* and *W* manifolds are 212–242 K, and 254–326 K respectively, over the parameter range. Note, due to the different temperature ranges the colour scaling of Fig. 10.3a and b is a factor of 4 different, with both scales starting from the same lowest value. The temperature range of the bistable region in the *SB* and *W* regimes are 218–242 K and 254–300 K respectively, meaning that the rate of change of surface temperature over the same range of S^* and $[CO_2]$ is approximately double in the *W* regime,

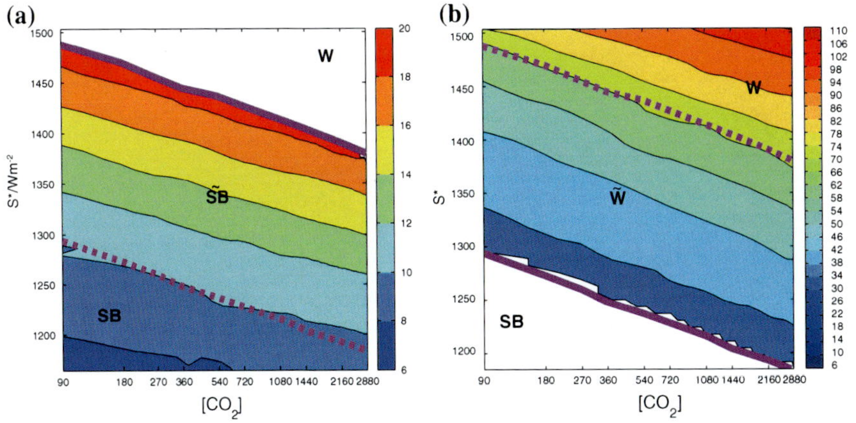

Fig. 10.4 Contour plot of $\overline{\dot{S}_{mat}}$ (mW m^{-2} K^{-1}) as a function of S^* and [CO_2] for the *lower SB* (**a**) and *upper W* (*b*) manifolds. The *SB* → *W* and *W* → *SB* transitions are shown by the *top* and *bottom purple lines* respectively

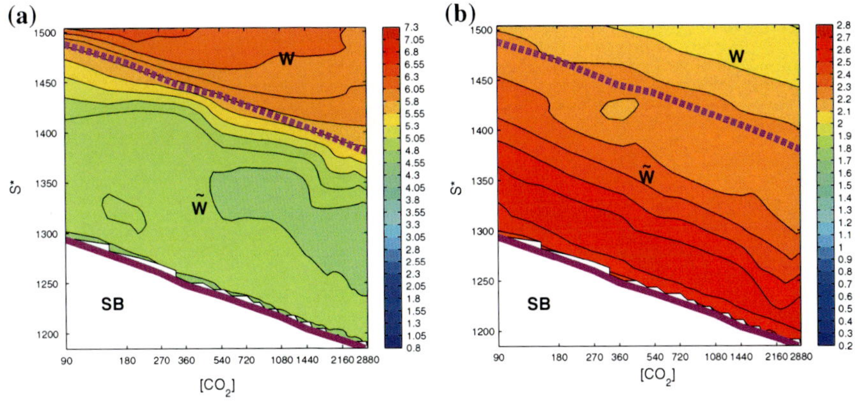

Fig. 10.5 Contour plot of: (**a**) meridional heat transport and (**b**) Lorenz energy cycle or dissipation as a function of S^* and [CO_2] for the *W* states

with respect to the *SB* regime and that the surface temperature difference between the two manifolds ranges between 40 and 60 K.

The *W* states (upper manifold) exists only in the region of the CS space above the *W* → *SB* transition line (the position of this line is expresses as $S^* = S^*_{wsb}$) whereas the *SB* states (lower manifold) only in the CS region below the *SB* → *W* transition line (the position of this line is expressed as $S = S^*_{sbw}$). Such lines, which are well separated and approximately parallel, are illustrated as solid and dashed purple lines on Figs. 10.3, 10.4, 10.5, 10.6, 10.7, 10.8 and 10.9 and have been found within an accuracy of 2 Wm^{-2} of the solar constant. The solid

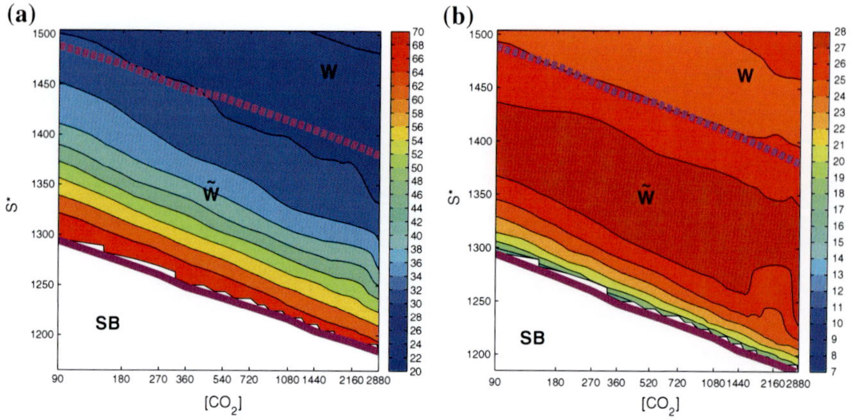

Fig. 10.6 Contour plot of: (**a**) meridional temperature gradient (*K*) and (**b**) midlatitude vertical temperature difference as a function of S^* and $[CO_2]$ for the *W* states

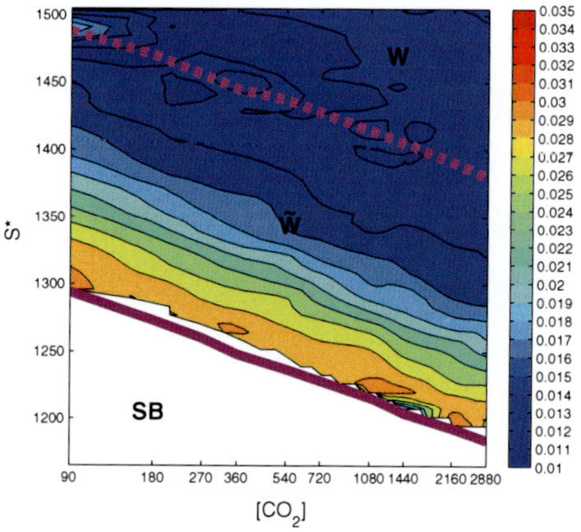

Fig. 10.7 Contour plot of the efficiency as a function of S^* and $[CO_2]$ for the *W* states

purple lines indicate the 'active' transition, dependent on which manifold the climate system lies in at that moment i.e. the active transition for the *SB* and *W* states are $SB \rightarrow W$ and $W \rightarrow SB$ respectively. The dashed lines illustrate the location of the 'inactive' transition, when the climate system exists in the alternative state. The bistable region is therefore located between the dashed and solid purple lines. As a result, a property of the system is that regardless of which combination of $[CO_2]$ and S^* is used, the transition from one state to another always occurs at almost exactly the same temperature. This indicates that the climate system has a low sensitivity to the mechanism of forcing.

The position of the two boundaries can be parameterised in terms of S^* and $[CO_2]$ as:

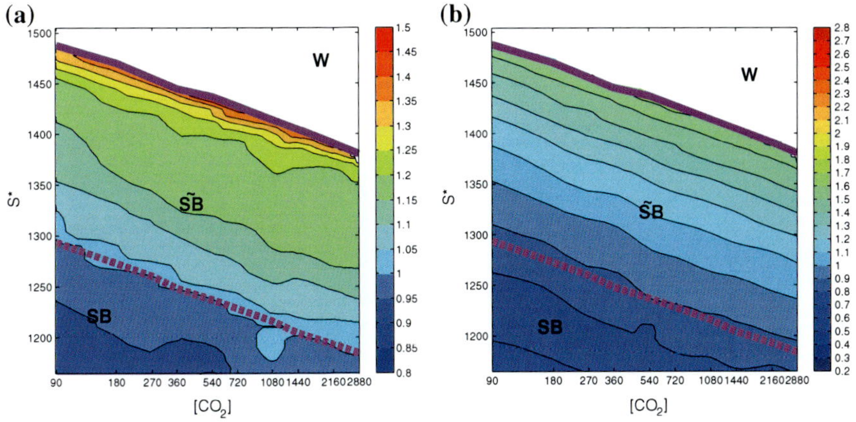

Fig. 10.8 Contour plot of: (**a**) meridional heat transport (*K*) and (**b**) Lorenz energy cycle or dissipation as a function of S^* and $[CO_2]$ for the *SB* states

$$S^*_{sbw} = a_{sbw} \log_{10}[CO_2] + C_{sbw}, \quad S^*_{wsb} = a_{wsb} \log_{10}[CO_2] + C_{wsb}, \tag{10.8}$$

where $a_{sbw} \approx a_{wsb} \approx -72\ \text{Wm}^{-2}$, $C_{sbw} \approx 1{,}629\ \text{Wm}^{-2}$ and $C_{wsb} \approx 1{,}438\ \text{Wm}^{-2}$ for the transition $SB \to W$ and $W \to SB$ respectively and $[CO_2]$ is expressed in ppm. The size of the bistable region, which we define as B, along $S*$ can therefore be defined by the difference between C_{sbw} and C_{wsb}:

$$B = C_{sbw} - C_{wsb}. \tag{10.9}$$

It is found that B is approximately 200 Wm^{-2}. The displacement between the position of the boundaries gives a precise measure of the hysteretic properties of the climate [2, 37, 51, 56] since it indicates the size of the overlap between the two manifolds in the CS plane.

The presence of a bistable region implies that when we change the values of S^* and $[CO_2]$ from an initial to a final value, the final steady state depends on the initial steady state and on the change of path in S^* and $[CO_2]$.

Let us assume that we start from an initial point $\left([CO_2]_0, S^*_0\right)$ in the bistable region of the W state. Let us also assume we perform a closed path when varying S^* and $[CO_2]$, so that $S^*_0 = S^*_f$ and $[CO_2]_0 = [CO_2]_f$. If the path does not cross S^*_{wsb} the final state will be identical to the initial one, that is, in an averaged sense:

$$T_s\left([CO_2]_0, S^*_0\right) = T_s\left([CO_2]_f, S^*_f\right). \tag{10.10}$$

On the other hand if the closed path crosses the transition line to the second manifold, the final state will be different from the initial:

$$T_s\left([CO_2]_0, S^*_0\right) \neq T_s\left([CO_2]_f, S^*_f\right). \tag{10.11}$$

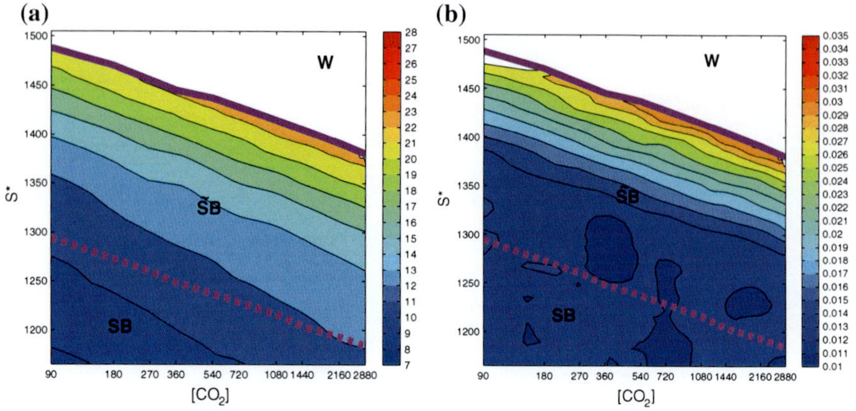

Fig. 10.9 Contour plot of: (**a**) mid-latitude vertical temperature gradient and (**b**) the Carnot-like efficiency as a function of S^* and $[CO_2]$ for the *SB* states

Furthermore, if the closed path crosses first S^*_{wsb} and then S^*_{wsb}, then once again we find $T_s([CO_2]_0, S^*_0) = T_s\left([CO_2]_f, S^*_f\right)$, since the system has performed first a $W \rightarrow SB$, and then a $SB \rightarrow W$ transition. The same applies starting from a *SB* state and exchanging *SB* with *W* in the previous discussion. This is true for any climate diagnostic. More specifically, in the case of T_s, for $W \rightarrow SB$ and $SB \rightarrow W$ transitions,

$$T_s([CO_2]_0, S^*_0) > T_s\left([CO_2]_f, S^*_f\right) \text{ and } T_s([CO_2]_0, S^*_0) < T_s\left([CO_2]_f, S^*_f\right) \quad (10.12)$$

respectively. Note that for $S^* > 1{,}438\ \mathrm{Wm}^{-2}$, even if $[CO_2]$ is 0 ppm, no transition to *SB* state can occur.

Figure 10.4a and b show the analogous behaviour for the material entropy production, $\dot{S}_{mat}$ in CS space, computed directly as described in [11]. As with temperature, $\dot{S}_{mat}$ increases monotonically with increasing S^* and $[CO_2]$ on both manifolds. In the *SB* state, $\dot{S}_{mat}$ is mostly generated by dissipation of kinetic energy and irreversible sensible heat transport, because the planet is almost entirely dry. For the *W* manifold the main contribution to $\dot{S}_{mat}$ comes from latent heat due to large scale and convective precipitation. In the bistable region the range of $\dot{S}_{mat}$ is (10, 19) $\mathrm{Wm}^{-2}\ \mathrm{K}^{-1}$ and (34, 62) $\mathrm{Wm}^{-2}\ \mathrm{K}^{-1}$ for the *SB* and *W* respectively, therefore a factor of 3 larger in the W regime respect to the *SB* regime. This confirms that $\dot{S}_{mat}$ may be a better indicator than temperature for discriminating between the *SB* and *W* states as already discussed in [37].

It is possible to parameterise $\dot{S}_{mat}$ as a function of both the solar constant, S^* and the logarithm of the $[CO_2]$ (the response of the climate to $[CO_2]$ increase is logarithmic, [40]) as:

$$\overline{\dot{S}_{mat}} = C + aS^* + b\log_{10}(CO_2) \quad (10.13)$$

Table 10.1 The coefficients of the empirical relationship (10.13) relating T_s, $\dot{S}_{mat}$ and the meridional heat transport to S^* and [CO_2]

Variable	Coefficients		
	C	b	a
T_S SB [K]	227 (K)	0.498 (K W^{-1} m^2)	0.00899 (K)
T_S W [K]	283 (K)	1.364 (K W^{-1} m^2)	0.01427 (K)
Mer. heat transport SB (PW)	2 (PW)	0.011 (m^2)	0.00026 (PW)
Mer. heat transport W (PW)	10 (PW)	0.031 (m^2)	0.00059 (PW)
$\dot{S}_{mat}$ *SB* (mW $m^{-2}K^{-1}$)	13 (mW m^{-2} K^{-1})	0.147 (K^{-1})	0.00358 (mW m^{-2} K^{-1})
$\dot{S}_{mat}$ *W* (mW m^{-2} K^{-1})	49 (mW m^{-2} K^{-1})	1.0 (K^{-1})	0.00492 (mW m^{-2} K^{-1})

where the coefficients are indeed different for the *W* and *SB* state and are reported in Table 10.1.

In the bistable region, the *SB* and *W* states are quantitatively very different with respect to their physical properties. Both disjoint attractors can thus be thought of as representing two different worlds, with completely different dynamical and thermodynamical properties. Therefore we treat them separately and then describe how the system makes a transition between both states. For this reason in the following two sub-sections dynamical and thermodynamical properties of the manifolds will be analysed individually in terms of the vertical and horizontal surface temperature differences, Carnot efficiency, meridional heat transport and dissipation of kinetic energy. Furthermore we shall relate these properties to the average mean global temperature and the material entropy production. As is illustrated by the solid and dashed purple lines, in the CS space Figures, each manifold is divided up into two sub regions: *W*, *W/Bistable* $(\tilde{W})$ on the upper manifold and *SB*, *SB/Bistable* $(\widetilde{SB})$ on the lower. Then, in a third section, the transitions between both manifolds occurring in the bifurcation regions are analyzed.

10.4.1.1 The Warm State

The meridional heat transport profile is worked out as explained in [36] and from this an index is defined as the mean of the magnitude between the Northern and Southern hemisphere maxima. In a moist atmosphere, the average global temperature and the meridional surface temperature difference (see Fig. 10.6a), defined as the difference between the mean surface temperature of the tropics (30°S, 30°N) and the polar regions ([90°S, 60°S] and [60°N, 90°N]), are the main contributing factors for controlling the meridional heat transport (see Fig. 10.5a). This is due to the fact that temperature controls the latent heat released in the atmosphere due to the Clausius-Clapeyron effect [17] and the meridional temperature gradient controls baroclinicity of the atmosphere [55]. Additionally, another modulating factor is the vertical stratification of the atmosphere, as conditions of low stratification in the mid-latitudes support stronger baroclinic activity for a given meridional temperature gradient [20]. In the bistable region of the warm sector, the meridional heat transport has a flat response to increasing S^* and $[CO_2]$ and therefore T_s. With increased T_s, water vapour concentration of the atmosphere increases, thus leading to the strengthening of the poleward latent heat fluxes. In addition, the increased T_s causes sea and continental ice as well as seasonal snow cover to retreat towards the poles, thus lowering the surface albedo gradient. This contributes negatively to changes in the meridional heat transport, through a decrease in the baroclinicity.

In the *W* regime, the boundary between the bistable and the monostable regime approximately marks the point at which the Earth surface loses its permanent sea-ice cover, thus supporting the idea that the presence of bistability is intrinsically linked with the powerful ice-albedo feedback. For T_s larger than approximately 300 K, the meridional temperature gradient decreases at a far slower rate with increasing T_s. Therefore, in this region the meridional heat transport is controlled only by the availability of water vapour in the atmosphere. This means, at temperatures above 300 K, the meridional temperature difference becomes decoupled from the surface meridional heat transport. This analysis consequently shows the importance of the hydrological cycle as a major contributing factor to the magnitude of the meridional heat transport and moreover, it indicates that when going from warm to very warm climates the hydrological cycle becomes the dominant climatic feature, leading to strong positive dependence of the meridional heat transport on the surface temperature. Our results agree with the findings of [4], who found in aqua planet simulations, there is little scope for reducing the meridional temperature gradient further once the sea-ice and snow have melted. For a constant meridional temperature gradient, an increase in the meridional latent heat fluxes with increasing global mean temperature was shown.

The conclusions drawn above find further support when looking at the mid-latitudes vertical temperature difference, ΔT_v, defined as the mean temperature difference between the surface and the 500 hPa level (see Fig. 10.6b). The vertical temperature difference is largest along a band of the CS space centered half way through the bistable region in the direction of S^*. Our current climate conditions would appear to be positioned at the centre along a band where such temperature differences are at their largest, implying conditions of reduced vertical stratification. For colder climates, increasing surface temperature causes the melting of sea-ice and of seasonal snow cover, so that the ensuing decrease in surface albedo (leading to increased surface absorption) accounts for the increasing vertical temperature difference. Instead, in warmer climates, the decrease in equatorial vertical temperature difference with T_s can be understood in terms of increased moist convection from warmer surface temperatures, resulting in an increase of moisture fluxes to the upper atmosphere, which then condense and release latent heat.

In the bistable region, for the reasons discussed above, we expect to find a pronounced weakening of the dynamics of the climate system with increasing surface temperature. We test this hypothesis by computing the strength of the Lorenz energy cycle (Fig. 10.5b), which is equal to the average rate of dissipation of kinetic energy, and the Carnot-like efficiency of the system (Fig. 10.7), which measures, instead, how far the system is from equilibrium. The dissipation, similarly to the meridional temperature gradient, decreases monotonically with the T_s, and reaches its largest value just before the $W \rightarrow SB$ transition boundary of the W manifold. The efficiency is maximized before the $W \rightarrow SB$ transition and decreases monotonically with increasing [CO_2] or S^*: warmer climates are characterised by smaller temperature differences, since the transport of water vapour acts as a very efficient means for homogenising the temperature across the system. Therefore, the system has lower ability to produce mechanical work and is characterised by very strong irreversible processes, as described by the very large values of $\dot{S}_{mat}$ shown in the W region of the CS space on Fig. 10.4b. A high value in the efficiency is closely related to the maximization of the temperature gradients. Comparison of the CS space figures show the Carnot efficiency to align much more closely with changes in the meridional rather than the vertical temperature differences, implying it is the meridional temperature gradients which contribute most actively in dictating the amount of work done by the climate system in the $\tilde{W}$ region.

10.4.1.2 The Snowball State

The SB state is intrinsically simpler than the W state because the hydrological cycle has a negligible influence. This is due to the fact that atmospheric temperatures are so low that the atmosphere in all cases is almost dry. Moreover, in the SB state the meridional gradients (and not only the globally averaged values) of albedo are very low and depend weakly on S^* and the [CO_2], because under all conditions the sea surface is frozen almost everywhere and the continents are covered by ice and snow. The meridional heat transport (Fig. 10.8a) is much smaller than for the corresponding W states, because in the SB state a large fraction of the radiation is reflected back to space and the albedo gradients are small, so that the energy imbalance between low and high latitudes is small.

Throughout the SB state, increases in S^* and [CO_2], which lead to an increase in the surface temperature and in the rate of entropy production are accompanied by increases in the meridional heat transport and in the dissipation of kinetic energy (Fig. 10.8b). This tells us that with increased meridional heat transport the intensity of the circulation also increases. However we find that the meridional temperature difference (not shown) has a very weak dependence on changing [CO_2] and S^*, as it varies by only 4K across the explored parameter range, so that the changes in the baroclinicity of the system cannot be due to changes in the meridional heat transport and in the intensity of the Lorenz energy cycle. The lack of large variations in the meridional surface temperature profile are essentially due to the fact that the net input of shortwave radiation is rather fixed by the constant surface albedo and minimal cloud cover. The system therefore can be seen as rather rigid, as changes in the absorbed radiation are almost exactly compensated by changes in the meridional heat transport. The re-equilibration mechanism must then contain elements, which are not present in the classic baroclinic adjustment [54]. We find that the mid-latitudes vertical temperature gradient (Fig. 10.9a) increases substantially with increased value of S^* and [CO_2], as an effect of the increased absorption of radiation near the surface. The reduced vertical stratification leads to more pronounced baroclinic activity even for a fixed meridional temperature gradient, thus leading to an increase in the meridional heat transport and the intensity of the Lorenz energy cycle. The argument is made more straightforward because we are considering an almost dry atmosphere. The efficiency has a dependence on S^* and [CO_2] which, as in the *W* case, is most closely related to the meridional temperature gradient field. In the case of the *SB* state, the main signature is given by the vertical gradient (Fig. 10.9a). This further reinforces the idea that the investigation of meridional temperature gradients is not enough to grasp the mechanisms through which the system generates available potential energy and material entropy production [36].

10.4.1.3 Transition and Comparison Between Manifolds

Up to now, the *W* and *SB* states have been characterized as two entirely distinct climate regimes, and have underlined that the basic mechanisms of re-equilibration are rather different. In this subsection we would like to present some ideas aimed at making sense of the transitions between the two states occurring when we get close to the boundary of the upper or lower manifold. As has been seen, the $W \rightarrow SB$ transition is associated with a large decrease in surface temperature, rate of material entropy production, and meridional heat transport. This is intimately related to the fact that whereas in the *W* state the hydrological cycle is a major contributor to the climate dynamics, in the *SB* state the hydrological cycle is almost absent. Nonetheless, this does not say much about the processes leading from one state to the other, or better still, describing how one of the attractors disappears.

We have also discovered that the usual dynamical indicators of the atmospheric state, i.e. the meridional temperature gradient and the vertical stratification do not necessarily indicate whether or not we are close to an irreversible transition of the system, e.g. by signaling something equivalent to a loss of "elasticity" of the system. In this regard, it is much more informative to observe how the efficiency behaves near the transitions. We find that, as a general rule, each transition is associated with a notable decrease (more than 30 %) of the efficiency of the system (see Figs. 10.7 and 10.9b), and the closer the system gets to the transition in the CS space, the larger is the value of the efficiency. This can be interpreted as follows. If the system approaches a bifurcation point, its positive feedbacks become relatively strong compared to the negative feedbacks, which act as re-equilibrating mechanisms, and become less efficient. As a result, the differential heating driving the climate is damped less effectively, and the system is further from equilibrium, since larger temperature differences are present. Therefore, the system produces more work, thus featuring an enhanced Lorenz energy cycle and a stronger circulation. At the bifurcation point, the positive feedbacks prevail and the circulation, even if rather strong, is not able to cope with the destabilising processes, and transition to the other manifold is realised. The new state is, by definition, more stable, and thus closer to equilibrium. The decreased value of the efficiency is exactly the marker of this property. This confirms what has been proposed by [37] in much greater generality and can be conjectured to be a rather general property of non-equilibrium systems featuring structural instabilities.

10.5 Summary and Conclusions

Motivated by paleoclimatic evidence and the goal to attain a parametric theory of climate change, which extends beyond the usual analysis of sensitivities around the present climate, in this contribution, we have studied how the stability properties of the climate system depend on the modulation of the two main parameters describing the radiative forcing, i.e. the solar constant S^* and the [CO_2]. In our analysis we propose that the point of view of non-equilibrium thermodynamics is especially useful for understanding the global properties of the climate system and for interpreting its global instabilities.

We have discovered that in a rather wide parameter range, which includes the present climate conditions, the climate is multistable, i.e. there are two coexisting attractors, one characterised by warm conditions, where the presence of sea ice and seasonal snow cover is limited (*W* state), and one characterised by a virtually completely frozen sea surface, the so-called snowball (*SB*) state. These qualitative and structural properties, obtained using the PlaSim climate model, confirm and extend what has been previously found in various studies using models of varying degrees of complexity. We point the reader to [49] and [37] for an extensive discussion. In this regard, the main improvement of this work is that a two dimensional parameter space is explored (whereas usually variations in the solar constant or in the opacity of the atmosphere are considered separately), which allows the gathering of more complete information on the possible states of the climate system that seem relevant for the actual mechanisms which play a crucial role in a paleoclimatological context, as explained in [49].

For all considered values of the [CO_2], which range from 90 to 2,880 ppm, the width of the bistable region is about 200 Wm^{-2} in terms of the value of the solar constant, and its position depends linearly on the logarithm of the [CO_2], being centered around smaller values of the solar constant for increasing opacity of the atmosphere, shifting by about 15 Wm^{-2} per doubling of [CO_2]. The *W* state is characterized by surface temperatures 40–60 K higher than in the *SB* state, and by values of material entropy production which are larger by a factor of 3–4 (order of

40–60 mW m^{-2} vs. 10–15 mW m^{-2}). The boundaries of the bistable region are approximately isolines of the globally averaged surface temperature, and in particular, the warm boundary, beyond which the *SB* state cannot be realized, is characterized by vanishing permanent sea ice cover in the *W* regime. This reinforces the idea that the ice-albedo feedback is the dominant mechanism for the multistability properties.

The thermodynamical and dynamical properties of the two states are very different, as if we were discussing two entirely different planets. In the *W* state the climate is dominated by the hydrological cycle and latent heat fluxes are prominent in terms of redistributing the energy in the system and as contributors to the material entropy production. The *SB* state is predominantly a dry climate, where heat transport is realized through sensible heat fluxes and entropy is mostly generated through the dissipation of the kinetic energy. The dryness of the *SB* atmospheres also explains why the climate sensitivity is much smaller.

In the *W* state, the meridional heat transport is rather constant throughout the bistable region, as the contrasting effect of the enhancement of latent heat fluxes driven by increasing surface temperature and the reduction in the baroclinicity due to the decrease in the meridional temperature gradient compensate almost exactly. In the warm range, beyond the bistability region, the meridional heat flux increases with the surface temperature, the reason being that the compensating albedo mechanisms are shut off as sea ice is completely removed from the surface.

In the *SB* state, increased incoming radiation or increased [CO_2] lead to increases in the meridional heat transport. In this case, the water vapor plays no role, and, somewhat surprisingly, the meridional temperature difference has also a rather flat response to the parameter modulation. In this case, the dominant mechanism determining the properties of the meridional heat transfer is the change in the vertical stratification, which becomes weaker for warmer climate conditions. This implies that the atmospheric circulation strengthens for increasing values of the solar constant and of the [CO_2]. In fact, the strength of the Lorenz energy cycle becomes stronger for warmer climate conditions, and the Carnot-like efficiency of the climate system has an analogous behaviour.

The opposite holds for the *W* state, where the intensity of the Lorenz energy cycle and the efficiency decrease for warmer conditions, the reason being that the water vapor becomes more and more efficient in homogenizing the system and destroying its ability to generate available potential energy.

A general property we have found is that, in both manifolds, the efficiency increases when we get closer to the bifurcation point and at the bifurcation point the transition to the newly realized stationary state is accompanied by a decrease in the efficiency. This can be framed in a rather general thermodynamical context: the efficiency gives a measure of how far the system is from equilibrium. The negative feedbacks tend to counteract the differential heating due to the sun's insolation pattern, thus leading the system closer to equilibrium. At the bifurcation point, the negative feedbacks are overcame by the positive feedbacks, so that the system makes a global transition to a new state, where, in turn, the negative feedbacks are more efficient in stabilizing the system.

The results discussed in this chapter support the adoption of new diagnostic tools for validating climate models based on the second law of thermodynamics. The next step in this direction is a more quantitative understanding of the global relationships between surface temperature, material entropy production, meridional heat fluxes and Carnot-like efficiency for the *SB* and *W* states and to propose possible parameterisations for the *SB* and *W* attractors. Another line of research will explore the dependence of these quantities on other fundamental parameters, e.g. the rotation rate and the surface drag, relevant for planetary atmospheres of terrestrial planets.

Acknowledgments *RB, VL and SP acknowledges the financial support of KlimaCamus (Hamburg), CLISAP and the EU-ERC project NAMASTE "Thermodynamics of the climate system". The authors thank F. Ragone, F. Lunkeit for help and insightful comments.*

References

1. Arnol'd, V.I.: Catastrophe Theory, 3rd edn. Springer, Berlin (1992)
2. Budyko, M.I.: The effect of solar radiation variations on the climate of the Earth. Tellus **21**, 611–619 (1969)
3. Caldeira, K., Kasting, J.F.: Susceptibility of the early Earth to irreversible glaciation caused by carbon dioxide clouds. Nature **359**, 226–228 (1992)
4. Caballero, R., Langen, P.L.: The dynamic range of poleward energy transport in an atmospheric general circulation model. Geophys. Res. Lett. **32**, L02705 (2005). doi:10.1029/2004GL021581
5. de Groot, S.R., Mazur, P.: Non-equilibrium Thermodynamics, Dover, New York (1984)
6. Donohoe, A., Battisti, D.S.: What determines meridional heat transport in climate models? J. Clim. **25**, 3832–3850 (2012)
7. Eliasen, E., Machenhauer, B., Rasmussen, E.: On a numerical method for integration of the hydrodynamical equations with a spectral representation of the horizontal fields. Report no. 2, Institute for Theoretical Meteorology, Copenhagen University, Denmark (1970)
8. Fraedrich, K.: Catastrophes and resilience of a zero-dimensional climate system with ice-albedo and greenhouse feedback. Q. J. R. Meteorol. Soc. **105**, 147–167 (1979)
9. Fraedrich, K.: A suite of user-friendly global climate models: hysteresis experiments. Eur. Phys. J. Plus 127 (2012). doi:10.1140/epjp/i2012-12053-7
10. Fraedrich, K., Lunkeit, F.: Diagnosing the entropy budget of a climate model. Tellus A. **60**, 921–931 (2008)
11. Fraedrich, K., Jansen, H., Kirk, U., Luksch, U., Lunkeit, F.: The planet simulator: towards a user friendly model. Meteor. Z. **14**, 299–304 (2005)
12. Gallavotti, G.: Nonequilibrium statistical mechanics (stationary): overview. In: Francois, J.-P., Naber, G.L., Tsun, T.S. (eds.) Encyclopedia of Mathematical Physics, pp. 530–539. Elsevier, Amsterdam (2006)
13. Ghil, M.: Climate stability for a sellers-type model. J. Atmos. Sci. **33**, 3–20 (1976)
14. Goody, R.: Sources and sinks of climate entropy. Q. J. R. Meteor. Soc. **126**, 1953–1970 (2000)
15. Gough, D.O.: Solar interior structure and luminosity variations. Sol. Phys. **74**, 21–34 (1981)
16. Grinstein, G., Linsker, R.: Comments on a derivation and application of the 'maximum entropy production' principle. J. Phys. A: Math. Theor. **40**, 9717 (2007). doi:10.1088/1751-8113/40/31/N01

17. Held, I.M., Soden, B.J.: Robust responses of the hydrological cycle to global warming. J. Climate **19**, 5686–5699 (2006)
18. Hoffman, P.F., Schrag, D.P.: The snowball Earth hypothesis: testing the limits of global change. Terra Nova **14**, 129–155 (2002)
19. Hoffman, P.F., Kaufman, A.J., Halverson, G.P., Schrag, D.P.: A Neoproterozoic snowball Earth. Science. **281**, 1342–1346 (1998). doi:10.1126/science.281.5381.1342
20. Holton, J.R.: 2004. An introduction to dynamic meteorology, IV edn. Elsevier, New York
21. Johnson, D.R.: General coldness of the climate models and the second law: implications for modelling the Earth system. J. Climate **10**, 2826–2846 (1997)
22. Johnson, D.R.: Entropy, the Lorenz energy cycle, and climate. In: Randall, D.A. (ed.) General Circulation Model Development: Past, Present and Future, Internation Geophysics Series, vol. 70, pp. 659–720. Academic Press, New York (2000)
23. Kennedy, M.J., Runnegar, B., Prave, A.R., Hoffmann, K.H., Arthur, M.A.: Two or four neoproterozoic glaciations? Geology **26**, 1059–1063 (1998)
24. Kleidon, A., Lorenz, R.D. (eds.): Nonequilibrium Thermodynamics and Maximum Entropy Production: Life, Earth and Beyond, pp. 260. Springer, Berlin (2005)
25. Kondepudi, D., Prigogine, I.: Modern Thermodynamics: From Heat Engines to Dissipative Structure. Wiley, England (1998)
26. Kuo, H.L.: On formation and intensification of tropical cyclones through latent heat release by cumulus convection. J. Atmos. Sci. **22**, 40–63 (1965)
27. Kuo, H.L.: Further studies of the parameterization of the influence of cumulus convection. J. Atmos. Sci. **22**, 40–63 (1974)
28. Lacis, A.A., Hanson, J.: A parameterization for the absorption of solar radiation in the Earth's atmosphere. J. Atmos. Sci. **31**, 118–133 (1974)
29. Landau, L.D., Lifshitz, E.M.: Statistical Physics, Part 1. Pergamon, Oxford (1980)
30. Laursen, L., Eliasen, E.: On the effects of the damping mechanisms in an atmospheric general circulation model. Tellus A. **41**, 385–400 (1989)
31. Lorenz, E.N.: Available potential energy and the maintenance of the general circulation. Tellus **7**, 157–167 (1955)
32. Lorenz, E.N.: The nature and theory of the general circulation of the atmosphere. WMO Bullettin, April 1967. World Meteorological Organization, Geneva (1967)
33. Louis, J.-F.: A parametric model of vertical eddy fluxes in the atmosphere. Bound.-Layer Meteorol. **17**, 187–202 (1979)
34. Louis, J.-F., Tiedtke, M., Geleyn, J.-F.: A short history of the PBL parameterization at ECMWF. In: Proceedings of the ECMWF Workshop on planetary boundary layer parameterization, Reading, 25–27 November 1981. pp. 59–80 (1982)
35. Lucarini, V.: Thermodynamic efficiency and entropy production in the climate system. Phys. Rev. E **80**, 021118 (2009)
36. Lucarini, V., Ragone, F.: Energetics of climate models: net energy balance and meridional enthalpy transport. Rev. Geophys. **49**, RG1001 (2011). doi:10.1029/2009RG000323
37. Lucarini, V., Fraedrich, K., Lunkeit, F.: Thermodynamic analysis of snowball Earth hysteresis experiment: efficiency, entropy production, and irreversibility. Q. J. R. Meteorol. Soc. **136**, 2–11 (2010)
38. Lucarini, V., Fraedrich, K., Lunkeit, F.: Thermodynamics of climate change: general sensitivities. Atmos. Chem. Phys. **10**, 9729–9737 (2010)
39. Marotzke, J., Botztet, M.: Present-day and ice-covered equilibrium states in a comprehensive climate model. Geophys. Res. Lett. **34**, L16704 (2007). doi:10.1029/2006GL028880
40. Myhre, G., Highwood, E.J., Shine, P.K., Stordal, F.: New estimates of the radiative forcing due to well mixed greenhouse gas. Geo. Res. Letters **25**(14), 2715–2718 (1998)
41. Orszag, S.A.: Transform method for the calculation of a vector-coupled sums: application to the spectral form of the vorticity equation. J. Atmos. Sci. **27**, 890–895 (1970)
42. Pascale, S., Gregory, J., Ambaum, M., Tailleux, R.: Climate entropy budget of the HadCM3 atmosphere-ocean general circulation model of FAMOUS, its low-resolution version. Clim. Dyn. **36**(5–6), 1189–1206 (2011)

43. Pascale, S., Gregory, J., Ambaum, M., Tailleux, R.: A parametric sensitivity study of the entropy production and kinetic energy dissipation using the FAMOUS AOGCM. Clim. Dyn. (2011). doi:10.1007/s00382-011-0996-2 (in press)
44. Pauluis, O., Held, I.M.: Entropy budget of an atmosphere in radiative–convective equilibrium. Part I: maximum work and frictional dissipation. J. Atmos. Sci. **59**, 125–139 (2002)
45. Pauluis, O., Held, I.M.: Entropy budget of an atmosphere in radiative–convective equilibrium. Part II: latent heat transport and moist processes. J. Atmos. Sci. **59**, 140–149 (2002)
46. Peixoto, J.P., Oort, A.H., de Almeida, M., Tome, A.: Entropy budget of the atmosphere. J. Geophys. Res. **96**(D6), 10981–10988 (1991)
47. Pierrehumbert, R.T.: High levels of atmospheric carbon dioxide necessary for the termination of global glaciations. Nature **429**, 646–649 (2004)
48. Pierrehumbert, R.T.: Climate dynamics of a hard snowball Earth. J. Geophys. Res. **110**, D01111 (2005). doi:10.1029/2004JD005162
49. Pierrehumbert, R.T., Abbot, D.S., Voigt, A., Koll, D.: Climate of the Neoproterozoic. Ann. Rev. Earth Planet Sci. **39**, 417–460 (2011)
50. Sasamori, T.: The radiative cooling calculation for application to general circulation experiments. J. Appl. Meteorol. **7**, 721–729 (1968)
51. Sellers, W.D.: A global climatic model based on the energy balance of the Earth-atmosphere system. J. Appl. Meteorol. **8**, 392–400 (1969)
52. Slingo, A., Slingo, J.M.: Response of the national center for atmospheric research community climate model to improvements in the representation of clouds. J. Geophys. Res. **96**, 15341–15357 (1991)
53. Stephens, G.L.: Radiation profiles in extended water clouds. 2: parameterization schemes. J. Atmos. Sci. **35**, 2123–2132 (1978)
54. Stephens, G.L., Ackerman, S., Smith, E.A.: A shortwave parameterisation revised to improve cloud absorption. J. Atmos. Sci. **41**, 687–690 (1984)
55. Stone, P.H.: Baroclinic adjustment. J. Atmos. Sci. **35**, 561–571 (1978)
56. Voigt, A., Marotzke, J.: The transition from the present-day climate to a modern snowball Earth. Clim. Dyn. **35**, 887–905 (2010)

Chapter 11
Entropy Production in Planetary Atmospheres and Its Applications

Yosuke Fukumura and Hisashi Ozawa

Abstract Distributions of temperature and longwave radiation are predicted from a state of maximum entropy production (MaxEP) due to meridional heat flux in the atmospheres of the Earth, Mars, Titan and Venus, and the predicted distributions are compared with observational results. In the predictions, we use a multi-box energy balance model that takes into account the effects of obliquity and latitudinal variation of albedo on shortwave absorption. It is found that the predicted distributions are generally in agreement with observations of the Earth, Titan and Venus, suggesting the validity of the MaxEP state for these planets. In the case of Mars, the predicted distributions do not agree well with the observations when compared with those predicted from a state of no meridional heat flux. A simple analysis on advective heat flux using a two-box model shows that the Martian atmosphere is so scant that it cannot carry the heat energy that is necessary for the MaxEP state by advection. These results suggest that the validity of the MaxEP state for a planetary atmosphere is limited when the total amount of atmosphere is not enough to sustain the advective heat flux that is necessary for the MaxEP state.

11.1 Introduction

Planetary atmospheres are known to be inherently turbulent because of their large length-scales and velocities, resulting in difficulty in estimating heat fluxes and temperature distributions from dynamic equations. In this respect, the maximum entropy production (MaxEP) principle is expected to be valid for the mean states of planetary atmospheres in this chapter. The MaxEP principle was first suggested by Zeigler [1] as a thermodynamic variational principle for nonlinear, non-equilibrium

Y. Fukumura · H. Ozawa (✉)
Graduate School of Integrated Arts and Sciences, Hiroshima University,
Higashi-Hiroshima 739-8521, Japan
e-mail: hozawa@hiroshima-u.ac.jp

R. C. Dewar et al. (eds.), *Beyond the Second Law*,
Understanding Complex Systems, DOI: 10.1007/978-3-642-40154-1_11,

systems. Later studies show that the MaxEP state is consistent with steady states of a variety of natural phenomena, including the global climate of the Earth [2–4], those of other planets [5], thermal convection [6], turbulent shear flow [7], oceanic general circulation [8, 9], and granular flows [10]. While the underlying physical mechanism is still debated, the MaxEP state is shown to be identical to a state of maximum generation of available energy [4, 11]. Moreover, several theoretical studies suggest that the MaxEP state is the most probable state that is realized by nonlinear, non-equilibrium systems [12–14]. Among phenomenological aspects of the MaxEP principle, the work by Lorenz et al. [5] is attractive because of the inherent simplicity of their model. The remarkable feature of this model is that it consists of only two regions (equator and pole) and it contains only a few model parameters. One can easily estimate the atmospheric conditions (temperature, radiation, and heat flux) at the MaxEP state using this simple model. However, the spatial resolution is crucially limited in their model, resulting in difficulty in comparing the estimated results with the actual observed distributions of the atmospheric conditions.

In order to estimate more accurate distributions of the atmospheric conditions of a planet, we developed a new one-dimensional multi-box model. In this model, the number of boxes can be increased to an arbitrary value, and the effects of axial tilt (obliquity) and latitudinal variation of albedo on the absorption of shortwave radiation are taken into account. By using this model, we estimate surface temperature, longwave and shortwave radiation, and then compare the estimated distributions with those observed for the Earth, Mars, Venus and Titan. The validity of MaxEP for the planetary atmospheres is thereby examined. We also discuss the role of the amount of atmosphere in advective heat flux, and examine the applicability of MaxEP to the planetary atmospheres from an advective parameter suggested by Jupp and Cox [15]. Finally, dependency of the MaxEP state on the optical properties of the atmospheres is investigated.

In what follows, the details of the multi-box model and the method of estimating the effects of obliquity and albedo variation are explained in Sect. 11.2. The results obtained from this model study are presented in Sect. 11.3. In Sect. 11.4, we discuss the validity of the MaxEP principle for planetary atmospheres with respect to the advective efficiency of each planet. The effects of the optical properties of each planet on the MaxEP states are also examined. The conclusions drawn from this study are contained in Sect. 11.5.

11.2 Multi-box Model

The multi-box energy balance model used in this study is shown in Fig. 11.1. A planet is divided into n latitude zones of equal surface area. We consider shortwave absorption SW_i (W m^{-2}) and longwave emission LW_i (W m^{-2}) from the top of the atmosphere of the ith zone, as well as meridional heat flux F_i (W) through the boundary between the ith and $(i + 1)$th zones; the heat flux is defined as

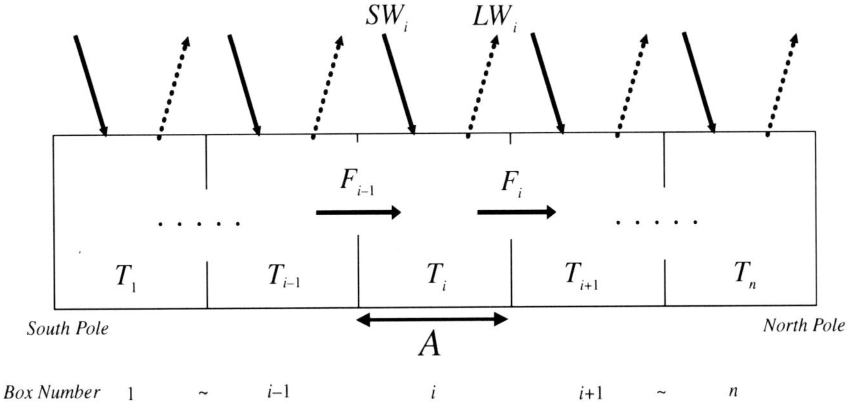

Fig. 11.1 A schematic representation of a multi-box energy balance model

positive northward. The advantage of this multi-box model over previous two-box models is that it can predict latitudinal distributions of temperature, heat flux and radiation, which can be compared with observations more quantitatively.

The entropy production rate due to the meridional heat flux in the model system can be expressed as a sum of entropy production in each zone:

$$\dot{\sigma} = \sum_{i=1}^{n-1} F_i \left(\frac{1}{T_{i+1}} - \frac{1}{T_i} \right), \tag{11.1}$$

where T_i is the surface air temperature of the ith zone. The steady-state energy balance condition for each ith zone is given by

$$(SW_i - LW_i)A = F_i - F_{i-1}, \tag{11.2}$$

where A is the surface area of each zone. Using Eq. (11.2), we can rewrite the entropy production rate (Eq. 11.1) in a different form:

$$\dot{\sigma} = \sum_{i=1}^{n-1} F_i \left(\frac{1}{T_{i+1}} - \frac{1}{T_i} \right) = \sum_{i=1}^{n} \frac{F_{i-1} - F_i}{T_i} = \sum_{i=1}^{n} \frac{LW_i - SW_i}{T_i} A. \tag{11.3}$$

In this manipulation, we have used no-flux boundary conditions at the polar ends: $F_0 = F_n = 0$. The right-hand side of the third equality represents the export rate of entropy from the planetary system. Equation (11.3) shows that the internal entropy production rate (Eq. 11.1) is balanced by the entropy export rate when the steady-state energy balance (Eq. 11.2) is maintained in each zone.

The emission rate of longwave radiation from the top of each zone is assumed to be a linear function of the surface air temperature [5, 16], expressed as:

$$LW_i = a + bT_i, \tag{11.4}$$

where a and b are empirical parameters depending on the optical properties of the atmosphere of a planet. In this study, these parameters are determined through the observational relation between LW_i and T_i using a least square analysis.

The absorption rate of shortwave solar radiation in the ith zone is expressed by the solar constant S, the local albedo α_i, and the solar elevation angle β_i as:

$$SW_i = \begin{cases} S(1-\alpha_i)\sin\beta_i & 0<\beta_i\leq\pi/2, \\ 0 & \beta_i\leq 0, \end{cases} \tag{11.5}$$

where $\beta_i \leq 0$ indicates a night time condition and therefore $SW_i = 0$. Applying spherical trigonometry to a zenith-pole-sun spherical triangle (e.g. [17]) we can find

$$\sin\beta_i = \sin\phi_i \sin\delta + \cos\phi_i \cos\delta \cos h, \tag{11.6}$$

where ϕ_i is the latitude, δ is the solar declination, and h is the hour angle from the local meridian (where $h = 0$). At sunrise ($h = -h_0$) and sunset ($h = h_0$), $\beta_i = 0$ so that

$$\sin\phi_i \sin\delta + \cos\phi_i \cos\delta \cos h_0 = 0, \tag{11.7a}$$

or

$$\cos h_0 = -\tan\phi_i \tan\delta. \tag{11.7b}$$

Daily mean shortwave absorption is thereby evaluated by integration of (11.5) from sunrise to sunset[1]:

$$\begin{aligned} SW_i(\text{day}) &= \frac{1}{2\pi}\int_{-h_0}^{h_0} S(1-\alpha_i)\sin\beta_i \, dh \\ &= \frac{S(1-\alpha_i)}{\pi}(h_0 \sin\phi_i \sin\delta + \cos\phi_i \cos\delta \sin h_0). \end{aligned} \tag{11.8}$$

Yearly mean shortwave absorption is then obtained by integration of (11.8) over a year:

$$\overline{SW_i} = \frac{S(1-\alpha_i)}{\pi\, t_0}\int_0^{t_0} (h_0 \sin\phi_i \sin\delta + \cos\phi_i \cos\delta \sin h_0)\, dt, \tag{11.9}$$

where t_0 is the period of the year (orbital revolution). The declination δ is a function of the obliquity (axial tilt) θ and time t from vernal equinox[2]:

$$\sin\delta = \sin\theta \sin\left(\frac{2\pi t}{t_0}\right). \tag{11.10}$$

[1] We have assumed that the declination δ is constant with respect to daily change of h.

[2] A cyclic orbit is assumed in this study. For a more general case with eccentricity see, e.g., [17].

Table 11.1 Astronomical parameters for Venus, Earth, Mars and Titan adapted from NASA Planetary Fact Sheet [18]

Parameter	Symbol	Units	Venus	Earth	Mars	Titan
Radius	R	km	6,052	6,371	3,390	2,575
Solar constant	S	W m^{-2}	2,614	1,360	589	14.9
Albedo (bond)[a]	α_B	–	0.90	0.306	0.25	0.265[b]
Albedo (geometric)	α_g	–	0.67	0.367	0.17	0.22
Orbital period	t_0	Days	224.7	365.2	687.0	10,747
Obliquity	θ	Degrees	177.4	23.4	25.2	27.0
Optical parameter	a	W m^{-2}	–	–302	–226	–5.92
Optical parameter	b	W m^{-2} K^{-1}	–	1.87	1.65	0.094
Surface gravity	g	m s^{-2}	8.87	9.80	3.69	1.35
Atmospheric mass	M	Mg m^{-2}	1,037	10.3	0.17	108.5
Velocity ratio	r	–	130,300	501	8.0	156,300

[a] Bond albedo is used for calculation of shortwave absorption unless otherwise noted
[b] Parameter value from Li et al. [30]

The declination $\delta = 0$ for $t = 0$ (vernal equinox) or $t = t_0/2$ (autumnal equinox) whereas $\delta = \theta$ for $t = t_0/4$ (summer solstice) and $\delta = -\theta$ for $t = 3t_0/4$ (winter solstice). If we assume $\theta = 0$ (no obliquity), then $\delta = 0$ and $h_0 = \pi/2$. In this case, Eq. (11.9) reduces to

$$\overline{SW_i} = \frac{S(1-\alpha_i)\cos\phi_i}{\pi} \quad \text{for } \theta = 0. \tag{11.11}$$

Equation (11.11) shows the situation of no obliquity or a permanent equinox planet, which has been assumed for simple box models of the Earth [2, 3]. However, this assumption may not be justified for planets with large obliquity. In this study, we therefore implement numerical integration of Eq. (11.9), and examine the effect of obliquity on the MaxEP state.

We selected the Earth, Mars, Titan (the largest satellite of Saturn), and Venus for evaluation of the MaxEP state. Astronomical parameters of these planets as used in our estimations are adapted from the NASA Planetary Fact Sheet [18] and are listed in Table 11.1. The box number N is set at 72, 60, 10 and 100, respectively, corresponding to spatial resolution of observations available for each planet. Using these parameters we calculate the entropy production rate due to meridional heat flux, and compare the MaxEP state with observations of each planet. First, we assume an arbitrary distribution of the heat flux F_i at the ith zone and evaluate the corresponding temperature distribution by the local energy balance requirement (Eq. 11.2) with the longwave radiation (Eq. 11.4). Second, the rate of entropy production is calculated with Eq. (11.3). Third, the heat flux F_i is modified by a finite small amount so that the rate tends to increase. This process is applied to successive zones and repeated until the rate reaches a maximum value. It is found that, starting from any initial distribution of F_i, the calculation converges into a single maximum point. This result suggests the existence of a single maximum in the entropy production rate in the model climate system.[3] The numerical calculations are carried out with the aid of a maximization routine in Mathematica (for details, see [19]).

11.3 Results

11.3.1 Earth

The results obtained for the Earth from the multi-box model are shown in Fig. 11.2a–c. Figure 11.2a is the result for the surface temperature, Fig. 11.2b is that for the longwave emission from the top of the atmosphere, and Fig. 11.2c is that for shortwave absorption. The solid lines are predictions from the MaxEP state (Fig. 11.2a, b) and the numerical calculation (Fig. 11.2c), and the dotted lines are observational results [20–22]. The optical parameters (a and b) are determined through the observational relation between the surface temperature and longwave radiation [23]. For comparison, estimations from a state with no entropy production ($\dot{\sigma} = 0$) and that with no obliquity ($\theta = 0$) are indicated by dashed lines in Fig. 11.2a, b and c, respectively.

The predicted distributions are generally in agreement with the observational results although we can find some discrepancies in the longwave emission. The longwave emission is overestimated in the central tropical region whereas it is underestimated in the northern polar region. We attribute the reason for this to cloud effects. In the central tropics, a large amount of high-level cloud associated with strong convection exists along the inter-tropical convergence zone. High-level cloud is known to reduce longwave emission from the atmosphere (e.g. [24]). In the northern polar region, a large amount of low-level stratus cloud exists over the Arctic Ocean. This low-level cloud is known to enhance longwave emission [25]. These cloud effects have not been taken into account in our simple atmosphere (Eq. 11.4). We therefore expect that the discrepancies can be improved by including the correct cloud effects, as shown by Paltridge [2, 3]. While some discrepancies exist, the predicted distributions show certain resemblance to the observations; the resemblance is much better than those with no entropy production (dashed lines in Fig. 11.2a, b). These results suggest that the MaxEP state tends to be realized in the mean state of the Earth's atmosphere under the cloud radiative forcing condition.

The predicted shortwave absorption shows a close resemblance to the observations, indicating the validity of the estimation method (Eq. 11.9). The prediction with no obliquity shows overestimation of 5 % in the tropics and underestimation of up to 50 % at the poles (Fig. 11.2c). This result suggests that the effect of

---

[3] In this sense, there do not seem to be multiple maxima for this simple model system with a fixed albedo distribution (cf. [16]).

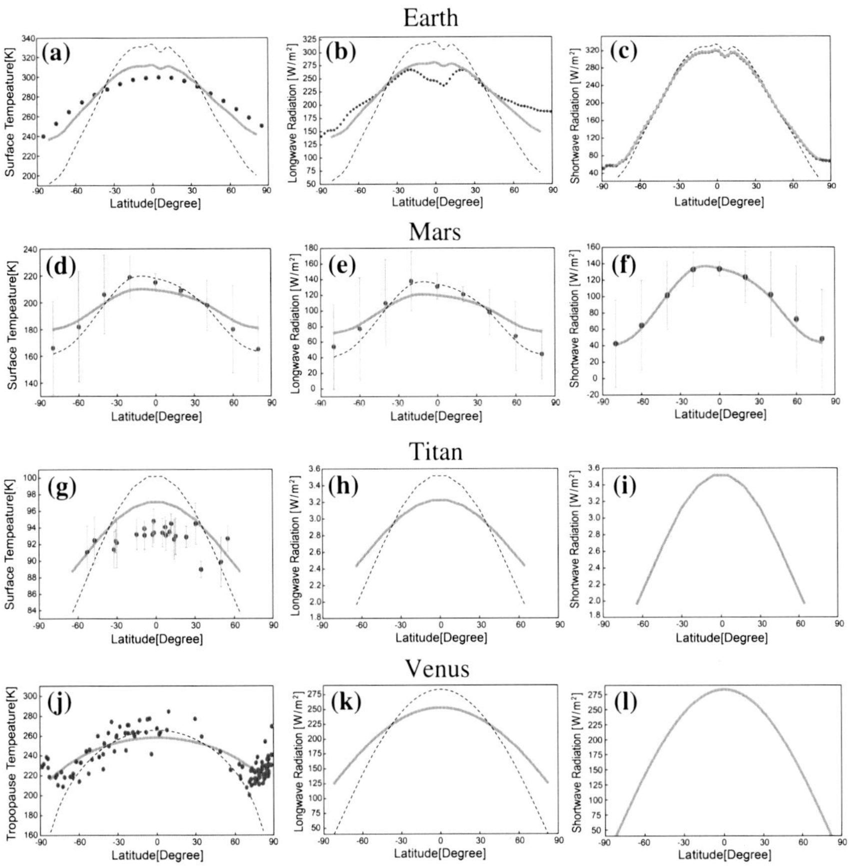

Fig. 11.2 Results for the Earth (**a–c**), Mars (**d–f**), Titan (**g–i**), and Venus (**j–l**). Each figure shows distribution of surface air temperature (**a**, **d**, **g**), effective radiation temperature (**j**), longwave emission (**b**, **e**, **h**, **k**), and shortwave absorption (**c**, **f**, **i**, **l**). *Solid lines* indicate those predicted from a state of maximum entropy production. *Dotted lines* and *dots* indicate those from observations. *Dashed lines* (**a**, **b**, **d**, **e**, **g**, **h**, **j**, **k**) indicate predictions from a state of no entropy production, and a *dashed line* (**c**) indicates prediction with no obliquity. Error bars (**d–f**) show standard deviations due to seasonal changes

obliquity cannot be neglected in the polar regions when the obliquity is larger than 20°. However, it is also confirmed that the effect of no obliquity is less significant on the predicted distributions of longwave emission and surface temperature, since the meridional heat flux is enhanced to compensate the over- and under-estimated shortwave absorption at the MaxEP state. The reason for this will be discussed with a simple box model in Sect. 11.4.

11.3.2 Mars

The results obtained for Mars are shown in Fig. 11.2d–f. Figure 11.2d is the result for the surface temperature, Fig. 11.2e is that for the longwave emission from the top of the atmosphere, and Fig. 11.2f is that for shortwave absorption. The solid lines are predictions from the MaxEP state (Fig. 11.2d, e) and the numerical calculation (Fig. 11.2f), and the dots are the observational results compiled from the Mars Climate Database [26–28]. The optical parameters (a and b) are determined through the relation between the surface temperature and longwave radiation in the database. For comparison, estimations from a state with no entropy production ($\dot{\sigma} = 0$) are indicated by dashed lines in Fig. 11.2d, e.

The predicted distributions of temperature and longwave emission are somewhat lower than those observed in the tropics and are higher in the polar regions. Although these discrepancies are roughly in the range of standard deviation of seasonal change (error bars), the predictions by the state of no entropy production show better agreement with the observations than those of the MaxEP state. This means that the observed state is better represented by the state of no meridional heat flux than by the MaxEP state. The reason for this will be discussed in relation to the total mass of the Martian atmosphere and the associated heat advection in Sect. 11.4. The predicted distribution of shortwave absorption shows reasonable agreement with the observed state and is within the range of standard deviation (Fig. 11.2f).

11.3.3 Titan

The results obtained for Titan are shown in Fig. 11.2g–i. Figure 11.2g is the result for the surface temperature, Fig. 11.2h is that for the longwave emission from the top of the atmosphere, and Fig. 11.2i is that for shortwave absorption. The solid lines are predictions from the MaxEP state (Fig. 11.2g, h) and the numerical calculation (Fig. 11.2i), and the dots in surface temperature are observational results [29]. The optical parameters (a and b) are taken from the values in Lorenz et al. [5], and the planetary (Bond) albedo is set at 0.265 from recent estimation [30]. In this case, the surface temperature distribution is compared with the observational data that are currently available. For comparison, estimations from a state with no entropy production ($\dot{\sigma} = 0$) are indicated by dashed lines in Fig. 11.2g, h.

The predicted temperature distribution shows a reasonable agreement with the observed results within the limit of accuracy of the observational data. Although a slight deviation can be found in the northern polar region, the agreement between the prediction and the observations is reasonable, considering the simplicity of the prediction method. This result suggests that the MaxEP state is realized in the mean state of Titan's atmosphere, being consistent with the earlier result obtained from a two-box model by Lorenz et al. [5]. More detailed study is, however, needed to verify the validity of the MaxEP state when more precise observational data are collected from Titan.

11.3.4 Venus

The results obtained for Venus are shown in Fig. 11.2j–l. Figure 11.2j is the result for the effective radiation temperature, Fig. 11.2h is that for the longwave emission from the top of the atmosphere, and Fig. 11.2i is that for shortwave absorption. The solid lines are predictions from the MaxEP state (Fig. 11.2j, k) and the numerical calculation (Fig. 11.2l), and the dots are temperatures estimated at the top of the troposphere [31]. In the case of Venus, no detailed surface temperature data is available because of the presence of the thick atmosphere. For this reason, we estimate the effective radiation temperature at the top of the atmosphere from the predicted longwave emission by using the Stefan–Boltzmann law:

$$T_{e,i} = \left(\frac{LW_i}{\sigma_B}\right)^{\frac{1}{4}}, \tag{11.12}$$

where $T_{e,i}$ is the effective radiation temperature of the ith zone and σ_B is the Stefan–Boltzmann constant. This effective radiation temperature is compared with the temperature distribution estimated at the top of the troposphere in Fig. 11.2j. Since no detailed surface temperature data is available, we cannot determine the optical parameters (a and b) of this planet. We therefore assume an arbitrary value of b (1, 10 and 100) and calculate the corresponding value of a that satisfies the total energy balance of the planet with the mean surface temperature of 735 K. The sensitivity of b to the predicted distributions of the surface temperature and the effective radiation temperature is thereby examined. For comparison, the effective temperature and longwave emission estimated from a state with no entropy production ($\dot{\sigma} = 0$) are indicated by dashed lines in Fig. 11.2j, k.

The predicted temperature distribution shows an agreement with the observational results within the limit of accuracy of the observational data. While we can find some discrepancies in the polar regions (overestimation at the poles and slight underestimation around the sub-polar regions), these discrepancies are within the range of accuracy of the observational data. This result suggests that the MaxEP state is realized in the mean state of the Venusian atmosphere. It is therefore interesting to conduct more detailed study when more precise observational data are obtained from the Venusian atmosphere.

It is found that the predicted surface temperature is sensitive to the change of the parameter b, that is, the temperature contrast between the tropics and poles tends to decrease with the increase of b. However, the predicted effective radiation temperature at the top of the troposphere (Eq. 11.12) shows no dependency on the change in b. This means that the change in the optical parameter b affects the surface temperature distribution whereas it does not affect the temperature at the top of the atmosphere at the MaxEP state. This result suggests the existence of a unique characteristic that is present at the top of the atmosphere in the MaxEP state and that is independent of the optical properties of the atmosphere. The reason for this characteristic will be discussed in the next section.

11.4 Discussion

We have seen in the previous section that the predicted distribution of temperature and longwave emission by the MaxEP state are generally in agreement with the observed mean states of the Earth, Titan and Venus. While some discrepancies exist, they are within the limit of accuracy of the observational data. These results suggest that the MaxEP state is realized in the mean states of these planets. In the case of Mars, the predicted distributions do not agree well with the observational results when compared with those predicted by the state of no meridional heat flux. This result suggests that the MaxEP state is not realized in the Martian atmosphere. In the case of Venus, we found that the distribution of the effective radiation temperature at the top of the atmosphere predicted by the MaxEP state is almost independent of the optical properties of the atmosphere. In order to clarify the reasons for these issues, we implement a simple analysis on entropy production in a simplified model planet.

For simplicity, let us assume a two-box model consisting of an equatorial zone and a polar zone (Fig. 11.3a). The energy balance requirement (Eq. 11.2) for each zone is then expressed as

$$\begin{cases} [SW_e - (a + bT_e)]A = F, \\ [SW_p - (a + bT_p)]A = -F, \end{cases} \tag{11.13}$$

where the suffix e and p denote the equatorial and polar regions, and F is the meridional heat flux from the equatorial zone to the polar zone. The entropy production rate (Eq. 11.3) is then given by

$$\dot{\sigma} = F\left(\frac{1}{T_p} - \frac{1}{T_e}\right). \tag{11.14}$$

Substituting Eq. (11.13) into Eq. (11.14) and eliminating T_e and T_p, and differentiating $\dot{\sigma}$ with respect to F, we find a maximum in the entropy production rate ($d\dot{\sigma}/dF = 0$) at

$$F_{\mathrm{MaxEP}} = \frac{\Delta SW}{2}\left[1 - \frac{1}{1 + \sqrt{1 - \left(\frac{\Delta SW}{2b\bar{T}}\right)^2}}\right]A \approx \frac{\Delta SW A}{4}, \tag{11.15}$$

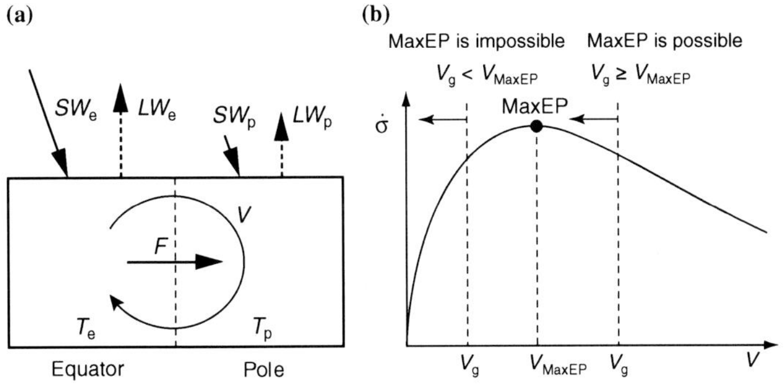

Fig. 11.3 **a** A two-box model of a planet. **b** entropy production rate as a function of mean velocity of circulation V. The MaxEP state is possible when the velocity of a steady gravity flow exceeds the velocity needed for the MaxEP state ($V_g \geq V_{MaxEP}$), whereas it is otherwise impossible ($V_g < V_{MaxEP}$)

where $\Delta SW = SW_e - SW_p$ is the difference in the shortwave radiation and $\bar{T} = (T_e + T_p)/2$ is the mean temperature of the system. In Eq. (11.15), we have assumed $\Delta SW/(2b\,\bar{T}) \approx \Delta T/\bar{T} \ll 1$. Similarly, the difference in the surface temperature $\Delta T = T_e - T_p$ at the MaxEP state is given by

$$\Delta T_{MaxEP} = \frac{\Delta SW}{b\left[1 + \sqrt{1 - \left(\frac{\Delta SW}{2b\bar{T}}\right)^2}\right]} \approx \frac{\Delta SW}{2b}, \tag{11.16}$$

and the corresponding difference in the longwave emission $\Delta LW = LW_e - LW_p$ is

$$\Delta LW_{MaxEP} = b\Delta T_{MaxEP} = \frac{\Delta SW}{1 + \sqrt{1 - \left(\frac{\Delta SW}{2b\bar{T}}\right)^2}} \approx \frac{\Delta SW}{2}. \tag{11.17}$$

We can see in Eqs. (11.16) and (11.17) that the surface temperature difference at the MaxEP state depends on the optical parameter b, whereas the difference in the longwave emission (as well as the effective radiation temperature) is independent of b, as long as the approximation ($\Delta SW/(2b\,\bar{T}) \approx \Delta T/\bar{T} \ll 1$) is valid. This explains the reason why the predicted distribution of the effective radiation temperature T_e at the MaxEP state for Venus is almost constant and is independent of the optical properties of the atmosphere (Sect. 11.3.4). We can also see in Eqs. (11.16) and (11.17) that each equation includes a factor of 1/2. This means that the sensitivity of ΔLW or ΔT to a change in ΔSW at the MaxEP state is about half of that expected at a state without meridonal heat flux (i.e. pure radiation balance: $\Delta LW = \Delta SW$). This explains the reason for the moderate sensitivity (or the stability) of the MaxEP climate state to the change in the shortwave radiative forcing that we have discussed in Sect. 11.3.1. This result is also consistent with an analytical result obtained by

Rodgers [32] who shows that the longwave emission at the MaxEP state is approximately proportional to the square root of the shortwave absorption: $LW_{\mathrm{MaxEP}} \propto SW^{1/2}$, resulting in $\mathrm{d}LW_{\mathrm{MaxEP}}/\mathrm{d}SW \approx 1/2$ at around $\overline{LW} \approx \overline{SW}$.

The meridional heat flux F is related to the mean velocity of circulation of the atmosphere (Fig. 11.3a). For a slowly rotating planet, we can write (e.g. [33])

$$F = cMV\frac{\Delta T}{R}A, \tag{11.18}$$

where c is the specific heat capacity, M is the mass of a unit atmospheric column, V is the mean velocity of circulation, and R is the radius of the planet (characteristic length of latitudinal zone). Substituting Eqs. (11.15) and (11.16) into Eq. (11.18), we find the mean velocity at the MaxEP state:

$$V_{\mathrm{MaxEP}} = \frac{bR}{2\,cM}\sqrt{1-\left(\frac{\Delta SW}{2b\bar{T}}\right)^2} \approx \frac{bR}{2\,cM}. \tag{11.19}$$

Equation (11.19) represents the mean circulation velocity that is needed to sustain the MaxEP state. The circulation should be driven dynamically in the atmosphere. Assuming a steady gravity flow, the mean velocity is at most of the order of the square root of potential energy of the atmosphere (e.g. [15]):

$$V_{\mathrm{g}} = \sqrt{gH} = \sqrt{\frac{gM}{\rho_{\mathrm{s}}}}, \tag{11.20}$$

where g is the acceleration due to gravity, $H = M/\rho_{\mathrm{s}}$ is the scale height, and ρ_{s} is the density at the surface. We can expect that the MaxEP state is dynamically attainable when $V_{\mathrm{g}} \geq V_{\mathrm{MaxEP}}$ whereas it may not be attainable when $V_{\mathrm{g}} < V_{\mathrm{MaxEP}}$. The attainability of the MaxEP state can then be determined by the ratio of the two velocities:

$$r \equiv \frac{V_{\mathrm{g}}}{V_{\mathrm{MaxEP}}} = \frac{2\,c}{bR}\sqrt{\frac{gM^3}{\rho_{\mathrm{s}}}} = \frac{2\,cM}{bR}\sqrt{R'\bar{T}}, \tag{11.21}$$

where R' is the gas constant for the atmosphere. In this manipulation, we have used the relation: $p_{\mathrm{s}} = gM$ and the equation of state: $p_{\mathrm{s}} = \rho_{\mathrm{s}} R' \bar{T}$, where p_{s} is the surface pressure. It should be noted that this ratio is proportional to the advective capability derived from a theoretical study by Jupp and Cox [15], although their derivation method is quite different from ours.[4] The calculated ratio r for each planet is listed in Table 11.1. We can see that the ratio is large for Venus, Earth and Titan ($r > 10^2$), because of their large amounts of atmosphere ($M > 10$ Mg m^{-2}).

[4] The difference can been seen in the numerical factor in each formulation: 2 in Eq. (11.21) whereas $2(3\gamma)^{1/2} \approx 1.62$ in their formulation, with $\gamma = 3^{1/2}/\pi - 1/3$ [15]. Their analysis also includes the planetary rotation rate, which turns out to be not important for slowly rotating planets where Eq. (11.18) is approximately valid.

By contrast, the ratio is relatively small for Mars ($r \approx 8$) because of its small amount of atmosphere ($M \approx 0.2$ Mg m^{-2}). This seems to explain the reason why the MaxEP state is not realized in the Martian atmosphere. The atmosphere is so scant that it cannot carry the heat energy that is needed for the MaxEP state by advection (Fig. 11.3b). This result is consistent with the theoretical analysis by Jupp and Cox [15] who showed that the advective capability of Mars is on the critical line below which a planet is prevented from achieving the MaxEP state by dynamical constraints.

11.5 Conclusion

In this chapter, we investigated the distributions of temperature and longwave radiation from a state of maximum entropy production (MaxEP) due to meridional heat flux in the atmospheres of the Earth, Mars, Titan and Venus, and compared the predicted distributions with observational distributions. In the predictions, we used a multi-box planetary model that took into account the effects of obliquity and latitudinal variation of albedo on shortwave radiation.

We have shown that the predicted distributions are generally in agreement with observations for the Earth, Titan and Venus, but not for Mars. While some discrepancies exist, they are within the limit of accuracy of the observational data. In the case of Mars, the predicted distributions do not agree well with the observational results when compared with those predicted by a state of no meridional heat flux. These results suggest that the MaxEP state is realized in the mean states of the Earth, Titan and Venus, whereas it is not realized in the Martian atmosphere. A simple analysis on advective heat transport using a two-box model shows that the Martian atmosphere is so scant that it cannot carry the amount of heat needed for the MaxEP state by advection. It is suggested that the validity of the MaxEP state for a planetary atmosphere is limited when the total amount of atmosphere is not enough to sustain the advective heat flux necessary for the MaxEP state. It was also shown from this analysis that the distributions of longwave emission and effective radiation temperature at the top of the atmosphere predicted by the MaxEP state are determined by the distribution of shortwave absorption, and is almost independent of the optical properties of a planet. The last result will provide a useful insight into the validity of MaxEP in planetary atmospheres when top-of-the-atmosphere temperature data are accumulated through future extra-planetary explorations.

Acknowledgments The authors wish to express their cordial thanks to the organizers of the MaxEP Workshop, Canberra, Sept. 2011, where the motivation for this work has been stimulated. Acknowledgment is also given to Dr. Ralph D. Lorenz for valuable comments on Titan's atmosphere, to Dr. Robert K. Niven for stimulating our interest in thermodynamics of planetary atmospheres, and to Dr. Ehouarn Millour for sending us the original Mars Climate Database. Valuable comments from two anonymous reviewers are also gratefully acknowledged.

References

1. Ziegler, H.: Zwei Extremalprinzipien der irreversiblen Thermodynamik. Ing. Arch. **30**, 410–416 (1961)
2. Paltridge, G.W.: Global dynamics and climate – a system of minimum entropy exchange. Quart. J. R. Met. Soc. **101**, 475–484 (1975)
3. Paltridge, G.W.: The steady-state format of global climate. Quart. J. R. Met. Soc. **104**, 927–945 (1978)
4. Ozawa, H., Ohmura, A., Lorenz, R.D., Pujol, T.: The second law of thermodynamics and the global climate system: a review of the maximum entropy production principle. Rev. Geophys. **41**, 1018 (2003)
5. Lorenz, R.D., Lunine, J.I., Withers, P.G., McKay, C.P.: Titan, Mars, and Earth: entropy production by latitudinal heat transport. Geophys. Res. Lett. **28**, 415–418 (2001)
6. Schneider, E.D., Kay, J.J.: Life as a manifestation of the second law of thermodynamics. Math. Comp. Modell. **19**, 25–48 (1994)
7. Ozawa, H., Shimokawa, S., Sakuma, H.: Thermodynamics of fluid turbulence: a unified approach to the maximum transport properties. Phys. Rev. E **64**, 026303 (2001)
8. Shimokawa, S., Ozawa, H.: On the thermodynamics of the oceanic general circulation: irreversible transition to a state with higher rate of entropy production. Q. J. Roy. Meteorol. Soc. **128**, 2115–2128 (2002)
9. Shimokawa, S., Ozawa, H.: Thermodynamics of irreversible transitions in the oceanic general circulation. Geophys. Res. Lett. **34**, L12606 (2007). doi:10.1029/2007GL030208
10. Nohguchi, Y., Ozawa, H.: On the vortex formation at the moving front of lightweight granular particles. Physica D **238**, 20–26 (2009)
11. Lorenz, E.N.: Generation of available potential energy and the intensity of the general circulation. In: Pfeffer, R.L. (ed.) Dynamics of climate, pp. 86–92. Pergamon, Oxford (1960)
12. Sawada, Y.: A thermodynamic variational principle in nonlinear non-equilibrium phenomena. Prog. Theor. Phys. **66**, 68–76 (1981)
13. Dewar, R.: Information theory explanation of the fluctuation theorem, maximum entropy production and self-organized criticality in non-equilibrium stationary states. J. Phys. A **36**, 631–641 (2003)
14. Niven, R.K.: Steady state of a dissipative flow-controlled system and the maximum entropy production principle. Phys. Rev. E **80**, 021113 (2009)
15. Jupp, T.E., Cox, P.M.: MEP and planetary climates: insights from a two-box model climate model containing atmospheric dynamics. Phil. Trans. R. Soc. B **365**, 1355–1365 (2010)
16. Budyko, M.I.: The effect of solar radiation variations on the climate of the Earth. Tellus **21**, 611–619 (1969)
17. Iqbal, M.: An introduction to solar radiation. Academic Press, New York (1983)
18. NASA Planetary Fact Sheet. http://nssdc.gsfc.nasa.gov/planetary/factsheet/ (2010)
19. Fukumura, Y.: A study on entropy production in planetary atmospheres. Hiroshima Univ, Master thesis (2012)
20. Sellers, W.D.: Physical climatology. Univ. Chicago Press, Chicago (1965)
21. Barkstrom, B.R., Harrison, E.F., Lee III, R.B.: Earth radiation budget experiment – Preliminary seasonal results. Eos, Trans. Amer. Geophys. Union **71**, 297–305 (1990)
22. Masuda, K.: Personal communication. http//macroscope.world.coocan.jp/ja/edu/clim_sys/erb/lat_seas.html (2003)
23. North, G.R., Coakley, J.A.: Differences between seasonal and mean annual energy balance model calculations of climate and climate sensitivity. J. Atmos. Sci. **36**, 1189–1204 (1979)
24. Peixoto, J.P., Oort, A.H.: Physics of climate. Amer. Inst. Phys. New York (1992)
25. Sedlar, J., Shupe, M.D., Tjernström, M.: On the relationship between thermodynamic structure and cloud top, and its climate significance in the arctic. J. Clim. **25**, 2374–2393 (2012)
26. Mars Climate Database. http://www-mars.lmd.jussieu.fr/mars/access.html (2008)

27. Forget, F., Hourdin, F., Fournier, R., Hourdin, C., Talagrand, O., Collins, M., Lewis, S.R., Read, P.L., Huot, J.-P.: Improved general circulation models of the Martian atmosphere from the surface to above 80 km. J. Geophys. Res. **104**, 24155–24175 (1999)
28. Lewis, S.R., Collins, M., Read, P.L., Forget, F., Hourdin, F., Fournier, R., Hourdin, C., Talagrand, O., Huot, J.-P.: A climate database for Mars. J. Geophys. Res. **104**, 24177–24194 (1999)
29. Courtin, R., Kim, S.J.: Mapping of Titan's tropopause and surface temperatures from Voyager IRIS spectra. Planet. Space Sci. **50**, 309–321 (2002)
30. Li, L., Nixon, C.A., Achterberg, R.K., et al.: The global energy balance of Titan. Geophys. Res. Lett. **38**, L23201 (2011)
31. Kliore, A.J.: Recent results on the Venus atmosphere from Pioneer Venus radio occultations. Adv. Space Res. **5**, 41–49 (1985)
32. Rodgers, C.D.: Comments on Paltridge's 'minimum entropy exchange' principle. Quart. J. R. Met. Soc. **102**, 455–457 (1976)
33. Golitsyn, G.S.: A similarity approach to the general circulation of planetary atmospheres. Icurus **13**, 1–24 (1970)

Chapter 12
Entropy Production-Based Closure of the Moment Equations for Radiative Transfer

Thomas Christen and Frank Kassubek

Abstract Heat radiation in gases or plasmas is usually out of local thermodynamic equilibrium (LTE) even if the underlying matter is in LTE. Radiative transfer can then be described with the radiative transfer equation (RTE) for the radiation intensity. A common approach to solve the RTE consists in a moment expansion of the radiation intensity, which leads to an infinite set of coupled hyperbolic partial differential equations for the moments. A truncation of the moment equations requires the definition of a closure. We recommend to use a closure based on a constrained minimum entropy production rate principle. It yields transport coefficients (e.g., effective mean absorption coefficients and Eddington factor) in accordance with the analytically known limit cases. In particular, it corrects errors and drawbacks from other closures often used, like the maximum entropy principle (e.g., the M1 approximation) and the isotropic diffusive P1 approximation. This chapter provides a theoretical overview on the entropy production closure, with results for an illustrative artificial example and for a realistic air plasma.

12.1 Introduction

Energy transfer by heat radiation in hot gases and plasmas is encountered in many different contexts like stellar and terrestrial atmospheres [1] and electric arcs [2], to mention a few examples. Although the energy carriers, the photons, do not interact with each other, the effective interaction due to scattering, emission, and absorption via the underlying matter makes the theoretical treatment of radiative

T. Christen (✉) · F. Kassubek
ABB Corporate Research, 5405 Dättwil, Switzerland
e-mail: thomas.christen@ch.abb.com

F. Kassubek
e-mail: frank.kassubek@ch.abb.com

R. C. Dewar et al. (eds.), *Beyond the Second Law*,
Understanding Complex Systems, DOI: 10.1007/978-3-642-40154-1_12,

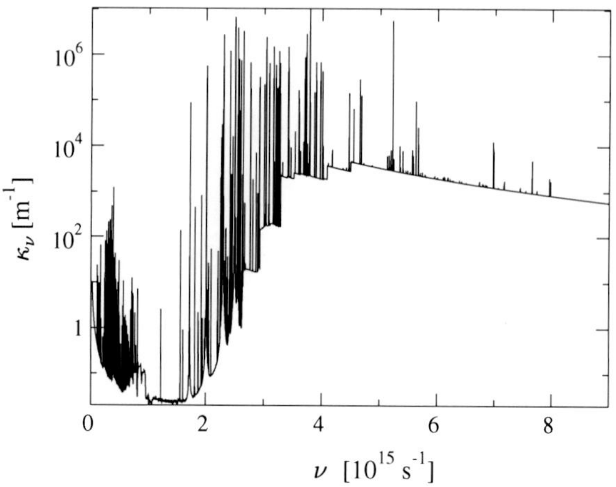

Fig. 12.1 Absorption spectrum of air plasma at 10,300 K and 2 bar [28], consisting of continuous bands (free-free, free-bound, bound-free transitions) superimposed to discrete peaks (bound-bound transitions) [29, 30]. The spectrum is not only a complicated function of frequency with huge variations ranging from 10^{-2} to 10^7 m^{-1}, but also strongly varies with temperature

transfer generally rather complicated [3]. In order to illustrate the possible complexity, the absorption spectrum κ_ν of air at about 10,000 K temperature is shown in Fig. 12.1. Here, κ_ν is the macroscopic spectral absorption coefficient in units of m^{-1}, and ν is the frequency. It consists of continuous bands and discrete peaks associated with electronic transitions of free-free, free-bound, and bound-bound states of the present air molecules, atoms, and ions.

An additional complication appears when the radiation is not in local thermal equilibrium (LTE). This is usually the case in gases and plasmas due to their partial transparency, even if the matter is in LTE. Non-LTE radiation refers to a photon distribution function n_ν that differs from the equilibrium Bose-Einstein or Planck distribution [4]

$$n_\nu^{(eq)} = \frac{1}{\exp(h\nu/k_B T) - 1}, \tag{12.1}$$

where h is the Planck constant, k_B the Boltzmann constant, and T the local temperature of the LTE matter.

For simplicity, we consider unpolarized radiation in an isotropic medium. The basic equation is then the radiative transfer equation (RTE) [1, 3] for the specific radiation intensity[1] [5]

$$I_\nu(\mathbf{x}, \Omega) = \frac{2h\nu^3}{c^2} n_\nu(\mathbf{x}, \Omega), \tag{12.2}$$

[1] In the following we will skip the term *specific*.

which describes the radiation flux at location $\mathbf{x}$ as a function of the direction Ω and frequency ν. The RTE reads

$$\frac{1}{c}\partial_t I_\nu + \Omega \cdot \nabla I_\nu = \mathscr{L}(B_\nu - I_\nu), \tag{12.3}$$

where $\mathscr{L}$ is linear in its argument $B_\nu - I_\nu$ (see discussion below) and can be expressed as

$$\mathscr{L}(B_\nu - I_\nu) = \kappa_\nu(B_\nu - I_\nu) + \sigma_\nu\left(\frac{1}{4\pi}\int_{S^2} d^2\tilde{\Omega}\, p_\nu(\Omega, \tilde{\Omega}) I_\nu(\tilde{\Omega}) - I_\nu\right). \tag{12.4}$$

The RTE can be transformed into a linear transport equation for n_ν by insertion of (12.2) in Eq. (12.3). It is a linear Boltzmann transport equation (see also [5] for further examples), where entropy production is caused uniquely by the term (12.4). Let us briefly explain the different terms of the RTE (see, e.g., [3]). The expression on the left hand side of Eq. (12.3) multiplied with c is the substantial derivative consisting of the explicit time derivative $\partial_t I_\nu$ plus the advection term $c\,\Omega \cdot \nabla$ due to the motion of the photons with speed c; $\Omega \cdot \nabla$ is the directional derivative. This net change of I_ν in direction of Ω must be equal to the sum of specific source and sink terms due to the radiation-matter interactions, written on the right hand side of Eq. (12.3) and detailed in Eq. (12.4). Photons are generated by emission and annihilated by absorption, expressed by $\kappa_\nu B_\nu$ and $\kappa_\nu I_\nu$, respectively. Here, B_ν is the Planck function for thermal equilibrium,

$$B_\nu = \frac{2h\nu^3}{c^2} n_\nu^{(eq)}. \tag{12.5}$$

By breaking time reversal symmetry, the "collision term" $\mathscr{L}(B_\nu - I_\nu)$ leads to the irreversibility that equilibrates nonequilibrium states, and is thus responsible for entropy production. The absorption coefficient κ_ν is generally a sum of products of particle densities, absorption cross-sections, and contains terms $1 - \exp(-h\nu/k_B T)$ [5]; it depends thus not only on frequency but also on the partial pressures of the present species, and the temperature. The expression (12.4) includes elastic scattering. Incoming photons of frequency ν from all directions $\tilde{\Omega}$ are scattered with probability $p_\nu(\Omega, \tilde{\Omega})$ into direction Ω. Among other properties [1], p_ν is assumed to be normalized according to $(4\pi)^{-1}\int_{S^2} d^2\tilde{\Omega}\, p_\nu(\Omega, \tilde{\Omega}) = 1$ with S^2 being the full solid angle 4π, and we denote by $d^2\Omega$ the (2-dimensional) volume angle increment. The strength of the scattering process is quantified by the spectral scattering coefficient σ_ν in units of m^{-1}. In the absence of any interaction, e.g., in vacuum or a fully transparent medium, the right hand side of Eq. (12.3) vanishes, which describes the so-called *(free) streaming limit*. In the particle picture it can be interpreted as the limit of ballistic propagation of the photons, i.e., propagation without any kind of scattering [6].

A number of procedures to solve the RTE exist [3]. In this chapter we discuss a simple but effective approach based on a truncated moment expansion with an entropy production minimization closure [7, 8]. Kohler has shown that entropy production optimization principles hold for linearized Boltzmann transport equations [9–12]. He especially discusses the Boltzmann transport equation for gases and the transport of electrons in a solid near equilibrium. Note that the expression "near equilibrium" is used in this case for the linear nonequilibrium regime, i.e. where the linearization of the Boltzmann transport equation is an appropriate approximation. Far from equilibrium, however, higher order terms in the deviation $n_\nu - n_\nu^{(eq)}$ from equilibrium have to be taken into account, and the entropy production principle is no longer applicable. Because the RTE (12.3) for heat radiation has the form of a linearized Boltzmann transport equation, Kohler's argument applies here analogously (for similar transfer equations, see also [5, 6]). The "collision term" $\mathscr{L}(B_\nu - I_\nu)$ in Eq. (12.3) is indeed a linear function of $B_\nu - I_\nu$: if one replaces $-I_\nu$ by $B_\nu - I_\nu$ in the large bracket of the right hand side in Eq. (12.4), the additional terms associated with B_ν add up to zero because B_ν is independent of direction Ω. Because photons do not interact with each other, the RTE is exactly[2] linear over the *whole* nonequilibrium range, i.e., for arbitrarily large deviation $|B_\nu - I_\nu|$ from equilibrium. It has been conjectured [7] that this exact linearity of the RTE is the reason for the success of our approach discussed below also far from equilibrium.

This chapter is organized as follows. Section 12.2 defines the moments and their governing equations. In order to truncate the system of equations, a closure based on entropy production rate is introduced in Sect. 12.3. The main results are discussed and illustrated in Sect. 12.4. Some remarks on boundary conditions for the moments are provided in Sect. 12.5.

12.2 The Moment Equations for Radiative Transfer

The macroscopic radiative properties of highest interest are related to those quantities that occur in the hydrodynamic equations of the underlying matter. Those are energy density, energy flux, and radiation pressure, and can be obtained from $I_\nu(\Omega)$ by integration over frequency ν and angle Ω. One thus introduces the moments

$$E = \frac{1}{c}\int d\nu\, d^2\Omega\, I_\nu, \tag{12.6}$$

$$F_k = \frac{1}{c}\int d\nu\, d^2\Omega\, \Omega_k I_\nu, \tag{12.7}$$

[2] Three and more photon processes are disregarded.

$$\Pi_{kl} = \frac{1}{c}\int d\nu\, d^2\Omega\, \Omega_k \Omega_l I_\nu, \qquad (12.8)$$
$$\dots = \dots,$$

where $k, l = 1, \dots, 3$ denote the three space directions. The integrations run from zero to infinity for ν and over the whole solid angle (sphere S^2) for Ω. The list of moments continues with higher order moments $\Omega_k \Omega_l \dots \Omega_N$ to infinite order. Multiplication of the RTE (12.3) with products and/or powers of Ω_k's, and integration over frequency and solid angle leads to the infinite set of equations

$$\frac{1}{c}\partial_t E + \nabla \cdot \mathbf{F} = P_E, \qquad (12.9)$$

$$\frac{1}{c}\partial_t \mathbf{F} + \nabla \cdot \Pi = \mathbf{P}_F, \qquad (12.10)$$
$$\dots = \dots,$$

where $\mathbf{F}$ and Π denote the vector and the tensor with components given by Eqs. (12.7) and (12.8), respectively. The right hand sides are given by

$$P_E = \frac{1}{c}\int d\nu d^2\Omega \;\; \mathcal{L}(B_\nu - I_\nu) = \kappa_{\mathrm{E}}^{(\mathrm{eff})}(E^{(eq)} - E), \qquad (12.11)$$

$$\mathbf{P}_F = \frac{1}{c}\int d\nu d^2\Omega \;\; \Omega\, \mathcal{L}(B_\nu - I_\nu) = -\kappa_{\mathrm{F}}^{(\mathrm{eff})}\mathbf{F}, \qquad (12.12)$$
$$\dots = \dots,$$

where

$$E^{(eq)} = \frac{4\pi}{c}\int_0^\infty d\nu\, B_\nu = \frac{4\sigma_{SB}}{c}T^4 \qquad (12.13)$$

is the LTE radiation energy density (σ_{SB} is the Stefan-Boltzmann constant), and for convenience the effective absorption coefficients $\kappa_{\mathrm{E}}^{(\mathrm{eff})}$ and $\kappa_{\mathrm{F}}^{(\mathrm{eff})}$ are introduced. In full equilibrium P_E and $\mathbf{P}_F$ vanish. The transport coefficients ($\kappa_{\mathrm{E}}^{(\mathrm{eff})}$, $\kappa_{\mathrm{F}}^{(\mathrm{eff})}$, ...) are still functionals of the unknown function I_ν. Once they are known, the moments (E, $\mathbf{F}$, ...), which are the variables of the (still infinite) set of partial differential equations [(12.9), (12.10), ...], can be determined in principle by solving the latter, provided appropriate initial and boundary conditions are given.

For practical purposes, one has to truncate the set of equations and to restrict the model to a finite number N of moments. The equation for the highest order moment will then contain the moment of the subsequent order, which is not a variable, but an additional unknown quantity that depends on I_ν. A closure method is a procedure that prescribes how to determine all these unknowns, which may eventually depend on all moments that are variables. In the following, we restrict ourselves to the two first moment equations (12.9) and (12.10) with variables E

and **F**, but we emphasize that our procedure is general and applicable to any order N of truncation.

Because the second rank tensor $\boldsymbol{\Pi}$ depends only on the scalar E and the vector **F**, it can be written by tensor symmetry reasons in the form

$$\Pi_{nm} = E\left(\frac{1-\chi}{2}\delta_{nm} + \frac{3\chi - 1}{2}\frac{F_n F_m}{F^2}\right), \tag{12.14}$$

where the *variable Eddington factor* (VEF) χ is in general a function of E and $F = |\mathbf{F}|$, and δ_{kl} ($= 0$ if $k \neq l$ and $\delta_{kl} = 1$ if $k = l$) is the Kronecker delta. In thermal equilibrium all fluxes vanish, and the stress tensor is proportional to the unit tensor with diagonal elements $E^{(eq)}/3$. This follows from Eqs. (12.6) and (12.8), $|\Omega| = 1$ and the isotropy of the equilibrium radiation. Because we assume that the underlying matter is isotropic, the only distinguished direction is given by **F**, and $\kappa_E^{(\mathrm{eff})}(E, \mathrm{v})$, $\kappa_F^{(\mathrm{eff})}(E, \mathrm{v})$, and $\chi(E, \mathrm{v})$ can be expressed as functions of E and

$$\mathrm{v} = \frac{F}{E}. \tag{12.15}$$

Note that $0 \leq \mathrm{v} \leq 1$, with $\mathrm{v} = 1$ corresponding to the free streaming limit. v can be roughly understood as the dimensionless average velocity of the photon gas, where $\mathrm{v} = 1$ is associated with the speed of light c, which cannot be surpassed.

12.3 Closure by Entropy Production Rate Minimization

The task of the closure is to determine the transport coefficients, i.e., the effective or mean absorption coefficients $\kappa_E^{(\mathrm{eff})}$ and $\kappa_F^{(\mathrm{eff})}$, and the VEF χ as functions of E and v (or F). A closure that is often considered is based on entropy maximization [13–15] (and is in the present context sometimes named "M1-model"). However, Kohler [9] has proved validity of entropy production rate principles for the linearized Boltzmann transport equation. According to his results, near equilibrium the distribution function optimizes the entropy production rate under certain constraints, which are associated with fixed moments or fluxes. The type of the optimum, i.e., whether the optimum is a maximum or a minimum, depends on the specific choice of constraints. Kohler's proof has been re-discussed several times in the literature [10–12]. We mention also three additional works which indicate the relevance of entropy production principles for radiative transfer. Firstly, Essex [16] has shown that the entropy production rate is minimum in a grey atmosphere in local radiative equilibrium. Secondly, Würfel and Ruppel [17, 18] discussed entropy production rate maximization by introducing an effective chemical potential of the photons, related to their interaction with matter. Finally, Santillan et al. [19] showed that for a constraint of fixed radiation power, black bodies maximize the entropy production rate.

The closure procedure based on entropy production minimization has been outlined for photons in [7, 8], and for a gas of independent electrons in [20]. The receipt, in a nutshell, is to minimize the entropy production rate, which is a functional of I_ν, subject to the constraints of fixed moments (given by Eqs. (12.6), (12.7) etc.). The result of this optimization problem will then be a function $I_\nu(E, \mathbf{F})$, from which all unknowns (P_E, $\mathbf{P}_F$, ...) can be determined. In order to derive the expression for the entropy production rate, we start with the entropy per volume of the photon gas [4, 21, 22]

$$S_{\text{rad}}[I_\nu] = -k_B \int d^2\Omega\, d\nu\, \frac{2\nu^2}{c^3}\left(n_\nu \ln n_\nu - (1+n_\nu)\ln(1+n_\nu)\right), \tag{12.16}$$

where Eq. (12.2) relates n_ν to I_ν. The total entropy production rate, $\sum$, consists of the two contributions $\sum_{\text{rad}}$ and $\sum_{\text{mat}}$ associated with entropy production in the photon gas and in the matter, respectively (cf. [23]). The contribution $\sum_{\text{rad}}$ is obtained from the time-derivative of Eq. (12.16), by making use of Eq. (12.3), and writing the result in the form $\partial_t S_{\text{rad}} + \nabla \cdot \mathbf{J}_S = \sum_{\text{rad}}$, which yields

$$\sum{}_{\text{rad}}[I_\nu] = -k_B \int d\nu\, d^2\Omega\, \frac{1}{h\nu} \ln\left(\frac{n_\nu}{1+n_\nu}\right) \mathscr{L}(B_\nu - I_\nu); \tag{12.17}$$

here $\mathbf{J}_S$ is the entropy current density.

The second contribution, the entropy production rate of the LTE matter, $\sum_{\text{mat}}$, can be derived from the fact that the matter can be considered locally as an equilibrium bath with temperature $T(\mathbf{x})$. Energy conservation implies that the local power production W of the matter is related to the radiation power density in Eq. (12.11) by $W = -cP_E$. The entropy production rate (associated with radiation) in the local heat bath is thus $\sum_{\text{mat}} = W/T = -cP_E/T$. Equation (12.1) implies $h\nu/k_B T = \ln(1 + 1/n_\nu^{(eq)})$, and one obtains with Eq. (12.11)

$$\sum{}_{\text{mat}}[I_\nu] = -k_B \int d\nu\, d^2\Omega\, \frac{1}{h\nu} \ln\left(\frac{1+n_\nu^{(eq)}}{n_\nu^{(eq)}}\right) \mathscr{L}(B_\nu - I_\nu) \ . \tag{12.18}$$

The total entropy production rate $\sum = \sum_{\text{rad}} + \sum_{\text{mat}}$ becomes

$$\sum[I_\nu] = -k_B \int d\nu\, d^2\Omega\, \frac{1}{h\nu} \ln\left(\frac{n_\nu(1+n_\nu^{(eq)})}{n_\nu^{(eq)}(1+n_\nu)}\right) \mathscr{L}(B_\nu - I_\nu) \ . \tag{12.19}$$

This quantity has to be minimized by varying I_ν and considering the constraints given by Eqs. (12.6) and (12.7) with E, $\mathbf{F}$ kept fixed. One has thus to solve

$$\frac{\delta}{\delta I_\nu}\left[\sum[I_\nu] - \lambda_E\left(E - \frac{1}{c}\int d\nu\, d^2\Omega\, I_\nu\right) - \lambda_{\mathbf{F}} \cdot \left(\mathbf{F} - \frac{1}{c}\int d\nu\, d^2\Omega\, \Omega\, I_\nu\right)\right] = 0 \tag{12.20}$$

for I_ν, where the Lagrange multipliers λ_E and λ_F can be eliminated with the help of Eqs. (12.6) and (12.7), which leads then to $I_\nu(\Omega, E, \mathbf{F})$. We mention that in the entropy maximization closure Σ is replaced by S_{rad} given by Eq. (12.16).

12.4 Results

In the following, we will not re-iterate the analytical calculations reported in [7, 8], but immediately discuss the results $\kappa_E^{(\mathrm{eff})}(E, \mathrm{v})$, $\kappa_F^{(\mathrm{eff})}(E, \mathrm{v})$, and $\chi(E, \mathrm{v})$ as functions of E and v, and explain their properties for simple illustrative cases.

12.4.1 Equilibrium Limit

If the radiation field (or photon gas) is in LTE with the matter ($I_\nu = B_\nu$), all transport properties can be obtained by considering the leading order deviations from LTE, $\delta I_\nu = I_\nu - B_\nu$, $\delta E = E - E^{(eq)}$, and $\delta F = F$. A corresponding expansion and subsequent solution of the minimization problem (12.20) leads then to [7, 8]

$$\kappa_E^{(\mathrm{eff})} = \langle \kappa_\nu \rangle_{\mathrm{Ro}}, \tag{12.21}$$

$$\kappa_F^{(\mathrm{eff})} = \langle \kappa_\nu + \sigma_\nu \rangle_{\mathrm{Ro}}, \tag{12.22}$$

$$\chi = \frac{1}{3}, \tag{12.23}$$

where the Rosseland average of any spectral function h_ν is defined by

$$\langle h_\nu \rangle_{\mathrm{Ro}} := \frac{\int_0^\infty d\nu\, \nu^4 \partial_\nu n_\nu^{(eq)}}{\int_0^\infty d\nu\, \nu^4 h_\nu^{-1} \partial_\nu n_\nu^{(eq)}}, \tag{12.24}$$

with ∂_ν being frequency differentiation. If h_ν is a physical rate (per time or per length) the Rosseland mean is the inverse of an average of inverse rates. Thus, frequencies with small κ_ν-values are dominating the Rosseland average, due to a macroscopic number of absorption-emission events (on the considered length scale). In order to establish LTE with matter, the medium must behave optically dense. The result (12.23) means that the stress tensor $\boldsymbol{\Pi}_{kl} = (E/3)\delta_{kl}$ is associated with isotropic radiation.

These LTE results are well-known and can be obtained also with other procedures [3]. Every reasonable closure to the moment equations should provide Eqs. (12.21)–(12.23). The often considered maximum entropy closure is incorrect near equilibrium as has been pointed out by Struchtrup [23]. For LTE, it is obvious from Kohler's work [9], that the entropy production rate is the appropriate quantity to be optimized.

12.4.2 Emission Limit

In this limit emission strongly predominates absorption. It is characterized by very low radiation intensity, $I_\nu \ll B_\nu$, such that $E \ll E^{(eq)}$, and is thus far from equilibrium. One can also derive analytical expressions for the transport coefficients [8] by an expansion in terms of the small quantities I_ν, E, and F. Entropy production minimization gives [7, 8]

$$I_\nu = \frac{2k_B}{c} \frac{\nu^2 \kappa_\nu}{\lambda_E + \lambda_\mathbf{F} \cdot \Omega} n_\nu^{(eq)}, \tag{12.25}$$

where the Lagrange multipliers are related to E and F by

$$E = \frac{k_B \mathcal{T}(\kappa_\nu)}{c^2 \lambda_F} \ln\left(\frac{\lambda_E + \lambda_F}{\lambda_E - \lambda_F}\right), \tag{12.26}$$

$$F = \frac{k_B \mathcal{T}(\kappa_\nu)}{c^2 \lambda_F} \left(2 - \frac{\lambda_E}{\lambda_F} \ln\left(\frac{\lambda_E + \lambda_F}{\lambda_E - \lambda_F}\right)\right). \tag{12.27}$$

Here, we introduced the integral

$$\mathcal{T}(h_\nu) = 4\pi \int_0^\infty \mathrm{d}\nu\, \nu^2 h_\nu n_\nu^{(eq)} \tag{12.28}$$

for frequency dependent functions h_ν. The transport coefficients are given by [8]

$$\kappa_E^{(\mathrm{eff})} = \langle \kappa_\nu \rangle_{\mathrm{Pl}}, \tag{12.29}$$

$$\kappa_F^{(\mathrm{eff})} = \frac{\mathcal{T}(\kappa_\nu(\kappa_\nu + \sigma_\nu))}{\mathcal{T}(\kappa_\nu)}, \tag{12.30}$$

$$\chi(\mathrm{v}) = -\frac{\lambda_E}{\lambda_F} \mathrm{v}, \tag{12.31}$$

where

$$\langle h_\nu \rangle_{\mathrm{Pl}} = \frac{\int_0^\infty \mathrm{d}\nu\, \nu^3 h_\nu n_\nu^{(eq)}}{\int_0^\infty \mathrm{d}\nu\, \nu^3 n_\nu^{(eq)}} \tag{12.32}$$

is the so-called Planck average of a frequency dependent function h_ν. Contrary to the opaque near-equilibrium limit [see Eq. (12.24)] from the previous subsection,

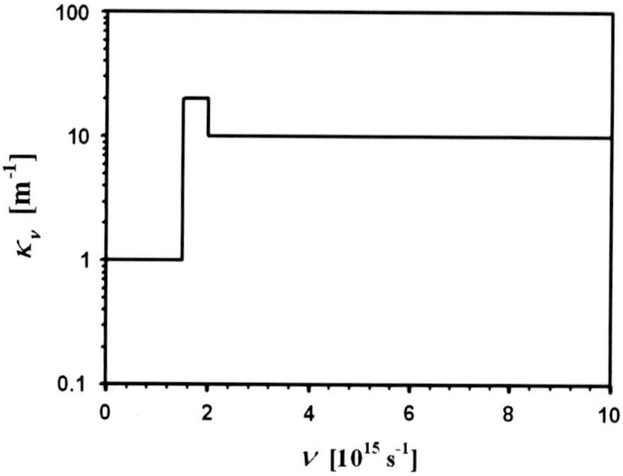

Fig. 12.2 Artificial spectrum with low absorption below a threshold frequency (1.5 PHz) and high absorption above, with an intermediate maximum below 2 PHz

in the transparent emission limit the effective absorption coefficients are averages (12.32) of the direct rates, rather than averages of inverse rates.

The VEF can easily be numerically calculated from the above equations. For small v, an expansion of Eqs. (12.26) and (12.27) gives $\lambda_E/\lambda_F = -1/(3\text{v})$, in accordance with the isotropic limit. In the free streaming limit, $\text{v} \to 1$ from below, one can show that $\lambda_F \to -\lambda_E$ [8], as one expects that $\chi \to 1$.

12.4.3 General Case

For arbitrary values of E and v (or F) the radiation intensity and the transport coefficients must be numerically computed. For use in radiation simulations, it is thus necessary to calculate the transport coefficients for real gases and plasmas and tabulate them as functions of all variables, including temperature and pressure of the LTE matter. In the following, we first consider an illustrative artificial example with negligible scattering ($\sigma_\nu \equiv 0$) and an absorption spectrum shown as in Fig. 12.2. In a frequency band below a certain threshold absorption is low, while at the threshold frequency absorption strongly increases to a maximum, beyond which it again decays or remains constant. The entropy production approach then leads to radiation intensities I_ν plotted in Fig. 12.3. The equilibrium radiation associated with $E = E^{(eq)}$ and $\text{v} = 0$ corresponds to the well-known Planck distribution (solid curve). Nonequilibrium occurs if $E \neq E^{(eq)}$ or $\text{v} \neq 0$. Consider first an isotropic nonequilibrium state where the energy of the radiation is smaller than the equilibrium energy, for instance $E = E^{(eq)}/2$ and $\text{v} = 0$. According to the figure, the radiation (dashed curve) is the closer to equilibrium the larger the absorption constant is. The same holds for the contrary case where the radiation energy is above the equilibrium value ($E = 2E^{(eq)}$ and $\text{v} = 0$, dotted curve). Because the magnitude of the absorption constant is a measure for the interaction

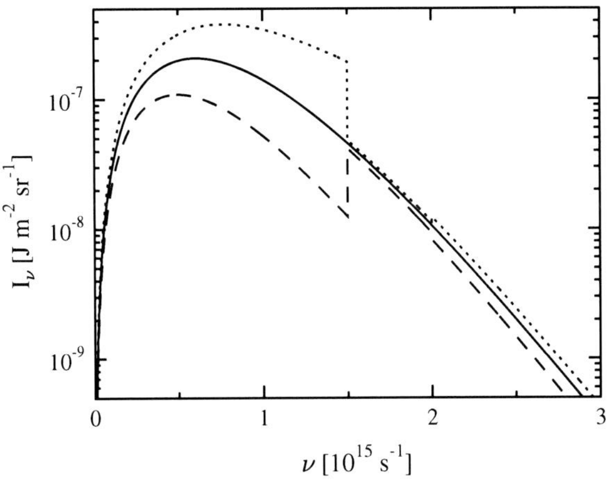

Fig. 12.3 Radiation intensities I_ν for v = 0, $T = 10,300$ K and different values of $E/E^{(eq)}$ and the spectrum given in Fig. 12.2. For $E/E^{(eq)} = 1$ (*solid curve*), equilibrium radiation is established (Planck distribution). For other $E/E^{(eq)}$ values (*dotted* $E/E^{(eq)} = 2$; *dashed* $E/E^{(eq)} = 0.5$), non-equilibrium occurs with a strength that is related to the magnitude of κ_ν. The larger κ_ν, the stronger is equilibration

Fig. 12.4 Effective absorption coefficients $\kappa_E^{(\mathrm{eff})}$ (*solid*) and $\kappa_F^{(\mathrm{eff})}$ (*dashed-dotted*) as functions of the radiation energy for v = 0, for the spectrum shown in Fig. 12.2. The Planck mean, Rosseland mean, and minimum of κ_ν are indicated by the *dotted*, *dashed*, and *thin solid lines*

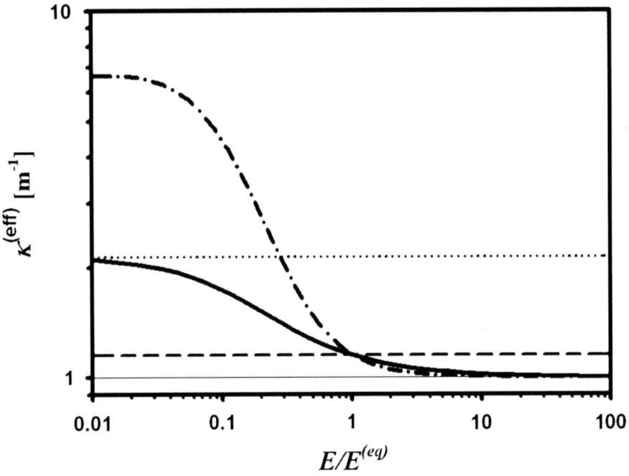

strength between radiation and matter, this behaviour reflects the fact that photons with more intensive interaction with LTE matter are more efficiently equilibrated. Entropy production rate optimization principles inherently take this general tendency into account [24]. For v $\neq$ 0, the intensity I_ν depends on Ω [see, e.g., Eq. (12.25)]; details will not be discussed here.

The mean absorption coefficients are shown in Fig. 12.4. The different limit cases discussed in the previous subsections are indicated by horizontal lines. At equilibrium, all effective absorption coefficients equal the Rosseland mean.

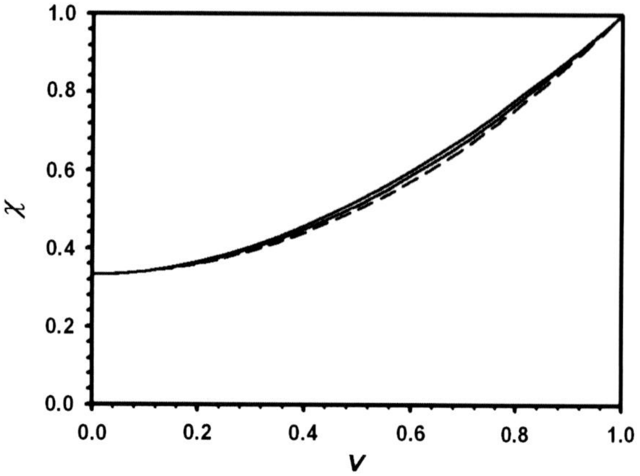

Fig. 12.5 *Solid curves*: The Eddington factor $\chi(\mathrm{v})$ for the spectrum in Fig. 12.2 for two different values $E/E^{(eq)} = 1$ (*lower solid curve*) and 0.5 (*upper solid curve*). *Dashed*: Kershaw approximation Eq. (12.33)

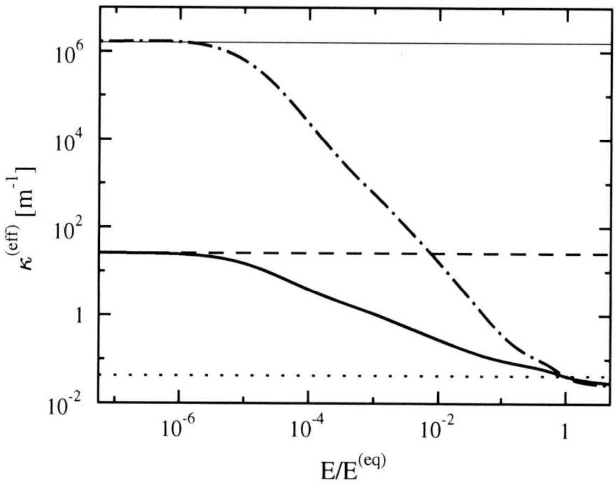

Fig. 12.6 Effective absorption coefficients $\kappa_E^{(\mathrm{eff})}$ (*solid*) and $\kappa_F^{(\mathrm{eff})}$ (*dashed-dotted*) for an air plasma at 10,000 K and 2 bar (see Fig. 12.1) as functions of the radiation energy for $\mathrm{v} = 0$. The Planck mean, Rosseland mean, and emission limit for $\kappa_F^{(\mathrm{eff})}$ from Eq. (12.30) are indicated by the *dashed*, *dotted*, and *thin solid lines*

A closure by entropy maximization would provide a wrong result [7, 8]. We note also that for $E/E^{(eq)} \to \infty$, the entropy production closure leads to a mean absorption dominated by the minimum absorption coefficient. In this limit the overwhelming amount of photons will occupy states with low photon-matter interaction, while the states with stronger interaction (large absorption) will be near the equilibrium distribution.

The VEF as a function of v is shown in Fig. 12.5 for two different E-values. It can be shown that the VEF satisfies a number of conditions [14]. For instance, as mentioned $\chi = 1/3$ for $\mathrm{v} = 0$ (isotropic radiation) and $\chi = 1$ for $\mathrm{v} = 1$ (free streaming limit). Furthermore, the dependence of χ on E is weak, and for many practical purposes $\chi(\mathrm{v})$ is well approximated by Kershaw's VEF [25]

$$\chi = \frac{1 + 2\mathrm{v}^2}{3}, \tag{12.33}$$

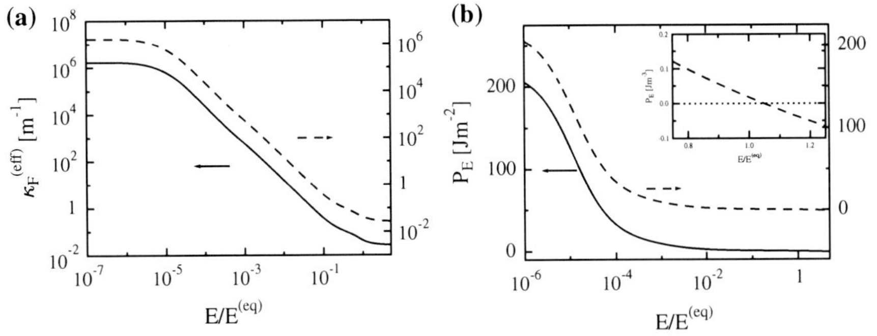

Fig. 12.7 *Left*: Effective absorption coefficient $\kappa_F^{(\text{eff})}$ as a function of radiation energy for $\text{v} = 0$ (*solid line*, left axis) and $\text{v} = 0.2$ (*dashed line*, right axis) *Right*: Absorption power P_E as a function of radiation energy for $\text{v} = 0$ (*solid line*, left axis) and $\text{v} = 0.2$ (*dashed line*, right axis). The inset shows the shift of the zero of P_E from $E/E^{(eq)} = 1$; because of this shift it is inconvenient to directly discuss κ_E, as it is defined here, in a graph

Fig. 12.8 Variable Eddington factor (VEF) for the air spectrum as a function of v for $E/E^{(eq)} = 1$ (*solid*) and $E/E^{(eq)} = 1/4$ (*dashed*). Kershaw's VEF is given as a reference (*dotted*)

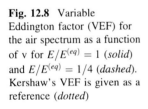

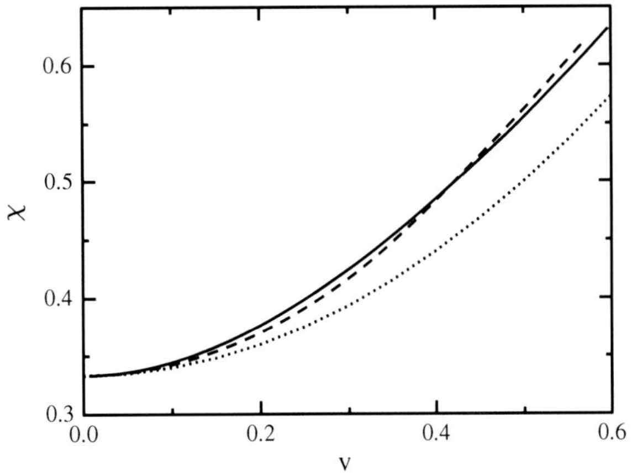

as is illustrated also in Fig. 12.5.

This general behaviour of the transport coefficients observed for the toy example is also valid for more complex absorption spectra. A calculation with a spectrum as given by Fig. 12.1 (air plasma at 10,000 K and 2 bar) has to take into account the small structures from the individual spectral lines and requires a rather high-frequency resolution. As a side remark, we mention an additional difficulty as there is a critical value of the Lagrange multiplier λ_F (for given λ_E) for which the solution of the variational equation (12.20) becomes singular. Even for this value, however, $\text{v} < 1$ and hence in order to go to the streaming limit, the intensity distribution acquires a δ-function contribution: part of the photons then concentrate (condense) at the frequency with minimal κ_ν in the streaming direction.

Details of the behaviour at large v will be described elsewhere, here we restrict ourselves to sufficiently small v-values. The effective absorption coefficients for $v = 0$ are shown in Fig. 12.6. The span over several orders of magnitude between the emission limit and a dense medium is remarkable. We mention that the detailed slightly wavy structures of $\kappa_E^{(\mathrm{eff})}$ and $\kappa_F^{(\mathrm{eff})}$ are not due to numerical inaccuracy but due to the specific frequency dependence of the absorption spectrum.

The numerical solutions show that the v dependence of the effective absorption coefficients is rather weak. In Fig. 12.7, a comparison between $v = 0$ and $v = 0.2$ is shown as an example. For practical use it is much more convenient to depict P_E instead of κ_E, because for finite v the zero of P_E is shifted away from $E = E^{(eq)}$ (cf. definition of κ_E by Eq. (12.11) and inset in Fig. 12.7).

On the other hand, the VEF depends relatively weakly on the energy of the radiation field. Fig. 12.8 shows χ for two different energies as an example.

12.5 Boundary Conditions

In order to have a well-defined hyperbolic problem associated with the partial differential equations (PDEs) (12.9) and (12.10), appropriate boundary conditions on E and $\mathbf{F}$ (or $\mathbf{v}$), at solid surfaces, at certain symmetry planes, and/or at infinity must be added. The qualitative nature of the boundary depends not only on the radiative behavior of the matter but also on the direction of the characteristics of the basic PDEs. This is analogous to gas dynamics, where a boundary condition at an outlet is needless if the Mach number of the flow is larger than one, because no information can travel from the boundary back into the system. For the Eqs. (12.9) and (12.10), this appears if v is larger than a critical value v_c. This value depends on the functional dependence of χ on v, but is typically around 0.7 [8]. If boundary conditions are needed for moment equations, they can be derived by projection of $I_\nu(\Omega)$, expanded in terms of the moments, onto a weight function. Often, the Marshak boundary condition is considered, which can be generalized in the present case to [8]

$$F = \frac{\varepsilon}{2(2-\varepsilon)}\left(E_{\mathrm{w}} - \frac{(3+15\chi)E}{8}\right), \qquad (12.34)$$

where ε is the surface emittance, and E_{w} is the equilibrium radiation energy density associated with the wall temperature. In the equilibrium limit ($\chi = 1/3$), Eq. (12.34) reduces to the usual Marshak boundary condition, as applied, for instance, in the diffusive P1-model [26].

In cases where the surface response to radiation is relevant and a good modelling of the surface behavior is crucial (e.g., if radiation-induced material ablation occurs), it may be more appropriate to include a solid surface layer in the simulation domain with realistic absorption and scattering coefficients [8].

12.6 Summary and Conclusion

We conclude that the entropy production rate is an appropriate variational functional for the closure of the moment equations of radiative transfer. Within the formalism, effective absorption coefficients and variable Eddington factors are calculated that have the correct limiting behavior in the analytically known cases, like the Planck and Rosseland mean absorption, and the VEF in the diffusive and free streaming limits. It turns out that the entropy production principle is superior to the often considered entropy maximization principle, which disregards the specific equilibration mechanisms and yields in general wrong results even in the Rosseland (equilibrium) limit, as has been extensively discussed in general in [23] and for specific examples in [7, 8]. As demonstrated with a toy example and a calculation for a realistic spectrum of air, the effective absorption coefficients can vary over several orders of magnitude in the physically relevant region. This shows that good models are necessary.

The success of the entropy production approach is related to Kohler's principle [9], because the linearity of the RTE is exact and not restricted to a linearization region near equilibrium. From the formalism discussed it is obvious that the approach is not limited to a specific number of moments, and it is applicable to other types of mutually non-interacting particles like neutrinos [27] or independent electrons [20].

References

1. Chandrasekhar, S.: Radiative Transfer. Dover Publ. Inc, New York (1960)
2. Jones, G.R., Fang, M.T.C.: The physics of high-power arcs. Rep. Prog. Phys. **43**, 1415 (1980)
3. Siegel, R., Howell, J.R.: Thermal Radiation Heat Transfer. Washington, Philadelphia (1992)
4. Landau, L.D., Lifshitz, E.M.: Statistical Physics. Elsevier, Amsterdam (2005)
5. Tien, C.L,: Radiation properties of gases. In: Irvine Jr, T.F, Hartnett, J.P (eds.) Advances in heat transfer, vol. 5. Academic Press, Inc., New York, 253 (1968)
6. Chen, G.: Nanoscale Energy Transport and Conversion: A Parallel Treatment of Electrons, Molecules, Phonons, and Photons. Oxford University Press, USA (2005)
7. Christen, T., Kassubek, F.: Minimum entropy production closure of the photo-hydrodynamic equations for radiative heat transfer. J. Quant. Spectrosc. Radiat. Transfer. **110**, 452 (2009)
8. Christen T, Kassubek F, and Gati R.: Radiative heat transfer and effective transport coefficients. In: Aziz Belmiloudi (ed.) Heat Transfer—Mathematical modelling, numerical methods and information technology, InTech, Rijeka, Croatia (2011)
9. Kohler, M.: Behandlung von Nichtgleichgewichtsvorĝangen mit Hilfe eines Extremalprinips. Z. Physik **124**, 772 (1948)
10. Ziman, J.M.: The general variational principle of transport theory. Can. J. Phys. **34**, 1256 (1956)
11. Ziman, J.M.: Electrons and Phonons. Clarendon Press, Oxford (1967)
12. Martyushev, L.M., Seleznev, V.D.: Maximum entropy production principle in physics, chemistry, and biology. Phys. Rep. **426**, 1 (2006)
13. Minerbo, G.N.: Maximum entropy Eddington factors. J. Quant. Spectrosc. Radiat. Transfer. **20**, 541 (1978)

14. Levermore, C.D.: Moment closure hierarchies for kinetic theories. J. Stat. Phys. **83**, 1021 (1996)
15. Turpault, R.: A consistent multigroup model for radiative transfer and its underlying mean opacities. J. Quant. Spectrosc. Radiat. Transfer. **94**, 357 (2005)
16. Essex, C.: Minimum entropy production in the steady state and radiative transfer. The Astrophys. J. **285**, 279 (1984)
17. Würfel, P., Ruppel, W.: The flow equilibrium of a body in a radiative field. J. Phys. C: Solid State Phys. **18**, 2987 (1985)
18. Kabelac, S.: Thermodynamik der Strahlung. Vieweg, Braunschweig (1994)
19. Santillan, M., de Parga, G.A., Angulo-Brown, F.: Black-body radiation and the maximum entropy production regime. Eur. J. Phys. **19**, 361 (1998)
20. Christen, T.: Nonequilibrium distribution function and generalized hydrodynamics for independent electrons from an entropy production rate principle. Europhys. Lett. **89**, 57007 (2010)
21. Oxenius, J.: Radiative transfer and irreversibility. J. Quant. Spectrosc. Radiat. Transfer **6**, 65 (1966)
22. Kröll, W.: Properties of the entropy production due to radiative transfer. J. Quant. Spectrosc. Radiat. Transfer **7**, 715 (1967)
23. Struchtrup H.: Rational extended thermodynamics. Müller I and Ruggeri T. (ed.) p. 308. Springer, New York, Second Edition (1998)
24. Christen, T.: Modeling electric discharges with entropy production rate principles. Entropy **11**, 1042 (2009)
25. Kershaw D, *Flux limiters nature's own way* Lawrence Livermore Laboratory UCRL-78378 (1976)
26. Nordborg, H., Iordanidis, A.: Self-consistent radiation based modelling of electric arcs: I. Efficient radiation approximations. J. Phys. D Appl. Phys. **41**, 135205 (2008)
27. Essex C and Kennedy DC J. Stat. Phys. **94**, 253 (1999). (or Minimum entropy production of neutrino radiation in the steady state, Report No: DOE/ER/40272-280 UFIFT-HEP-97- 7)
28. Data for the air spectrum has been provided by R. Gati (ABB Corporate Research) with use of tools by V. Aubrecht
29. Aubrecht, V., Lowke, J.J.: Calculations of radiation transfer in SF6 plasmas using the method ofpartial characteristics. J. Phys. D Appl. Phys. **27**, 2066 (1994)
30. Chaveau, S., et al.: Radiative transfer in LTE air plasmas for temperatures up to 15,000 K. J. Quant. Spectrosc. Radiat. Transfer. **77**, 113 (2003)

Chapter 13
MaxEP and Stable Configurations in Fluid–Solid Interactions

Ashwin Vaidya

Abstract We review the experimental and theoretical literature on the steady terminal orientation of a body as it settles in a viscous fluid. The terminal orientation of a rigid body is a classic example of a system out of equilibrium. While the dynamical equations are effective in deriving the equilibrium states, they are far too complex and intractable as of yet to resolve questions about the nature of stability of the solutions. The maximum entropy production principle is therefore invoked, as a selection principle, to understand the stable, steady state patterns. Some on-going work and inherent complexities of fluid solid systems are also discussed.

13.1 Introduction

This chapter is concerned with examining the thermodynamic principles behind fluid structure interactions. Variational principles abound in physics and much has been written about the mathematical, physical and philosophical aspects of these theories. Perhaps the earliest of such explanations came from Fermat through the principle of least time followed by Maupertius in his formulation of the principle of least action. One of the most noteworthy versions of these principles is Hamilton's principle of least action which is still a part of the standard physics curriculum and has found use in various applications (Fig. 13.1).

In addition, one can find several other optimal principles such as the principle of minimum potential, principle of minimum energy and principle of maximum entropy among others. In this chapter, we discuss the maximum entropy production principle (MaxEP) which has been gaining ground, particularly in the past decade.

A. Vaidya (✉)
Department of Mathematical Sciences, Montclair State University,
Montclair, NJ 07043, USA
e-mail: vaidyaa@mail.montclair.edu

R. C. Dewar et al. (eds.), *Beyond the Second Law*,
Understanding Complex Systems, DOI: 10.1007/978-3-642-40154-1_13,

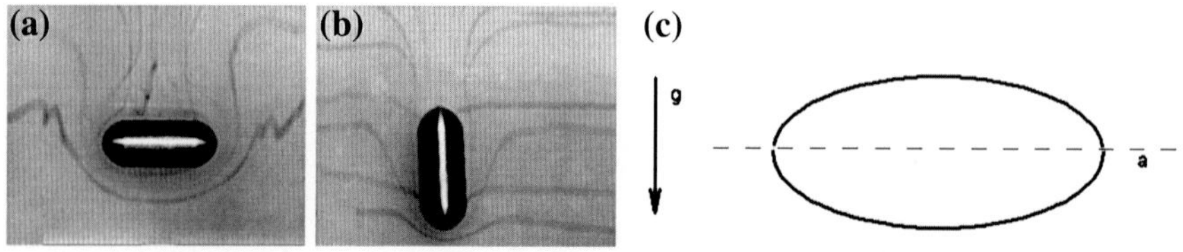

Fig. 13.1 This figure shows the terminal orientation of a sedimenting cylinder in a **a** Newtonian fluid and **b** viscoelastic fluid (Polyoxide) (Courtsey of Prof. D.D. Joseph). The cartoon in panel **c** helps clarify the meaning of the axis *a* which has been used to describe the terminal orientation of a sedimenting body

The purpose of the chapter is to point to some evidence for the utility of this 'variational principle' in certain complex problems in fluid mechanics while also reviewing its successes in other fields with the goal of examining its role as an overarching optimal principle of nature. The physical sciences have been slow in accepting this principle despite its many successes in various disciplines for over half a century now. This reluctance can perhaps be attributed to looming questions about the teleological nature of the theory which has been a subject of some controversy in regards to variational arguments [1]. Even historically, the now universally accepted 'Least action principle' and other such variational arguments have been questioned by philosophers of science who attribute an 'end purpose' to such theories [2] thereby diminishing their value over causal arguments. While addressing the issue of teleology and science is outside the scope of this chapter, we approach the issue of an optimal principle in the 'Popperian' sense that the theory is only as good as its ability to withstand the test of falsification. In this article we lend some credence to this argument by pointing to some interesting examples where the MaxEP argument can provide valuable insight.

13.2 Orientation of a Sedimenting Body

It is well established that homogeneous bodies of revolution around an axis, a, with fore-aft symmetry, when dropped in a quiescent liquid, will orient themselves in certain ways with respect to the direction of gravity. The orientation is seen to depend upon the shape of the body and also upon the nature of the fluid in which they are immersed.

13.2.1 The Steady State Case

In a highly viscous fluid, in creeping flow regimes, the body is seen to keep its initial orientation as it falls [3]. In a Newtonian fluid when the inertia of the fluid exceeds the viscous forces, the body falls with *a* eventually becoming perpendicular to the direction of gravity. If the same body falls in a viscoelastic fluid, such as a polymer, where the inertial and elastic effects compete, then, a will eventually become parallel to the direction of gravity. In fact, the orientation behavior becomes very complex in viscoelastic fluids[1] since, at critical concentrations of the polymer, it can also allow for some intermediate angles, referred to in the literature as tilt angles [4–6]. In fact the tilt angle varies continuously with the polymer concentration thereby allowing the particle to fall anywhere between the horizontal and vertical state. Theoretical explanations of these observations have been provided in a variety of fluid models, Newtonian and non-Newtonian by considering that in the terminal state, the net torque imposed by the body on the fluid, due to viscosity (constant and shear dependent), inertia and viscoelasticity (or normal stress) must be in equilibrium. Hence, in its steady state, the terminal angle can be obtained from the vanishing of the net torque [7–10] $\mathcal{M}$, which can be decomposed as

$$\mathcal{M}(\theta) = \mathcal{M}_v + Re\,\mathcal{M}_I + We\,\mathcal{M}_{NN}. \qquad (13.1)$$

Here, $\mathcal{M}_v$ refers to the viscous component of the torque, $\mathcal{M}_I$ is the inertial component and $\mathcal{M}_{NN}$ is the non-Newtonian part of the torque. The previous, mechanical approach successfully explains the orientation phenomena in various cases at first order in Re and We (where the Reynolds number $Re = \frac{UL}{\mu}$ measures the relative strength of fluid inertia to viscosity and the Weissenberg number, $We = \frac{\alpha U}{L\mu}$, measures the relative strengths of elasticity to viscosity, where L and U are the characteristic length and velocity, μ is the viscosity and α is the normal stress coefficient of the fluid). In particular previous computations have been successful at obtaining the equilibrium states of a falling body. The question of stability is however, not easily addressed; it involves the resolution of complex, coupled, nonlinear integro-differential equations which are currently being addressed. In this chapter, we argue that the MaxEP principle serves as a useful alternative selection principle to identify the appropriate stable steady configuration in this problem.

13.2.2 The Time Dependent Case

The orientational dynamics becomes even more interesting in the unsteady regime where vortex shedding effects become significant and give rise to oscillations of the body . In the context of sedimentation, several relevant studies have been conducted, both experimental and numerical to document the highly nonlinear dynamics of disk like bodies, i.e. bodies whose aspect ratios (length to diameter ratio, denoted τ) are much less than 1, which typically represent disks or flat plates.

[1] Viscoelastic fluids are a class of non-Newtonian fluids which display viscosity and elasticity and normal stresses which give rise to ‘memory’ effects, as in elastic solids.

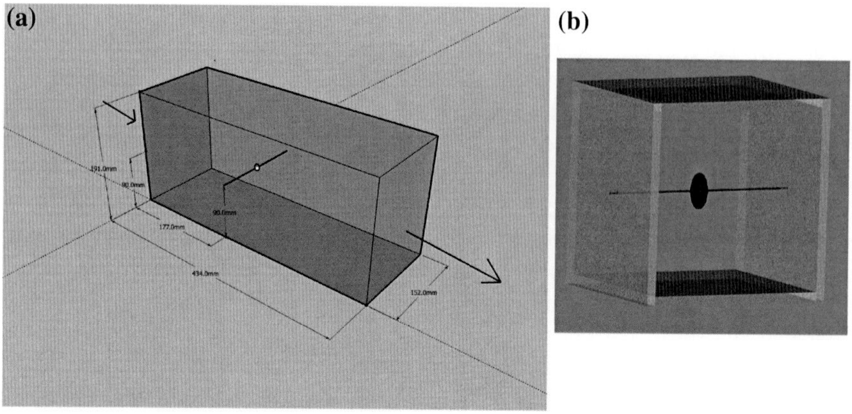

Fig. 13.2 The Panel (**a**) shows a schematic of the experimental setup for the time dependent experiments. The *arrows* indicate the flow direction of water in the flow chamber. The Panel (**b**) shows the details of the particle suspension mechanism which restricts the motion of the body (*cylinder* or *prolate spheroid*) to rotation about the suspension axis alone

See for instance [11–14] and references therein. It is seen that a sedimenting disk or flat plate can exhibit (1) fluttering, (2) tumbling and (3) chaotic motions, depending upon the Re . Attempts have been made to classify these different phenomena by means of non-dimensional parameters such as particle aspect ratio, reduced inertia (defined by $I^* = \frac{I}{\rho_f d^5}$, where I is the moment of inertia of the body with respect to the symmetry axis and ρ_f stands for the fluid density and d is the characteristic length of the cylinder), Froude number and Strouhal number ($Sr = \frac{fd}{U}$ where f is the frequency of oscillation and U is the characteristic velocity).

Our preliminary time-dependent experimental study was carried out in a horizontal recirculating water tunnel (see Fig. 13.2). Most of the previous studies have been conducted for sedimenting bodies (spheroids, cylinders, spheres etc.) which are allowed to fall through a fixed height. In such experiments, however the only way to achieve Re in the intermediate range is by increasing the density of the particle. This results in a reduction of the observation time with the only possible solution being the design of a large tank which could be rather expensive. Therefore we consider a better alternative whereby the particle is fixed in the center of a recirculating flow tank in which the fluid flow can be controlled, thus changing the Re (see Fig. 13.2). The advantage of this experiment is that it allows for very long observation times when compared to the case of sedimentation while displaying the same orientation dynamics in the range of Re explored. In this case, however, the particle only has one degree of rotational freedom around its axis of suspension. For more details about the experiments, the readers are referred to our earlier chapter on the subject [15]. We investigated the dynamics of cylindrical bodies in flows with Reynolds numbers based on particle dimensions in the range $0 < Re < 6{,}000$ and the non-dimensional inertia, $0 < I^* < 0.6$ by changing the

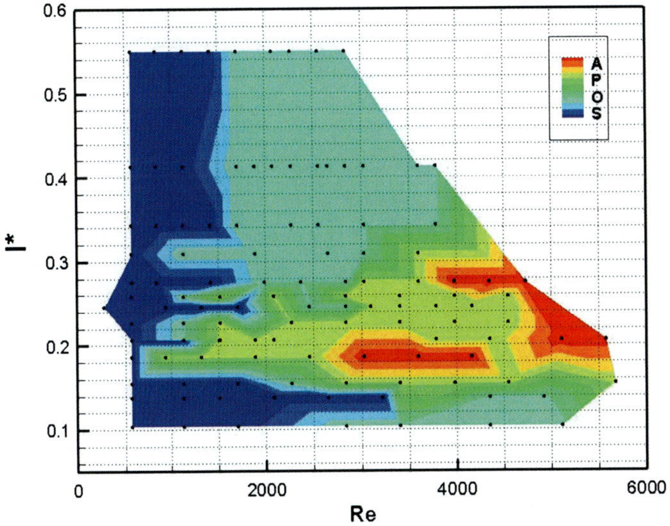

Fig. 13.3 The figure shows a bifurcation diagram displaying the variety of orientation dynamics displayed by a cylinder in a flow—based on the experiments in a flow tank—as a function of the non-dimensional inertia and Reynolds number. In the figure *A* represents autorotation (or tumbling of the cylinder), *P*: periodic oscillations (where the cylinder oscillates with a consistent frequency), *O*: flutter (where the oscillations are random and very small) and *S*: steady state [15]

flow speed, particle density and aspect ratio [15]. We examined the various possible motions based upon variations in *Re* and I^* which range from *steady orientation* (*S*), *random fluctuations* (*O*), *periodic oscillation* (*P*) *to autorotation* (*A*). Figure 13.3 summarizes the multiple bifurcations observed in this phenomena and the appropriate steady state case. More details about the oscillatory state are discussed in our recent chapter on the subject [16] and we refer the readers to a movie [17] documenting the experiments under discussion here.

13.3 The MaxEP Principle

It is very often found that problems concerning pattern formation are intricately related to optimal principles and conservation laws such as the principle of minimum potential energy, principle of least action, Fermat's principle of least time etc. It has always seemed to us therefore that some such quantity must be optimized in the orientation problem since the problem that we are studying is one of pattern selection. The message of this article are two-fold: (1) The first is to establish that the problem of terminal orientation of a symmetric body in a fluid is governed by an optimal principle which is related to the entropy production of the system, i.e. an optimal principle exists and (2) secondly, we want to show that the nature of the extremum is also an important issue and is related to the choice of the extremizing variable.

In this section, we review some essential points concerning dissipative structures in fluid mechanics.[2] The application of non-equilibrium thermodynamics to fluid mechanics has been a subject of some importance in the past [21–26] but seems to have been sidelined in the recent literature. The greatest relevance of this subject to fluid motion, of course, lies in the regime of turbulence. However, it is observed that even in the case of slow flows, thermodynamics plays an essential role, since fluid motion is inherently dissipative in nature. Motivated by the work of Zeigler [26], there have been several studies, in particular, concerning the application of thermodynamic principles to the constitutive modeling of complex fluids (non-Newtonian fluids). Specifically, the requirement that viscous energy dissipation be non-negative is very useful in obtaining restriction on the material parameters of the constitutive models [27].

The second law of thermodynamics leads us to the local entropy equation which describes a system out of equilibrium, namely

$$\frac{\partial(\rho_f s)}{\partial t} + \operatorname{div} j_s = \sigma_s \tag{13.2}$$

where ρ_f represents the fluid density, s represents the entropy density, $\mathbf{j}_s$ is the entropy flux density and σ_s is the local entropy production. It has been established [28] that the equation for entropy production can be given by the product of forces[3] (denoted X) and fluxes (denoted Y)

$$\sigma_s = \sum_i X_i Y_i + \sum_j X_j Y_j + \sum_{kl} X_{kl} Y_{kl} \tag{13.3}$$

which may be represented as scalars, vectors or second order tensors. Onsager suggested that for looking at near equilibrium phenomena, we may represent the fluxes as a linear function of the forces, namely

$$Y_i(X_1, X_2, \ldots, X_n) \; = \; L_{ij} X_j, Y_i(0, 0, \ldots, 0) \; = \; 0$$

where L_{ij} represent phenomenological constants which satisfy the well known Onsager reciprocity relations [28, 29],

$$L_{ii} \geq 0, \; L_{ii} L_{kk} \geq \frac{1}{4}(L_{ik} + L_{ki})^2 \tag{13.4}$$

In the case of motion of an incompressible fluid, $\sigma_s = \sigma_s(x)$ takes the form [19]

$$\sigma_s = \frac{1}{T}\mathbf{T} : \mathbf{D} + \mathbf{j}_q \cdot \nabla\left(\frac{1}{T}\right) - \frac{1}{T}(\rho - \rho_f)\mathbf{g} \cdot \mathbf{U} \tag{13.5}$$

[2] See [18–20] for an introduction to the subject of non-equilibrium thermodynamics.

[3] The forces may originate from hydrodynamic viscosity, chemical reactions, thermal gradients etc.

where T is the temperature, $\mathbf{j}_q$ is the heat flux, $\mathbf{T}$ is the Cauchy stress tensor, $\mathbf{D}$ is the symmetric part of the velocity gradient [30] $\mathbf{D} = \frac{1}{2}(\nabla \mathbf{u} + (\nabla \mathbf{u})^T)$ and $\rho = \rho(\mathbf{x})$ refers to the density of the material occupying the position X. This choice of σ_s is in line with the Curie principle and we will remark on its structure a little more below once it is written in its integral form. The first term on the right hand side represents the viscous dissipation term while the second term refers to heat conduction due to a temperature gradient and the third term refers to the rate of work done by the system. The integral of the local entropy production over the entire (unbounded) domain, Ω_∞ yields the entropy production $\mathcal{P}$. In order to obtain a simple form of the total entropy production, we write $\Omega_\infty = \Omega \cup \mathcal{B}$ where Ω refers to the fluid domain and $\mathcal{B}$ refers to the region occupied by the rigid body.[4] It is easy to see that the dissipation term vanishes over $\mathcal{B}$ since there is no flow in this region while the integral of $\rho - \rho_f$ vanishes over Ω since the density difference is zero in this region; the integral of this term over $\mathcal{B}$ yields the effective mass of the body. In the rest of this chapter we assume, as in [23] that the effect of the heat conduction term is negligible and also that the ambient temperature $T = T_0$ is a constant (see also [31]). Therefore the integral form of entropy production functional is (see also [32, 33])

$$\mathcal{P} = \frac{1}{T_0} \int_\Omega \mathbf{T} : \mathbf{D} \, dV - \frac{m_e}{T_0} \mathbf{g} \cdot \mathbf{U} \tag{13.6}$$

where and $m_e = (\rho_b - \rho_f)|\mathcal{B}|$ is the effective mass, where $|\mathcal{B}|$ represents the volume of the body and $\mathbf{g}$ refers to the acceleration due to gravity. We decompose the stress tensor in general into a Newtonian and a non-Newtonian component $\mathbf{T} = \mathbf{T}_N + \alpha \mathbf{T}_{NN}$ [34, 35] where $\mathbf{T}_N = -p\mathbf{I} + 2\mu \mathbf{D}$, p being the isotropic pressure and α represent the viscoelastic material parameter. The entropy production is therefore given by

$$\mathcal{P} = \frac{1}{T_0} \int_\Omega \Big(\mathbf{T}_N : \mathbf{D} + \frac{\alpha}{T_0} \int_\Omega \mathbf{T}_{NN} : \mathbf{D} \Big) dV - \frac{m_e}{T_0} \mathbf{g} \cdot \mathbf{U} \tag{13.7}$$

$$= \frac{2\mu}{T_0} \int_\Omega \mathbf{D} : \mathbf{D} \, dV + \frac{\alpha}{T_0} \int_\Omega \mathbf{T}_{NN} : \mathbf{D} \, dV - \frac{m_e}{T_0} \mathbf{g} \cdot \mathbf{U} \tag{13.8}$$

where in the last equation the pressure term drops out of the stress tensor due to the divergence free nature of the velocity field.

The Eq. (13.6) is not merely a mathematical construct but based on physically sound reasoning. In order to interpret this equation appropriately, one must see the problem now from the point of view of the fluid. We interpret the entropy of the system as the energy dispersed by the system, which cannot be employed to do useful work, which we also identify with the inertial term i.e. the term involving the *Re* [33] in the linear momentum equation. In the Eq. (13.6), we take the first

---

[4] Note that $\int_{\Omega_\infty} = \int_\Omega + \int_\beta$.

term on the right hand side to refer to the rate of change of energy entering the system and the second term to mean the rate at which the fluid performs useful work (i.e. work done in changing its potential energy). Therefore the excess energy in the system, which cannot be used to perform useful work, must be dispersed to the surroundings at a rate given by the rate of change of kinetic energy of the system [36] which, as stated above is dictated by the *Re* which can also be interpreted as a measure of irreversibility. Our definition of entropy production can be seen to be closely related to the concept of available or excess potential energy that has been discussed earlier by Lorenz [37], Ozawa [38] and Kleidon [39].

In order to simplify the equations further, we consider a rigid body of any shape moving in a fluid in its steady state. We consider a frame attached to the body[5] with the origin at the center of gravity and hence we can decompose the translational and rotational motion of the fluid from that of the body in the following manner:

$$\mathbf{u} = \sum_{i=1}^{3} \left(U_i \mathbf{h}^{(i)} + \Omega_i \mathbf{H}^{(i)} \right) \tag{13.9}$$

where $\mathbf{h}^{(i)}$ and $\mathrm{H}^{(i)}$ (i=1, 2, 3) are the translational and rotational auxiliary incompressible fields satisfying the steady Stokes equations with no slip conditions and respectively equal to $\mathbf{e}_i$ and $\mathbf{x} \times \mathbf{e}_i$ as $x \to \infty$ (see [9, 40] for a discussion about these auxiliary fields). It must be kept in mind that this approximation is valid for very small Reynolds numbers, *Re*. We justify such a linearization using the fact that experimentally, the orientation phenomena is valid for extremely small *Re*, when the system is longer in equilibrium. In fact, the Reynolds numbers in the experiments performed can be as small as 0.016 (see [9] and references cited therein). In the rest of this section, we examine the problem of particle orientation for three different cases (1) Stokes case, (2) Newtonian case in the presence of small inertial effects and (3) viscoelastic case.

Using the Eq. (13.5), Horne et al. [23], ignoring the heat conduction term, have shown that steady flow of a viscous Newtonian fluid in some simple geometries is seen to coincide with the minimum of the entropy production, subject to a constant pressure gradient. They observe that this principle may be invalid when the system reaches a state which is far from equilibrium. The extrema (minimization and maximization) of entropy production in irreversible processes have been found to be valid in several physical contexts [21, 24, 25, 31, 38, 41–48]. There has been considerable confusion on whether the appropriate extrema to apply is a maxima or a minima and how to distinguish between the two states. We refer the readers to an up to date review about this issue and its relevance to fluid flow, written by Niven [46]. By analyzing flow in single and parallel pipes, Niven discusses appropriate conditions under which flows satisfy the minimum or the maximum

5 While the results of our calculations are frame independent, the body frame, if appropriately chosen to align with the natural symmetries of the body, can make the computations considerably simple.

EP principle, while maintaining that the latter is a more fundamental one. Martyushev and Seleznev [43] state that the two are equivalent concepts and the MinEP which is limited in its scope is only a special case of the MaxEP principle. Our own work on particle sedimentation (see [32, 33] and below) seems to indicate that one possible difference in the maximum and minimum principles arises simply from the variable with respect to which we choose to extremize $\mathcal{P}$. We provide more comments on this issue in the concluding section.

13.3.1 Stokes Case

We first consider the problem of sedimentation in a Newtonian fluid, i.e. $\alpha = 0$. Putting Eq. (13.9) into Eq. (13.7), we may write $\mathcal{P}$ as

$$T_0 \mathcal{P} = U_i\left(K_{ij}U_j - m_e\, g_i\right) + 2U_iC_{ij}\Omega_j + \Omega_i M_{ij}\Omega_j \tag{13.10}$$

where we define [49]

$$K_{ij} = 2\mu \int_\Omega \mathbf{D}(\mathbf{h}^{(i)}) : \mathbf{D}(\mathbf{h}^{(j)})\, dV, \tag{13.11}$$

$$C_{ij} = 2\mu \int_\Omega \mathbf{D}(\mathbf{h}^{(i)}) : \mathbf{D}(\mathbf{H}^{(j)})\, dV \tag{13.12}$$

$$M_{ij} = 2\mu \int_\Omega \mathbf{D}(\mathbf{H}^{(i)}) : \mathbf{D}(\mathbf{H}^{(j)})\, dV. \tag{13.13}$$

We employ the Einstein summation convention with i, j=1, 2, 3. The appropriate Onsager's phenomenological constants in this problem are K_{ij}, M_{ij} and the gravity term which are easily seen to satisfy the reciprocity relations given by Eq. (13.4). In the Stokes regime, when $Re = 0$, since inertial effects are absent, the balance of linear and angular momentum yield [49]

$$K_{ij}U_j + C_{ij}\Omega_j = m_e g_i \tag{13.14}$$

$$C_{ij}^T U_j + M_{ij}\Omega_j = 0. \tag{13.15}$$

Putting Eqs. (13.14) and (13.15) into the entropy equation automatically yields $\mathcal{P} = 0$. The zero entropy production case is therefore identified with the creeping flow regime and the vanishing of $\mathcal{P}$ indicates that sedimentation if slow enough, is a reversible process [23]. Also, since $\mathcal{P}$ is independent of $\mathbf{U} = (U_1, U_2, U_3)$ and hence is also independent of the orientation of the falling body. In other words, we see that in the creeping motion regime, the sedimenting body can fall with any orientation, which is consistent with the observations of Leal [3].

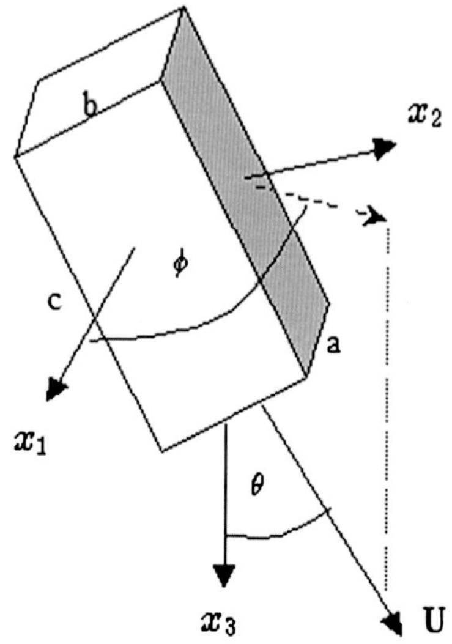

Fig. 13.4 A schematic of the reference frame attached to the body and the velocity vector. The origin of the body frame coincides with the center of gravity of the body

13.3.2 Irreversible Case

In the case where inertial effects appear, $\mathcal{P} > 0$ (while still letting $\alpha = 0$). The corresponding expression for force displays the emergence of an inertial force term for non-vanishing *Re* [32, 33]. As a consequence, there is an interesting transition in the behavior of the sedimenting body from the earlier case as indicated in Fig. 13.3 which can be predicted from applying the MaxEP principle to the form of $\mathcal{P}$ in Eq. (13.6). Before we do so, we will follow along the lines of Happel and Brenner [49] who have studied the various forms that the tensor K_{ij} can take corresponding to different symmetries of the falling body.

In order to see the relation between the terminal orientation of the body with $\mathcal{P}$ clearly, we write $\mathbf{U} = (U\,\cos\phi\sin\theta, -U\,\sin\theta\sin\phi, U\,\cos\theta)$ in polar coordinates (see Fig. 13.4) with $U = |\mathbf{U}|$. As a result, we have

$$\begin{aligned} T_0\mathcal{P} = {} & U^2(K_{11}\cos\phi^2\sin\theta^2 + K_{22}\sin\theta^2\sin\phi^2 + K_{33}\cos\theta^2 - K_{12}\sin2\phi\sin\theta \\ & + K_{13}\sin2\theta\cos\phi - K_{23}\sin2\theta\sin\phi) + 2U(C_{11}\Omega_1 + C_{12}\Omega_2 \\ & + C_{13}\Omega_3)\cos\phi\sin\phi - 2U(C_{21}\Omega_1 + C_{22}\Omega_2 + C_{23}\Omega_3)\sin\phi\sin\theta \\ & + \sum_{i,j=1}^{3} M_{ij}\Omega_i\Omega_j - m_e g \cdot \mathbf{U} \end{aligned} \quad (13.16)$$

Note that the last term, $\mathbf{U} \cdot \mathbf{g}$, in fact depends only on the angle between the velocity field and direction of gravity and is hence independent of θ and ϕ. Therefore, to estimate the extremum of $\mathcal{P}$, we consider derivatives now with respect to (θ, ϕ) which gives us

$$\frac{\partial \mathcal{P}}{\partial \theta} = U^2 \sin 2\theta (K_{11} \cos\phi + K_{22} \sin\phi - K_{33}) - K_{12} \sin 2\phi \cos\theta \\ + 2K_{13} \cos 2\theta \cos\phi - K_{23} \cos 2\theta \sin\phi + 2U(C_{11}\Omega_1 + C_{12}\Omega_2 \\ + C_{13}\Omega_3) \cos\phi \cos\theta + 2U(C_{21}\Omega_1 + C_{22}\Omega_2 + C_{23}\Omega_3) \sin\phi \cos\theta \tag{13.17}$$

$$\frac{\partial \mathcal{P}}{\partial \theta} = U^2(-K_{11} \sin 2\phi \sin\theta^2 + K_{22} \sin 2\phi \sin\theta^2 - 2K_{12} \cos 2\phi \sin\theta \\ - K_{13} \sin 2\theta \sin\phi - K_{23} \sin 2\theta \cos\phi) - 2U(C_{11}\Omega_1 + C_{12}\Omega_2 \\ + C_{13}\Omega_3) \sin\phi \sin\theta + 2U(C_{21}\Omega_1 + C_{22}\Omega_2 + C_{23}\Omega_3) \cos\phi \sin\theta \tag{13.18}$$

The solutions to (13.17) and (13.18) can be obtained from $\frac{\partial P}{\partial \theta} = 0$, $\frac{\partial P}{\partial \phi} = 0$ which we denote (θ_0, ϕ_0). Among the immediate observations that we can make include the following special cases:

13.3.2.1 Isotropic Symmetry

In the case of isotropic bodies such as a sphere, $K_{11} = K_{22} = K_{33} = K$ and $\mathcal{P} = KU^2 - m_e g \cdot U$. Since both these terms are independent of the angles θ and ϕ, we can say that *for the sphere all angles are permitted.*

13.3.2.2 Axis or Revolution and Fore-aft Symmetry

Secondly, for the case of bodies with an axis of revolution and fore-aft symmetry we can choose the reference frame such that $U = (U\cos\phi, -U\sin\phi, 0)$, i.e. $\theta = \pi/2$. Also, for such bodies $K_{12} = 0$. Therefore the derivative of $\mathcal{P}$ with respect to ϕ gives us

$$\frac{\partial \mathcal{P}}{\partial \phi} = U^2(K_{22} - K_{11}) \sin 2\phi \tag{13.19}$$

yielding two possible equilibrium states $\phi = 0$ and $\phi = \pi/2$ [32, 33]. Numerical computations for prolate spheroids indicate that $K_{22} > K_{11}$ [32]. In our more recent study [33], we also observed that this relationship is true for a variety of other bodies such as oblate spheroids, cylinders and tori provided the symmetry resulted in $K_{12} = 0$ [32]. It is now simple enough to verify that $\mathcal{P}$ has a maximum when $\phi = \pi/2$ corresponding with experimental observations. The MaxEP principle therefore serves as a selection principle to find the stable configuration from the possible equilibrium states.

13.3.2.3 Orthotropic Symmetry

Consider now an orthotropic body i.e. one with 3 planes of symmetry but without an axis of revolution such as a tri-axial ellipsoid or a rectangular block with all three dimensions different. In such cases, choosing the origin at the center of reaction [49], we have

$$0 = K_{12} = K_{21} = K_{13} = K_{31} = K_{23} = K_{32} \tag{13.20}$$

$$= M_{12} = M_{21} = M_{13} = M_{31} = M_{23} = M_{32} \tag{13.21}$$

$$= C_{ij}(i,j = 1,2,3). \tag{13.22}$$

Hence

$$\frac{\partial \mathcal{P}}{\partial \theta} = U^2 \sin 2\theta (K_{11}\cos\phi + K_{22}\sin\phi - K_{33}) \tag{13.23}$$

$$\frac{\partial \mathcal{P}}{\partial \phi} = U^2 (K_{22} - K_{11}) \sin^2\theta \sin 2\phi \tag{13.24}$$

The equilibrium states corresponding to Eqs. (13.23) and (13.24) are $(\theta_0, \phi_0) = (0,0), (0, \pi/2), (\pi/2, 0), (\pi/2, \pi/2), (0, \phi*)$ where $\phi*$ satisfies the equation

$$K_{11}\cos\phi * + K_{22}\sin\phi * - K_{33} = 0.$$

Therefore the second variation of $\mathcal{P}$, gives us an idea about the nature of the extremum.

$$\begin{aligned} \mathcal{D}(\theta, \phi) &= \frac{\partial^2 \mathcal{P}}{\partial \theta^2}\frac{\partial^2 \mathcal{P}}{\partial \phi^2} - \left(\frac{\partial^2 \mathcal{P}}{\partial \theta \partial \phi}\right)^2 \\ &= 4U^4 (K_{22} - K_{11})(K_{11}\cos\phi + K_{22}\sin\phi - K_{33})\cos 2\phi \cos 2\theta \sin^2\theta. \end{aligned} \tag{13.25}$$

Evaluating $\mathcal{D}(\theta, \phi)$ at the different equilibrium points yields inconclusive results (i.e. $\mathcal{D}$ vanishes) at $(0,0), (0, \pi/2), (0, \phi*)$. However,

$$\mathcal{D}(\pi/2, 0) = 4U^4 (K_{22} - K_{11})(K_{33} - K_{11}) = \mathcal{D}_1 \neq 0 \tag{13.26}$$

$$\mathcal{D}(\pi/2, \pi/2) = 4U^4 (K_{22} - K_{11})(K_{22} - K_{33}) = D_2 \neq 0. \tag{13.27}$$

The sign of $\mathcal{D}_1, \mathcal{D}_2$ along with that of $\frac{\partial^2 \mathcal{P}}{\partial \theta^2}$ at the two equilibria determines the appropriate stable configuration. While we have not evaluated the values of the drag coefficients we make an educated guess below. It is well known that the drag coefficient is larger for bodies with larger frontal areas. Therefore considering the rectangular block as in Fig. 13.4 such that the areas of the x_2x_3, x_1x_2 and x_1x_3 planes are bc, ac and ab respectively $(c > b > a)$, then it follows by geometric considerations that $K_{11} > K_{22} > K_{33}$. As a result $\mathcal{D}_2 < 0$ indicating a saddle point at $(\pi/2, \pi/2)$. Further, $\mathcal{D}_1 > 0$ and $\frac{\partial^2 \mathcal{P}}{\partial \theta^2}|_{(\pi/2,0)} = -2U^2(K_{11} - K_{33}) < 0$. Therefore $(\pi/2, 0)$ represents a maxima for $\mathcal{P}$ corresponding to experimental observations.

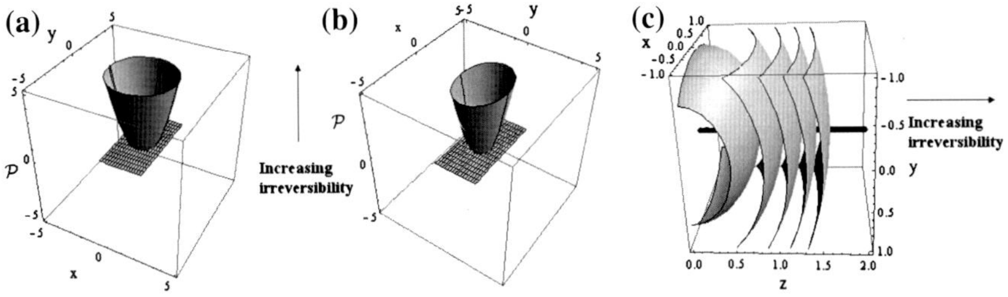

Fig. 13.5 This figure shows the path of maximum entropy production for the case of a sedimenting **a** sphere, **b** spheroid and **c** tri-axial ellipsoid. It represents the phase space showing the various states of the system with increasing *Re*

13.3.2.4 The MaxEP Path

Perhaps one of the most succinct definitions of the MaxEP principle is due to Swenson [50] which states: *A system will select the path or assemblage of paths out of available paths that minimizes the potential or maximizes the entropy at the fastest rate given the constraints.* While this statement indicates that it is the maximum rate of entropy production (with respect to time) which underlies non-equilibrium behavior, things have not been quite as clear. There has been some confusion regarding the nature of the extrema of $\mathcal{P}$ such as due to Prigogine [18] who argues for a minimum of entropy production. Martyushev and Seleznev [43] have quite well articulated this issue and tried to reconcile the problem in their review chapter. Our own results indicate that the issue of the nature of extrema could be a 'red herring'; the extremum could be determined by the form of the independent variable that is considered. In Eq. (13.16), taking the derivative with respect to U gives a minimum (although the minimum may not be physical as we will see soon) while the derivatives with respect to θ or ϕ can yield more complicated results as we have seen in the previous section. However, perhaps a more appropriate way to view this result denoted 'the path of steepest ascent' has been suggested in a series of papers by Beretta [51–54] who has successfully applied this idea in the area of Quantum Thermodynamics. While the examples considered by Beretta have no direct bearing upon the problem considered in this article, we consider it a useful analogy. Let us plot the contours of the equation $\mathcal{P} =$ constant where the value of the constant depends upon the *Re* or magnitude of irreversibility in the system (see Fig. 13.5). Therefore increasing *Re* would be reflected by an increasing value of the right hand side with $\mathcal{P} = 0$ referring to the reversible case. On each contour one may identify one or more equilibrium points. The trace of identical equilibria on increasing level curves (or surfaces) defines a path. *The essential idea behind this argument is that of all the paths allowed by the system upon the contour 'surface' of $\mathcal{P}$, the equilibrium state corresponding to the steepest one tends to be the most stable.* In Fig. 13.5a we indicate the case of isotropic bodies where the underlying manifold of $\mathcal{P}$ is a paraboloid whose cross section is a circle. Therefore all radial paths along this surface increase at the same rate and so all orientations are equally likely. In the case of a spheroid or cylinder for instance, the manifold is a paraboloid with an elliptical cross section. In this case there is a clear path along this surface which is steepest and corresponds to the one where the body falls along $\phi = \pi/2$. In the case of orthotropic bodies, the contours are slightly more complicated due to the higher number of degrees of freedom of this problem. In Fig. 13.5c, we plot the level surfaces of $\mathcal{P} = \mathcal{P}(x, y, z)$; the bold line indicates the maxEP path along the level surfaces and corresponds to the state of mechanical equilibrium ($\theta = \pi/2, \phi = 0$).

13.3.3 Open Problems and Difficulties

In this section we briefly discuss some on-going issues that we are currently grappling with. In order to set the MaxEP principle on a firm and rigorous footing, we need to resolve several loose ends and clarify the looming issues. At the same time we also need to be able to consider more sophisticated examples than the ones considered so far allowing us to build the success of the MaxEP. This can help point to a general overarching argument. The particular problems discussed below are meant to indicate some such issues related to the sedimentation problem and the particular complexities that we have encountered.

13.3.3.1 Non-Newtonian Case

Consider the case of a non-Newtonian fluid represented by the Reiner-Rivlin model where $\mathbf{T}_{NN} = \alpha \mathbf{D}^2$, the entropy production equation can be given by

$$\begin{aligned} T_0 \mathcal{P} =& U_i(K_{ij}U_j - m_e g_i) + U_i C_{ij} \Omega_j + \Omega_i M_{ij} \Omega_j + U_i U_j U_k R_{ijk} \\ &+ U_i U_j \Omega_k S_{ijk} + \Omega_i \Omega_j U_k T_{ijk} + \Omega_i \Omega_j \Omega_k Z_{ijk} \end{aligned} \tag{13.28}$$

where

$$R_{ijk} = \alpha \int_\Omega \mathbf{D}(\mathbf{h}^{(i)})\mathbf{D}(\mathbf{h}^{(j)}) : \mathbf{D}(\mathbf{h}^{(k)})\, dV, \tag{13.29}$$

$$S_{ijk} = \alpha \int_\Omega \mathbf{D}(\mathbf{h}^{(i)})\mathbf{D}(\mathbf{h}^{(j)}) : \mathbf{D}(\mathbf{H}^{(k)})\, dV, \tag{13.30}$$

$$T_{ijk} = \alpha \int_\Omega \mathbf{D}(\mathbf{H}^{(i)})\mathbf{D}(\mathbf{H}^{(j)}) : \mathbf{D}(\mathbf{h}^{(k)})\, dV, \tag{13.31}$$

$$Z_{ijk} = \alpha \int_\Omega \mathbf{D}(\mathbf{H}^{(i)})\mathbf{D}(\mathbf{H}^{(j)}) : \mathbf{D}(\mathbf{H}^{(k)})\, dV. \tag{13.32}$$

In the case of bodies with an axis of revolution and fore-aft symmetry, we can show that the terms C_{ij}, R_{ijk}, Z_{ijk} vanish for all $i, j, k = 1, 2, 3$ [55]. Therefore the Eq. (13.28) can be rewritten as

$$T_0\mathcal{P} = U_i\left(K_{ij}U_j + U_j\Omega_k S_{ijk} - m_e g_i\right) + \Omega_j\left(\Omega_i U_k T_{ijk}\right) + \Omega_i M_{ij}\Omega_j \qquad (13.33)$$

In order to evaluate the appropriate stable configurations, we would now be required to evaluate several more coefficients, which are not available in the literature. Our preliminary computations have also shown [56] that the appropriate creeping flow field that is required is not the Stokes' flow but the creeping flow for the viscoelastic fluid under consideration which remains an open problem.

13.3.3.2 Time Dependent Case

As indicated in the earlier section, when *Re* exceeds a threshold, the particle undergoes a transition from a steady to an oscillatory state, about the previous stable configuration. In Fig. 13.6 we indicate this transition on the contour surface of the corresponding $\mathcal{P}$ which can be interpreted as the phase space of $\mathcal{P}$. As indicated by the figure, the path of maximum Entropy Production transitions from a monotonic curve to a fluctuation about the stable path (indicated by the shaded region in the paraboloid), with the oscillations getting larger, as the system is pushed further and further out of equilibrium.[6] The increasing amplitude and frequency of oscillations with *Re* is validated by recent experiments and theory [16]. In such a state it is perhaps not incorrect to say that the system is 'far from equilibrium', a term that is ill-defined despite its frequent use. Oscillatory behavior far from equilibrium has been noticed in other examples as well [57–61] but the universality of such behavior and its details require further investigation. For one, it is not even clear if the oscillation of the body is indicative of an oscillation in $\mathcal{P}$[7] and the depiction in Fig. 13.6 is based upon the universally observed Hopf bifurcation from steady state which is the central theme of the work by Prigogine [18]. Could this be a macroscopic manifestation of the microscopic fluctuations in the system as predicted by the fluctuation theorem [57]?

13.3.3.3 Equivalence to the Dynamical Approach

The relationship between the MaxEP principle and the dynamic equations (i.e. force and torque equations in our case) merit serious attention. At present, we note that the derivative of $\mathcal{P}$ with respect to U_i and Ω_i gives

---

[6] The intersection of the dark line and shaded region on the paraboloid with a plane parallel to its cross-section in Fig. 13.6 gives the number of allowable states at any given *Re*.

[7] The simplistic depiction of $\mathcal{P}$ in Fig. 13.6 obviously leaves out 'memory' terms which become dominant at large *Re* [9].

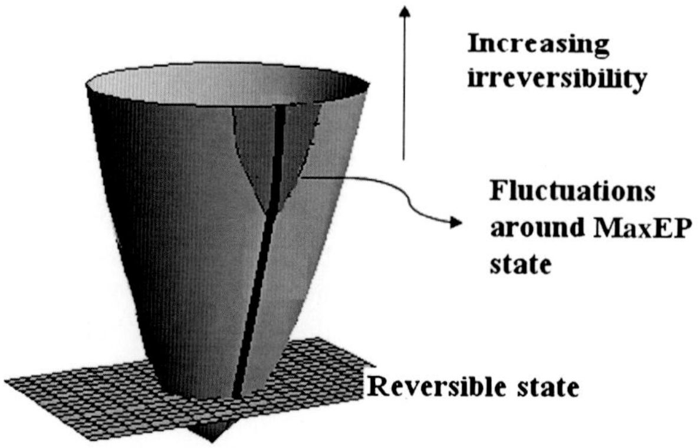

Fig. 13.6 The figure depicts a schematic of the path of maximum entropy production as the particle goes from steady state orientation to a state of periodic oscillations. The figure displays the appropriate orbits corresponding to steady state (or MaxEP) and the state of fluctuation in the phase space under a quasi-steady assumption, made by ignoring inertial effects upon entropy production at larger *Re*. Each cross section of the paraboloid represents the set of all allowed states at a fixed *Re* which is the measure of irreversibility of the system

$$T_0 \frac{d\mathcal{P}}{dU_i} = K_{ij}U_j + U_j\Omega_k S_{ijk} - m_e g_i + \Omega_j C_{ij} + \Omega_i \Omega_k T_{ijk} \tag{13.34}$$

$$T_0 \frac{d\mathcal{P}}{d\Omega_i} = U_j U_k S_{ijk} + U_j C_{ij} + U_i \Omega_k T_{ijk} \tag{13.35}$$

respectively. We recognize the Eqs. (13.34) and (13.35) to be the linearized versions of the linear momentum and angular momentum equations i.e. the nonlinear terms involving *Re* are not present on the right hand side of the above equations. The entropy production, by definition, contains merely the internal energy contributions and therefore does not include the inertial terms. Does the entropy production have an equivalent extremum in the force and torque? In the examples considered in Sect. 13.3.2, we find that at least for bodies with fore-aft and orthotropic symmetries, the bodies sediment in the orientation corresponding to maximum drag. We are looking into the possibility of generalizing this result for bodies of other symmetries.

13.4 Conclusions

In conclusion, we note that the problem of fluid structure interaction, in particular the terminal, stable orientation of a body in a fluid during freefall is a problem well suited to be tackled by the laws of non-equilibrium thermodynamics. It remains to be seen whether this approach allows for similar successes in other more complex problems in fluid mechanics. Specifically, we see that the extrema of entropy production of the system, determines the allowed stable states of a fluid–solid system where dynamics based techniques fail at present. In the creeping flow regime, when inertia is absent, the system has already reached a state of maximum entropy which the second law of thermodynamics dictates and also the rate of entropy production is in fact zero. This suggests that the body can take on any orientation which is determined by its initial state. When inertial effects emerge i.e. $0 < Re < Re_1$ where Re_1 refers to a critical value of Reynolds number, the terminal state corresponds to one of maximum entropy production or one corresponding to maximum drag in the examples of steady orientations of bodies in a flow.[8] Re_1 marks the threshold below which the equilibrium orientation (or maximum entropy production state) is achieved and beyond which the particle undergoes fluctuations about the MaxEP state.

While our results points to the maximum as the appropriate extrema of $\mathcal{P}$, we are conscious, in light of the insightful argument made by Niven [46], of the fact that the nature of the extrema can reverse under different constraints. This subject requires serious attention and could help in the resolution of the debate concerning the "correct" nature of the extrema of entropy production. Oddly enough, our previous and current work [32, 33, 62] on sedimentation suggests that stable configurations can show up as a maximum or a minimum of $\mathcal{P}$ depending upon the nature of the variable used to compute the variation. As can be seen easily, the derivative of $\mathcal{P}$ with respect to U in Eq. (13.19), for instance, yield a minimum while the derivatives with respect to θ, in the case of a spheroidal body, displays maxEP corresponding to the stable state. While, this example by no means resolves the min–max issue, it points to questions that need closer investigation. We take up this issue in some detail in our forthcoming work [62] on this subject.

All in all, we view this work as another success of the MaxEP principle. Optimal principles such as the MaxEP have a natural appeal due to their inherent simplicity, elegance and universality. A rigorous scientific justification is however slightly harder to provide and a completely convincing one is still forthcoming. While scientific explanations are not required to be simplistic, they must be the simplest of the set of all possible correct explanations [63] and must be subject to experimental verifiability or falsification. We strongly believe that the MaxEP approach possesses these features. Its survival is based on evidentiary success in various independent problems. It should also be pointed out that the MaxEP theorists by no means seek to replace dynamical arguments by this extremum principle. Instead, they seek to find the relationship between MaxEP and the causal arguments. The hope is that the MaxEP concept, if validated, can serve to augment the explanatory power of science and extend its sphere of influence; due to its natural simplicity, it can provide essential clues into very complex non-linear phenomena where traditional and causal arguments have been unable to make headway so far. It has also been suggested in a recent article by Kleidon [64] that

[8] The MaxEP does not always seem to correspond with the maximum drag state [33] and this is an issue that needs further attention.

the maximum power principle which describes the rate of energy transfer is in fact a sufficient practical condition to describe out of equilibrium states. This statement is indeed true in our case as well; the examples that we treat here do not ultimately rely on the potential energy term but merely on the dissipation. The entropy production however allows us to verify the reversible cases when $\mathcal{P} = 0$. No matter what function is adopted for optimization, the underlying connections between them all must be explored and clearly understood. In our own case, we fully realize that MaxEP, as we understand it now, fails to capture several aspects of the physics, especially when the system is very far from equilibrium, such as when the *Re* in the problems becomes very large. In such cases, however dynamical arguments based on linear momentum equations also become analytically intractable and one must resort to full scale numerical simulations. We are optimistic that the growing body of work on MaxEP will point to solutions to these unresolved issues in the very near future.

Acknowledgments The author wishes to thank the organizing committee of the 2011 MaxEP Workshop in Canberra. This afforded a wonderful opportunity to clarify several of the ideas that have appeared in this article and will serve as fodder for future work. Many thanks to the reviewers for their valuable suggestions to improve the article.

References

1. Kincaid, H.: Routledge encyclopedia of philosophy, Version 1.0. Routledge, London and New York (1998)
2. Schoemaker, P.J.H.: The quest for optimality: A positive heuristic of science? Behav. Brain Sci. **14**, 205–245 (1991)
3. Leal, L.G.: The slow motion of slender rod-like particles in a second-order fluid. J. Fluid Mech. **69**, 305–337 (1975)
4. Cho, K., Cho, Y.I., Park, N.A.: Hydrodynamics of a vertically falling thin cylinder in non-newtonian fluids. J. Non-Newtonian Fluid Mech. **45**, 105–145 (1992)
5. Chiba, K., Song, K., Horikawa, A.: Motion of a slender body in quiescent polymer solutions. Rheol. Acta **25**, 280–388 (1986)
6. Joseph, D.D., Liu, Y.J.: Orientation of long bodies falling in a viscoelastic fluid. J. Rheol. **37**, 961–983 (1993)
7. Galdi, G.P., Vaidya, A.: Translational steady fall of symmetric bodies in Navier-Stokes liquid, with application to particle sedimentation. J. Math. Fluid Mech. **3**, 183–211 (2001)
8. Galdi, G.P., Pokorny, M., Vaidya, A., Joseph, D.D., Feng, J.: Orientation of bodies sedimenting in a second-order liquid at non-zero reynolds number. Math. Models and Methods in the Appl. Sci. **12**(11), 1653–1690 (2002)
9. Galdi, G, P.: On the motion of a rigid body in a viscous fluid: A mathematical analysis with applications, handbook of mathematical fluid mechanics, pp. 653–791. Elsevier Science, Amsterdam (2002)
10. Vaidya, A.: A note on the terminal orientation of symmetric bodies in a power-law fluid. Appl. Math. Lett. **18**(12), 1332–1338 (2005)
11. Belmonte, A., Eisenberg, H., Moses, E.: From flutter to tumble: Inertial drag and Froude similarity in falling paper. Phys. Rev. Lett. **81**(2), 345–349 (1998)
12. Fields, S.B., Klaus, M., Moore, M.G., Nori, F.: Chaotic dynamics of falling disks. Nature **388**, 252–254 (1997)

13. Tanabe, Y., Kaneko, K.: Behavior of falling paper. Phys. Rev. Lett. **73**(10), 1372–1376 (1994)
14. Willmarth, W.W., Hawk, N.E., Galloway, A.J.,Roos, F.W. J. Fluid Mech. 27, 177–207 (1967)
15. Camassa, R., Chung, B.J., Howard, P., McLaughlin, R.M., Vaidya, A.: Vortex induced oscillations of cylinders at low and intermediate Reynolds numbers. Sequeira, A., Rannacher, R. (ed.) Advances in mathematical fluid mechanics: A tribute to Giovanni Paolo Galdi, pp. 135–145. Springer Verlag (2010)
16. Cohrs, M., Ernst, W., Galdi, G.P., Vaidya, A., Theory and experiments on oscillating cylinders in a flow, submitted for publication (2012)
17. Camassa, R., Chung, B., Gipson, G., McLaughlin, R., Vaidya, A.: Vortex induced oscillations of cylinders, http://ecommons.library.cornell.edu/handle/1813/11484/ (2008)
18. Prigogine, I.: Introduction to thermodynamics of irreversible processes. Interscience Publishers, New York (1955)
19. Kreuzer, H.J.: Non-equilibrium thermodynamics and its statistical foundations. Clarendon Press, Oxford (1981)
20. Ottinger, H.C.: Beyond equilibrium thermodynamics. Wiley Interscience, USA (2005)
21. Biot, M.A.: Variational principles and irreversible thermodynamics with applications to viscoelasticity. Phys. Rev. **97**(6), 1463–1469 (1955)
22. Ghesselini, R.: Elastic free energy of an upper convected maxwell fluid undergoing fully developed planar poiseuille flow: a variational result. J. Non-Newtonian Fluid Mech. **46**, 229–241 (1993)
23. Horne, C., Smith, C.A., Karamcheti, K.: Aeroacoustic and aerodynamic applications of the theory of non-equilibrium thermodynamics, NASA technical paper 3118, June (1991)
24. Woo, H.-J.: Variational formulation of non-equilibrium thermodynamics for hydrodynamic pattern formation. Phys. Rev. E **66**(066104–1), 066104–066105 (2002)
25. Kawazura, Y., Yoshida, Z.: Entropy production rate in a flux-driven self-organizing system. Phys. Rev. E **82**, 066403 (2010)
26. Ziegler, H.: An introduction to thermodynamics. North-Holland, Amsterdam (1983)
27. Coleman, B.D., Noll, W.: On the thermodynamics of continuous media. Arch. Rat. Mech. Anal. **3**, 289–303 (1959)
28. Onsager, L.: Reciprocal relations in irreversible processes. I. Phys. Rev. **37**, 405–426 (1931)
29. Onsager, L.: Reciprocal relations in irreversible processes. II. Phys. Rev. **38**, 2265–2279 (1931)
30. Gurtin, M.E., Fried, E., Anand, L.: The mechanics and thermodynamics of continua. Cambridge University Press, Cambridge, UP (2010)
31. Christen, T.: Application of the maximum entropy production principle to electrical systems. J. Phys. D Appl. Phys. **39**, 4497–4503 (2006)
32. Chung, B.J., Vaidya, A.: An optimal principle in fluid-structure interaction. Physica D **237**(22), 2945–2951 (2008)
33. Chung, B.J., Vaidya, A.: Non-equilibrium pattern selection in particle sedimentation. Appl. Math. Comput. **218**(7), 3451–3465 (2011)
34. Morrison, F.A.: Understanding rheology, Oxford University Press, Oxford (2001)
35. Larson, R.: The structure and rheology of complex fluids, Oxford University Press, Oxford (1999)
36. Massoudi, M.: On the heat flux vector for flowing granular materials Part I: Effective thermal conductivity and background. Math. Meth. Appl. Sci. **29**, 15851598 (2006)
37. Lorenz, E.: Generation of available potential energy and the intensity of the general circulation. Scientific Report No. 1, UCLA, Dept. of Meteorology, July (1955)
38. Ozawa, H., Ohmaru, A., Lorenz, R.D., Pujol, T.: The second law of thermodynamics and global climate system: A review of the maximum entropy production principle. Rev. of Geophys. **41**(4), 1–24 (2003)
39. Kleidon, A.: The atmospheric circulation and states of maximum entropy production, Geophys. Res. Lett., **30**(23) 2223 (2003)

40. Chwang, A.T., Wu, T.Y.: Hydromechanics of low-reynolds-number flow. Part 2. singularity method for stokes flows. J. Fluid Mech. **45**, 105–145 (1992)
41. Jaynes, E.T.: The minimum entropy production principle. Ann. Rev. Phys. Chem. **31**, 579–601 (1980)
42. Malek, J., Rajagopal, K.R.: On the modeling of inhomogeneous incompressible fluid-like bodies. Mech. Mater. **38**, 233–242 (2006)
43. Martyushev, L.M., Seleznev, V.D.: Maximum entropy production principle in physics, chemistry and biology. Phys. Rep. **426**, 1–45 (2006)
44. Martyushev, L.M.: Some interesting consequences of the maximum entropy production principle. J. Exper. Theor. Phys. **104**, 651654 (2007)
45. Niven, R.K.: Steady state of a dissipative flow-controlled system and the maximum entropy production principle. Phys. Rev. E **80**, 021113 (2009)
46. Niven, R.K.: Simultaneous extrema in the entropy production for steady-state fluid flow in parallel pipes. J. Non-Equilib. Thermodyn. **35**, 347–378 (2010)
47. Paltridge, G.W.: Global dynamics and climate—a system of minimum entropy ex-change, Q. J. R. Meteorol. Soc. **101**, 475 (1975); Nature (London) **279**, 630 (1979)
48. Paulus Jr, D.M., Gaggioli, R.A.: Some observations of entropy extrema in fluid flow. Energy **29**, 28472500 (2004)
49. Happel, V., Brenner, H.: Low reynolds number hydrodynamics. Prentice Hall, New Jersey (1965)
50. Swenson, R.: Autocatakinetics, yes-autopoiesis, no: steps toward a unified theory of evolutionary ordering. Int. J. General Syst. **21**, 207–228 (1992)
51. Beretta, G.P., Sc.D. thesis, M.I.T., (un published), e-print quant-ph/0509116 (1981)
52. Beretta, G.P.: A nonlinear model dynamics for closed-system, constrained, maximal-entropy-generation relaxation by energy redistribution. Phys. Rev. E **73**, 026113 (2006)
53. Beretta, G.P.: On the relation between classical and quantum-thermodynamic entropy. J. Math. Phys. **25**, 1507 (1984)
54. Beretta, G.P.: Quantum thermodynamics: a new equation of motion for a single constituent of matter. Nuovo Cimento B **82**, 169 (1984)
55. Vaidya, A.: Steady fall of bodies of arbitrary shape in a second-order fluid at zero Reynolds numbers. Japan J. Ind. Appl. Math., **21**(3), 299–321 (2004)
56. Chung, B.J., Vaidya, A.: On the slow motion of a sphere in fluids with non-constant viscosities. Int. J. Eng. Sci. **48**(1), 78–100 (2010)
57. Evans, D.J., Cohen, E.G.D., Morriss, G.P.: Probability of second law violations in shearing steady states. Phys. Rev. Lett. **71**(15), 24012404 (1993)
58. Glansdorff, P., Prigogine, I.: Thermodynamics of structure, Stability and Fluctuations. Wiley-Interscience, New York (1971)
59. Morita, H.: Collective oscillation in two-dimensional fluid, arXiv:1103.1140, March (2011)
60. Nicolis, G., Prigogine, I.: Fluctuations in non-equilibrium systems. Proc. Nat. Acad. Sci. USA **68**(9), 2102–2107 (1971)
61. Pearson, J.E.: Complex patterns in a simple system. Science **261**(5118), 189–192 (1993)
62. Chung, B.J., McDermid, K., Vaidya, A.: On the affordances of the MaxEP principle, in preparation (2012)
63. Feuer, L.S.: The principle of simplicity. Philos. Sci. **24**(2), 109–122 (1957)
64. Kleidon, A.: How does the earth system generate and maintain thermodynamic disequilibrium and what does it imply for the future of the planet?, arXiv:1103.2014v2 (2011)

Chapter 14
Can the Principle of Maximum Entropy Production be Used to Predict the Steady States of a Rayleigh-Bérnard Convective System?

Iain Weaver, James G. Dyke and Kevin Oliver

Abstract The principle of Maximum Entropy Production (MaxEP) has been successfully used to reproduce the steady states of a range of non-equilibrium systems. Here we investigate MaxEP and maximum heat flux extremum principles directly via the simulation of a Rayleigh-Bèrnard convective system implemented as a lattice gas model. Heat flux and entropy production emerges in this system via the resolution of particle interactions. In the spirit of other related works, we use a reductionist approach, creating a lattice-Boltzmann model to produce steady-convective states between reservoirs of different temperatures. Convection cells emerge that show meta-stability where a given lattice size is able to support a range of convective states. Slow expansion and contraction of the model lattice, implemented by addition and subtraction of vertices, shows hysteresis loops where stable convection cells are expanded to regions wherein they become meta-stable, and eventually transition into more stable configurations. The maximally stable state is found to be that which maximises the rate of heat transfer, which is only equivalent to maximum internal entropy production in a strong forcing regime, while it is consistent with minimising entropy production in a weak forcing case. These results demonstrate the utility of lattice-Boltzmann models for future studies of non-equilibrium systems, and highlight the importance of dissipation and forcing rates in disambiguating proposed extremum principles.

I. Weaver (✉) · J. G. Dyke
School of Electronics and Computer Science, University of Southampton, Southampton, UK
e-mail: isw1g10@soton.ac.uk

K. Oliver
National Oceanography Centre, Southampton University of Southampton, Southampton, UK

R. C. Dewar et al. (eds.), *Beyond the Second Law*,
Understanding Complex Systems, DOI: 10.1007/978-3-642-40154-1_14,

14.1 Introduction

The power of equilibrium statistical mechanics stems from the very large number of particles in the systems of study. At equilibrium, deviations from predictions are so small as to be safely discounted. With unlimited computational resources and unlimited time it is, in principle, possible to construct numerical simulations of equilibrium systems from a 'bottom up' approach by explicitly modelling the kinetics of the system. Given that non-equilibrium systems often defy exact analysis because equilibrium assumptions no longer hold then such an approach is attractive. For example, a model convective system would be one in which particles would be resolved and their interactions lead to the bulk properties of patterns of circulation, heat transport and entropy production. The behaviour of such simulations could be assessed in terms of the principle of Maximum Entropy Production (MaxEP). For example, given the known boundary conditions of the system, would assuming that the system self-organised into a state of MaxEP lead to predictions in agreement with simulation? There are reasons to think that it would. The proposed MaxEP principle states that complex-dissipative systems are characterised by a non-equilibrium thermodynamic state in which the rate of thermodynamic entropy production is maximised [1, 2]. A number of examples have demonstrated the utility of the MaxEP principle. For example, the prediction of atmospheric heat transport from simple considerations [3–5] and geological process within the Earth [6, 7]. These studies typically proceed on the basis of formulating a model of the system in question, and using constrained optimisation to produce a MaxEP state. The resultant model output is then compared to empirical data. While the success of these studies are both fascinating and very promising, there is a significant paucity of experimental or simulated validation which would otherwise shed light on the types of system for which this approach may be successful [8, 9], and while in progress, a theoretical grounding for MaxEP is yet to be established [10, 11]. Here we present a series of simulations of Rayleigh-Bènard convective systems, produced in order to evaluate the applicability of MaxEP and other competing principles. In particular, we ask "would the formation of convective cells within the system correlate to entropy production in a way that would allow certain bulk properties of the system be predicted without recourse to simulation of the internal kinetics?''

Convection cells are ubiquitous [12]. They occur in a vast range of systems across an enormous range of scales, from currents in the core of early stars to a heated pan of soup, an observation which caused Getling to remark "Convection due to nonuniform heating is, without overstatement, the most widespread type of fluid motion in the Universe''. In this work, we do not interest ourselves with the specifics of scale and the fluid properties, rather we are interested in the driving principles behind convection cells as a generic example of self-organisation. A specific type of convection cell, known as *Rayleigh-Bénard* convection cell, occur where fluids with positive thermal expansion coefficient are heated from below. Warm parcels of fluid experience an upwards buoyancy force from the induced vertical density gradient. At the surface, they are able to deposit, or radiate heat, eventually becoming more dense than the rising fluid below. Being unable to descend here, currents form outward from the up-welling point until the inverse density gradient is encountered, and the parcel descends. Through this mechanism, a fluid subject to a temperature gradient is able to self-organise to a dissipative state, transporting energy from the hot to cold reservoir. The internal entropy production is easily computed for this phenomena, and the associated fluid mechanics are irrefutably complex, making an ideal case study for the investigation of extremum principles for non-equilibrium systems.

In modelling this type of system from the bottom up, rather than prescribe system-level properties such as heat flux or rates of convection, we allow these to emerge via the interaction of the simulated particles within the system. We implement a lattice gas cellular automata (LGCA) as our approach for modelling our convecting fluid system [13]. Furthermore, in order to be able to produce statistically significant results, a novel computational method was developed that allowed large scale simulations to be conducted.

We introduce the LGCA class of models in more depth, particularly our choice of lattice gas in Sect. 14.2 along with the necessary modifications to produce convection cells. Results are presented and discussed in detail in Sects. 14.3 and 14.4.

14.2 A Model of Rayleigh-Bénard Convection

The challenge is to develop a simulation of a convective system that captures the important processes over sufficient temporal and spatial scales whilst remaining computationally feasible as well as intuitive. With a number of modifications (detailed in the following sections) LGCA represent an ideal trade-off in terms of efficiency and fidelity.

14.2.1 Lattice Gas Cellular Automata

LGCA consist of discrete particles traversing a regular lattice, characterised both by the choice of lattice vectors, that is the vector describing the edges attached to each vertex, $\vec{c}_i$ and collision rules. While in terms of implementation of the model we talk of *particles*, we are not constrained to modelling on the molecular scale. Indeed from this abstraction, the equations of motion of a continuous fluid can be extracted by application of Chapman-Enskog theory [14]. A commonality shared by all lattice gas models is that they advance in alternating *collision* and *propagation* steps:

- **Collision**
  Particles at a vertex participate in mass and momentum conserving collisions. Importantly, certain configurations allow for scattering collisions in which

particles redistribute their momenta. It is these collisions which essentially provide viscosity, and enable us to recover the Navier–Stokes equations.

- **Propagation**
  Between collisions, particles travel along lattice vectors to neighbouring vertices separated by the characteristic lattice spacing ΔL. Particles propagating across a periodic boundary are re-injected at the corresponding boundary in the same momentum state, while particles which would travel across a solid surface instead have their momentum state inverted, known as a *no-slip boundary* due to the result of zero net velocity at a boundary.

The original LGCA was developed in 1973 used a simple 2D square lattice and is commonly named the HPP lattice gas after it's authors, Hardy et al. [15, 16]. These early LGCA suffered greatly from anisotropy, resulting for example in curious square vortices, although a similar method was implemented later in 1986 on a hexagonal lattice called the FHP lattice gas, again named after the authors, Frisch et al. [17]. The increased symmetry of this lattice ensures it does not suffer so strongly from anisotropy. A thorough account of the zoo of lattice gas models can be found in Wolf-Gladrow [18].

14.2.2 The FHP Lattice Gas

As a hexagonal lattice with six lattice vectors, each vertex in the FHP lattice has six momentum states, whose occupancy is denoted by the binary variable n_i with $i = 1 \ldots 6$. The state $n_i = 1$ corresponds to an occupied vertex while 0 represents a hole. Almost all collisions are trivial since there is only one mass and momentum conserving outcome to a collision, such that the state of the vertex $\vec{n}$ is unchanged. The exceptions to this are mainly zero net-momentum collisions which have multiple mass and momentum conserving outcomes. The choice proves extremely important to the model dynamics, and *scattering* (momentum-state altering) collisions are chosen, breaking what would otherwise be unwanted invariants. The collision rules are summarised in Fig. 14.1 to highlight this distinction. The final element of the model is the propagation step. Between collision steps, momentum states travel along lattice vectors to participate in collisions at a neighbouring vertex.

14.2.3 FHP Buoyancy Modification

The traditional FHP model does not have a defined temperature since for most hydrodynamical problems thermal anomalies are assumed small, and temperature is therefore not important [19]. However, the temperature gradient serves as the driving force for convection for which Schaffenberger et al. [20] modify the usual

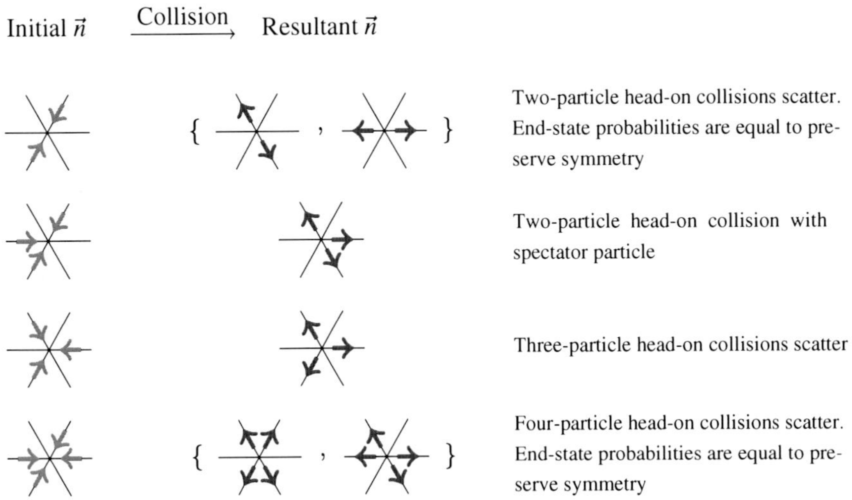

Fig. 14.1 Explicit definitions of the FHP lattice gas collision rules which change the momentum states at a vertex. The FHP lattice has six momentum states at each vertex, where arrows indicate the presence of a particle in momentum state n_i and the vector $\vec{n}$ holds all the momentum state occupancies for a vertex. Two- and four-particle head-on collisions are probabilistic, and the resulting states occur with equal probability

FHP model by introducing an effective particle temperature, along with an associated buoyancy force modelled by spontaneous momentum-state flipping. As well as a momentum-state, particles have a temperature state θ_i which can take the value ± 1 for a hot or cold particle respectively. The temperature state is a *passive scalar*; it follows particles passively during the propagation step. Hot particles can be thought of as being less dense than the surrounding medium and experience a positive buoyancy force. Cold particles are relatively dense and experience a negative buoyancy force. The inclusion of θ states modifies the lattice gas in the following ways:

- **Collision**
  In the collision step the total particle thermal energy, $\sum_i \theta_i$, is conserved. That is to say, the number of 1 and 0 temperature particles are conserved. However, the temperature states are randomly distributed amongst the participating particles. The thermal diffusion coefficient then becomes a function of density only.
- **Thermal flipping**
  A force can be simulated by spontaneous flipping of particle momentum states, resulting in a change of net-momentum [14]. The buoyancy force is simulated by inverting particle y-momentum, without altering x-momentum. This is only possible between corresponding pairs of momentum states. Particles in the negative y-momentum state with $\theta = 1$ flip into the corresponding positive y-momentum state with some small probability, here labelled by ω. The thermal flipping rule is illustrated in Fig. 14.2. While flipping causes spontaneous

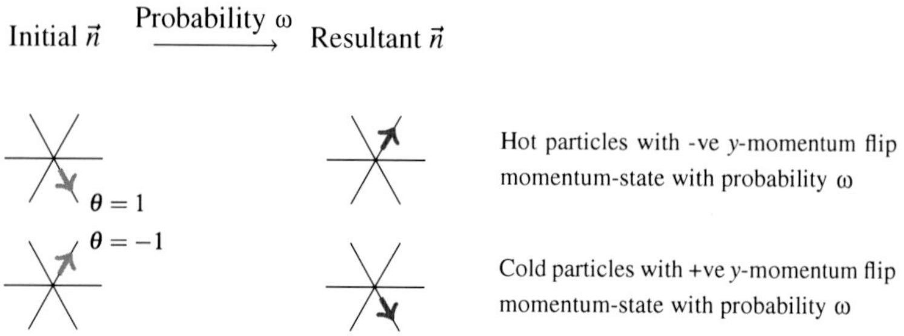

Fig. 14.2 Explicit definitions of FHP buoyancy rules. Hot particles have positive buoyancy and flip to positive y-momentum states without altering their x-momentum, and vice versa for negatively buoyant cold particles. This can be thought of as equivalent to applying a force in $\hat{y}$ where the strength of the force is determined by the rate of such transitions

changes in local momentum, symmetry of the thermal flipping rule with θ ensures that on the average, global momentum is conserved.

- **Propagation**
  As a passive scalar, θ is propagated between lattice points along with particles. Additionally particles reflected from a boundary set θ to ± 1 depending on the simulated temperature of the boundary.

14.2.4 A Lattice Boltzmann Model

The simulation results of Schaffenberger et al. [20] included measurement of some common hydrodynamical quantities, and required coarse-graining and long time averages to be taken over a very large lattice. Even then, they exhibit large amounts of noise, and rapid transitions can be seen between different convective states for parameter values where, say, period-1 (where the lattice is occupied by a single convection cell) and period-$\frac{1}{2}$ (where two cells of half the size occupy the lattice) are similarly favourable. While the effects of such fluctuations may be important in real systems, we are interested in the thermodynamic properties of steady-state convection cells, a very difficult parameter to extract from this type of simulation.

Fortunately, the Boltzmann transport equation grants us reprieve from many of these problems. The very simple and intuitive lattice gas rules can be transformed from a binary model of lattice site occupations n_i to a continuum model of mean momentum-state occupancies N_i. Through some careful manipulation, a continuum model can be made to approximate the real lattice gas dynamics with the advantage of much smaller lattices, and removing the need for large space and time averages. In the following steps, we write $N_i(\vec{x}, t)$ to be the mean occupancy of momentum-state i of the vertex at $\vec{x}$ at simulation time t. The collision and propagation steps can be written as

$$N_i(\vec{x} + \vec{c}_i\Delta t, t + \Delta t) = N_i(\vec{x}, t) + \Delta_i \tag{14.1}$$

where $\vec{c}_i\Delta t$ is a vector of length ΔL, the lattice spacing, and therefore the left side of this equation can be thought of as the propagation step, advancing mean momentum occupancy along lattice vectors. The right side can be thought of as the collision step, and includes the collision term, Δ_i. This operator adjusts the mean occupancy propagated between lattice sites according to the collision rules. It includes all the information in Fig. 14.1. Rather than ennumerating all possible momentum-state altering collisions here, we can call on the much earlier work of Bhatnagar et al. and the BGK approximation [21]. The large and highly non-linear collision operator can be replaced by a less computationally intensive linear relaxation towards $N_i^{(eq)}$, the local mass and momentum conserving equilibrium with characteristic timescale τ_1

$$\Delta_i = -\frac{1}{\tau_1}\left[N_i(\vec{x}, t) - N_i^{(eq)}(\vec{x}, t)\right]. \tag{14.2}$$

Frisch et al. [14] proved the equilibrium momentum-state occupancies $N_i^{(eq)}$ of a vertex with a given density ρ and momentum $\vec{v}$, given by

$$\rho = \sum_{i=1}^{6} n_i(\vec{x}, t) \qquad \vec{v} = \sum_{i=1}^{6} N_i(\vec{x}, t)\vec{c}_i \tag{14.3}$$

obeys Fermi–Dirac statistics, since the particles are subject to the exclusion principle; each momentum state may be occupied by at most one particle, and therefore $0 \leq N \leq 1$. The proof itself is somewhat involved [18, p. 64–67], so here we present only the result;

$$N_i^{(eq)}(\rho, \vec{v}) = \frac{1}{1 + e^{(h + \vec{c}_i \cdot \vec{q})}} \tag{14.4}$$

where h and $\vec{q}$ are the Lagrange multipliers associated with our conserved quantities. The full derivation of these Lagrange multipliers is cumbersome [18, p. 248–251] and again we give only the result, which can be substituted into the linearised collision operator of Eq. (14.2)

$$N_i^{(eq)}(\rho, \vec{v}) = d + d(d-1)q_1\vec{c} \cdot \vec{v} + \frac{1}{2}d(d-1)(2d-1)q_1^2 c_{i,\alpha}^2 v_\alpha^2 + d(d-1)h_2\vec{v}^2 \tag{14.5}$$

where the sub-scripted α notation represents a sum over the x and y compenents, and

$$d = \frac{\rho}{6}, \quad q_1 = \frac{2}{d-1} \quad \text{and} \quad h_2 = \frac{1-2d}{(d-1)^2}. \tag{14.6}$$

In a similar way, we re-implement the buoyancy force as linear relaxation towards thermal equilibrium occupancy, $N_i^{(eq\prime)}$, where the rate of flow between momentum states due to buoyancy is zero. This necessitates the addition of another timescale, τ_2 which parameterises the strength of this buoyancy force in a similar way to ω in the discrete model of Schaffenberger et al. [20]. Momentum states with $c_{i,y} = 0$ are unchanged, while the rest are relaxed towards

$$N_i^{(eq\prime)}(\vec{n}, \theta) = \frac{\text{sign}(c_{i,y}) + \theta\Sigma_i^{(\text{pair})} + \sqrt{\theta^2\left(\Sigma_i^{(\text{pair})} - 2\right)\Sigma_i^{(\text{pair})} + 1}}{2\theta} \tag{14.7}$$

where $\text{sign}(c_{i,y})$ denotes the sign of the y-momentum of momentum state i, and $\Sigma_i^{(\text{pair})}$ is the sum of mean-occupancy of corresponding momentum states between which flipping can occur.

14.2.5 Boundary Conditions

To impose a temperature gradient across the lattice gas, we heat the top and bottom boundaries at different rates, allowing their temperatures to diverge to T_c and T_h respectively. These temperatures evolve according to our imposed external heating, and the dissipation of energy between them. This differential heating can be understood as capturing a number of different real-world processes. For example, on Earth equatorial regions receive more incident radiation from the sun than polar regions. T_c and T_h would then represent the polar and equatorial temperatures in the absence of any latitudinal heat flux. Newtonian-relaxation refers to a forcing scheme where the strength of forcing is linearly related to surface temperature by some characteristic relaxation timescale λ. Here, the rate of heating or cooling of the boundaries is proportional to the difference in temperature between a surface, T, and its temperature in the absence of any additional dissipation, T^*. Such schemes can be used to model sensible heat transfer from thermostats, or in the case of small temperature anomalies ($T^* - T \ll T^*$) it can be used to approximate flux driven forcing by linearising the Stefan-Boltzmann law. In this scheme, the rate of change of the boundary temperatures are given by;

$$\frac{\mathrm{d}T_h}{\mathrm{d}t} = \frac{1}{\lambda}\left(T_h^* - T_h\right) - Q_h$$

$$\frac{\mathrm{d}T_c}{\mathrm{d}t} = \frac{1}{\lambda}\left(T_c^* - T_c\right) + Q_c$$

where T_h^* and T_c^* are the steady-state temperatures of the lower and upper boundaries respectively, and Q_h and Q_c are rates of heat exchange between the lower and upper boundaries and the simulated fluid, normalised by the boundary

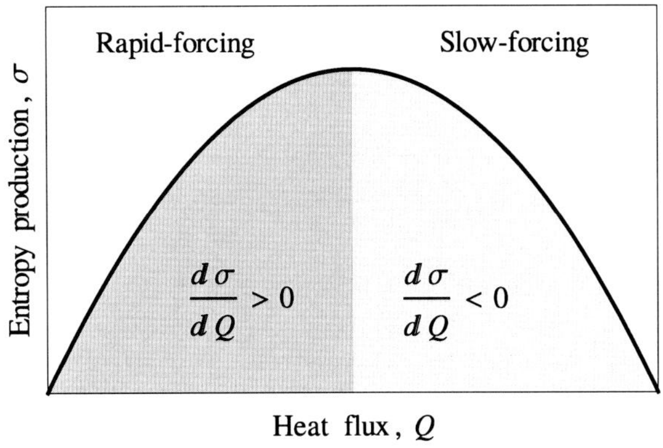

Fig. 14.3 Entropy production σ as a function of heat flux Q for our Newtonian relaxation scheme in the limit $\Delta T = T_h^* - T_c^* \ll T_h^*$. In the rapid-forcing regime, maximising entropy production and maximising heat flux are equivalent, while in the slow-forcing regime they are easily distinguished

width L_x. We define the corresponding normalised rate of entropy production, σ, by considering only internal entropy production;

$$\sigma = Q\left(\frac{1}{T_c} - \frac{1}{T_h}\right). \tag{14.8}$$

In the limit $\Delta T = T_h^* - T_c^* \ll T_h^*$, the rate of entropy production, Eq. (14.8) becomes a parabolic function of Q [22], shown in Fig. 14.3 with a maxima at $Q_{\text{crit.}} = \frac{\Delta T}{4\lambda}$.

We define a rapid-forcing regime to be one where λ is sufficiently small such that relaxation of the boundaries to their steady-state temperature occurs on much shorter timescales than the dissipation of heat between them, and $Q < Q_{\text{crit.}}$. Here, MaxEP is indistinguishable from maximising heat flux Q, while in a slow-forcing regime, where λ is large and $Q > Q_{\text{crit.}}$, entropy production and heat flux are anti-correlated, and the maximisation principles could be easily distinguished. A steady-state can be detected where these boundary variables are constant in time and $Q = Q_h = Q_c$, simply measured to be the rate of exchange of heat between the reservoirs and particles reflecting from the boundary.

14.3 Model Results

We present results from two sets of simulations, one with rapid forcing, one with slow forcing. For both sets we find a relationship between the vertical heat flux, Q, and the stability of the emergent convection cells, where the system favours configurations that produce the greatest flux of heat from hot to cold reservoir. This phenomena is responsible for producing hysteresis loops. With rapid forcing, this maximum rate of heat flux is equivalent to maximum entropy production. With slow forcing, the maximum rate of heat flux is equivalent to *minimum* entropy production. For all simulated results, the model mean particle density

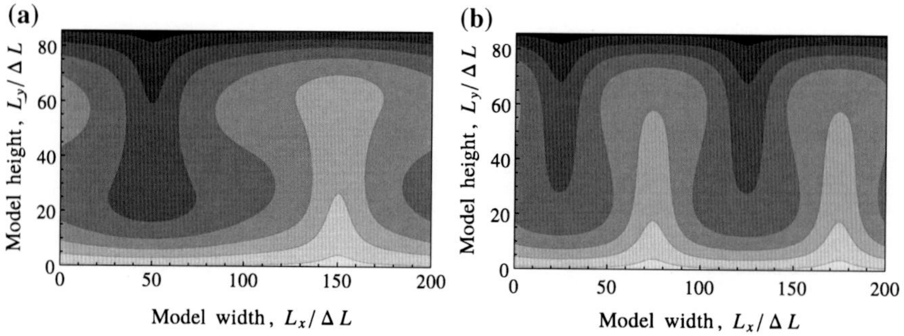

Fig. 14.4 Example convection cell configurations produced by the modified lattice Boltzmann model with $\tau_1 = \tau_2 = 20$, $L_x = 200$. Contours show isotherms and are *shaded light to dark* with increasing temperature. **a** Period-1 convection cell. **b** Period-1/2 convection cell

$\bar{\rho} = 1.2$, upper and lower boundary steady-state temperatures $T_c^* = 273\,\text{K}$, $T_h^* = 293\,\text{K}$, lattice height $L_y = 100\sqrt{3}/2\Delta L$ and characteristic model relaxation times $\tau_1 = \tau_2 = 20$.

We examine the rate of heat flux and corresponding entropy production of a range of convection cell configurations outside the dimensions in which they might typically occur. To begin with, we initialise a small cavity and evolve the model to a steady-state convection such as that shown in Fig. 14.4. From here, we increase the lattice width, L_x, some small fraction, injecting new particles to maintain the gas density at a constant. Since the gas is already in a steady-state of a specific period, we expect it will be attracted to a steady-state of the same period in the new, slightly larger cavity. This process can be used to expand and contract the lattice for a range of convection configurations.

14.3.1 Meta-stability and Hysteresis

A snapshot of two separate simulations are displayed in Fig. 14.4, highlighting the ability of the cavity to support a range of convection cell configurations. By slow expansion, we are able to maintain prescribed convective configurations over a large range of model dimensions. Additionally, perturbation experiments can be carried out (simply achieved by applying small, global mass- and momentum-conserving fluctuations to lattice momentum states), and find transitions to higher Q states are strongly favoured. This can be seen clearly where the fluctuations caused by expanding the model lattice produce clean hysteresis loops, illustrated in Fig. 14.5 for the case of a transition between period-1 and period-$\frac{1}{2}$ convection cells. This phenomena is illustrated more generally in Fig. 14.6a. From this, we infer that maximal Q states are the most stable.

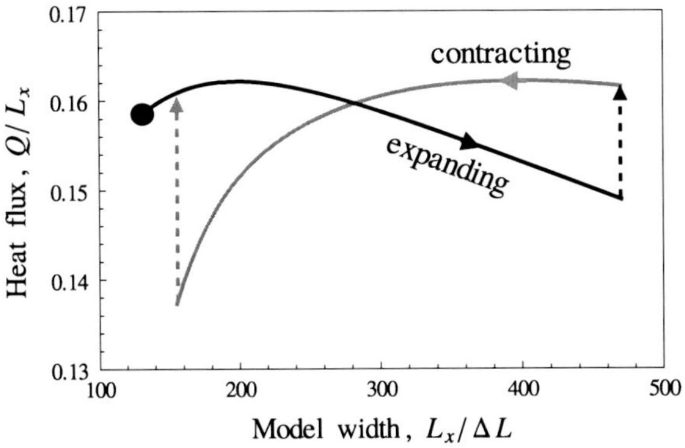

Fig. 14.5 A hysteresis loop formed from expansion and contraction model with $\lambda = 20$. Expanding a period-1 convection cell results in a metastable convection regime. Fluctuations caused by the slow expansion cause a phase transition to a more stable, more efficient period-$\frac{1}{2}$ configuration. The same effect is found to exist for the reverse process

14.3.2 Heat Flux and Maximum Entropy Production

MaxEP is but one of a plethora of suggested extremum principles, albeit one which has experienced a wide range of successes. Still the challenge remains of being able to identify the appropriate principle to apply to a given system a priori. In part, this is down to the difficulty in disentangling the competing principles for specific systems. Here, we attempt to disambiguate between a principle of maximum heat flux, and MaxEP by tuning λ such that dissipation through the fluid is more rapid than forcing, entering the slow-forcing regime of Fig. 14.3 where the two principles are mutually exclusive.

- **Rapid-forcing Regime**
  In the rapid-forcing regime, the thermal forcing on the model boundaries occurs on much shorter timescales than the dissipation of the established gradient. This exists for small λ where forcing is sufficiently strong that a principle of MaxEP and maximum heat flux would be indistinguishable, as illustrated in Fig. 14.3. In our model, $\lambda = 20$ results in $Q < Q_{\text{crit.}}$, shown in Fig. 14.6.
- **Slow-forcing Regime**
  In contrast, $\lambda = 100$ finds $Q > Q_{\text{crit.}}$, where MaxEP and maximum heat flux are mutually exclusive. The results of simulation are shown in Fig. 14.7. The heat flux curves of Fig. 14.7a are characteristically identical to those in the strong-forcing case, Fig. 14.6a, while the corresponding entropy production curves are inverted. Here, maximum heat flux is equivalent to a minimum entropy production principle, and we conclude that the model as described appears to be represented by a principle of maximum heat flux, rather than of MaxEP considering only internal entropy production.

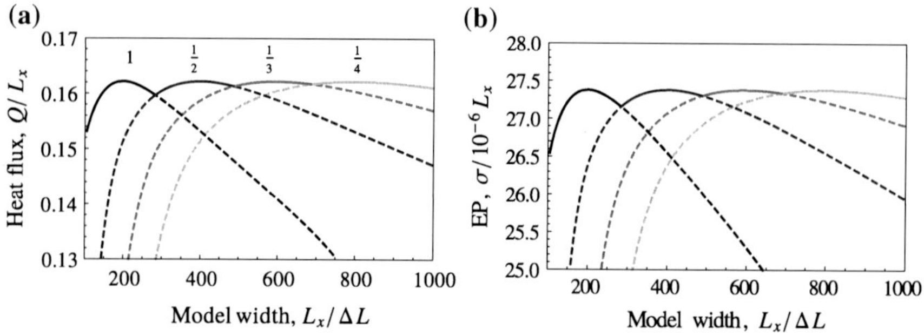

Fig. 14.6 Specific heat flux (Fig. 14.6a) and entropy production (Fig. 14.6b) with model width L_x for $\lambda = 20$. Labels show the period of the emerging convection cells in terms of L_x. *Dashed lines* indicate metastable states, where transitions to higher Q states are favoured. Crossover points between different maximally efficient convective regimes are in exactly the same position in both cases. Here, assuming that the systems self-organises into a maximum heat flux or MaxEP state accurately represents the system's behaviour. It is impossible to distinguish between a maximum heat flux and MaxEP principle, consistent with the expectation of the rapid-forcing regime. **a** Heat flux. **b** Entropy production

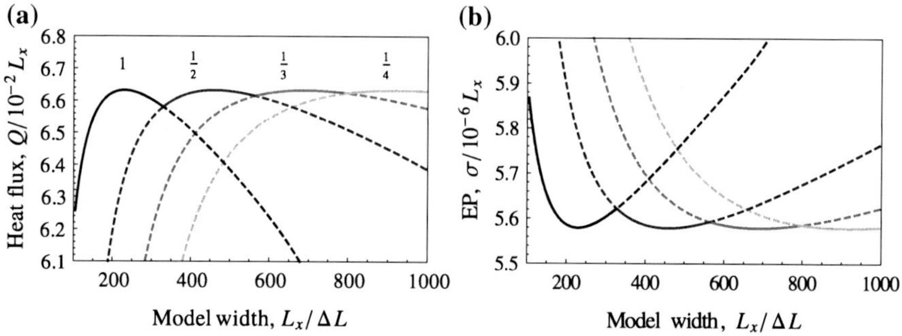

Fig. 14.7 Specific heat flux (Fig. 14.7a) and entropy production (Fig. 14.7b) with model width L_x for $\lambda = 100$. Labels show the period of the emerging convection cells in terms of L_x. Here we can see that MaxEP predictions would select inefficient convection cell configurations, while we find the model to favour maximally efficient configurations. **a** Heat flux. **b** Entropy production

14.4 Discussion

Through a principled and reductionist approach, we have implemented the modified FHP lattice gas of Schaffenberger et al. [20] as a lattice Boltzmann model, and mapped the heat flux and entropy production characteristics of a range of convection cell periods for a range of model dimensions. We find that more efficient dissipative convection configurations are universally more stable resulting in hysteresis loops between convection regimes. It is found that in the regime of rapid forcing, assuming the system self-organises to maximise heat flux or maximum entropy production would lead to accurate predictions of the systems behaviour. In regimes of slow forcing the two assumptions are diametrically opposed: the system continues to self-organise into states of maximum heat flux which are associated with minimum entropy production. Consequently, having MaxEP emerge within a system is not sufficient evidence for us to conclude that the principle of MaxEP will accurately predict steady states for that system over a range of forcing or assumptions.

How are we to understand this result in the light of previous studies that have accurately reproduced the steady states of non-equilibrium systems by assuming them to be those characterised by maximum entropy production? It is important to note that the approach we have taken in allowing the proportion of heat that is transported from hot to cold to emerge is atypical. We do not tune the model's behaviour via a parameter that affects rates of heat flux (for example, convection or advection in the Earth's hydrosphere or mantle). We are able to reconstruct the parabolic entropy production function via altering the rate of thermal forcing acting on the hot and cold reservoirs. Next, we need to revisit the formulation of entropy production. We have considered only the aspect of the model which exhibits self-organisation—the lattice gas between the heated surfaces. However, if we were to include the entropy production in heating the surfaces, we would find σ to be an linear function of Q since T^* are constant—assumptions of maximum heat flux and entropy production would be equivalent. Finally we need to consider the epistemological basis of the principle of MaxEP. As has recently been argued, the principle of MaxEP may not be understood as a physical law, rather an information theoretic procedure or algorithm [11, 23]. It is when predictions deviate from observations that we gain information about those systems and processes we are studying.

Acknowledgments This work was supported by an EPSRC Doctoral Training Centre grant (EP/G03690X/1).

References

1. Ozawa, H., Ohmura, A.: J. Clim. **10**(3), 441 (1996). doi:10.1256/003590002320603566
2. Ito, T., Kleidon, A. (Springer Berlin/Heidelberg, 2005), pp. 93–106. doi:10.1007/11672906_8
3. Paltridge, G.W.: Q. J. Royal Meteorol. Soc. **101**(429), 475 (1975). doi:10.1002/qj.49710142906
4. E. Lorenz. Predictability–a problem partly solved, chap 3 in: Palmer, T., Hagedorn, R. (eds.) Predictability of Weather and Climate (2002)
5. Shimokawa, S., Ozawa, H.: Q. J. Royal Meteorol. Soc. **128**(584), 2115 (2002). doi:10.1256/003590002320603566
6. Lorenz, R.D.: Int. J. Astrobiol. **1**, 3 (2002). doi:10.1017/S1473550402001027
7. Dyke, J.G., Gans, F., Kleidon, A.: Earth Syst. Dyn. **2**, 139 (2011). doi:10.5194/esd-2-139-2011
8. Ozawa, H., Ohmura, A., Lorenz, R.D., Pujol, T.: Rev. Geophys. **41**, 1018 (2003). doi:10.1029/2002RG000113

9. Martyushev, L., Seleznev, V.: Phys. Rep. **426**(1), 1 (2006). doi:10.1016/j.physrep.2005.12.001
10. Dewar, R.C.: J. Phys. A: Math. Gen. **38**(21), 931 (2005). doi:10.1088/0305-4470/38/21/L01
11. Dewar, R.C.: Entropy **11**(4), 931 (2009). doi:10.3390/e11040931
12. Getling, A.: Rayleigh-Bénard Convection: Structures and Dynamics, vol. 11. World Scientific Pub Co Inc (1998)
13. Vichniac, G.Y.: Physica D: Nonlinear Phenomena **10**(1–2), 96 (1984). doi:10.1016/0167-2789(84)90253-7
14. Frisch, U.: Complex Syst. **1**, 649 (1987). http://ci.nii.ac.jp/naid/10013412080/en/
15. Hardy, J., Pomeau, Y., de Pazzis, O.: J. Math. Phys. **14**, 1746 (1973). doi:10.1063/1.1666248
16. Hardy, J., de Pazzis, O., Pomeau, Y.: Phys. Rev. A **13**, 1949 (1976). doi:10.1103/PhysRevA.13.1949
17. Frisch, U., Hasslacher, B., Pomeau, Y.: Phys. Rev. Lett. **56**(14), 1505 (1986). doi:10.1103/PhysRevLett.56.1505
18. Wolf-Gladrow, D.: *Lattice Gas Cellular Automata and Lattice Boltzmann Models* (Springer, 2005)
19. Landau, L., Lifshitz E.: *Fluid Mechanics. Course of Theoretical Physics* (Pergamon Press, 1987)
20. Schaffenberger, W., Hanslmeier, A., Messerotti, M.: Hvar Observatory Bull. **25**, 49 (2001)
21. Bhatnagar, P., Gross, E., Krook, M.: Phys. Rev. **94**(3), 511 (1954)
22. Axel, Kleidon: Phys. Life Rev. **7**(4), 424 (2010). doi:10.1016/j.plrev.2010.10.002
23. Dyke, J., Kleidon, A.: Entropy **12**(3), 613 (2010). doi:10.3390/e12030613

Chapter 15
Bifurcation, Stability, and Entropy Production in a Self-Organizing Fluid/Plasma System

Zensho Yoshida and Yohei Kawazura

Abstract The self-organization of macroscopic structure apparently contradicts the second law of thermodynamics. However, disorder can still develop on microscopic scales. In nonlinear systems, order and disorder may thus coexist on different scales. Here we study the self-organization of macroscopic structures in driven, nonlinear systems. Using a simple phenomenological transport model (of current in an electric circuit, or heat transport in a turbulent fluid/plasma) with linear and nonlinear impedances, we analyze the behavior of the rate of entropy production (σ) as a macroscopic system undergoes a bifurcation between linear and nonlinear operating points. Here σ acts as a generating function for a Legendre transformation between flux-driven and force-driven systems, the thermodynamic potential for which is a generalization of Onsager's dissipation function. We derive a duality relation that implies min/max-σ behavior depending on the connectivity of the impedances (series or parallel) and the type of driving.

List of Symbols
Symbol Meaning

Roman Symbols

I	Current in circuit
V	Voltage in circuit
T	Temperature at boundaries in turbulent layer model
T_{D}	Minimum temperature at inner boundary achieved by linear branch

Z. Yoshida (✉) · Y. Kawazura
Graduate School of Frontier Sciences, The University of Tokyo, Kashiwa, Chiba 277-8561, Japan
e-mail: yoshida@ppl.k.u-tokyo.ac.jp

Y. Kawazura
e-mail: kawazaura@ppl.k.u-tokyo.ac.jp

R. C. Dewar et al. (eds.), *Beyond the Second Law*,
Understanding Complex Systems, DOI: 10.1007/978-3-642-40154-1_15,

T_c Critical temperature for bifurcation
F Heat flux through boundaries
F_c Critical heat flux for bifurcation
P Free power available to create ordered flow

Greek Symbols
β Inverse temperature ($= T^{-1}$)
η Impedance
χ Conductance
α_S^V Thermodynamic amplification factor in current-driven series model
α_S^I Thermodynamic amplification factor in voltage-driven series model
α_P^V Thermodynamic amplification factor in current-driven parallel model
α_P^I Thermodynamic amplification factor in voltage-driven parallel model
Ψ Thermodynamic potential function as a function of V
Φ Thermodynamic potential function as a function of I

15.1 Introduction

The aim of this chapter is to describe and analyze a simple mathematical model that enables a nonlinear driven system to create bifurcated operating points (quasi-stationary states)—in energetics, the *entropy production rate* (σ) plays an interesting role. The relevant phenomena are taken from fluid or plasma physics. However, the thermodynamic relations to be derived are general, i.e., mechanism-independent.

Here we focus on how the behavior of σ in the model compares to various min/max-entropy production conjectures in the literature [1–15]. In this work, σ appears as a generating function of the *Legendre transformation* relating thermodynamic potentials (characterizing the operating points of a nonlinear driven system) for *flux-driven* and *force-driven* conditions [4]. The min/max duality of σ is then a natural consequence of the convexity of the thermodynamic potential.

Before discussing this new aspect of σ, we briefly review related (though distinct) variational principles. We often assume σ to be equivalent to *energy dissipation rate*. The history of the latter concept goes back to Helmholtz' minimum-dissipation principle [16], by which the velocity distribution of a slow stationary incompressible fluid can be determined. The viscous dissipation of the fluid kinetic energy is an example of Rayleigh's *dissipation function* [17] which may be added to Lagrange's equations to formulate a dissipative equation of motion.[1]

---

[1] However, Rayleigh's dissipation function and a Lagrangian are not on an equal footing to formulate a unified target functional of a variational principle; in the latter, the velocity $\boldsymbol{v}$ is related to $\dot{\boldsymbol{q}}$, which must be perturbed as a function of the position $\boldsymbol{q}$, while the dissipation function deals the velocity $\boldsymbol{v}$ as a direct state vector.

Generalizing the notion of force and current (or flux), we may describe thermodynamic non-equilibrium in parallel with mechanical non-equilibrium; the gradient of an intensive quantity (e.g. temperature T or $\beta = T^{-1}$) acts as a force that causes a current of an extensive quantity (e.g. heat flux). Then the mechanical minimum-dissipation principle (Helmholtz' principle) extends to thermodynamic minimum-dissipation principle (Onsager's principle) [18, 19]. Prigogine's *minimum σ principle* has an independent origin; first it was applied to linear irreversible processes in discontinuous systems, and was then extended to some nonlinear and continuous systems [8, 20]. Interestingly, σ becomes equivalent to the energy dissipation function for some physical systems; the product $-\mathbf{F} \cdot \nabla\phi = -F^j\partial_j\phi$ measures σ by an irreversible current $\mathbf{F}$ driven by a non-equilibrium intensive quantity ϕ, which in the linear regime becomes a dissipation function $R_{jk}F^jF^k$ by an appropriate impedance R_{jk}, hence both minimum principles are equivalent [21]. In a nonlinear regime of dissipation (the subject of the present study), we need a different target of variation—the thermodynamic potential to be formulated in Sect. 15.2 is a nonlinear generalization of Onsager's dissipation function, which is no longer equivalent to σ [4] (instead, σ is a generating function of Legendre transformation).[2]

The dissipation function for a diffusion-type equation is a convex (and coersive, or higher order [22]) functional, thus the variational principle (known as Dirichlet's principle) must be a *minimum principle* [23, 24]. Physically, this is because diffusion dominates on smaller scales, and the measure of dissipation is unbounded for small-scale fluctuations [22]. A "maximum" σ principle, then, means something different—it is not the maximum of the dissipation function for a diffusion process. In a general nonlinear system (like fluid), σ may be changed by a process on large scales, even if the ultimate dissipation (conversion of ordered flow to random motion) occurs on small scales.[3] Then, σ may have some different (bifurcated) modes that are determined by self-organized coherent flows. Each mode of σ may be modeled by a different impedance [15]. Moreover, the macroscopic operating point depends on how the system is driven. We define a max-σ selection principle according to which the actual operating point (as determined by

[2] As noticed by some authors, the formula dissipation function $= \sigma$ fails to apply to a strikingly simple and important example: the stationary distribution of temperature T on a thermal conductor minimizes the dissipation function $\int D|\nabla T|^2\,dx$ (Dirichlet's principle) if the conductivity D is given and Fourier's law applies. The corresponding σ is, however, $\int (D\nabla T) \cdot \nabla T^{-1}\,dx = \int D|\nabla \ln T|^2\,dx$ whose minimizer gives a different temperature distribution; see also [1].

[3] Shock is a well-known example: let us consider a one-dimensional compressible flow $u(x)$ with boundary conditions $u(0) = u_0$ and $u(1) = u_1$. Obviously the viscous dissipation $\int \nu(\partial_x u)^2\,dx$ is minimized when $u(x)$ has a linear distribution: $u(x) = u_0(1 - x) + u_1 x$. Creation of a shock will increase the variation in $(\partial_x u)^2$, resulting in an enhanced dissipation [24].

a stability analysis) has the greatest σ among all possible operating points. Similar discussions have been made by different authors, which are known as *optimum theory* or *upper bound theory* [25–27] (see also Chapters by Ozawa & Shimokawa and Dewar & Maritan). For example, in a Bénard convection, Ozawa [28] pointed out that a finite depth of the thermal boundary layer yields a bound for σ. The minimum depth is determined by the critical Rayleigh number. Here we inquire further into the root cause of such bounds; there must be an energy (power) bound.[4] Our phenomenological model of turbulent heat transport (Sect. 15.3) considers a nonlinear impedance (or conductance) that is a function of a *free power* available to cause a macroscopic structure.

In the next section, we analyze simple nonlinear electric circuits that illustrate the basic idea of *operating points*, their *bifurcations*, and *stability*—in the latter part of discussion, the current and voltage in the circuit model will be replaced, respectively, by a heat flux and (inverse) temperature in a plasma or fluid. In Sect. 15. 3, we will study a phenomenological model of turbulent heat transport in a plasma (or fluid) layer—an annular thin layer surrounding a high-temperature core plasma/fluid and surrounded by a cold heat bath. When the plasma/fluid self-organizes into an ordered shear flow (so-called zonal flow [29, 30]; see also [31, 32] for zonal flow in a plasma), the heat transport through the layer is reduced; this transition in a plasma is called the H-mode [33]. There is another type of self-organization of an ordered flow—in fluid mechanics, known as Bénard convection [34, 35], and in plasma physics, often called a *streamer* [36–38]—which brings about an "opposite" effect on heat transport; such a flow is parallel to the direction of heat flow and causes convective heat transport, resulting in an increased heat transport. While the zonal flow blocks the thermal conduction, the streamer opens up a new channel of heat transfer. These two examples allow us to analyze min/max-σ behavior in the context of a simple model.

15.2 Nonlinear Impedance and Equivalent Circuits

15.2.1 Electric Circuit Model

We consider a nonlinear impedance (resistance) η_1 which varies as a function of some power P available in the system (the power is used for self-organization of some internal structure that affects the impedance). We may consider a more complex model of η_1 that depends on other parameters [4]. The nonlinear impedance is connected to a linear impedance η_0 ($\equiv$ constant) that determines a baseline dissipation in the system. We will consider two different topologies of

[4] Evidently, the averaged dissipation in a quasi-stationary state cannot exceed the energy input. In the example of Bénard convection, the critical Rayleigh number measures the temperature difference in a non-equilibrium state, which is primarily caused by a thermal energy input.

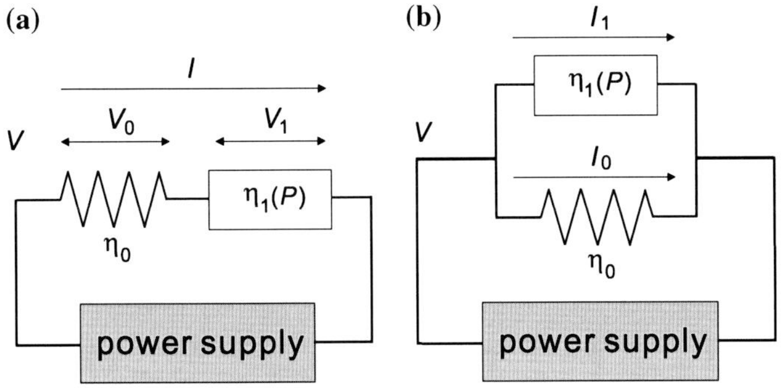

Fig. 15.1 Two different topologies of equivalent circuits connecting linear and nonlinear impedances: **a** series connection and **b** parallel connection. In **a**, the nonlinear impedance works as a blocking impedance when $\eta_1(P)$ increases. In **b**, the nonlinear impedance opens a new channel of current when $\eta_1(P)$ decreases [$\chi_1(P)$ increases]. The operating point of the circuits depends on how it is driven; two different modes are represented by the constant-current and constant-voltage power supplies

connections [see (Fig. 15.1)]; one is the series connection of η_1 and η_0, and the other is the parallel connection (for convenience, we will also use conductances $\chi_1 = \eta_1^{-1}$ and $\chi_0 = \eta_0^{-1}$).

In the series system, we consider a nonlinear impedance $\eta_1(P)$ that is a positive increasing function of P such that $\eta_1(0) = 0$ (a finite value of $\eta_1(0)$ can be absorbed in η_0). And, in a parallel system, $\chi_1(P)$ is a positive increasing function of P such that $\chi_1(0) = 0$. The series system is a model of the self-organization of zonal flow [29–32] that is directed perpendicular to a heat flux; its strong shear suppresses the turbulence, resulting in an increased impedance of heat transport (for example, see [39]). There is another type of self-organization of an ordered flow—Bénard convection or streamer—which brings about an opposite effect on heat transport; such a flow is parallel to the direction of heat flow and causes convective heat transport, resulting in a reduced impedance of heat transport [36–38]. While the zonal flow blocks the thermal conduction, the streamer opens up a new channel of heat transfer.

The other essential element of the circuit model is the power supply. The operating point of the circuit depends on whether it is driven by a *constant-current supply* or a *constant-voltage supply*. In a more general context, a system driven by the former will be called a *flux-driven system*, and the latter a *force-driven system*. The power σ dissipated by a resistance η is, if the current I is fixed, $\sigma = \eta I^2$, and, if the voltage V is fixed, $\sigma = V^2/\eta$. If η increases with σ (as it may occur in a nichrome-wire resistance), one must be careful in connecting a constant-current power supply, since an instability (so-called thermal runaway) occurs:

$$\sigma \nearrow \Rightarrow \eta \nearrow \Rightarrow \sigma \nearrow \cdots .$$

This well-known precaution of electric engineering is, in fact, the basis of the present study; a flux-driven system tends to maximize entropy production. To make this statement more precise and solid, we need to formulate an appropriate variational principle by which we can study the bifurcation of operating points and their stability.

15.2.2 Model of Nonlinear Impedance and Circuit Equations

We formulate a mathematical model of nonlinear impedance. The change of the impedance (a macroscopic property of the device) is due to some transition in the device, which is driven by a free power (the time derivative of a free energy to be denoted by P) available for the device; hence we denote $\eta_1(P)$. Here we assume a general form in terms of the current I_1 and the voltage V_1 on the device, which we write as

$$P = \wp(I_1, V_1). \tag{15.1}$$

First, we consider the series circuit connecting impedances η_0 and $\eta_1(P)$; see (Fig. 15.1a). The determining equation is Ohm's law, which reads as

$$V = [\eta_0 + \eta_1(P)]I. \tag{15.2}$$

Substituting (15.1) with $I_1 = I$ and $V_1 = \eta_1(P)I = V - \eta_0 I$ into (15.2), we obtain

$$V = [\eta_0 + \eta_1(\wp(I, V - \eta_0 I))]I. \tag{15.3}$$

When this system is driven by a constant-current power supply, we solve (15.3) for the voltage V with a given current I; this determines V as an implicit function of I

$$V = \mathscr{V}_S(I). \tag{15.4}$$

Depending on the specific form of $\wp$ and η_1, $\mathscr{V}_S$ (I) may have a singularity[5] or bifurcated branches; in Sect. 15.3 we will see an example of physical system that has bifurcated solutions. When the same series circuit is driven by a constant-voltage power supply, we solve (15.3) for the current I with a given voltage V;

$$I = \mathscr{I}_S(V). \tag{15.5}$$

Next, we consider the parallel circuit (Fig. 15.1b). Ohm's law reads as

$$I = [\chi_0 + \chi_1(P)]V, \tag{15.6}$$

where $\chi_0 = \eta_0^{-1}$ and $\chi_1(P) = \eta_1(P)^{-1}$. Substituting (15.1) with $V_1 = V$ and $I_1 = I - \chi_0 V$ into (15.6), we obtain

---

[5] Thermal runaway occurs if $P = W = I_1 V_1$ and $\eta_1(P) = Z_1 + aP$ (Z_1 and a are positive constants): then, from (15.3) we obtain, $V = I[\eta_0 + Z_1/(1 - aI^2)]$, which blows up at $I = 1/\sqrt{a}$.

$$I = [\chi_0 + \chi_1(\wp(I - \chi_0 V, V))]V, \tag{15.7}$$

by which we may determine implicit functions

$$V = \mathscr{V}_{\mathrm{P}}(I), \tag{15.8}$$

or

$$I = \mathscr{I}_{\mathrm{P}}(V); \tag{15.9}$$

the former applies to the constant-current drive, and the latter to the constant-voltage drive.

15.2.3 Stability

Here we formulate a general stability condition; explicit examples of stability analysis will be given in Sect. 15.3 for a plasma/fluid model.

We start by analyzing the series system; see (Fig. 15.1a). Let us first consider the constant-current drive. When a perturbation in the voltage δV occurs, it causes a chain of events:

$$\begin{aligned} \delta V \Rightarrow \delta P &= \frac{\partial \wp(I, V)}{\partial V}\delta V \Rightarrow \delta\eta_1 = \frac{\partial \eta_1(P)}{\partial P}\delta P \\ \Rightarrow \delta V' = I\delta\eta_1 &= I\frac{\partial \eta_1(P)}{\partial P}\frac{\partial \wp(I, V)}{\partial V}\delta V =: \alpha_{\mathrm{S}}^{\mathrm{V}}(I)\delta V, \end{aligned} \tag{15.10}$$

where the derivatives are evaluated at the given I, around $V = (V)_{\mathrm{S}}(I)$, and $P = \wp(I, \mathscr{V}_{\mathrm{S}}(I))$. When $\alpha_{\mathrm{S}}^{\mathrm{V}}(I) > 1$, the corresponding operating point is unstable because the fluctuation is amplified. If this cycle of perturbations takes a period of time τ, the evolution of the perturbation may be written as $\delta V(t) = \exp(\gamma t)\delta V(0)$ with $\gamma = \log \alpha_{\mathrm{S}}^{\mathrm{V}}/\tau$.

When the same series system is driven by a constant-voltage supply, we consider a perturbation of the current around $I = \mathscr{I}_{\mathrm{S}}(V)$ evaluating all coefficients at V:

$$\begin{aligned} \delta I \Rightarrow \delta P &= \frac{\partial \wp(I, V)}{\partial I}\delta I \Rightarrow \delta\chi_1 = \frac{\partial \chi_1(P)}{\partial P}\delta P \\ \Rightarrow \delta I' = V_1\delta\chi_1 &= V\frac{\eta_1}{\eta_0 + \eta_1}\frac{\partial \chi_1(P)}{\partial P}\frac{\partial \wp(I, V)}{\partial I}\delta I =: \alpha_{\mathrm{S}}^{\mathrm{I}}\delta I. \end{aligned} \tag{15.11}$$

When $\alpha_{\mathrm{S}}^{I}(V) > 1$, the corresponding operating point is unstable.

The amplification factor of the parallel system is evaluated by a similar procedure: for the constant-current drive, $\delta V' = \alpha_{\mathrm{P}}^{\mathrm{V}}(I)\delta V$ with [evaluating all coefficients at I and $V = \mathscr{V}_{\mathrm{P}}(I)$]

$$\alpha_{\mathrm{P}}^{\mathrm{V}}(I) = I\frac{\chi_1}{\chi_0+\chi_1}\frac{\partial\eta_1(P)}{\partial P}\frac{\partial\wp(I,V)}{\partial V}, \tag{15.12}$$

and, for the constant-voltage drive, $\delta I' = \alpha_{\mathrm{P}}^{\mathrm{I}}(V)\delta I$ with [evaluating all coefficients at V and $I = \mathscr{I}_{\mathrm{P}}(V)$]

$$\alpha_{\mathrm{P}}^{\mathrm{I}}(V) = V\frac{\partial\chi_1(P)}{\partial P}\frac{\partial\wp(I,V)}{\partial I}. \tag{15.13}$$

15.2.4 Thermodynamic Potential and Entropy Production Rate

We now introduce a *thermodynamic potential* by which we may characterize the operating point as its extremal (in fact, a maximum, as shown below).

The following analysis applies to both series and parallel systems. For the flux-driven (or, constant-current) case, we invoke the dual solution (15.5) for series system or (15.9) for parallel system to construct a potential function[6]:

$$\Psi(V) = \int_0^V \mathscr{I}(V)dV, \tag{15.14}$$

and

$$\Phi(V,I) = IV - \Psi(V). \tag{15.15}$$

The operating point of the system is the extremal of this thermodynamic potential, because, $\partial\Phi(V,I)/\partial V = 0$ gives $I - \mathscr{I}(V) = 0$.

Evaluating $\Phi(V,I)$ at the operating point (i.e., maximizing $\Phi(V,I)$ with respect to V), we define

$$\Phi(I) = \max_V \Phi(V,I) = \max_V[IV - \Psi(V)], \tag{15.16}$$

which reads as the Legendre transformation of $\Psi(V)$. The generator IV of this transformation is σ. The total derivative of σ is

$$\mathrm{d}(IV) = I\mathrm{d}V + V\mathrm{d}I. \tag{15.17}$$

On operating points, we may evaluate I as a function of V, i.e., $I = \mathscr{I}(V)$. Then the first term on the right-hand side of (15.17) may be written as $\mathscr{I}(V)\mathrm{d}V$. Similarly, evaluating $V = \mathscr{V}(I)$ [see (15.4)], the second term reads as $\mathscr{V}(I)\mathrm{d}I$. Hence, σ (on an operating point) splits into two terms:

[6] For a linear impedance $\eta \equiv R$ (constant) and $\mathscr{I}(V) = V/R$, thus, $\Psi(V) = V^2/(2R)$. If nonlinearity increases η, the corresponding $\Psi(V)$ gets smaller. Provided $\eta > 0$, however, $\Psi(V)$ is a monotonically increasing convex function.

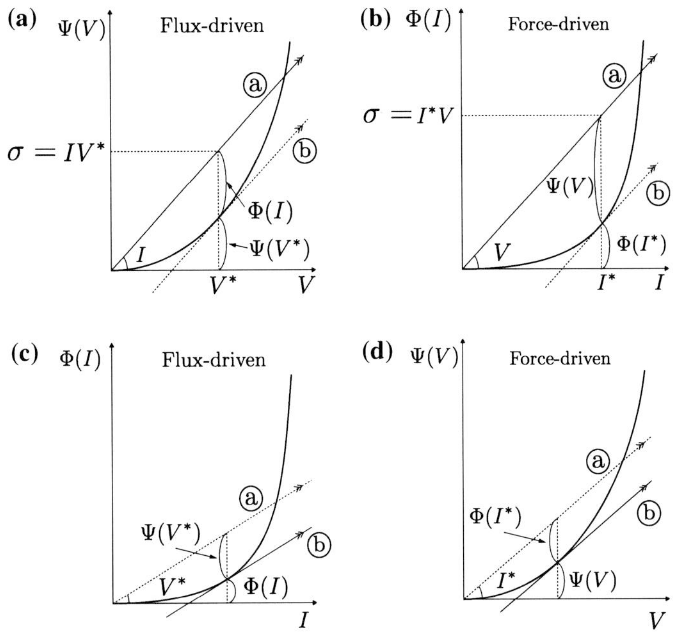

Fig. 15.2 Graphical method of evaluating σ in the flux-driven (**a** and **c**) and force-driven (**b** and **d**) systems. ⓐ Linear graph of IV. For the flux-driven system, the gradient is I, and for the force-driven system, V. ⓑ Tangent line to the graph of $\Psi(V)$ (flux-driven system) or $\Phi(I)$ (force-driven system)

$$IV = \Psi(V) + \Phi(I). \tag{15.18}$$

The evaluation of σ by (15.18) must be done with care. In the flux-driven system, I is the independent variable. For a given I, we maximize $\Phi(V, I)$ to find the operating point—graphically, we draw a line of gradient I which is tangent to the graph of $\Psi(V)$, and find a tangent point, which we denote by $V = V^*$ [see (Fig. 15.2a)]. In terms of the operating point V^*, we can evaluate σ as

$$IV^* = \Psi(V^*) + \Phi(I). \tag{15.19}$$

The force-driven (or, constant-voltage) system is dictated by the *dual potential* $\Phi(I) = \int \mathscr{V}(I)\,dI$, the Legendre transformation of $\Psi(V)$. For a given V, we find the tangent point $I = I^*$ of a line of gradient V (Fig. 15.2b). Then σ is given by

$$I^*V = \Psi(V) + \Phi(I^*). \tag{15.20}$$

Alternatively, we may use the graph of $\Psi(V)$ to evaluate σ of the force-driven system. For a given V, we evaluate the gradient $I^* = d\Psi(V)/dV$ and put $I^*V = \Psi(V) + \Phi(I^*)$ (Fig. 15.2d). Similar method applies to derive σ of the flux-driven system from $\Phi(I)$ (Fig. 15.2c).

In the linear regime, our potential $\Psi(V)$, defined by the integral (15.14), reduces to Onsager's dissipation function (see footnote 5). As pointed out by Gyarmati [21], σ is equivalent to the dissipation function in the linear regime (i.e., $d(IV) = 2d\Phi = 2d\Psi$), thus σ determines of the operating point. After the generalization of Onsager's dissipation functions to our potentials (which encompass nonlinear relations), however, σ is no longer equivalent to the potentials Φ and Ψ, and it ceases to be the single determinant of the operating point. In writing the determining equation of the operating point [$I = \mathscr{I}(V)$ for the flux-driven system] as a variational principle (15.16), we maximized $\Phi(V, I)$; in this case $\sigma(= IV)$ is not the target function to be minimized or maximized Instead, it is the generating function of the Legendre transformation between Φ and Ψ. The flux-driven and force-driven systems are connected by this Legendre transformation. In (15.19) and (15.20), we have given the formal estimates of σ for both systems.

15.2.5 Bifurcation and Min/Max Duality of Entropy Production Rate

As remarked in Sect. 15.2.2, the implicit function $\mathscr{V}_S(I)$ [or its inverse $\mathscr{I}_S(V)$; similarly the parallel-connection counterparts $\mathscr{V}_P(I)$ or $\mathscr{I}_P(V)$] may have bifurcated branches. Here we assume that $\mathscr{V}_S(I)$ [thus, $\mathscr{I}_S(V)$] has two branches (as to be shown in Sect. 15.3, a plasma/fluid model does have two-branch solutions [15]). Using the thermodynamic potential, we show that the difference between the entropy productions $\sigma = IV$ of these two branches changes sign when the drive is switched from a constant-current supply to a constant-voltage supply.

Here we consider the series system. Figure 15.3 depicts bifurcated thermodynamic potentials; the branches of larger and smaller values, respectively labeled by ① and ②, correspond to linear and nonlinear impedances. Since the nonlinearity increases the total impedance, branch-② represents the nonlinear state of the system. The bifurcation of the nonlinear branch occurs at $I = I_c$ ($V = V_c$).

First, we show that, in the flux-driven system, σ of the branch- ① is smaller than that of the branch- ②, i.e., for a given $I > I_c$ (the bifurcation point),

$$IV_1^* < IV_2^*, \tag{15.21}$$

where $I = \Psi_1'(V_1^*) = \Psi_2'(V_2^*)$. It suffices to show that $V_1^* < V_2^*$. By definition (15.14), $\Psi_1'(V_c) = \Psi_2'(V_c)$ and $\Psi_1'(V) > \Psi_2'(V) > 0$ ($V > V_c$). Therefore, $V_1^* < V_2^*$.

In the force-driven system, we obtain the dual relation, i.e., for a given $V > V_c$,

$$I_1^* V > I_2^* V, \tag{15.22}$$

where $V = \Phi_1'(I_1^*) = \Phi_2'(I_2^*)$. This follows from $\Phi_1'(I_c) = \Phi_2'(I_c)$, $\Phi_1'(I) < \Phi_2'(I)$ ($I > I_c$), thus $I_1^* > I_2^*$.

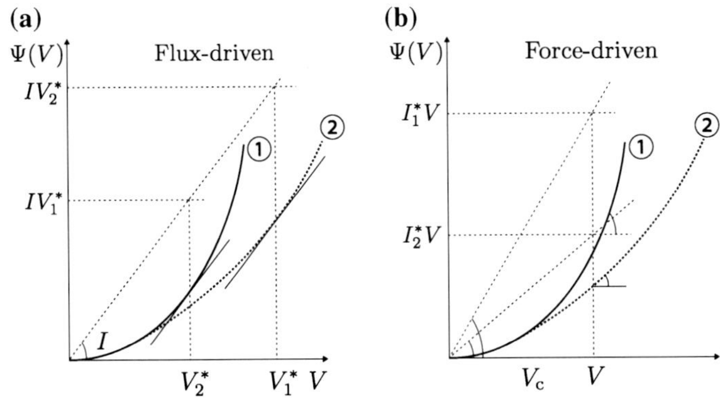

Fig. 15.3 Thermodynamic potential $\Psi(V)$ evaluated for bifurcated branches; ① linear impedance branch, ② nonlinear impedance branch. Graphical method to estimate of σ in **a** the flux-driven system and **b** the force-driven system

Figure 15.3 gives a graphical explanation of the min/max duality of σ in flux-driven and force-driven systems. As shown in Sect. 15.2.4, the operating point is characterized by the critical point of $IV - \Psi(V)$ (for the flux-driven system) or $F\beta - \Phi(F)$ (for the force-driven system); the criticality condition reads as the *Legendre transformation* between $\Psi(V)$ and $\Phi(I)$; see (Fig. 15.2). Remembering that the Legendre transformation seeks tangent lines, with a given gradient I, to the graph of $\Psi(V)$ [see (Fig. 15.2a)], it is evident from (Fig. 15.3a) that the branch-② yields a larger $\sigma = IV_2^*$. We can use the same graph to derive σ in the force-driven system [see (Fig. 15.2d)]; for a given V, we read the gradient $d\Psi(V)/dV$ from the graph, which evaluates I^*. Since branch-② (larger-impedance nonlinear branch) has a smaller gradient, it yields a smaller $\sigma = I_2^* V$; see (Fig. 15.3b).

As noted in Sect. 15.2.1, a flux-driven system tends to increase σ for $I > I_c$. In fact, we will prove, in Sect. 15.3, that the larger σ branch is stable, and smaller one is unstable. By the aforementioned *duality* of the Min/Max relation of σ, we may also expect that a smaller σ branch is chosen in the force-driven condition. This will be also proven by stability analysis.

15.3 Phenomenological Model of Turbulent Heat Transport

15.3.1 Layer Model

In this section, we describe a thermodynamic model of self-organizing plasma/fluid turbulence [15]. The force (voltage) and flux (current) in the previous electric circuit model will be translated into inverse-temperature β (or temperature $T = \beta^{-1}$) and heat flux F. The series-connection model is at the core of the present discussion —we will compare it with a parallel-connection model in Sect. 15.4.

We consider a thin layer which encloses a high-temperature core plasma. The outer boundary of the layer domain is connected to a low-temperature (T_0) heat bath. At the inner boundary, heat flux F is given, while the temperature (T) is an unknown variable. We note that, in usual plasma experiments (as well as in most natural processes, like atmospheric heat transfer driven by solar heat), the input power is controlled (or, given as a determining parameter), which must be transferred to some heat sink in a steady state, thus the flux is the controlling parameter. On the other hand, theoretical analysis or a computational study often uses a force-driven model giving a boundary condition on the intensive quantities (like the temperature). Mathematically, a flux-driven condition and a force-driven condition correspond to a Neumann boundary condition and a Dirichlet boundary condition, respectively (one may consider a mixed-type boundary condition if appropriate).

The series-system Ohm's law (15.2) is now rewritten as

$$T = T_0 + [\eta_0 + \eta_1(P)]F. \tag{15.23}$$

Here we note the analogy between $T - T_0$ and V (and also between F and I). To define the model, we have to specify the functions $\eta_1(P)$ and $\wp(F, T)$. For $\eta_1(P)$, we assume the simplest relation:

$$\eta_1(P) = aP, \tag{15.24}$$

where $a(>0)$ is a constant. In the analysis of the bifurcation point and stability, the first derivative $\partial\eta_1/\partial P$ is essential [15], hence, this minimum model suffices (a more complicated graph may cause, for example, hysteresis [4]). A positive a yields an increased impedance of heat conduction when a positive P creates an ordered flow (zonal flow) and suppresses turbulent heat transport. In a thermodynamic system, the free power $P = \wp(F, T)$ must be smaller than the Carnot-cycle's power $[P_c = F(1 - T_0/T)]$ (the ideal conversion of the internal energy to a mechanical energy). The difference between P_c and the actually available free power P is the power dissipation in the background linear impedance, which we write $F(1 - T_0/T_D)$ introducing $T_D = T_0 + \eta_0 F$, i.e.,

$$P = \wp(F, T) = F\left(1 - \frac{T_0}{T}\right) - F\left(1 - \frac{T_0}{T_D}\right) = F\left(\frac{T_0}{T_D} - \frac{T_0}{T}\right). \tag{15.25}$$

The *non-organized state* is such that $T = T_D$ and $P = 0$; this state may be called a *linear branch* because (15.23) reduces to a linear relation between T and F. In the next subsection, we will show how an *organized state* (or, *nonlinear branch*) emerges with $T > T_D$.

15.3.2 Bifurcation

First, we consider the flux-driven system in which F is a given parameter and T is an unknown variable. We solve (15.23)–(15.24)–(15.25) for T to obtain a set of bifurcated solutions:

$$T = \begin{cases} T_1 = T_D & (F < F_c), \\ T_1 \text{ or } T_2 = aF^2 T_0 / T_D & (F \geq F_c), \end{cases} \tag{15.26}$$

where

$$F_c = \frac{T_0}{\sqrt{T_0 a} - \eta_0} \tag{15.27}$$

is the *critical flux* at which the nonlinear branch bifurcates. Stability analysis (Sect. 15.3.3) will show that the organized state (or, nonlinear branch) $T = T_2 = aF^2 T_0 / T_D \ (\geq T_D)$ is stable, while the non-organized state (or, linear branch) $T = T_1 = T_D$ destabilizes beyond the bifurcation point [15].

For a force-driven system, where the inner-boundary temperature T is a given parameter, we solve (15.23)–(15.24)–(15.25) for F to obtain

$$F = \begin{cases} F_1 = (T - T_0)/\eta_0 & (T < T_c), \\ F_1 \text{ or } F_2 = \frac{\eta_0 T + \sqrt{(\eta_0 T)^2 + 4aT_0^2 T}}{2aT_0} & (T \geq T_c), \end{cases} \tag{15.28}$$

where

$$T_c = \frac{T_0 \sqrt{aT_0}}{(\sqrt{aT_0} - \eta_0)} \tag{15.29}$$

is the *critical temperature* at which the bifurcation occurs. In Sect. 15.3.3, we will show that the nonlinear branch $F = F_2$ is stable, while the linear branch $F = F_1$ destabilizes beyond the critical temperature.

15.3.3 Thermodynamic Stability

Invoking the method of Sect. 15.2.3, we study the stability of the foregoing bifurcated solutions.

In a flux-driven system (F is given), we consider a chain of perturbations $\delta T \Rightarrow \delta\eta_1 \Rightarrow \delta T' = F\delta\eta_1 = \alpha^T \delta T$. The amplification factor on the temperature perturbation is

$$\alpha^T = aF^2 \frac{T_0}{T^2}. \tag{15.30}$$

When $\alpha^F > 1$, the system is unstable because the fluctuation amplifies. For $F \leq F_c$, there exists only the linear branch. We denote by α_1^T the amplification factor of this branch. By (15.27), the condition $F \leq F_c$ reads as $T_D > F\sqrt{T_0 a}$. We thus find

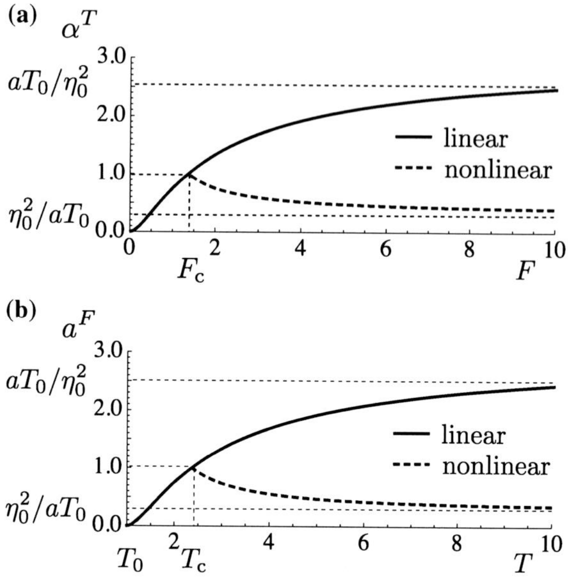

Fig. 15.4 The amplification factor α^{T} or α^{F} of each branch of solutions (*solid line* linear branch, *dashed line* nonlinear branch) [4]. **a** Flux-driven system: α^{T} is evaluated as a function of the flux F. **b** Temperature-driven system: α^{F} is evaluated as a function of temperature T. In these graphs, parameters are normalized so that $\eta_0 = 1$ and $T_0 = 1$. We assume $a = 2$

$$\alpha_1^{\mathrm{T}} = \frac{aT_0F^2}{T_{\mathrm{D}}^2} < \frac{aT_0F^2}{(F\sqrt{T_0 a})^2} = 1 \tag{15.31}$$

before the bifurcation. Beyond the bifurcation point $(F > F_c)$, the linear branch is unstable, because

$$\alpha_1^{\mathrm{T}} = \frac{aT_0F^2}{T_1^2} = \frac{T_{\mathrm{D}}}{T^2}\frac{aT_0F^2}{T_{\mathrm{D}}} = \frac{T_2}{T_1} > 1, \tag{15.32}$$

while the nonlinear branch is stable, because

$$\alpha_2^{\mathrm{T}} = \frac{aT_0F^2}{T_2^2} = \frac{T_{\mathrm{D}}}{T_2^2}\frac{aT_0F^2}{T_{\mathrm{D}}} = \frac{T_1}{T_2} < 1. \tag{15.33}$$

Hence, the nonlinear branch (or, organized state), if it exists, is always stable [15].

For the force-driven case (T is given), we consider a chain of perturbations $\delta F \Rightarrow \delta\chi_1 \Rightarrow \delta F' = (T - T_0)\delta\chi_1 = \alpha^{\mathrm{F}}\delta F$. The amplification factor of the flux perturbation is

$$\alpha^{\mathrm{F}} = -\frac{(T - T_0)}{\left[\eta_0 + aT_0F\left(\frac{1}{T_{\mathrm{D}}} - \frac{1}{T}\right)^2\right]^2}\left[aT_0\left(\frac{1}{T_{\mathrm{D}}} - \frac{1}{T}\right) - aT_0F\frac{\eta_0}{T_{\mathrm{D}}^2}\right], \tag{15.34}$$

by which we find that the nonlinear branch, if it exists, is always stable (for the detail, see [4]).

In Fig. 15.4, we plot the amplification factor of each branch for both flux and force drives.

Table 15.1 Selection rules of operating points in terms of σ. The nonlinear branch, whenever it exists, is always stable, while the linear branch destabilizes beyond the bifurcation point. However, the min/max relation of the selected (nonlinear) branch exhibits two-by-two duality. These selection rules apply whenever $d\eta_1(P)/dP|_{P=0} > 0$ for series connection and $d\chi_1(P)/dP|_{P=0} > 0$ for parallel connection

	Flux-driven	Force-driven
Series system	Maximum	Minimum
Parallel system	Minimum	Maximum

15.4 Concluding Remarks

We have developed a thermodynamic model of bifurcation and stability of self-organizing driven systems; introducing a nonlinear impedance of heat transfer, the energetics of self-organization has been analyzed. Creation of a macroscopic ordered structure affects the heat transport in a *layer* hemmed by inner and outer boundaries; on the other hand, the creation of the structure (represented by a change of the impedance) requires power that is supplied by the heat flux coming from the inner boundary. The boundary condition on the inner boundary is an important factor in determining the operating point of the system.

As formulated in Sect. 15.2.1, we have a two-by-two matrix of series/parallel and flux/force-drive systems. In Sect. 15.3, we have analyzed the series-connection model of nonlinear impedance pertinent to the self-organization of zonal flow and suppression of turbulence—the nonlinearity of the impedance works to increase the impedance of the heat transport. We have shown that the nonlinear branch (or, the organized state), whenever it exists, is always stable, while the linear branch (or, the baseline turbulent state) destabilizes beyond the bifurcation point; this selection rule holds both in flux-driven and force-driven cases. As we have shown in Sect. 15.2.5, σ of a higher-impedance state (nonlinear branch) is larger when it is flux-driven, is smaller when force-driven.

While we have omitted detailed calculations about the parallel-connection model pertinent to the Bénard convection or streamer, we reach the same conclusion about the bifurcation and stability in the parallel system; the nonlinear branch (in the parallel model, the nonlinearity works to increase the conductance, creating a new channel of heat transport), whenever it bifurcates, is always selected as the stable branch [5]. However, in terms of entropy production, the selection of this stable branch is characterized by min-σ or max-σ depending on how the system is driven; see Sect. 15.2.5.

Table 15.1 summarizes the selection rule of operating points in terms of σ on the two-by-two matrix of series/parallel connections and flux/force drivers. In the series system, the nonlinear effect is assumed to increase the impedance as a function of the free power scale by Carnot's efficiency; in the parallel system, the nonlinear effect decreases the impedance.

We end this chapter with a short comment on turbulence theory. In two dimensional ideal fluid, the energy and entrophy conserve simultaneously. When a finite viscosity is added, the entrophy decays faster than the energy (selective decay), because the entrophy includes higher-order derivatives. The relaxed state is, then, the minimizer of the entrophy under the constraint on the energy (as well as the total angular momentum) [40]. The minimum enstrophy principles, naturally, implies minimum entropy production, because the viscous dissipation is proportional to the enstrophy. However, one may consider the *dual* variational principle, i.e., maximization of energy under a given entrophy [22], which is mathematically equivalent but physically more plausible than the former setting in the context of turbulence theory; the key assumption in Kolmogorov's cascade model is that the energy dissipation rate (= energy transfer rate = energy injection rate), is a control parameter of the turbulence, and the so-called inverse cascade, resulting in the energy transfer to a large scale, is the maximization of the energy in the coherent structure. This min/max duality of dissipation is, in principle, similar to the duality of σ delineated here. However, our problem is somewhat more complex than the argument about turbulent cascade, because the energy injection is not directly mechanical, but it is given as a heat flow; σ is not directly the constraint or the target of minimization/maximization, but it is the generating function of the thermodynamic potentials to be maximized.

References

1. Bertola, V., Cafaro, E.: A critical analysis of the minimum entropy production theorem and its application to heat and fluid flow. Int. J. Heat Mass Transf. **51**, 1907–1912 (2008)
2. Dewar, R.C.: Information theory explanation of the fluctuation theorem, maximum entropy production and self-organized criticality in non-equilibrium stationary states. J. Phys. A **36**, 631–641 (2003)
3. Dewar, R.C.: Maximum entropy production and the fluctuation theorem. J. Phys. A **38**, L371–L381 (2005)
4. Kawazura, Y., Yoshida, Z.: Entropy production rate in a flux-driven self-organizing system. Phys. Rev. E **82**(066403), 1–8 (2010)
5. Kawazura, Y., Yoshida, Z.: Comparison of entropy production rates in two different types of self-organized flows: Bénard convection and zonal flow. Phys. Plasmas **19**(012305), 1–5 (2012)
6. Kleidon, A., Lorenz, R. D.: Non-equilibrium Thermodynamics and the Production of Entropy–Life, Earth, and Beyond. Springer, Berlin (2004) Chap. 8
7. Li, J.: On the extreme of internal entropy production. J. Phys. A: Math. Theor. **42**(035002), 1–15 (2009)
8. Nicolis, G., Prigogine, I.: Self-Organization in Nonequilibrium Systems—From Dissipative Structures to Order through Fluctuations. Wiley, New York (1977)
9. Niven, R.K.: Steady state of a dissipative flow-controlled system and the maximum entropy production principle. Phys. Rev. E **80**(021113), 1–15 (2009)
10. Niven, R.K.: Simultaneous extrema in the entropy production for steady-state fluid flow in parallel pipes. J. Non-Equilib. Thermodyn. **35**, 347–378 (2010)
11. Ozawa, H., Ohmura, A., Lorenz, R.D., Pujol, T.: The second law of thermodynamics and the global climate system: A review of the maximum entropy production principle. Rev. Geophys. **41**, 1–24 (2003)

12. Paltridge, G. W.: Global dynamics and climate - a system of minimum entropy exchange, Q. J. R. Meteorol. Soc. **101**, 475–484 (1975)
13. Paltridge, G.W.: Climate and thermodynamic systems of maximum dissipation. Nature **279**, 630–631 (1979)
14. Shimokawa, S., Ozawa, H.: On the thermodynamics of the oceanic general circulation: entropy increase rate of an open dissipative system and its surroundings. Tellus Ser. A **53**, 266–277 (2001)
15. Yoshida, Z., Mahajan, S.M.: "Maximum" entropy production in self-organized plasma boundary layer: A thermodynamic discussion about turbulent heat transport. Phys. Plasmas **15**(032307), 1–6 (2008)
16. Helmholtz, H.: Zur Theorie der stationären Ströme in reibenden Flüssigkeiten. Wiss. Abh **1**, 223–230 (1868)
17. Rayleigh, Lord: On the motion of a viscous fluid. Phil. Mag. **26**, 776–786 (1913)
18. Onsager, L.: Reciprocal relations in irreversible processes. I. Phys. Rev. **37**, 405–426 (1931)
19. Onsager, L.: Reciprocal relations in irreversible processes. II. Phys. Rev. **38**, 2265–2279 (1931)
20. Gransdorff, P., Prigogine, I.: Thermodynamic Theory of Structure. Stability and Fluctuations, Wiley-Interscience, New York (1971)
21. Gyarmati, I.: Nonequilibrium Thermodynamics. North-Holland, Amsterdam (1962)
22. Yoshida, Z., Mahajan, S.M.: Variational principles and self-organization in two-fluid plasmas, Phys. Rev. Lett. **88**(095001), 1–4 (2002)
23. Evans, L.C.: Partial Differential Equations, (American Mathematical Society, Providence (1998)
24. Yoshida, Z.: Nonlinear Science—The Challenge of Complex Systems. Springer, Heidelberg (2010)
25. Busse, F.H.: Bounds for turbulent shear flow. J. Fluid Mech. **41**, 219–240 (1970)
26. Howard, L.N.: Heat transport by turbulent convection. J. Fluid Mech. **17**, 405–432 (1963)
27. Malkus, W.V.R.: The heat transport and spectrum of thermal turbulence. Proc. R. Soc. London, Ser. A **225**, 196 (1954)
28. Ozawa, H., Shimokawa, S., Sakuma, H.: Thermodynamics of fluid turbulence: A unified approach to the maximum transport properties. Phys. Rev. E **64**(026303), 1–8 (2001)
29. Chapman, C.R.: Jupiter's zonal winds: Variation with latitude. J. Atmos. Sci. **26**, 986–990 (1969)
30. Pedlosky, J.: Geophysical Fluid Dynamics, 2nd edn. Springer-Verlag, New York (1987)
31. Hasegawa, A.: Self-organization processes in continuous media. Adv. Phys. **34**, 1–42 (1985)
32. Hasegawa, A., Wakatani, M.: Self-organization of electrostatic turbulence in a cylindrical plasma. Phys. Rev. Lett. **59**, 1581–1584 (1987)
33. Wagner, F., et al.: Regime of improved confinement and high beta in neutral-beam-heated divertor discharges of the ASDEX tokamak. Phys. Rev. Lett. **49**, 1408–1412 (1982)
34. Bénard, H.: Les tourbillons cellulaires dans une nappe liquide. Rev. Gén. Sci. pures et appl. **11**, 1261–1271, 1309–1328 (1900)
35. Rayleigh, Lord: On the convective currents in a horizontal layer of fluid when the higher temperature is on the under side. Phil. Mag. **32**, 529–546 (1916)
36. Champeaux, S., Diamond, P.H.: Streamer and zonal flow generation from envelope modulations in drift wave turbulence. Phys. Lett. A **288**, 214–219 (2001)
37. Das, A., Sen, A., Mahajan, S., Kaw, P.: Zonal and streamer structures in magnetic-curvature-driven Rayleigh-Taylor instability. Phys. Plasmas **8**, 5104–5112 (2001)
38. Sagdeev, R.Z., Shapiro, V.D., Shevchenko, V.I.: Convective cells and anomalous plasma diffusion. Sov. J. Plasma Phys **4**, 306–314 (1979)
39. Terry, P.W.: Suppression of turbulence and transport by sheared flow. Rev. Mod. Phys. **72**, 109–165 (2000)
40. Kraichnan, R.H.: Two-dimensional turbulence. Rep. Prog. Phys. **43**, 547 (1980)

Chapter 16
MaxEnt and MaxEP in Modeling Fractal Topography and Atmospheric Turbulence

Jingfeng Wang, Veronica Nieves and Rafael L. Bras

Abstract Recent investigations on scale-invariant processes such as topography and modeling evapotranspiration demonstrate the usefulness and potential of the principles of maximum entropy (MaxEnt) and maximum entropy production (MaxEP) in the study of Earth systems. MaxEnt allows theoretical predictions of probability distributions of geophysical multifractal processes based on a small number of geometric parameters. MaxEP leads to the prediction of evapotranspiration and heat fluxes using fewer input variables than existing process based models. Encouraging progress in the application of MaxEnt and MaxEP, viewed as unified principles of inference, paves the way for more approaches of understanding, characterizing and predicting the behavior of the complex Earth systems.

List of Symbols

Symbol Meaning (SI Units)

Roman Symbols

$A_{r/r'}$	Multiscaling parameter (–)
B	Dimentionless function (–)
c_p	Specific heat of air at constant pressure $\left(\mathrm{J\ kg^{-1}\ K^{-1}}\right)$
D	Dissipation function
E	Latent heat flux $(\mathrm{W\ m^{-2}})$
G	Ground heat flux $(\mathrm{W\ m^{-2}})$

J. Wang (✉) · R. L. Bras
Georgia Institute of Technology, Atlanta, GA 30332-0355, USA
e-mail: jingfeng.wang@ce.gatech.edu

R. L. Bras
e-mail: rlbras@gatech.edu

V. Nieves
Jet Propulsion Laboratory, California Institute of Technology, Pasadena, CA 91109, USA
e-mail: veronica.nieves@jpl.nasa.gov

R. C. Dewar et al. (eds.), *Beyond the Second Law*,
Understanding Complex Systems, DOI: 10.1007/978-3-642-40154-1_16,

H	Sensible heat flux $(\mathrm{W\ m^{-2}})$
$I_0,\ I_1,\ I_2$	Electric currents (amp)
$I_s,\ I_a,\ I_e$	Thermal inertia parameters $\left(\mathrm{J\ m^{-2}\ K^{-1}\ s^{-1/2}}\right)$
K	Moment scaling function (–)
$\boldsymbol{L}^2$	Functions with continuous derivatives (–)
L_v	Latent heat of vaporization of liquid water $\left(2.5\times10^6\right)\left(\mathrm{J\ kg^{-1}}\right)$
m_A	Ignorance prior of A (–)
$n,\ N$	Integer (–)
p_A	Probability of A (–)
q_s	Specific humidity $\left(\mathrm{kg\ kg^{-1}}\right)$
$r,\ r'$	Scale parameters (–)
$\mathbb{R}^2$	Two-dimensional real domain (m)
$R_1,\ R_2$	Electrical impedance (Ohm)
R_n	Net radiation $(\mathrm{W\ m^{-2}})$
R_v	Gas constant for water vapor (461) $\left(\mathrm{J\ kg^{-1}\ K^{-1}}\right)$
S_{SJ}	Shannon-Jaynes information entropy (–)
T_s	Surface temperature (K)
$U_0,\ U_1,\ U_2$	Voltage (Volt)
$\boldsymbol{x}$	Location vector (m)
z	Elevation (m)
Δz	Incremental elevation (m)
Z_A	Partition function (–)

Greek Symbols

$\alpha_{l,j,\boldsymbol{k}}$	Wavelet coefficients (–)
β	Ratio of eddy diffusivity of water vapor to that of heat (–)
$\eta_{l,j,\boldsymbol{k}}$	Wavelet coefficients (–)
$\lambda_n,\ \lambda_I,\ \lambda_U$	Lagrangian multipliers for the nth (–), current (Volt) and voltage (amp) constraint, respectively
σ	Dimensionless parameter (–)

16.1 An Overview

The complexity of Earth systems results from interactions between its interwoven geo-eco-hydro-atmospheric components. Prediction of the evolution of an earth system depends on both our understanding of the natural laws governing the physical processes and our knowledge of the past and current states of the system over a wide range of scales. The challenges of understanding, characterizing and predicting the behavior of earth systems have also been related to the fact that we either lack full understanding of the mechanisms behind the processes or lack data. Consequently, earth system science is, in essence, about making inferences, i.e. reasoning with incomplete information guided by certain principles of inference.

Since its proposition in the late 1950s as an application of information theory in statistical mechanics [1], the maximum entropy (MaxEnt) principle has been accepted as a general and powerful inference algorithm in the context of Bayesian probability theory for finding solutions of ill-posed problems when complete information is not available [2, 3]. Application of MaxEnt in the study of earth systems started in early 1960s. One of the earliest works is that of Leopold and Langbein [4] on river networks. There have been sporadic publications pursuing this idea since then. Lienhard [5] derived the dimensionless unit hydrograph through maximizing the Boltzmann entropy based on the argument that such derived unit hydrograph is the most probable one. Sonuga [6] directly applied the MaxEnt theory in hydrologic frequency analysis. Fiorentino et al. [7] and Claps and Fiorentino [8] attempted to explain the fractal structure of river basin networks based on the MaxEnt formalism. Rodriguez-Iturbe and Rinaldo [9] showed that some structures of river networks might be explained by the MaxEnt. Niedda [10] studied the problem of linking measurements at small scales to model output at large scales from the perspective of information entropy although MaxEnt was not used explicitly. Recent efforts by the authors expanded these earlier works to common phenomena in earth system, namely self-similar and multifractal processes, for deriving their probability distributions at different spatial scales [11, 12]. Compared to other existing methods of statistical analysis, the MaxEnt method has evident advantages: (1) it highlights essential information while leaving out unimportant details of the processes under study; (2) it provides statistics of self-similar and multifractal processes covering scales beyond those of observations; (3) it allows derivation of probability distributions without computing histograms that requires a large volume of data.

The principle of maximum entropy production (MaxEP) evolved out of the classical treatise of non-equilibrium thermodynamics [13] dealing with thermodynamic entropy production, defined as the ratio of exchange rate of heat to temperature. An earlier successful application of MaxEP in numerical modeling of climate [14] has motivated a number of later studies to pursue the idea further. Recent years have witnessed rapid progress of the application of MaxEP in the study of earth systems and beyond [15, 16]. Among these efforts was the ground breaking work of Dewar [17] that laid a theoretical foundation under the intuitive concept of thermodynamic entropy production by deriving a formalism of MaxEP from that of MaxEnt. Dewar further emphasized that MaxEP, like MaxEnt, is a principle of inference in the context of Bayesian probability [18]. The MaxEP formalism [17] paves the way to modeling a wider class of transport processes that are not necessarily related to the production of thermodynamic entropy. A recent study on modeling energy balance over the earth's surface [19] demonstrated how MaxEP leads to prediction of evapotranspiration and heat fluxes using a small number of surface hydro-meteorological variables that would be considered insufficient according to the classical bulk transfer equation based models.

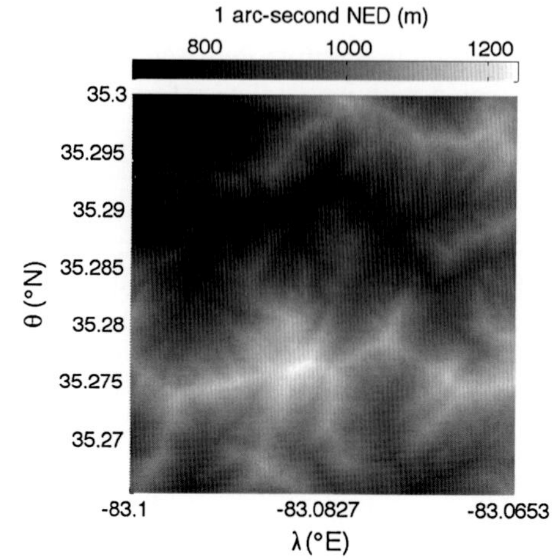

Fig. 16.1 The multifractal pattern of topography of Anza-Borrego Desert State Park in California from the USGS National Elevation Dataset (NED) map of elevation in meters with 1 arc second resolution. The domain is covered with 128 × 128 pixels with longitude λ and latitude θ in degrees

16.2 Application of MaxEnt in Modeling Scale-Invariant Fractal Topography

Natural processes in earth systems often manifest "scaling" behavior, i.e. a part resembles the whole as quantified by certain statistical properties known as scaling laws. Familiar examples include rough coast lines [20], meandering river channels [9] and topographic landscapes [21] that appear to have infinite similar structures without a characteristic length scale. Processes with such geometric properties are often classified as self-similar and multifractal. Exact definition of self-similarity and multifractality may vary. Herein, self-similarity refers to scaling laws associated with constant parameters, while multifractality to scaling laws associated with random-variable parameters (or multiscaling parameter) described by probability distributions. A graphical illustration of the multifractal processes is given in Fig. 16.1.

A common tool for studying multifractal behavior is the cascade model that relates the process at different scales according to some geometric characteristics, i.e. scale-invariant fractal statistics [22], which can be solved recursively through the multiscaling parameter (the cascade variable). The cascade model is not considered a complete characterization of a stochastic process without specifying the corresponding probability distribution. Identification of probability distributions through empirical histograms requires a large amount of data that are not always available. This is the typical situation of incomplete information we always face in the study of earth systems. The theory of MaxEnt, as an inference algorithm, can help in deriving the probability distribution of a multiscaling parameter of multifractal process.

Consider the incremental elevation Δz of a field of topographic elevation z satisfying the multifractal condition [23],

$$\Delta_r z \doteq A_{r/r'} \Delta_{r'} z, \tag{16.1}$$

where $\doteq$ indicates "equal in probability", $\Delta_r z \equiv |z(r\boldsymbol{x}_2) - z(r\boldsymbol{x}_1)|$ is the rescaled incremental elevation at the location vector $\boldsymbol{x}$, $A_{r/r'}$ is the multiscaling parameter independent of z, and r, r' are a pair of given scales with $r' > r$ for a downscaling cascade process. In the study of multifractal processes, $A_{r/r'}$ is commonly characterized in terms of several empirical moments, whose estimation requires a large amount of data. Yet the probability distributions of $A_{r/r'}$ are often needed to fully describe the multifractal processes such as erosion shaped topography [22], mesoscale rainfall [24], oceanic chlorophyll [25], etc.

MaxEnt allows an analytical expression of the probability distribution of $A_{r/r'}$ to be derived based on a given number (N) of moments of $A_{r/r'}$ by selecting the most probable configuration among all possibilities. Define the multiscaling moments of $A_{r/r'}$,

$$\int p_A(A_{r/r'}) A^n_{r/r'} dA_{r/r'} = \left(\frac{r}{r'}\right)^{K(n)}, \quad 1 \le n \le N, \tag{16.2}$$

where $K(n)$ is the moment scaling function independent of r/r', and N a fixed integer [26]. Following the standard MaxEnt formalism [2] to maximize the Shannon-Jaynes information entropy S_{SJ} of $p_A(A_{r/r'})$ for the continuous variable $A_{r/r'}$,

$$S_{SJ} \equiv \int p_A(A_{r/r'}) \ln\left[\frac{p_A(A_{r/r'})}{m_A(A_{r/r'})}\right] dA_{r/r'} \tag{16.3}$$

where m_A is a prescribed ignorance prior distribution [27] under the constraint as in Eq. (16.2) leads to,

$$p_A(A_{r/r'}) = \frac{m_A(A_{r/r'})}{Z_A} \exp\left\{-\sum_{n=1}^{N} \lambda_n A^n_{r/r'}\right\}, \tag{16.4}$$

with the partition function

$$Z_A(\lambda_1, \ldots, \lambda_N) = \int m_A(A_{r/r'}) \exp\left\{-\sum_{n=1}^{N} \lambda_n A^n_{r/r'}\right\} dA_{r/r'}. \tag{16.5}$$

The Lagrangian multipliers λ_n are uniquely related to the given multiscaling moments on the RHS of Eq. (16.2) through the Legendre transformation,

$$\frac{\partial \ln Z_A}{\partial \lambda_n} = \left(\frac{r}{r'}\right)^{K(n)}, \quad 1 \le n \le N. \tag{16.6}$$

The probability distribution of $\Delta z_{r'}$ for given multiscaling moments and geometric mean can be obtained in the same way [12] according to Eqs. (16.1) and (16.4).

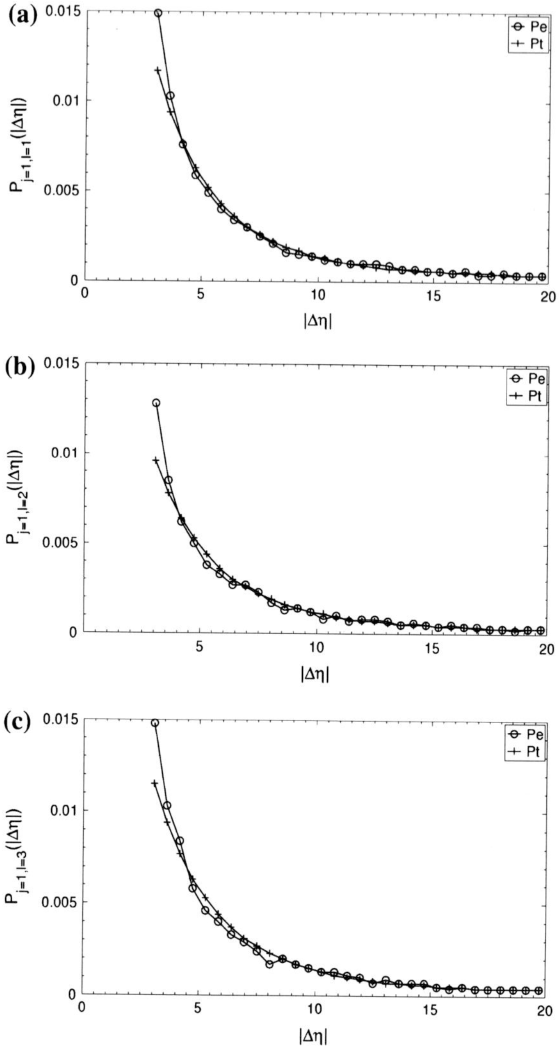

Fig. 16.2 Observed (Pe) versus MaxEnt predicted (Pt) probability distribution p_A for $N = 2$ as in Eq. (16.4) in the wavelet domain assuming a uniform ignorance prior m_A [12]. According to the wavelet decomposition, $|\Delta\eta| \equiv \eta_{j,j,\boldsymbol{k}}$ corresponds to $A_{r/r'}$ at the scale of observation $j = 1$ to $j = 2$ (downscale cascade) using order 3 Battle-Lemarié basis. **a** $l = 1$, **b** $l = 2$ and **c** $l = 3$ corresponding to horizontal, vertical, and diagonal spatial orientation, respectively

Figure 16.2 compares the MaxEnt predicted p_A as in Eq. (16.4) with that estimated from the histogram of Δz at various spatial scales using the tree-structure of wavelet-based statistical analysis [28–33]. A multi-resolution decomposition of $z(\boldsymbol{x}) \in \boldsymbol{L}^2(\mathbb{R}^2)$ can be expressed in terms of a dyadic wavelet transform at orientations $l = 1, 2, 3$ (representing the horizontal, vertical and diagonal, respectively), scale 2^j, and translational vector $\boldsymbol{k} = (k_1, k_2)$ for integers j, k_1 and k_2,

$$z(\mathbf{x}) = \sum_{j} \sum_{l=1,2,3} \sum_{\mathbf{k}} \alpha_{j,l,\mathbf{k}} \psi_{j,l,\mathbf{k}}(\mathbf{x}) \tag{16.7}$$

where $\psi_{lj\mathbf{k}}(\mathbf{x}) = 2^{-j}\psi_l(2^{-j}\mathbf{x} - \mathbf{k})$ are the mother wavelets and $\alpha_{lj\mathbf{k}} = \langle \psi_{lj\mathbf{k}}, z \rangle$ the wavelet coefficients [34, 35] where $\langle \cdot, \cdot \rangle$ stands for inner product. The wavelet representations of $\Delta_r z$ and $\Delta_{r'} z$ are given by,

$$\Delta\alpha_{j,l,\mathbf{k}} = |\alpha_{j,l,\mathbf{k}}(r\mathbf{x}_1) - \alpha_{j,l,\mathbf{k}}(r\mathbf{x}_2)|, \tag{16.8}$$

$$\Delta\alpha_{j-1,l,\left[\frac{\mathbf{k}}{2}\right]} = |\alpha_{j-1,l,\left[\frac{\mathbf{k}}{2}\right]}(r'\mathbf{x}_1) - \alpha_{j-1,l,\left[\frac{\mathbf{k}}{2}\right]}(r'\mathbf{x}_2)|. \tag{16.9}$$

where $r > 0$. Hence, Eq. (16.1) in the wavelet domain can be expressed as

$$\Delta\alpha_{j,l,\mathbf{k}} \doteq \eta_{j,l,\mathbf{k}} \Delta\alpha_{j-1,l,\left[\frac{\mathbf{k}}{2}\right]} \tag{16.10}$$

where $\eta_{j,l,k}$ is the coefficient relating the wavelet transformation of $\Delta_r z$ and $\Delta_{r'} z$ in Eq. (16.1). More details can be found in [12].

The close agreement between the observed and MaxEnt predicted p_A for different types of wavelet basis (e.g. Haar, Daubechies, Symmlet, Coiflet, Battle-Lemarié) suggests that p_A is adequately quantified by the first two moments of $A_{r/r'}$ as long as sufficient number of wavelet coefficients are included. Knowing that the statistics of the incremental elevation is well captured by its first and second moment provides more clues about the underlying physical mechanisms [23]. Discrepancies, if any, between the observed and MaxEnt predicted distribution would indicate that p_A must include higher moment(s) or other parameters. Thus, MaxEnt as an inference algorithm sheds more light on the physical mechanisms underlying the multifractal processes in particular and other processes of the Earth systems in general.

16.3 Application of MaxEP in Modeling Atmospheric Turbulent Transport

Exchange of energy, water, and carbon between the earth and the atmosphere is driven by radiative energy from the Sun. The partition of solar radiation into various heat fluxes over the Earth's surface is strongly influenced by the surface conditions, e.g. land versus ocean, bare soil versus vegetation, etc. Among the processes of exchange of energy and mass occurring in the interwoven energy-water-carbon cycles, evapotranspiration (i.e. the phase change of water between solid/liquid/vapor state) is presumably the most important one as it couples the energy, water and carbon cycles. Monitoring and modeling these surface fluxes at local, regional and global scales using bulk transfer equations based approaches have been challenging [36, 37]. The difficulty results from lack of either complete understanding of the underlying mechanisms or observations at desired space–time coverage and resolution necessary for accurate estimation and prediction. The theory of maximum entropy production (MaxEP), as an application of MaxEnt to non-equilibrium systems, offers alternatives to the classical transport models, particularly in the representation of evapotranspiration. The added element in the MaxEP formalism (see [17] for mathematical details) to the MaxEnt formalism is the concept of a dissipation (or entropy production) function satisfying certain equations known as the orthogonality conditions. In the information based entropy production theories [17], the "orthogonality conditions" are the mathematical relationships between constraints and the corresponding Lagrangian multipliers. To our knowledge, a rigorous derivation of the orthogonal conditions under the most general case [38] is underway at the time of writing this chapter. Nonetheless, there is a plethora of empirical evidence justifying its usefulness. The thermal dissipation function of an electric circuit [39] is an example. Below we re-analyze the electric circuits in the context of the MaxEP formalism [17] from a different angle than that of [40] to provide a unified interpretation of MaxEP as an inference algorithm and a physical principle.

Consider the problem of finding the electric currents I_1 and I_2 of an electric circuit of two parallel resistors R_1 and R_2 given a total current I_0. The solution is given by the well-known Ohm's law and Kirchhoff's laws. This state of the electric circuit corresponds to minimum heat dissipation for the resistors at different temperatures or equivalently minimum thermodynamic entropy production for the resistors at the same temperature [41–44]. Therefore, depending on the temperature homogeneity of the circuit elements given that Ohms law and Kirchhoff's laws are temperature independent, two sets of physical principles are needed to describe the states of electric circuits. The same solution obtained using the concept of maximum information entropy [2, 17], avoids such dichotomy.

The thermal dissipation function D resulting from the heat generated by electric current passing through the resistors is expressed as,

$$D = \frac{I_1^2}{R_1^{-1}} + \frac{I_2^2}{R_2^{-1}}. \tag{16.11}$$

Minimizing D under the constraint of conservation of electric charge, according to Kirchhoff's current law, by introducing a Lagrangian multiplier λ_I,

$$I_0 = I_1 + I_2, \tag{16.12}$$

leads to,

$$I_1 R_1 = I_2 R_2 = \lambda_I, \tag{16.13}$$

where λ_I is recognized as the voltage. MaxEP predicts that two resistors connected in parallel have the same voltage, i.e. Ohm's law. The companion problem of finding the voltages U_1 and U_2 of an electric circuit of two serial resistors R_1 and R_2 given a total voltage U_0 can also be solved in the same way using the MaxEP method except that D in Eq. (16.11) is expressed as a function of voltages and the constraint due to conservation of electric charges in Eq. (16.13) is replaced by the conservation of (potential) energy according to Kirchhoff's voltage law,

$$D = \frac{U_1^2}{R_1} + \frac{U_2^2}{R_2}, \tag{16.14}$$

$$U_0 = U_1 + U_2, \tag{16.15}$$

Minimizing D in Eq. (16.14) under the constraint Eq. (16.15) by introducing a (different) Lagrangian multiplier λ_U leads to,

$$U_1 R_1^{-1} = U_2 R_2^{-1} = \lambda_U, \tag{16.16}$$

where λ_U is recognized as the current. MaxEP again predicts that two resistors connected in series have the same current, i.e. Ohm's law, independent of temperature. This example reveals that whether energy dissipation or entropy production rate of an electric circuit is maximum or minimum depends on whether the constraint in the formulation of the problem is linear or non-linear. A non-linear constraint as in the case of Županovíc et al. [40] corresponds to a maximum entropy production, while a linear constraint as in the above analysis corresponds to a minimum entropy production. MaxEP as an inference algorithm selects the state of an electric circuit (or a transport model) described by Ohm's law among all possible states allowed by the conservation of charge and energy represented by Kirchhoff's laws. Therefore, maximum and minimum entropy production are seen as two sides of the same coin from the perspective of information theory.

The thermodynamic dissipation functions defined in Eqs. (16.11)–(16.14) are formally identical to the dissipation functions associated with the information entropy introduced in the general formalism of MaxEP (see Eq. (16.17) of [17]). When applied to the physical system of electric circuit, MaxEP as an inference algorithm becomes a physical principle of minimum energy dissipation or power loss. The above analysis shows that the Legendre transform in the MaxEnt formalism relating the given constraints to the associated Lagrangian multipliers reduces to Ohm's law. Conversely, Ohm's law implies that D must be minimized under the conservation laws. Again, D being minimum instead of maximum results from the conservation laws expressed as linear functions of currents and voltages. Note that voltage λ_I turns out to be the equal Lagrangian multipliers associated with I_1 and I_2 (equivalent to F_1 and F_2, respectively, in Eq. (16.18) of [17]) due also to the conservation law as in Eq. (16.12). MaxEP not only predicts the macroscopic (i.e. observable and/or controllable) behavior of electric circuits described by Ohm's law under conservations of energy and charge represented by Kirchhoffs laws, but also reveals the underlying mechanisms of microscopic (i.e. unobservable and/or uncontrollable) variables: the observed electric current as a macroscopic transport phenomenon results from the most probable or macroscopically reproducible microscopic configurations of a non-equilibrium system (free electrons) under the (macroscopic or observable) constraint of conservation of electric charges. This is another example of a macroscopic flux-gradient equation interpreted as an "inference" instead of a "physical principle" in the sense that it cannot be derived from more fundamental physical principles governing the microscopic dynamics. An excellent demonstration of this point was the derivation of diffusion equation as an inference [45]. Consequently, all constitutive relationships in continuum mechanics (e.g. shear stress proportional to velocity gradient for Newtonian fluids) can be viewed as inference in the context of MaxEP.

While this example highlights the physical and mathematical significance of MaxEP, it is important to emphasize that the dissipation function (or entropy production function) defined in the MaxEP formalism of [17], contrary to that implied by the misleading name, is not limited to the cases of thermal energy dissipation (or thermodynamic entropy production when D defined in Eq. (16.11) is divided by a constant temperature of the resistors). When applied to the process of heat transfer for example, the dissipation function associated with heat fluxes [46] following the MaxEP formalism, is not the thermal energy dissipation as for the case of electric circuit or thermodynamic entropy production discussed in other studies [15, 16]. The generality of MaxEP for solving less intuitive inference problems (in terms of the physical meaning of the dissipation function) than the previous example can be demonstrated with a MaxEP model of evapotranspiration. Nonetheless, the example of electric circuits as an analogy gives a clue of how to formulate dissipation functions for non-equilibrium systems.

Let's now consider the partition of net radiative energy at the Earth's surface (from solar and atmospheric radiation) into fluxes of latent heat or evapotranspiration for phase change of water, sensible heat warming or cooling the air and ground heat into the Earth. The problem of finding evapotranspiration knowing only the surface thermal and water state is ill-posed because more information is needed to model the fluxes based on the classical flux-gradient relationships [47]. MaxEP as an inference algorithm offers an alternative approach to finding a solution by selecting the most probable partition of net radiation into evapotranspiration and heat fluxes among all possibilities allowed by the conservation of energy. Guided by the example of electric circuits, a dissipation function was formulated using an analogy between heat fluxes and electric currents with electrical impedances replaced by thermal inertia [19],

$$D = \frac{G^2}{I_s} + \frac{H^2}{I_a} + \frac{E^2}{I_e}, \tag{16.17}$$

where E, H and G are the unknown latent, sensible and ground heat flux, respectively. The thermal inertia of heat conduction I_s is a physical property of the soil matrix, and the thermal inertia of turbulent transport I_a and I_e can be formulated as functions of H using the Monin–Obukhov similarity theory [19, 46]. It turns out that the dissipation function D as in Eq. (16.17) is not thermal energy dissipation or thermodynamic entropy production function as in the case of electric circuit discussed above. Although the (thermal) dissipation functions defined in Eqs. (16.11) and (16.14) is not directly related to the dissipation function defined in Eq. (16.17), they are the specific formula of the generic dissipation function, defined using the concept of information entropy in the MaxEP formalism of [17], applied to different physical phenomena. D expressed in terms of E, H, and G is more non-linear than quadratic due to the dependence of I_a and I_e on H, representing a thermodynamic system far-from equilibrium. Minimizing D under the constraint of energy conservation, $G + H + E = R_n$ where R_n is the net radiation at the surface, leads to non-trivial algebraic equations from which E must be solved simultaneously with G and H, referred to as the MaxEP (or MEP) model of evapotranspiration,

$$\left[1 + B(\sigma) + \frac{B(\sigma)}{\sigma}\frac{I_s}{I_a}\right]H = R_n, \tag{16.18}$$

$$E = B(\sigma)H, \tag{16.19}$$

$$G = R_n - H - E, \tag{16.20}$$

with

$$B(\sigma) = 6\left(\sqrt{1 + \frac{11}{36}\sigma} - 1\right), \quad \sigma = \sqrt{\beta}\frac{L_v^2}{c_p R_v}\frac{q_s}{T_s^2}, \tag{16.21}$$

leading to the solution of H from Eq. (16.18) (a nonlinear algebraic equation for H due to the dependence of I_a on H) and that of E and G from Eqs. (16.19)–(16.20) where the variables are defined in the List of Symbols. Note that the MaxEP (MEP) solution of E, H and G is expressed in terms of given R_n, T_s and q_s. A field test of Eqs. (16.18)–(16.20) is shown in Fig. 16.3. The power of the MaxEP is not only about the MaxEP model predictions being in close agreement with observations, as illustrated in Fig. 16.3. Rather, it is the fact that the MaxEP model predictions are based on far less information than the classical models requiring only three surface hydro-meteorological variables, i.e. net radiation, temperature and humidity.

16.4 Conclusion

We have demonstrated the usefulness of MaxEnt and MaxEP as inference algorithms in modeling multifractality and evapotranspiration. The maximum entropy probability distributions of multifractal processes based on a limited number of physical constraints is a straightforward application of MaxEnt. The seemingly trivial example of simple electric circuits may serve as a prototype of MaxEP modeling framework of transport processes in Earth systems as illustrated by the MaxEP model of evapotranspiration. Our results add to a growing number of applications of MaxEP in modeling Earth systems; some of the new developments are reported in this volume. On-going research by the authors: (1) incorporates the distribution of scale-invariant processes in the design of an optimal hydro-meteorological network; (2) generalizes the MaxEP model of evapotranspiration to all types of surfaces including the ocean, snow and ice. Preliminary tests look promising. As unified principles of inference, MaxEnt and MaxEP are applicable

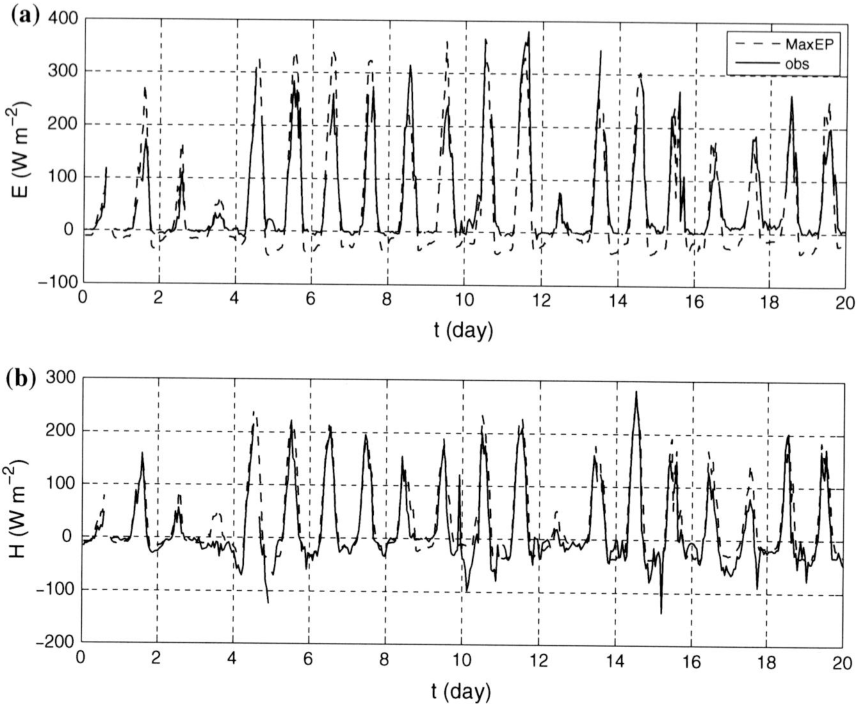

Fig. 16.3 The MaxEP model [19] predicted evapotranspiration E and sensible heat flux H (*broken*) given in Eqs. (16.18)–(16.20) versus the observed fluxes (*solid*) from the Harvard Forest experiment during 19 August–8 September 1994. Over vegetated surfaces, $I_s \rightarrow 0$ and $G \rightarrow 0$ [19]. Data courtesy of Steven Wofsy of Harvard University

to any situation where we lack complete information, especially in remote sensing of the environment and numerical simulations of climate change. The possibilities are unlimited and more exciting findings can be expected once the concept of MaxEnt and MaxEP as principles of inference is understood and accepted.

Acknowledgments This work was supported by the US National Science Foundation grant EAR-0943356 and the US Army Research Office grant W911NF-10-1-0236.

References

1. Jaynes, E.T.: Information theory and statistics mechanics. Phys. Rev. **106**(620), 630 (1957)
2. Jaynes, E.T., Bretthorst, G.L. (eds.): Probability Theory the Logic of Science, p. 755. Cambridge University Press, New York (2003)
3. Gregory, P.C.: Bayesian Logical Data Analysis for the Physical Sciences, p. 468. Cambridge University Press, New York (2005)

4. Leopold, L.B., Langbein, W.B.: The concept of entropy in landscape evolution. U.S Geological Survey professional paper no. 500-A (1962)
5. Lienhard, J.H.: A statistical mechanics prediction of the dimensionless unit hydrograph. J. Geophys. Res. **69**(24), 5231–5238 (1964)
6. Sonuga, J.O.: Principle of maximum entropy in hydrologic frequency analysis. J. Hydrol. **17**(3), 177–219 (1972)
7. Fiorentino, M., Claps, P., Singh, V.P.: An entropy-based morphological analysis of river basin network. Water Resour. Res. **29**(4), 1215–1224 (1993)
8. Claps, P., Fiorentino, M.: The information entropy of fractal river networks. In: 1st International Conference on Fractals in Hydroscience Hydrofractals 1993, Polytechnic of Milan, Ischia (1993)
9. Rodríguez-Iturbe, I., Rinaldo, A.: Fractal River Basins: Chance and self-organization. Cambridge University Press, Cambridge (1997)
10. Niedda, M.: Upscaling hydraulic conductivity by means of entropy of terrain curvature representation. Water Resour. Res. **40**, W04206 (2004)
11. Nieves, V., Wang, J., Bras, R.L.: Maximum entropy distributions of scale-invariant processes. Phys. Rev. Lett. **105**, 118701 (2010)
12. Nieves, V., Wang, J., Bras, R.L.: Statistics of multifractal processes using the maximum entropy method. Geophys. Res. Lett. **38**, L17405 (2011)
13. de Groot, S.R., Mazur, P.: Non-equilibrium thermodynamics. Dover Publications Inc., New York (1984)
14. Paltridge, G.W.: Global dynamics and climate—A system of minimum entropy exchange. Q. J. R. Meteorol. Soc. **101**, 475–484 (1975)
15. Kleidon, A., Lorenz, R.D. (eds.): Non-Equilibrium Thermodynamics and the Production of Entropy, p. 260. Springer, Heidelberg (2005)
16. Kleidon, A., Malhi, Y., Cox, P.M.: Maximum entropy production in ecological environmental systems: applications and implications. Phil. Trans. Roy. Soc. B **365**, 1295–1455 (2010)
17. Dewar, R.C.: Maximum entropy production and the fluctuation theorem. J. Phys. A **38**, L371–L381 (2005)
18. Dewar, R.C.: Maximum entropy production as an inference algorithm that translates physical assumptions into macroscopic predictions: dont shoot the messenger. Entropy **11**, 931–944 (2009)
19. Wang, J., Bras, R.L.: A model of evapotranspiration based on the theory of maximum entropy production. Water Resour. Res. **47**, W03521 (2011)
20. Mandelbrot, B.B.: The Fractal Geometry of Nature. W. H. Freeman, San Francisco (1983)
21. Stanley, H.E., Meakin, P.: Multifractal phenomena in physics and chemistry. Nature **335**, 405–409 (1988)
22. Newman, W.I., Turcotte, D.L.: Cascade model for fluvial geomorphology. Geophys. J. Int. **100**, 433–439 (1990)
23. Veneziano, D., Niemann, J.D.: Self-similarity and multifractality of fluvial erosion topography 1. Mathematical conditions and physical origin. Water Resour. Res. **36**(7), 1923–1936 (2000)
24. Gupta, V.K., Waymire, E.C.: A statistical analysis of mesoscale rainfall as a random cascade. J. Appl. Meteorol. **32**, 251–267 (1993)
25. Pottier, C., Turiel, A., Garcon, V.: Inferring missing data in satellite chlorophyll maps using turbulent cascading. Remote Sens. Environ. **112**, 4242–4260 (2008)
26. Schertzer, D., Lovejoy, S.: Resolution Dependence and Multi-fractal in Remote Sensing and Geographical Information System. McGill University Press, Montreal (1996)
27. Wang, J., Bras R.L.: On ignorance priors and the principle of indifference. In: 31st International Workshop on Bayesian Inference and Maximum Entropy Methods in Science and Engineering, AIP Conference Proceedings, vol. 1443, p. 40, (2012)
28. Arneodo, A., Grasseau, G., Holschneider, M.: Wavelet transform of multifractals. Phys. Rev. Lett. **61**, 2281 (1988)

29. Vergassola, M., Frisch, U.: Wavelet transforms of self-similar processes. Physica D **54**, 5864 (1991)
30. Arneodo, A.: Wavelets: Theory and applications. Oxford University Press, New York (1996)
31. Turiel, A., Parga, N.: Multifractal wavelet filter of natural images. Phys. Rev. Lett. **85**, 3325–3328 (2000)
32. Lashermes, B., Foufoula-Georgiou, E., Dietrich, W.E.: Channel network extraction from high resolution topography using wavelets. Geophys. Res. Lett. **34**, 04 (2007)
33. Nieves, V., Turiel, A.: Analysis of ocean turbulence using adaptive CVE on altimetry maps. J. Mar. Syst. **77**, 482–494 (2009)
34. Daubechies, I.: Ten Lectures on Wavelets. CBMS-NSF Series in Applied Mathematics Society for Industrial and Applied Mathematics (1992)
35. Mallat, S.: A wavelet tour of signal processing. Academic, New York (1999)
36. Desborough, C.E., Pitman, A.J., Irannejad, P.: Analysis of the relationship between bare soil evaporation and soil moisture simulated by 13 land surface schemes for a simple non-vegetated site. Global Planet. Change **13**, 47–56 (1996)
37. Henderson-Sellers, A., Irannejad, P., McGuffie, K., Pitman, A.J.: Predicting land-surface climates Better skill or moving targets? Geophys. Res. Lett. **30**(14), 1777 (2003)
38. Grinstein, G., Linsker, R.: Comments on a derivation and application of the maximum entropy production principle. J. Phys. Math. Theor. **40**, 9717–9720 (2007)
39. Ziegler, H.: An Introduction to Thermomechanics, p. 355. North-Holland, Amsterdam (1983)
40. Županovíc, P., Juretic, D., Botric, S.: Kirchhoff's loop law and the principle of maximum entropy production. Phys. Rev. E **70**, 056108 (2004)
41. Jean, J.H.: The Mathematical Theory of Electricity and Magnetism, p. 633. Cambridge University Press, Cambridge (1925)
42. Landauer, R.: Stability and entropy production in electrical circuits. J. Stat. Phys. **13**, 1–6 (1975)
43. Jaynes, E.T.: The minimum entropy production principle. Ann. Rev. Phys. Chem. **31**, 579–601 (1980)
44. Kondepudi, G., Prigogine, I.: Modern Thermodynamics. Wiley, Chichester (2002)
45. Jaynes, E.T.: Clearing up mysteries—the original goal. In: Skilling, J. (ed.) The Proceedings Volume Maximum Entropy and Bayesian Methods, Kluwer Academic Publishers, Dordrecht, pp. 1–27. (1989)
46. Wang, J., Bras, R.L.: A model of surface heat fluxes based on the theory of maximum entropy production. Water Resour. Res. **45**, W11422 (2009)
47. Arya, S.P.: Introduction to Micrometeorology, p. 307. Academic, New York (1988)

Chapter 17
Entropic Bounds for Multi-Scale and Multi-Physics Coupling in Earth Sciences

Klaus Regenauer-Lieb, Ali Karrech, Hui Tong Chua, Thomas Poulet, Manolis Veveakis, Florian Wellmann, Jie Liu, Christoph Schrank, Oliver Gaede, Mike G. Trefry, Alison Ord, Bruce Hobbs, Guy Metcalfe and Daniel Lester

Abstract The ability to understand and predict how thermal, hydrological, mechanical and chemical (THMC) processes interact is fundamental to many research initiatives and industrial applications. We present (1) a new Thermal–Hydrological–Mechanical–Chemical (THMC) coupling formulation, based on non-equilibrium thermodynamics; (2) show how THMC feedback is incorporated in the thermodynamic approach; (3) suggest a unifying thermodynamic framework for multi-scaling; and (4) formulate a new rationale for assessing upper and lower bounds of dissipation for THMC processes. The technique is based on deducing time and length scales suitable for separating processes using a macroscopic finite

K. Regenauer-Lieb (✉) · J. Liu · M. G. Trefry · A. Ord · B. Hobbs
School of Earth and Environment, The University of Western Australia, M004, 35 Stirling Highway, Crawley, WA 6009, Australia
e-mail: klaus.regenauer-lieb@uwa.edu.au

K. Regenauer-Lieb · T. Poulet · M. Veveakis · F. Wellmann
CSIRO Earth Science and Resource Engineering, PO Box 1130 Bentley, WA 6102, Australia

H. T. Chua
School of Mechanical and Chemical Engineering, The University of Western Australia, 35 Stirling Highway, Crawley, WA 6009, Australia

C. Schrank · O. Gaede
School of Earth Sciences, Queensland University of Technology, 2434 Brisbane, QLD 4001, Australia

G. Metcalfe · D. Lester
CSIRO Mathematics, Informatics and Statistics, Applied Fluid Chaos Group, Graham Rd, Highett, VIC 3190, Australia

M. G. Trefry
CSIRO Land and Water, Floreat Park, WA 6014, Australia

A. Karrech
School of Earth and Environment, The University of Western Australia, M051, 35 Stirling Highway, Crawley, WA 6009, Australia

R. C. Dewar et al. (eds.), *Beyond the Second Law*,
Understanding Complex Systems, DOI: 10.1007/978-3-642-40154-1_17,

time thermodynamic approach. We show that if the time and length scales are suitably chosen, the calculation of entropic bounds can be used to describe three different types of material and process uncertainties: geometric uncertainties, stemming from the microstructure; process uncertainty, stemming from the correct derivation of the constitutive behavior; and uncertainties in time evolution, stemming from the path dependence of the time integration of the irreversible entropy production. Although the approach is specifically formulated here for THMC coupling we suggest that it has a much broader applicability. In a general sense it consists of finding the entropic bounds of the dissipation defined by the product of thermodynamic force times thermodynamic flux which in material sciences corresponds to generalized stress and generalized strain rates, respectively.

17.1 Introduction

The Earth is a complex system in which non-linear feedbacks lead to critical point phenomena. Classical modelling and simulation approaches are based on forward prediction of basic scenarios without any, or with only limited, capability to comprehensively assess the multitude of dissipative patterns emerging from nonlinear, multi-scale non-equilibrium thermodynamic feedbacks. Moreover, in classical mechanical and fluid dynamic modelling of Earth processes the variable *time* is often neglected. Processes are deemed to occur at a very slow, geological pace, therefore in most cases kinetic energy does not play a role. Hence, it is thought that classical dynamics does not apply and the mechanical framework can be reduced to an isothermal, quasi-static case where time is replaced, for instance, by the position of a reference point. For fluid flow modelling, steady state solutions are often sought without considering chaotic time evolution or system transients. Chemical modelling likewise is classically reduced to equilibrium thermodynamics; that is, time is assumed to be infinite. Through these assumptions we are throwing overboard all insights into the natural processes underpinning localization phenomena, which require modelling of the slow dynamics underpinning the creation of dissipative patterns that we can see in nature. It follows that by explicitly solving for time-dependent dissipative processes, the newly emerging methods for thermo-hydro-chemo-mechanical-chemical THMC modelling have the potential to deliver a new class of predictive models that can accurately describe the slow dynamic processes that we see in nature. The next step would be to recognize the value of slow dynamic time-dependent data from dissipative processes at multiple scales in order to nudge THMC solutions forward into the actual dynamic state occupied by the modeled natural system. Such cross-scale inversion is beyond the current aim of the formulation presented here. An extension of the present theory for data compaction and data assimilation is found elsewhere [1]. In this chapter we aim to deliver the basic non-equilibrium thermodynamic framework for multi-scale THMC data assimilation. We also explain explicitly how recently postulated extremum principles for dissipative processes [2] can be applied to coupled THMC modelling.

17.2 Non-equilibrium Thermodynamic Approach for THMC Coupling

17.2.1 The Role of Irreversible Entropy Production

The theory of thermodynamics covers the energetics of THMC processes. In the case of classical thermodynamic equilibrium it is assumed that the energy fluxes have relaxed to some form of equilibrium, that is, the system is assumed to have been given sufficient time to home in on an energy attractor such that time-dependent terms in the energy balance equations can be neglected. Here, we summarize the key ingredient of such time-dependent terms that appear as additional feedback parameters in the energy equation of non-equilibrium thermodynamic approaches. We assume familiarity with basic thermodynamic concepts and refer to our earlier work [2] and references cited therein for more in-depth thermodynamic formulations.

We first define the irreversible entropy production by using the concept of generalized stress and generalized strain rate on a given volume element to calculate their product. This product is the internal power of a given volume element,

$$\tilde{S}_{ir} = \tilde{W}^{diss} = \int F_{ij}^{diss} v_{ij}^{diss} dV \geq 0 \tag{17.1}$$

where F_{ij}^{diss} is the generalized stress and v_{ij}^{diss} the generalized strain rate [3]. The generalized stress corresponds to a thermodynamic force and the generalized strain rate corresponds to a thermodynamic flux. In a purely mechanical problem generalized stresses and strain rates are the classical stresses and strain rates. In the more general case the thermodynamic force can for instance be a pressure difference, a temperature difference, a difference in chemical potentials, and electrical potential. The conjugate thermodynamic fluxes can be a volume change, a heat change, a change in chemical species and electrical current. The work rate is denoted by $\tilde{W}^{diss}$. The use of the tilde instead of the overdot for time derivatives is to emphasize that the work rate is an incomplete differential of time. This leads to uncertainties due to the inherent path dependence of the time integration.

According to the second law of thermodynamics the constraint for time evolution is that for the thermodynamic reference volume under consideration S_{ir} is always positive or equal to zero. It is possible that the system does not obey the laws of thermodynamics but rather obeys the laws of statistical mechanics, if the reference volume for the energy consideration is too small. For now we consider that the reference volume is large enough to employ the laws of thermodynamics where S_{ir} is bound by two entropic bounds, the minimum and maximum of irreversible entropy production, respectively, which we will come back to later.

We emphasize first the role of the irreversible entropy production on assessing the overall behavior of a dissipative system. The important role of the irreversible entropy production is that it appears as an additional source term in the energy equation. In Earth Sciences we are dealing with multi-physics mechanical problems for which the role of the irreversible entropy production is well understood [4–7]. We find that the deforming system forms a new dissipative pattern if it breaches a critical mechanical dissipation level or a critical rate of entropy production. This critical level of entropy production forms an attractor of the particular multi-physics feedback system [8] and we propose here a method for deriving estimates of entropic bounds of the system and thereby an estimate of the uncertainty of the thermo-mechanical system.

In order to derive these bounds we need to address the problem of time integration of Eq. (17.1). We consider entropy production of a far from equilibrium system over a predefined time scale. This time scale can be derived through consideration of the concept of finite time thermodynamics [9]. Finite-time thermodynamics is developed from a macroscopic point of view (with heat conductance, friction coefficients, overall reaction rates) rather than based on a microscopic knowledge (with phonons, yielding of asperities on contacts, local chemical reactions) of the processes involved. In order to assess this macroscopic behavior a critical aspect is that the system has to be given sufficient time for the development of the material attractor that is derived from the boundary value problem of the material volume under consideration. In Earth Sciences the multi-physics problem is often defined by a thermal (T), hydro (H), mechanical (M) and chemically (C) coupled system.

For THMC coupling, a natural time scale is defined by the availability of a thermodynamic force such as (T) a temperature difference, (H) a pressure difference, (M) an applied force or (C) a difference in chemical concentration. The so defined finite-time provides a link to a finite length scale of the dissipative mechanism and the possible emergence of a dissipative structure. In the geomechanical problem the time scale is often given by the inverse of the background strain rate and the corresponding length scale is given by the diffusive length scale. Since the diffusivities of THMC in many Earth Science problems are orders of magnitude apart, we can simplify fully coupled modelling of the non-equilibrium thermodynamic processes by the selection of spatial dimensions and their associated diffusive or convective time and length scales over which the dissipative structures develop.

For THMC coupling the *thermodynamic force* can either be:

- (T) a temperature difference
- (H) a pressure difference
- (M) an applied force
- (C) a difference in chemical species concentration

and the *thermodynamic flux* can be a:

- (T) a heat flow

- (H) a fluid flow
- (M) a velocity
- (C) a chemical flux.

The explicit forms of these equations will be described in the next two sections.

17.2.2 Balance Laws for Thermodynamic Forces

The general framework for thermodynamic forces and fluxes was described in Coussy [10], Poulet et al. [11], Regenauer-Lieb et al. [2] and Karrech et al. [12] to mention a few. In this section we summarize the basic local forms of these equations of conservation of thermodynamic force that read:

$$\rho C_p \dot{T} + \rho^f C_p^f v_i T_i + q_{i,i}^T = r^T \tag{17.2a}$$

$$\frac{1}{M}\dot{p} + b\dot{\varepsilon}_{ii} + q_{i,i}^f = r^f \tag{17.2b}$$

$$\sigma'_{ij,j} - bp_{,i} = 0 \tag{17.2c}$$

$$\varphi \dot{c}^\alpha + \varphi v_i \cdot c_i^\alpha + q_{i,i}^\alpha = \varphi r^\alpha \quad \alpha = 1, 2 \ldots \tag{17.2d}$$

where e.g.: $q_{i,i}^T = \sum_i \frac{\partial q}{\partial x_i}$.

We use compact index notation with implicit summation. The subscripts i and j denote the directions of space. The superscripts f, T and α denote respectively the fluid phase, temperature and chemical species. Equation (17.2a) spells out the conservation of energy and where ρC_P is the overall volumetric heat capacity of the solid-fluid mixture, ρC_P^f is the volumetric heat capacity of the fluid content, $v_i T_i$ accounts for the advective thermal contribution, q_i^T is the conductive heat flux and r^T is the volumetric heat source. In the geomechanical problems it is often identified as shear heating being a driving force for emergence of dissipative structures. Equation (17.2b) derives from the conservation of fluid mass and describes the pore pressure evolution. In the first transient term of this equation, $1/M = \Phi/K_f + 1/N$, where Φ is the matrix porosity, K_f is the fluid bulk modulus, and N is the Biot modulus. The second transient term describes the effect of the rate of volume change (denoted by $\dot{\varepsilon}_{ii}$) on the pressure variation, the third term represents the fluid flux and the right hand side term denotes the fluid source. Equation (17.2c) represents the conservation of momentum, which involves the effective stress tensor σ'_{ij} as well as the fluid pore pressure p, and the Biot coefficient b (see Coussy [10]). Equation (17.2d) is similar to (17.2b), it represents the mass transfer of chemical concentrations c^α with the corresponding concentration flux, q_i^α and sources r^α.

17.2.3 Balance Laws for Thermodynamic Fluxes

The system of Eq. (17.2a–d) is ill-posed and requires additional constraints which derived from the framework of thermodynamics [10]. These constraints are often identified experimentally in order to relate the thermodynamic fluxes to their corresponding forces. In this chapter, we limit ourselves to the following classical first order relationships for isotropic media:

$$q_i^f = -kT_i + hot \tag{17.3a}$$

$$q_i^f = -\frac{\kappa}{\mu_f} p_{,i} + hot \tag{17.3b}$$

$$\dot{\sigma}'_{ij} = C^{ep}_{ijkl}\dot{\varepsilon}_{ij} + hot \tag{17.3c}$$

$$q_i^\alpha = -\varsigma_\alpha c^\alpha_{,i} + hot \quad \forall\alpha \tag{17.3d}$$

Equation (17.3a) represents Fourier's law where k is the thermal conductivity. Higher order terms (*hot*) may need to be considered for special cases. In Fourier's law, for instance, higher order terms are necessary when the finite speed of the phonons must be considered leading to a relativistic heat wave equation. In Earth Sciences this is most often neglected owing to the large time scales considered and the lack of THMC cross coupling (e.g. Dufour effect and Sorret effect in Fick's law). Equation (17.3b) describes Darcy's law, where κ is the rock permeability and μ_f is the fluid viscosity. Equation (17.3c) represents the incremental relationship between effective stresses and matrix deformation with a Jacobian tensor C^{ep}_{ijkl}. Similar forms were used to describe the mass flux using Fick's law (17.3d), where ς_α is the chemical diffusivity of the chemical species c^α.

17.3 THMC Feedbacks and Dissipative Patterns

THMC feedbacks manifest themselves in the formation of dissipative patterns. A well known pattern is the one that is caused by chemical feedbacks [13]. However, dissipative patterns are much more widespread in geological applications and well known throughout all of the THMC couplings in geology [14]. There is a succinct relationship between the dissipative pattern and its length and time scale which can be identified from Eqs. (17.2a–d) and (17.3a–d). This relationship is known as the diffusive scaling length.

$$l_i = 2\sqrt{\eta_i t} \tag{17.4}$$

where the index i, refers to the diffusional process identified in Eq. (17.3a–d) and the diffusivity η to the corresponding diffusion coefficient, that is, the thermal, pressure, mechanical and chemical diffusivities, respectively. This equation offers the identification of space–time coupling for a particular feedback process and its associated dissipative pattern. An explicit example for the identification of time and length scales for a simple thermal–mechanical coupling problem can be found in [15]. It is a fundamental relationship that separates out the space–time continuum of THMC couplings of thermal, hydrological, mechanical and chemical feedbacks. We know that thermal diffusivities of rocks are of the order of 10^{-6} m^2/s while chemical diffusivities are easily ten orders of magnitude lower. Therefore, there is a hierarchy of scales for a given geological time scale. Even when considering the square root relationship between length and time scale there should be enough separation of scales of THMC feedbacks to clearly identify the dominant dissipative feedback mechanism for a given geological time scale. We will use this scale separation to recommend a multi-scale framework in the next chapter.

17.4 Thermodynamic Framework for Multi-Scaling

The link between time scale, length scale and diffusivity allows us to define a multi-scale framework for non-equilibrium thermodynamics processes. We first of all define a relevant time scale for THMC coupling, which by way of the diffusive length scale Eq. (17.4) directly defines the associated length scale.

For this, we need to extend the second law of thermodynamics so that it applies to finite systems observed for finite times. This formulation is known as the "fluctuation theorem" [16]. We can derive the time-scale necessary for separation of scales from the fluctuation theorem of thermodynamics:

$$\frac{P\{\tilde{S}_{ir} = A\}}{P\{\tilde{S}_{ir} = -A\}} = e^{At} \tag{17.5}$$

The fluctuation theorem simply states that for any dissipative process the probability P of positive entropy production over that of negative entropy production increases exponentially with time. That is to say, for infinite time, we can expect the process to evolve to a state of maximum dissipation characterized by a new dynamic quasi-steady state. Since we are dealing with finite systems we can use the fluctuation theorem to define a time scale where the system approaches that of an infinite time system for the particular dissipation mechanism under consideration. In the case of simple thermal–mechanical coupling the most probable dynamic attractor of the dissipative pattern is the maximum of entropy production [8]. An extension to all THMC diffusive processes is natural and allows us to define a complete multi-scale framework for non-equilibrium thermodynamics.

The simple assumption is that if the diffusivities of the individual feedback processes are far enough apart we can consider a macro-time/spatial scale for which the micro-system is approaching its maximum entropy, that is, it behaves in a similar manner as for infinite time (Fig. 17.1). For the micro-system we can obtain a solution by minimizing the stored (potential) energy, whereas for the

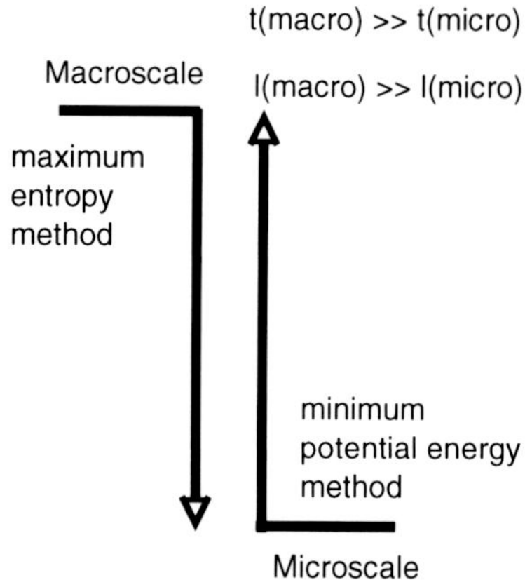

Fig. 17.1 The different diffusivities of the THMC feedback processes in geological applications allow a clear separation of scales. If such a separation of scales is given and one is interested in the macroscopic behavior, one can describe the micro-system by solving the minimum Helmholtz or Gibbs free energy (minimum potential energy method) or the macroscale by explicitly solving the time-dependent energy equation under the constraint of maximum dissipation (maximum entropy method) [10]

macro-system we still have to solve the full time dependent energy equation under the constraint of the second law of thermodynamics [9]. Repeating this process over the different length scales in nested calculations, we can in theory obtain a full assessment of the THMC dissipative pattern.

Complexities arise for similar diffusivities with overlapping scales, which prompt the requirement for an explicit fully coupled treatment of the various dissipation mechanisms. In forthcoming contributions, we will introduce a method to deal with the upscaling of such more general systems. In simple cases, we can assume that for the time scale of the longer-term geological process the smaller scale geological feedback process has converged and the basic dynamic attractor for each length scale can be found.

17.5 Upper and Lower Bound Approach

Before we deal with uncertainties from the time integration of Eq. (17.1), stemming from the dissipative force, we consider a simpler system where a conservative force does work on the thermodynamic system independent of its path. That implies that the system recovers it original state after removal of the conservative force. The time derivative of the work is hence a complete differential, which does not depend upon the path.

$$\dot{W}^{cons} = \int F_{ij}^{cons} v_{ij}^{cons} dV \tag{17.6}$$

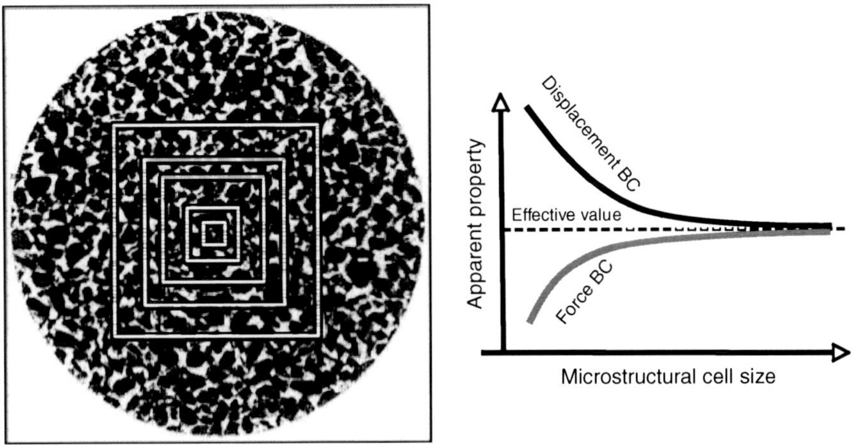

Fig. 17.2 Modified after a plenary lecture by Robert L. Taylor, University of Berkeley, California, on "Computational Mechanics Today" 2008. We have extended the approach by considering combinations of constant thermodynamic force and constant thermodynamic flux as shown in Table 17.1. We propose this method as a new thermodynamic homogenization procedure

where F_{ij}^{cons} denotes the conservative force and v_{ij}^{cons} the rate of displacement. The force can hence be described as a gradient of a potential function and be described in the framework of equilibrium thermodynamic since there is no time dependent dissipation. Equation 17.3a–d simply becomes:

$$\sigma_{ij} = C_{ijkl}^{elastic} \varepsilon_{\mathrm{ij}}^{elastic} \tag{17.7}$$

Let us now consider in addition a microstructure as shown in Fig. 17.2 where the matrix property is described by Eq. (17.7) and the pore space is assumed to have no mechanical strength. For this problem an asymptotic computational homogenization method has been devised [17] whereby through a stepwise increase in microstructural cell size the apparent material property can be derived asymptotically. The material property (e.g. Young's modulus) is found to be consistently overestimated (stronger) for the choice of a displacement boundary condition (constant thermodynamic flux) while the property is consistently underestimated through the choice of a traction boundary condition (constant thermodynamic force).

Using variational extremum principles the above described boundary value problem of the asymptotic homogenization technique can be shown to always deliver an absolute lower bound for a constant force boundary while a constant displacement boundary problem always gives an absolute upper bound of the work done. When the volume is sufficiently large the two bounds are the same [18] (Fig. 17.3).

It is curious to note that these limit theorems on upper and lower bounds initially have been developed for the more complicated case of dissipative elasto-

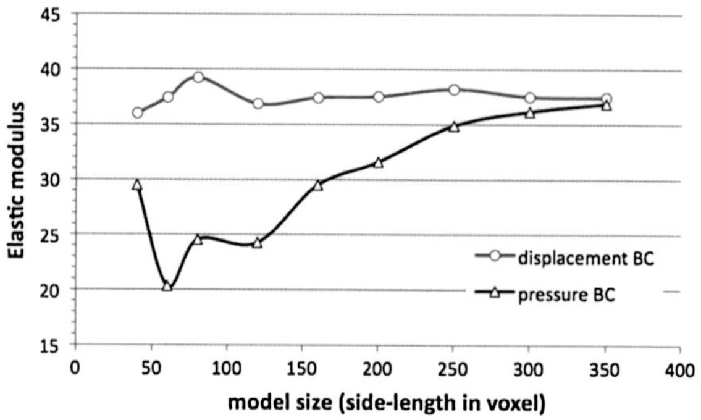

Fig. 17.3 A typical result for computational homogenization of the elastic modulus from our laboratory. Note that for small subvolumes the asymptotic trend oscillates and in severe cases (not shown here) the lower bound (pressure BC) predicts a higher modulus than the upper bound (displacement BC). When this is the case the laws of thermodynamics break down and the problem has to be dealt with by a statistical mechanics approach

plastic systems to overcome the problem of lack of uniqueness of predicted velocity fields in incomplete semi-analytical solutions [19] of rigid-plastic continua. They were later on generalized to elasto-plastic cases and the method was called limit analysis design. These elasto-plastic solutions were derived for the time independent, isothermal situation where the derivation of the constitutive laws were constrained by the postulate of maximum entropy production [20]. The interested reader can find a good review on the origin of the principle of maximum entropy production elsewhere [21].

17.6 Entropic Bounds for Time Dependent Processes

The choice of entropic bounds to deal with the non-uniqueness of the velocity field in elasto-plastic continua and the uncertainty of the material heterogeneity in elastic continua lays the theoretical ground for the time dependent non-isothermal Earth Science THMC problem. The generalized thermodynamic process according to Eq. (17.1) leads to an additional uncertainty owing to the path dependence of the integration of the deformational work. We have already discussed the definition of the relevant "finite time thermodynamic" time for the generalized process and have shown the importance of the coupling between the irreversible entropy production and the diffusion of the entropic source term in Eq. (17.4). We have also shown that the fluctuation theorem defines a maximum of the irreversible entropy production if the time scale is sufficiently long. It remains to show how to derive the minimum of the entropy production. This minimum of entropy production can also be derived by the theory of "finite time thermodynamics" through variational principles, or alternatively, optimal control theory. It is in fact defined by the optimal path where the system delivers the maximum rate of available work to its exterior $\dot{\mathrm{W}}^{ext}_{\max}$.

$$\dot{\mathrm{W}}^{ext}_{\max} = \max\left[\dot{W}^{ext}_{t(i)} - \dot{W}^{ext}_{t(f)} - T_0 \int_{t(i)}^{t(f)} \dot{S}_{tot} dt\right] \tag{17.8}$$

where $\dot{S}_{tot}$ is the total entropy production, i.e. the sum of the irreversible entropy production inside the control volume plus the entropy flux through its boundaries and the time interval is defined by $t(i) - t(f)$ where i and f stand for initial and final configurations. The entropic bounds for the generalized time dependent process is therefore given by the minimum entropy production (Eq. 17.8) and the maximum entropy production (Eq. 17.5).

17.7 Upper and Lower Bounds of Dissipation in THMC Processes

While Fig. 17.1 provides the basis for dealing with multi-scale THMC problems we have so far only used Eq. (17.4) to identify an empirical thermodynamic ruler, which is the length scale of the thermodynamic feedback underpinning the intrinsic dissipative patterns. This can at best be understood as semi-quantitative guidance of scale separation. This is because the fluctuation theorem does not allow us to specify quantitatively how close we are to the maximum entropy production for a given time scale and a certain number of degrees of freedom and the diffusion length-scale is only calculated in 1-D. In order to assess this quantitatively, we have to give up the macroscopic view and explicitly calculate the fully coupled THMC problem, that is, solve the five basic conservation equations, conservation of mass, conservation of linear momentum, conservation of angular momentum, conservation of energy, and the explicit formulation of the entropy production under the constraint of the second law of thermodynamics.

In order to quantify whether our calculations have converged such that they can deliver homogenized values as input parameters for the THMC coupled problems at the next larger scale we make use of the thermodynamic upper and lower bound principles first postulated in Regenauer-Lieb et al. [2]. Table 17.1 spells out what these boundary conditions are.

Note that our choice of neglect of higher order terms in the thermodynamic fluxes in Eq. (17.3a–d) was based on the vast separation of diffusional length scales in Earth Science problems. This does not intrinsically rule out complex nonlinear coupling required to identify dynamic attractors. We have been able to show that thermodynamic bounds (maximum and minimum entropy production defined by the product of thermodynamic force times flux) can be employed to address the non-uniqueness problem arising from non-linear interactions [2]. This

Table 17.1 Boundary conditions to obtain upper and lower bounds of dissipation

	Upper bound (constant thermodynamic flux)	Lower bound (constant thermodynamic force)
T	Constant heat flow	Constant temperature difference
H	Constant fluid flow	Constant pressure difference
M	Constant velocity	Constant force
C	Constant chemical flux	Constant difference in chemical concentration

approach has so far only been used as a method in continuum mechanics [22] for isothermal dissipative processes where stresses (resp. forces) and strain (resp. displacements) are applied alternatively to obtain converging solutions representing the responses of structures (Fig. 17.2). The extension to coupled non-equilibrium THMC problems is natural and involves the use of thermodynamic duals to estimate the responses of materials and structures as specified in Table 17.1.

17.8 Conclusions

We have presented here a basic non-equilibrium thermodynamic framework for multi-scale THMC coupling. We have derived from this formulation a basic 1D diffusive scaling length for the formation of dissipative patterns stemming from non-equilibrium thermodynamic feedback. Considering the separation of diffusivities of THMC systems we have identified a wide separation of scales of the dissipative structures beginning with chemical feedback at sub-mm level, to fluid flow feedbacks from meters to hundreds of meters, to thermal feedback from hundreds of meters to kilometer scale and mechanical feedbacks operating at all scales but particularly at coarse scales. The introduction of finite time thermodynamics and the fluctuation theorem allows an application of the second law of thermodynamics to finite systems observed for finite times. It spells out the probability of entropy production from a macroscopic view. Based on this formulation we have developed suitable energy solution methods for macro- and microstates for a given macroscopic process time scale (Fig. 17.1). We have also introduced a new homogenization method (Table 17.1) for coupled THMC solutions for convergence of the upper and lower bounds of dissipation.

Acknowledgments We wish to acknowledge discussions with Jeff Gordon and Ian Collins and support through CSIRO, the University of Western Australia and the Western Australian Government through the Premiers Science Fellowship Program and the Western Australian Geothermal Centre of Excellence. Some of this work was supported through ARC Linkage Grant number LP10020078: Multiscale Modelling of Ore Body Formation and ARC Discovery Grant number DP1094050: The dynamic strength of continents and how they break apart.

References

1. Regenauer-Lieb, K., Veveakis, M., Poulet, T., Wellmann, F., Karrech, A., Liu, J., Hauser, J., Schrank, C., Gaede, O., Trefry, M.: Multiscale coupling and multiphysics approaches in Earth sciences: theory. J. Coupled Syst. Multiscale Dyn. **1** (2013) (in press)
2. Regenauer-Lieb, K., Karrech, A., Chua, H., Horowitz, F., Yuen, D.: Time-dependent, irreversible entropy production and geodynamics. Philos. Trans. R. Soc. Lond. A **368**, 285–300 (2010)
3. Houlsby, G., Puzrin, A.: Principles of Hyperplasticity, an Approach to Plasticity Theory Based on Thermodynamic Principles, vol. XXIV. Springer, Berlin, p. 351 (2007)
4. Gruntfest, I.J.: Thermal feedback in liquid flow—plane shear at constant stress. Trans. Soc. Rheol. **7**, 195–207 (1963)
5. Yuen, D.A., Schubert, G.: Stability of frictionally heated shear flows in the asthenosphere. Geophys. J. Roy. Astron. Soc. **57**(1), 189–207 (1979)
6. Regenauer-Lieb, K., Yuen, D.A.: Modeling shear zones in Geological and Planetary sciences: solid- and fluid- thermal-mechanical Approaches. Earth-Sci. Rev. **63**, 295–349 (2003)
7. Veveakis, E., Vardoulakis, I., Di Toro, G.: Thermo-poro-mechanics of creeping landslides: the 1963 Vaiont slide, Northern Italy. J. Geophys. Res. **112**, F03026 (2007)
8. Regenauer-Lieb, K., Weinberg, R.F., Rosenbaum, G.: The role of elastic stored energy in controlling the long term rheological behaviour of the lithosphere. J. Geodyn. **55**, 66–75 (2012)
9. Andresen, B., Salamon, P., Berry, R.: Thermodynamics in finite time. Phys. Today **37**(9), 62–70 (1984)
10. Coussy, O.: Poromechanics. Wiley, Chichester (2004)
11. Poulet, T., Regenauer-Lieb, K., Karrech, A.: A unified multi-scale thermodynamical framework for coupling geomechanical and chemical simulations. Tectonophysics **483**, 178–189 (2010)
12. Karrech, A., Regenauer-Lieb, K., Poulet, T.: Frame indifferent elastoplasticity of frictional materials at finite strain. Int. J. Solids Struct. **48**(3–4), 407 (2011)
13. Glansdorff, P., Prigogine, I., Nyden Hill, R.: Thermodynamics theory of structure, stability and fluctuations. Am. J. Phys. **41**(1), 147–148 (1973)
14. Hobbs, B., Ord, A., Regenauer-Lieb, K.: The thermodynamics of deformed metamorphic rocks: a review. J. Struct. Geol. **33**(5), 758–818 (2011)
15. Regenauer-Lieb, K., Yuen, D.A.: Positive feedback of interacting ductile faults from coupling of equation of state, rheology and thermal-mechanics. Phy. Earth Planet. Inter. **142**(1–2), 113–135 (2004)
16. Evans, D., Searle, D.: The fluctuation theorem. Adv. Phys. **51**(7), 1529–1585 (2002)
17. Terada, K., Muneo, H., Kyoya, T., Kikuchi, N.: Simulation of the multi-scale convergence in computational homogenization approaches. Int. J. Solids Struct. **37**, 2285–2311 (2000)
18. Hill, R.: Elastic properties of reinforced solids: some theoretical principles. J. Mech. Phys. Solids **11**, 357–372 (1963)
19. Hill, R.: On the state of stress in a plastic-rigid body at the yield point. Phil. Mag. **42**(331), 868–875 (1951)
20. Prager, W.: An Introduction to Plasticity. Addison Wessley, Reading, Massachussets (1959)
21. Martyushev, L.M., Seleznev, V.D.: Maximum entropy production principle in physics, chemistry and biology. Phys. Rep.-Rev. Sect. Phys. Lett. **426**(1), 1–45 (2006)
22. Ostoja-Starzewski, M.: Microstructural Randomness and Scaling in Mechanics of Materials. Chapman &Hall/CRC, London (2008)

Chapter 18
Use of Receding Horizon Optimal Control to Solve MaxEP-Based Biogeochemistry Problems

Joseph J. Vallino, Christopher K. Algar, Nuria Fernández González and Julie A. Huber

Abstract The maximum entropy production (MaxEP) principle has been applied to steady state systems, but biogeochemical problems of interest are typically transient in nature. To apply MaxEP to biogeochemical reaction networks, we propose that living systems maximum entropy production over appropriate time horizons based on strategic information stored in their genomes, which differentiates them from inanimate chemistry, such as fire, that maximizes entropy production instantaneously. We develop a receding horizon optimal control procedure that maximizes internal entropy production over different intervals of time. This procedure involves optimizing the stoichiometry of a reaction network to determine how biological structure is partitioned to reactions over an interval of time. The modeling work is compared to a methanotrophic microcosm experiment that is being conducted to examine how microbial systems integrate entropy production over time when subject to time varying energy input attained by periodically cycling feed-gas composition. The MaxEP-based model agrees well with experimental results, and model analysis shows that increasing the optimization time horizon increases internal entropy production.

Accepted (July 2012) in: *Beyond the Second Law: Entropy Production and Non-Equilibrium Systems*. R. C. Dewar, C. H. Lineweaver, R. K. Niven and K. Regenauer-Lieb, Springer.

18.1 Introduction

In this chapter we examine the application of the maximum entropy production (MaxEP) principle for describing microbial biogeochemistry. Biogeochemistry enlists the fields of biology and geochemistry to understand chemical transformations and element cycling that occur in natural environments. Because the

J. J. Vallino (✉) · C. K. Algar · N. F. González · J. A. Huber
Marine Biological Laboratory, Woods Hole, MA, USA
e-mail: jvallino@mbl.edu

R. C. Dewar et al. (eds.), *Beyond the Second Law*,
Understanding Complex Systems, DOI: 10.1007/978-3-642-40154-1_18,

majority of biologically catalyzed reactions that occur on Earth, such as nitrogen fixation, denitrification, metal redox reactions, sulfate reduction, etc., are orchestrated by bacteria and archaea [12], we restrict our current focus to microbially catalyzed reactions. Microbes (including viruses, bacteria and archaea) are the simplest living organisms and are at the interface between chemistry and biology, because they catalyzed reactions that also occur abiotically, such as the oxidation of iron (rusting), oxidation of hydrogen sulfide and methane, fixing N_2 into NH_3 and HNO_3 (lightening and combustion). Since we can view bacteria and archaea as simple molecular machines [12], they are most likely amendable to thermodynamic description. They are critical for the support and functioning of all higher life on Earth, so it is particularly important to understand how their presence and growth controls the chemistry at local, regional and global scales. Our expectation is that by employing MaxEP we will be able to develop more robust models that can be used to study how biogeochemistry changes as the environment is altered by natural phenomena and human actions.

Biogeochemistry can be viewed from two extreme perspectives. In the classic perspective, organisms determine the overall biogeochemical processes that occur in an ecosystem. This organismal centric view derives naturally from reductionism, as biogeochemistry is by definition a product of organismal growth. However, the organismal centric view implies that changing species composition will likely produce different biogeochemistry. Furthermore, this approach requires detailed knowledge on organism growth kinetics, predator–prey interactions, as well as on how community composition may change as a result of internal dynamics or external drivers. Except for extremely simple systems, this information is usually lacking. Despite these short comings, the majority of biogeochemical models use the organismal perspective as a basis of their design [13].

The second perspective on biogeochemistry takes a systems approach and views ecosystems thermodynamically as open, non-equilibrium systems. In this case, it is free energy potential, resource availability and information that determine ecosystem biogeochemistry. While organisms ultimately carry out the process, thermodynamics determines which metabolic functions will dominate. Organisms are viewed as interchangeable components, similar to microstates that underlie macrostates in equilibrium thermodynamics [44]. It is this thermodynamic perspective that we will employ to describe ecosystem biogeochemistry, where MaxEP will serve as the governing principle. Because we will limit our analysis to microbial processes, we will remove the typical organismal emphasis and instead view a microbial community in functional terms as a collection of catalysts (or molecular machines [12]) that are synthesized and degraded to achieve MaxEP.

In this chapter we develop a MaxEP-based biogeochemical (BGC) model of a distributed metabolic network. Model degrees of freedom are determined by solving a receding horizon optimal control problem that maximizes entropy production over successive intervals of time. Results from the model are compared to data from two methanotrophic microcosm experiments, a control, and a treatment where energy input is cycled over a 20 day period.

18.2 MaxEP and Living Systems

The MaxEP conjecture [8, 10, 35, 36] states that steady state, non-equilibrium systems with many degrees of freedom will likely be found in a state that maximizes internal entropy production. If internal self-organization, such as vortices and macroscopic structures, facilitates internal entropy production, then those structures will likely develop [26]. Similar to equilibrium thermodynamics that requires systems to be found in the state of maximum entropy, MaxEP indicates that nonequilibrium systems will head towards equilibrium along the fastest possible pathway. That is, they will dissipate free energy as fast as possible within the constraints imposed on the system [28, 44]. As discussed in this book and elsewhere, several phenomena appear consistent with MaxEP, including planetary-scale heat transport [19, 27], laminar to turbulent flow transition [29], plant evapotranspiration [46], and many others (see [35] and references therein).

18.2.1 Living Systems as Catalysts

If MaxEP indicates that systems should race down free energy surfaces towards equilibrium as fast as possible, then why isn't the universe already at equilibrium? The answer is because systems often get trapped in metastable states. For instance, a mixture of methane and air at 20 °C, even within the combustible mixture envelope (5–15 % CH_4), will remain in this metastable state for a considerable length of time due to the high activation energy required to overcome the repulsive force of the electron cloud that prevents spontaneous reaction. Of course, if a spark is introduce, then the highly exothermic reaction proceeds in a MaxEP manner due to the increase in temperature. Another means in which the free energy can be released is by introducing a catalyst. By reducing the activation energy, the catalyst frees the system from its metastable state, so the reaction can proceed at room temperature even if the system lies outside the combustion envelope or the reactants are dissolved in water.

While most man-made catalysis are crude and exhibit poor selectivity, enzyme catalysts synthesized by bacteria, as well as all living organisms, achieve extreme reductions in activation energies along very selective reaction pathways. It is the presence of these enzyme catalysts that hastens the dissipation of free energy and entropy production through the destruction of chemical and electromagnetic potentials. However, the increase in reaction rates provided by catalysts is proportional to the amount of catalyst present. To maximize entropy production, it is necessary for a system to rely on autocatalytic reactions that not only dissipate chemical potential but also synthesize more catalyst in the process, such as the methane oxidation reaction given by,

$$CH_4 + 2O_2 + \Sigma \xrightarrow{\$} \$ + H_2CO_3 + H_2O \qquad (18.1)$$

where $\$$ is a catalyst, or *biological structure*, synthesized from available resources, such as C, N, P, Fe, in the environment, Σ. Because catalyst is produced as a product of methane oxidation in Eq. (18.1), the reaction will proceed exponentially provided resources, Σ, needed to build catalyst are not limiting. Of course, Eq. (18.1) also represents growth of methanotrophs (specialized bacteria that eat methane), but we are placing emphasis here on catalyst synthesis for the dissipation of chemical potential, not on the nature of bacterial growth. This distinction *represents a paradigm shift from 'we eat food' to 'food has produced us to eat it'* [25].

In order to calculate the rate of reaction, Eq. (18.1), we need to know the standard molar entropy associated with biological structure, $\$$. Unfortunately, there is considerable confusion associated with entropy calculations involving living organisms. It is often believed that living organisms represent extremely low entropy structures. This misconception can be attributed to confusion over the association between entropy and order. Order, as might be represented by a pattern, does contribute to entropy, but the entropy (or free energy) of the material the pattern is constructed from must also be accounted for in the entropy calculation. As Morrison [34] has shown, only when the pattern is written at the atomic scale does the entropy of the pattern become significant compared to the entropy of the material the pattern is written in.

Consider the words written on this page. Because the ink on the page forms a pattern that contains information, the entropy of the page is lower than a page with randomized letters [5]; however, the reduction of entropy is trivial compared to the entropy of the paper the ink is written on. If the paper is burned, it hardly matters in a thermodynamic context if the text contains the meaning of life or only jibberish; the difference in the amount of free energy dissipated, or entropy produced, between the two cases is virtually undetectable, because the pattern on this page is written at a macroscopic scale. Likewise, entropy associated with information contained in DNA/RNA or protein is small compared to the entropy associated with the nucleic or amino acid polymers the information is written in [45]. All too often the entropy of the material a pattern is written in is overlooked, which leads to incorrect assessments, such as the popular statement that a clean desk has lower entropy than a messy one; both have the same thermodynamic entropy or free energy. In terms of entropy and free energy calculations, a gram of freeze-dried yeast or bacteria, which are viable upon rehydration, has the same molar entropy and free energy of formation as an equivalent weight of a macromolecules in the appropriate proportions [3]. To paraphrase Morrison [34], the *élan vital* carries no thermodynamic burden.

While the entropy associated with the information content of a cell is trivial compared to the material of a cell, it is nevertheless of critical importance. It is useful information [1] contained in the genome that allows for the construction of complex macromolecules that gives rise to the catalytic nature of biological structure, $\mathfrak{S}$. Ultimately, then, it is information that releases systems from metastable states to flow down free energy surfaces and produce entropy. Evolution works to refine this information and thereby increase the rate of entropy production. Information and entropy are intimately coupled [17]. Philosophically, we postulate that free energy spawns the creation of information that hastens free energy's destruction.

18.2.2 MaxEP and Transient Systems

An element of time has been implied in the MaxEP description above for constructing biological structure to dissipate free energy; however, all MaxEP theories to date have been applied to steady state systems only, where time is not involved in the equations. There currently does not exist a MaxEP theory for transient systems where the state is allowed to vary with time, but it is transient systems we are often most interested in. The objective in modeling is usually to understand and predict how a system of interest will respond to perturbations or changes in external drivers. To build a transient biogeochemistry model based on MaxEP requires that we speculate as to how time may affect the MaxEP solution.

For any particular system we can define internal entropy production once the system boundaries have been defined [32, 35], as well as formulate an entropy balance equation, such as

$$\frac{dS}{dt} = J_S + \dot{\sigma} \tag{18.2}$$

where S is system entropy (kJ K^{-1}), J_S is the entropy flux into the system (from mass and heat transport) and $\dot{\sigma}$ is the entropy production rate due to irreversible processes occurring within the system. The second law requires that $\dot{\sigma} \geq 0$ [20]. We also define σ as $\int \dot{\sigma} dt$, which is the amount entropy that derives from internal irreversible processes over some interval of time. Throughout this manuscript we will only be concerned with σ or $\dot{\sigma}$, but not S, because MaxEP applies to internal entropy production only.

For a transient system, internal entropy production is a function of time, $\dot{\sigma}(t)$, so how can MaxEP be defined when $\dot{\sigma}$ varies with time? One special case would be to maximize $\dot{\sigma}$ at every instance in time, which would be equivalent to taking a steepest decent pathway along the free energy surface defined by the current state and all possible pathways leading from that point, similar to water flowing downhill. However, following a steepest decent pathway at each instance in time may not lead to the greatest internal entropy production over an interval of time. Consider Fig. 18.1 for example. Instantaneous internal entropy production at time t_n is greater along pathway PA than along pathway PB, since $\dot{\sigma}_A(t_n) > \dot{\sigma}_B(t_n)$. But taking the

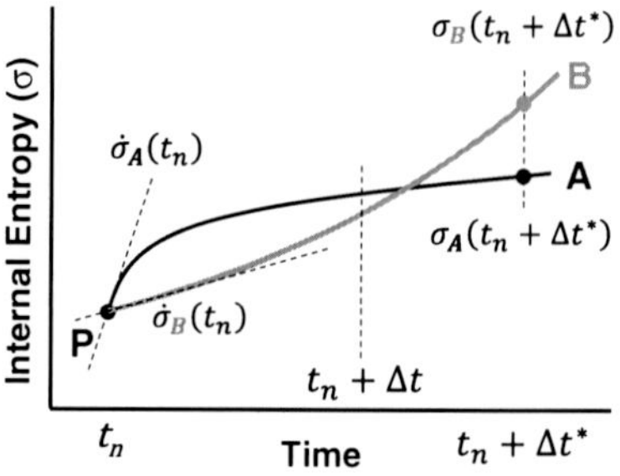

Fig. 18.1 Increases in internal entropy over two possible pathways starting from point P at time t_n. Here, σ is the part of system entropy S that is from internal irreversible processes and $\dot{\sigma}$ its production rate

steepest descent pathway at point P sets the system along a trajectory that ultimately produces less internal entropy than had the system followed pathway PB, since $\sigma_B(t_n + \Delta t^*) > \sigma_A(t_n + \Delta t^*)$. If the system had a means to explore all possible future pathways leading from P over Δt^* time, then the system could increase entropy produced over the steepest descent pathway, PA, by following pathway PB. That is, if the system has a way to generate predictions, then forgoing the steepest descent pathway can lead to greater internal entropy production over time. We postulate that this is precisely what living systems do.

Because living systems can store information in their genome, they can develop temporal strategies based on passed events that become refined via evolutionary selection. Genomic information not only allows organisms to access free energy trapped in metastable states, but also allows them to follow pathways that avoid the steepest decent route and produce more entropy over time. For instance, some bacteria form spores or dormant cells that increase their fitness when conditions become hostile [23, 24]. Likewise, many organisms will increase fat storage in the fall to survive the winter months. While temporal strategies are well recognized, they are often not accounted for in models. Instead, most biogeochemistry models view the system as a type of Markov process where system response only depends on the current state. We believe what differentiates abiotic systems from biotic ones, is the ability of the latter to store information that allows them to develop temporal strategies and out compete abiotic systems over time in internal entropy production [44]. Maximizing internal entropy produced over intervals of time is the basis of the model and associated experiment discussed in the next section.

18.3 Methods

Discussed below are descriptions of a microbial microcosm experiment and an associated mathematical model that are intended to explore the idea that living systems develop temporal strategies that increase entropy production when averaged over time. The experimental setup employs methanotrophic microcosms whose energy input is cycled over time, while the modeling of the microcosms is based on a distributed metabolic network of biochemical reactions that are controlled to maximize averaged entropy production over intervals of time using a receding horizon optimal control approach.

18.3.1 Experimental System

The experiment is designed to examine how microbial communities adapt and evolve to cope with periodic energy inputs using methane plus air as the sole source of energy. The experimental setup [44] consists of four 18 L microcosms that are operated in chemostat mode at a dilution rate of 0.1 d^{-1} and are sparged at a gas flow rate of 20 mL min^{-1} (0 °C, 101.3 kPa). Two control microcosms are sparged continuously with a gas mixture of 4.9 % CH_4, 19.6 % O_2, 0.03 % CO_2, balance N_2, while two other microcosms are cycled between the methane plus air mixture and just air (20.95 % O_2, 0.033 % CO_2, balance N_2) over a 20 d period (10 days with CH_4 on, 10 days with CH_4 off). All microcosms were inoculated approximately 4 years ago with whole water samples collected from a coastal pond and cedar bog (1 L each). A mineral salts medium (10 mM K_2HPO_4, 50 μM KNO_3, 100 μM $MgSO_4$, 100 μM $CaCl_2$, 100 μM NaCl, plus trace elements solution) adjusted to pH 6.8 is used as feed.

Output gas composition is analyzed on-line every 5 h for CH_4 (NDIR, California Analytical Instruments), O_2 and CO_2 (laser diode adsorption spectroscopy, Oxigraf) concentrations, and analyzer drift is compensated for by monitoring input gas composition. Dissolved oxygen and pH electrodes are measured and recorded every hour. Gas cycling and all data acquisition are under computer control and posted on-line (http://ecosystems.mbl.edu/MEP). Periodically, liquid samples are withdrawn for both nutrient analysis (NO_3^-, NH_3, particulate organic C (POC), N (PON), dissolved organic C (DOC), and N (DON)) and microbial community assessment via cell counts and 454-tag pyrosequencing of the V4-V6 hypervariable regions of the 16S rRNA gene [16].

18.3.2 Metabolic Network Model

The MaxEP-based biogeochemistry model uses a distributed metabolic network approach to simulate the functional attributes of a microbial community [43]. For the methanotrophic microcosms, four biological structures are used to capture methane oxidation to methanol $\mathfrak{S}_1$, methanol to CO_2 $\mathfrak{S}_2$, the turnover of biological structures $\mathfrak{S}_3$, and the consumption of recalcitrant (i.e., hard to decompose) organic C (dC) and N (dN), $\mathfrak{S}_4$ (Table 18.1 and Fig. 18.2). This metabolic network structure differs significantly from our previous approach [44]. Here, we use whole reactions instead of half reactions to represent metabolism, which has two main advantages: (1) since half reactions produce (or consume) electrons, we do not need equations and constraints to insure electron conservation and (2) biological

Table 18.1 Reaction stoichiometries and optimal control variables (*OCV*: ε_j and $\omega_{i,j}$ in Eq. 18.4) associated with the four biological structures used to represent methanotrophic communities

Reaction Rate	Stoichiometry	OCV
$r_{1,1}$	$CH_4 + \varepsilon_1\gamma_1 HNO_3 + a_{1,1}O_2 \xrightarrow{\mathbb{S}_1} \varepsilon_1\mathbb{S}_1 + (1-\varepsilon_1)CH_3OH + b_{1,1}H_2O$	$\varepsilon_1, \omega_{1,1}$
$r_{2,1}$	$CH_4 + \varepsilon_1\gamma_1 NH_3 + a_{2,1}O_2 \xrightarrow{\mathbb{S}_1} \varepsilon_1\mathbb{S}_1 + (1-\varepsilon_1)CH_3OH + b_{2,1}H_2O$	
$r_{1,2}$	$CH_3OH + \varepsilon_2\gamma_2 HNO_3 + a_{1,2}O_2 \xrightarrow{\mathbb{S}_2} \varepsilon_2\mathbb{S}_2 + (1-\varepsilon_2)H_2CO_3 + b_{1,2}H_2O$	$\varepsilon_2, \omega_{1,2}$
$r_{2,2}$	$CH_3OH + \varepsilon_2\gamma_2 NH_3 + a_{2,2}O_2 \xrightarrow{\mathbb{S}_2} \varepsilon_2\mathbb{S}_2 + (1-\varepsilon_2)H_2CO_3 + b_{2,2}H_2O$	
$r_{i,3}$	$\mathbb{S}_i + a_{i,3}O_2 \xrightarrow{\mathbb{S}_3} \varepsilon_3\mathbb{S}_3 + (1-\varepsilon_3)[\gamma_i(\varepsilon_3 NH_3 + (1-\varepsilon_3)dN) + \varepsilon_3 H_2CO_3 + (1-\varepsilon_3)dC] + \varepsilon_3(\gamma_i - \gamma_3)NH_3 + b_{i,3}H_2O \quad \text{for } i = 1,\ldots,4$	ε_3
$r_{1,4}$	$dCN + a_{1,4}O_2 \xrightarrow{\mathbb{S}_4} \varepsilon_4\mathbb{S}_4 + (1-\varepsilon_4)H_2CO_3 + b_{1,4}H_2O + d_{1,4}NH_3$	ε_4

Biological structure is unit carbon based and its composition is given by $CH_{\alpha_j}O_{\beta_j}N_{\gamma_j}$. The stoichiometric coefficients, $a_{i,j}$, $b_{i,j}$ and $d_{1,4}$ are determined from O, H and N elemental balances for each reaction as necessary

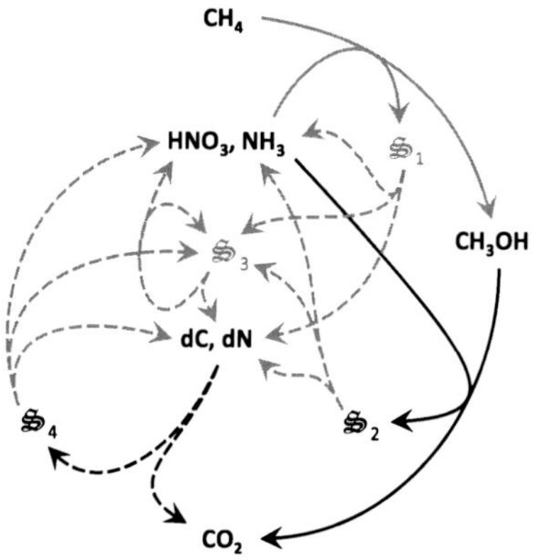

Fig. 18.2 Distributed reaction network for methanotrophic communities. dC and dN: recalcitrant organic C and N, respectively. Lines of similar color and style represent a single reaction group. H_2CO_3 not shown to improve readability. See Table 18.1 for stoichiometry

structure synthesis is directly coupled to its associated redox reaction pair. Nevertheless, networks based on half reactions are useful for discovering important reaction pairs that evade detection, such as anammox [21], because models based on half reactions build their own redox pair combinations.

Reaction stoichiometries are parameterized by two types of optimal control variables, ε_j and $\omega_{i,j}$, where the former controls the efficiency of biological structure synthesis, and the latter controls how biological structure is allocated to sub-reactions associated with each biological structure (Table 18.1). For instance, $\omega_{1,1}$ determines how $\mathfrak{S}_1$ is partitioned between nitrate uptake ($r_{1,1}$) and ammonia uptake ($r_{2,1}$). The value of ε_j plays a critical role in the model, because as ε_j approaches 0 the reaction behaves as pure combustion dissipating substantial amounts of free energy, while as ε_j approaches 1 biological structure is synthesized with minimum free energy dissipation and maximum conversation of C substrate into biomass. As discussed above and elsewhere [45], reaction free energy for biological structure synthesis as ε_j approaches 1 is still negative or within the neighborhood of 0, but in order to achieve a growth efficiency of near 100 %, reactions must proceed reversibly (i.e., infinitely slowly). This thermodynamic constraint explains why we do not find bacteria opting for an ε_j near 1 strategy.

The partitioning of labile (i.e., easily degraded) versus detrital C and N in the four reactions associated with biological structure decomposition, $r_{i,3}$, is solely determined by ε_3. While this is a crude approximation, it has the advantage that no additional parameters are needed. One of the objectives of the model is to limit the number of adjustable parameters and place as many degrees of freedom as possible in the optimal control variables ε_j and $\omega_{i,j}$. The detrital C (dC) and N (dN) pools are modeled separately, but are treated as a single molecule, dCN, in reaction $r_{1,4}$ with its concentration, c_{dCN}, set to c_{dC} and its N:C ratio given by $\gamma_{\mathrm{dCN}} = c_{\mathrm{dN}}/c_{\mathrm{dC}}$.

Total internal entropy produced by the microbial community (kJ K^{-1}), ignoring small contributions from mixing entropy [45], is readily calculated from the product of reaction rate ($r_{i,j}$) and the associated reaction free energy ($\Delta_r G_{r_{i,j}}$) summed over each reaction in the network, as given by,

$$\dot{\sigma}(t) = -\frac{V_L}{T}\sum_{j=1}^{n_{\mathrm{S}}}\sum_{i=1}^{n_{\mathrm{S}j}} r_{i,j}(t)\Delta_r G_{r_{i,j}}(t) \tag{18.3}$$

where V_L is the liquid volume of the microcosms (m^3), T is temperature (K), n_{S} is the number of biological structures (4 in this case), and n_{S_j} is the number of sub-reactions associate with $\mathfrak{S}_j$ (Table 18.1). We use Alberty's [2] approach for calculating reaction free energies, $\Delta_r G_{r_{i,j}}$, that accounts for species concentrations and activity coefficients, and Battley's [4] value for the free energy of formation of biological structure (see also [44, 45]).

Reaction rates are given by the following modified Monod kinetics expression [45]

$$r_{i,j} = \nu_j \varepsilon_j^2 (1 - \varepsilon_j^2) \prod_{k=1}^{n_c} \left(\frac{c_k}{c_k + \kappa_j \varepsilon_j^4} \right)^{\Lambda_{i,j,k}} \omega_{i-1,j} \prod_{l=i}^{n_{\mathrm{S}_j}-1} (1 - \omega_{l,j}) f_G(\Delta_r G_{r_{i,j}}) c_{\mathrm{S}_j}. \tag{18.4}$$

The parameters ν_j and κ_j were chosen to capture bacterial growth kinetics observed in nutrient deplete (i.e., oligotrophic) to nutrient abundant (i.e., eutrophic) conditions. That is, ν_j and κ_j are independent of community composition. The exponent $\Lambda_{i,j,k}$ is set to either 0 or 1 depending on reaction stoichiometry (Table 18.1) for the n_c state concentration variables, c_k, and $\omega_{l,j}$ determines how

$\mathfrak{S}_j$ is partitioned to its associated n_{S_j} sub-reactions, where $\omega_{0,j} = 1$ for all reactions. For all model runs, we assume decomposition of biological structure occurs indiscriminately, so that $\omega_{i,3} = c_{S_{i+1}} / \sum_{k=1}^{i+1} c_{S_k}$ for $i = 1, \ldots, 4$. To insure no reaction proceeds if its free energy of reaction, $\Delta_r G_{r_{i,j}}$, is greater than zero, the function f_G is set to,

$$f_G(\Delta_r G_{r_{i,j}}) = \begin{cases} 1 - e^{\chi_G \Delta_r G_{r_{i,j}}} & \Delta_r G_{r_{i,j}} \leq 0 \\ 0 & \Delta_r G_{r_{i,j}} > 0 \end{cases}; \tag{18.5}$$

where χ_G is chosen for numerical integration criteria, because the $(1 - \varepsilon_j^2)$ term in Eq. [18.4] imposes an empirical thermodynamic constraint as ε_j approaches 1.

Once again, the motivation for Eq. (18.4) is based on minimizing the number of free parameters. Since ν_j and κ_j have predetermined values for all reactions [45], except for reaction $r_{1,4}$ discussed below, reaction rates solely depend on the values of the optimal control variables and the concentration of the state variables.

A process that is difficult to model is biofilm formation in the MCs. After several hundred days of operation, considerable biomass accumulated on the reactor walls, even though the MCs were gently mixed. While we could have developed a sophisticated biofilm sub-model, this would result in numerous poorly defined additional parameters. Instead, we simply introduce one parameter, f_{PL}, to represent the fraction of particulate matter (both living and detrital) that is not subject to chemostat washout because it is associated with the biofilm (see Table A.1).

18.3.3 Optimization Model

To determine how ε_j and $\omega_{i,j}$ must vary over time in order to maximize internal entropy production, we formulate and solve a receding horizon optimal control (RHOC) problem [7, 30]. RHOC is used in many fields. For example, in economics RHOC is used to determine how short-term investments should be allocated to maximize long-term returns, such as in retirement fund management. Because long-term prediction of markets is not perfect, short-term strategies are updated periodically based on current market conditions. We implement a similar approach and maximize internal entropy production over successive intervals of time as given by,

$$\max_{\mathbf{u}(t_{n+1})} \frac{1}{\Delta t^*} \int_{t_n}^{t_n+\Delta t^*} \dot{\sigma}(\tau) e^{-k_w(\tau - t_n)} d\tau \quad \textit{where} \quad \mathbf{u} = \left[\boldsymbol{\varepsilon}^T \ \boldsymbol{\omega}^T\right]^T \tag{18.6a}$$

$$\textit{subject to}: \quad \frac{d\mathbf{x}(t)}{dt} = \mathbf{f}(\mathbf{x}(t), \mathbf{u}(t)) \quad \textit{and} \quad \mathbf{0} < \mathbf{u}(t) \leq \mathbf{1} \tag{18.6b}$$

where Δt^* is the long-term optimization interval from the current time, t_n, over which entropy production, $\dot{\sigma}$, is maximized. A conventional weighting function, $e^{-k_w(t-t_n)}$, discounts the importance of entropy production as time increases beyond t_n due to uncertainties in predicting future states. After the value of the optimal control variables ε_j and $\omega_{i,j}$, are determined over the optimization interval $[t_n, t_n + \Delta t^*]$, the state equations are updated only to $t_{n+1} = t_n + \Delta t$, where $\Delta t \leq \Delta t^*$, as illustrated in Fig. 18.1. The updating interval, Δt, is typically less than Δt^* to minimize discontinuities in state and control variables at the end of an interval. Average internal entropy production over the update interval, Δt, is given by,

$$\langle \dot{\sigma}(t_{n+1}) \rangle = \frac{1}{\Delta t} \int_{t_n}^{t_n+\Delta t} \dot{\sigma}(\tau) d\tau. \tag{18.7}$$

Total internal entropy produced over k intervals is given by $\sigma(t_n : t_{n+k}) = \Delta t \sum_{i=1}^{k} \langle \dot{\sigma}(t_{n+i}) \rangle$. Once the state and control variables are updated to $t_{n+1} = t_n + \Delta t$, Eq. (18.6) is used to solve the next optimization interval, $t_{n+1} + \Delta t^*$ to extend the solution to $t_{n+2} = t_{n+1} + \Delta t$; this iteration is repeated until the desired final simulation time is reached.

The optimization, Eq. (18.6a), is subject to box constraints on the control variables between 0 and 1, and by mass balance constraints on the state variables, $\mathbf{x}(t)$, given by the differential equations defined by $\mathbf{f}(\mathbf{x}(t), \mathbf{u}(t))$ in Eq. (18.6b). The state variables for the microcosm experiment consist of nutrient concentrations, $\mathbf{c}(t)$, gas partial pressures, $\mathbf{p}(t)$, and concentration of biological structures, $\mathbf{c}_S(t)$, so that $\mathbf{x}(t) = \left[\mathbf{c}^T(t), \mathbf{p}^T(t), \mathbf{c}_S^T(t)\right]^T$. The mass balance equations are listed in the Appendix (Table A.1). The differential equations were numerically integrated using a high precision method [6] and the optimization problem was solved using a derivative free algorithm (BOBYQA [39]). Control variables are discretized over the $[t_n, t_n + \Delta t^*]$ interval using n_{knots} grid points and linear interpolation is used to produce continuous control functions.

18.4 Results

Time zero of the microcosm experiments corresponds to 00:00 20 Aug 2010, and on day 210.5 gas cycling of microcosms (MC) 1 and 4 commenced after experimental operating conditions had been finalized, in particular nitrogen-limited growth was achieved. Numerical simulations using the MaxEP-based BGC model were initialized on day 100, which provided sufficient time to achieve steady state conditions prior to gas cycling. Both experimental and modeling results are compared over days 200–500.

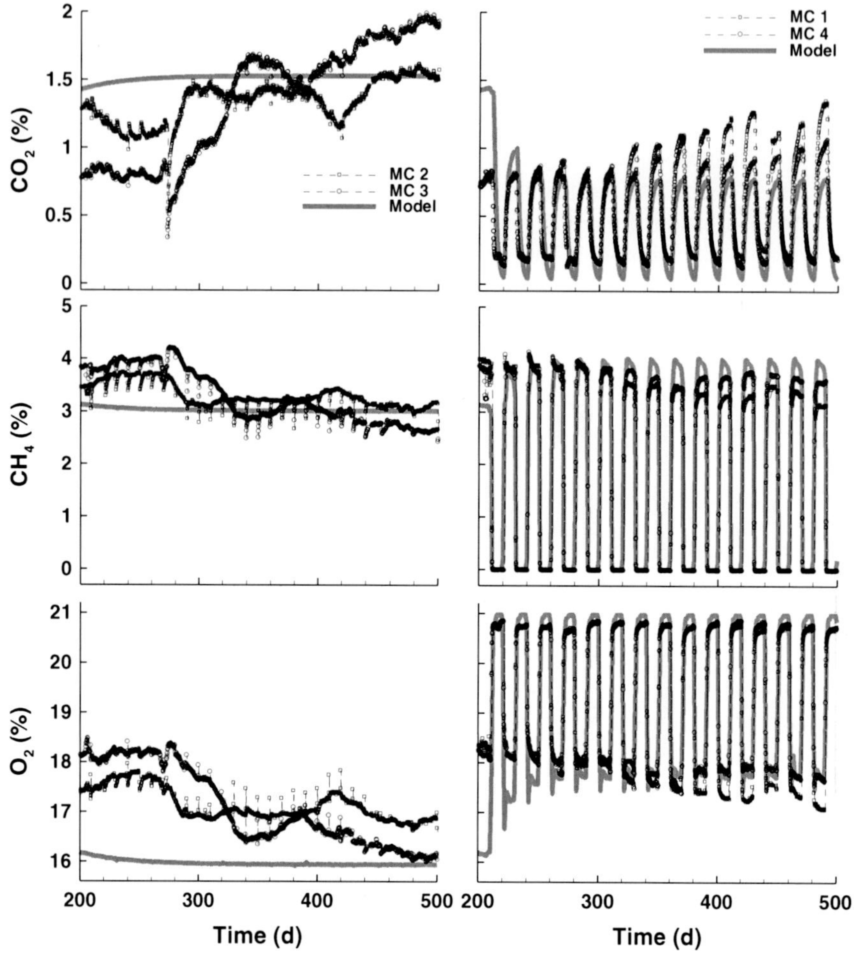

Fig. 18.3 Observed and modeled reactor exit gas concentrations for the controls (MC 2 and 3, *left column*) and cycled (MC 1 and 4, *right column*) microcosms. Modeled predictions are shown as the *orange* (or *grey* in BW) *solid line*, while experimental data are shown as open symbols connected by *dashed lines*. Model results are for $k_w = 0.230\ \mathrm{d}^{-1}$, $\Delta t = 10$ d and $\Delta t^* = 20$ d

Only two model parameters were qualitatively adjusted to achieve reasonable agreement between model results and observations for all four MCs (Figs. 18.3 and 18.4). Because detritus is a rather amorphous, non-polymeric material, its decomposition is difficult and is often the rate limiting step in microbial BGC [14]. Consequently, we reduced v_4 in Eq. (18.4) to 35 d^{-1} from the standard value of 350 d^{-1} [45]. We also tuned the biofilm parameter, f_{PL}, to 0.2. All other parameter values are well-defined constants, such as MC volume, dilution rate, feed concentrations, etc. All model degrees of freedom, other than v_4 and f_{PL}, reside in the 6

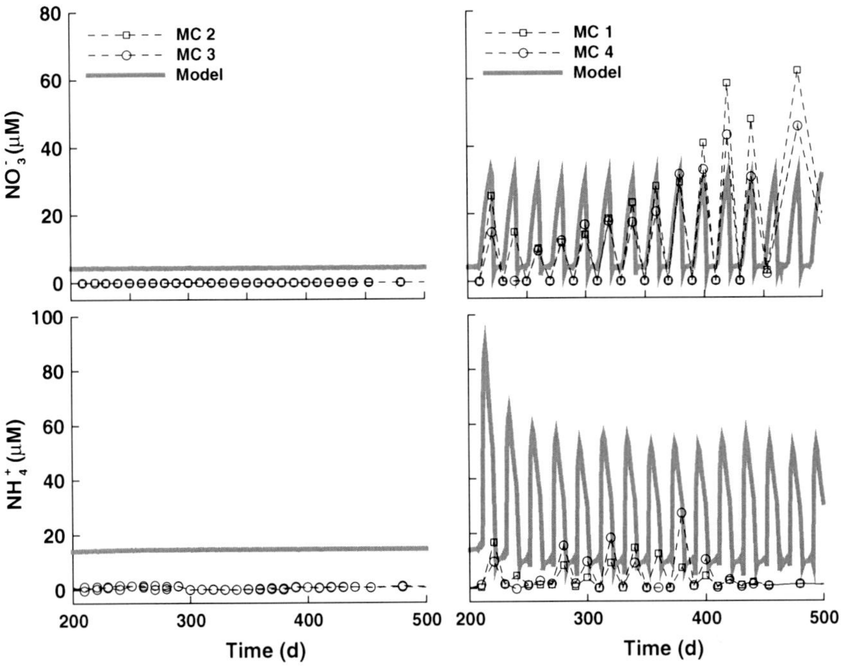

Fig. 18.4 Simulated data [*orange* (or *grey*) *lines*] compared to observations of nitrate and ammonium concentrations for the control MCs (MC 2 and 3, *left column*) and the methane cycled MCs (MC 1 and 4, *right column*). Also see caption to Fig. 18.3

Table 18.2 Internal entropy produced over 400 days for the control and gas-cycled simulations for different optimal interval parameters values: k_w, Δt and Δt^*

Δt	Δt^*	k_w	n_{knots}	$\sigma(100:500)$ (kJ K^{-1})	
(d)	(d)	(d^{-1})		Control	Cycled
0.1	0.1	0	1	2.07	1.45
0.1	1	3.00	5	18.56	7.71
1	1	0	1	16.82	6.94
1	5	0.921	5	22.85	9.05
10	20	0.230	15	24.19	10.53
20	40	0.115	20	24.64	14.55
20	50	0.0921	25	24.77	15.15

optimal control variables and the three interval optimization parameters, k_w, Δt and Δt^*.

To examine how the optimal interval parameters affect the solution and overall internal entropy production, we conducted several simulations by varying k_w, Δt and Δt^* for both the control and the gas-cycled simulations (Table 18.2). In general, these results show that as the optimization interval increases, total internal entropy produced (σ) over the 400 days of simulation increases, but asymptotes to approximately 25 and 15 kJ K^{-1} for the control and gas-cycled simulations, respectively. Except for very short intervals, entropy production in the control simulations is not strongly affected by choices of Δt or Δt^* (Table 18.2). However, for very short optimization intervals ($\Delta t^* = 0.1$ d), entropy production is significantly depressed (Table 18.2). A similar phenomenon occurs in the gas-cycled simulations, but the decrease in total entropy production as Δt^* decreases is more gradual.

As Δt^* becomes small, biological structures are allocated to maximize entropy production in a manner that resembles abiotic systems, such as fire. In particular, examination of the control simulations reveals that the system does not sufficiently allocate resources to biological structure turnover, $\mathfrak{S}_3$. The concentration of $\mathfrak{S}_3$ in the control simulation with $\Delta t^* = 0.1$ is approximately equal to $\mathfrak{S}_1$ and $\mathfrak{S}_2$, but in the simulation with larger Δt^* values, $\mathfrak{S}_3$ concentration is twice that of $\mathfrak{S}_1$ and $\mathfrak{S}_2$. The optimal controller attains higher concentration of $\mathfrak{S}_3$ by setting ε_3 to approximately 0.62, while in the low entropy producing case ε_3 is only set to 0.34. The higher concentration of $\mathfrak{S}_3$ allows the system to achieve much higher remineralization rates, so that reactions $r_{2,1}$ and $r_{2,2}$ can attain much higher rates due to the increase in NH_3 availability from $\mathfrak{S}_i$ turnover. However, under short optimization intervals, the system's time horizon is too short to realize a return on investment in $\mathfrak{S}_3$ with respect to entropy production or utilization of available chemical potential. When the time scale is short, the system does not make best use of available resources.

The gas-cycled simulations also generate interesting results when different $(\Delta t, \Delta t^*)$ values are examined. Figure 18.5a shows methanol dynamics over two gas cycling periods (40 d, beginning on day 300) for four selected simulations in Table 18.2 based on the $(\Delta t, \Delta t^*)$ values. When time scales are short, $(\Delta t, \Delta t^*) =$ (0.1, 1 d) and (1, 5 d), methanol (CH_3OH) accumulates immediately after methane gas is turned on (at 300.5 and 320.5 d; Fig. 18.5a, dashed lines). However, when the time scale specified by the optimization parameters approach the period length of the gas cycling, $(\Delta t, \Delta t^*) =$ (10, 20 d) and (20, 50 d), methanol accumulation occurs immediately before methane is switched off (at 310.5 and 330.5 days; Fig. 18.5a, solid lines). When the optimization time scales are long, the model develops an anticipatory control strategy, where methanol is produced as a storage compound that can be utilized during the phase when methane is absent. By storing some of the methane captured in the first half of the cycle as methanol, the system is able to oxidize more methane and produce more internal entropy compared to the simulations using short term optimization parameters. The strategy only accumulates methanol near the end of the period, because methanol is also lost due to dilution, which does not contribute to internal entropy production.

We can see how the control strategy achieves methanol accumulation by examining the concentration of biological structures and growth efficiencies for the case where $(\Delta t, \Delta t^*)$ equals (20, 50 d) over the two gas cycling periods

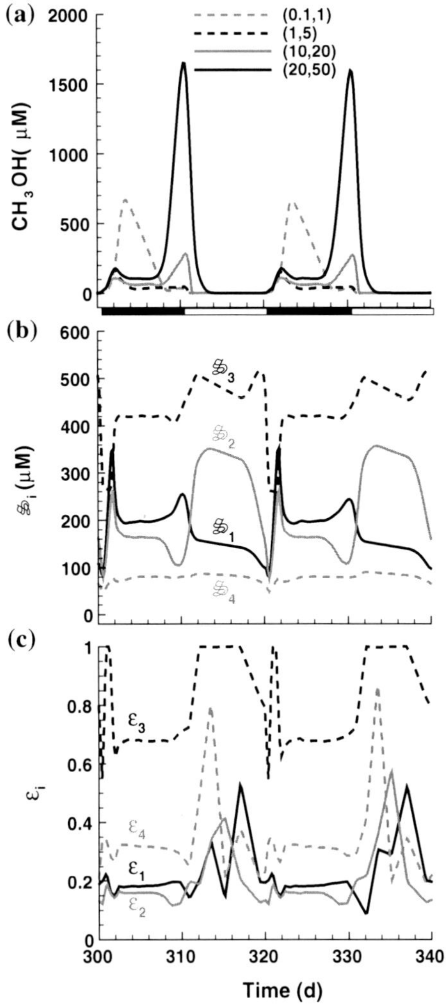

Fig. 18.5 (**a**) Methanol accumulation over two cycles for the optimization intervals given in the legend: $(\Delta t, \Delta t^*)$. The bar along the x axis shows when CH_4 gas is on (*black*) and off (*white*). Biological structures (**b**) and growth efficiencies (**c**) over two cycle periods for $(\Delta t, \Delta t^*) = (20\text{ d}, 50\text{ d})$. All data are from the gas-cycled simulations only

(Fig. 18.5b, c). Just prior to the loss of methane (310.5 and 330.5 d), there is an increase in $\mathfrak{S}_1$ and a decrease in $\mathfrak{S}_2$ (Fig. 18.5b). Based on the reaction network (Fig. 18.2, Table 18.1), this allocation of catalyst favors methanol overproduction, so methanol accumulates rapidly. Immediately following the addition of methane, there is a rapid rise in $\mathfrak{S}_2$ concentration and a decrease in $\mathfrak{S}_1$, which drives methanol consumption up. To attain these changes in $\mathfrak{S}_1$ and $\mathfrak{S}_2$ abundances, there are the expected changes in the associated growth efficiencies (Fig. 18.5c), but there is also a large change in ε_3. In particular, ε_3 is driven to 1 following the loss of methane feed, which allows all biological structures to remain at high concentrations in the absence of methane because $r_{i,3}$ is driven to zero (Eq. 18.4). Just prior to the introduction of methane, ε_3 is reduced significantly, which causes a large turnover of biological structure (Fig. 18.5b), but biological structure concentrations quickly rebound once methane is again made available. To examine if these changes are occurring in the actual microcosms, we are currently sampling for cell abundances, DNA/RNA, and methanol concentration.

18.5 Discussion

In this chapter we have shown that a microbial biogeochemistry model based on the MaxEP principle produces results that are comparable to those obtained experimentally from microbial methanotrophic microcosms (Figs. 18.3 and 18.4). Unlike most microbial biogeochemistry models, the MaxEP model contains very few adjustable parameters, because we have been able to place most of the model's degrees of freedom into the optimal control variables, ε_j and $\omega_{i,j}$, whose values are determined by maximizing internal entropy production. By placing emphasis on catalytic activity at the system level, rather than on competition of individuals, the MaxEP approach provides a unique perspective on how ecosystems may function and evolve. Due to the novelty of the MaxEP approach, many of the ideas and conjectures that derive from MaxEP need to be tested, or at least shown to be improvements over canonical approaches. Microbial microcosms provide excellent experimental systems for testing MaxEP-based approaches for describing living systems, as microbial systems have fast characteristic times scales, high population densities and high biodiversity, all of which can be readily manipulated and monitored.

The MaxEP-BGC model predicts a comprehensive suite of output variables that can be compared to observations, only some of which were presented here. In addition to providing concentration data and reaction efficiencies, the model predicts reaction rates through the metabolic network (Table 18.1), reaction free energies, and how biological structure partitioning among sub-reactions changes over time (i.e., $\omega_{i,j}$). We expect our on-going measurements of community composition from 454-tag pyrosequencing and quantitative PCR analysis of function gene levels and expression will assist in comparing model output to observations [15]. Preliminary molecular results show that very high microbial diversity is maintained in the microcosms ($\sim$600 operational taxonomic units); however, community composition of the methanotrophs changes substantially over time (microscopic behavior), but this does not alter methane oxidation rates (macroscopic behavior), a characteristic consistent with MaxEP [9].

Perhaps the most intriguing result from our implementation of MaxEP for describing microbial biogeochemistry is the proposed distinction between abiotic and biotic systems based on instantaneous versus averaged entropy production. When entropy production is maximized instantaneously, no biological structure is produced because some of the free energy would simply be converted to another form of chemical potential instead of being destroyed. This problem is solved by maximizing entropy produced over an interval of time, which leads to the hypothesis. Because biotic systems are able to store information in their genome, they can implement temporal strategies that can out-compete abiotic processes in some situations. Because of genomic complexities, we do not know a priori the nature of the temporal strategies at this time, but this lack of knowledge can be circumvented by assuming that evolution has produced systems that extract the greatest possible free energy from a system over some appropriate characteristic time scale. Our results indicate (Table 18.2) that the longer the time scale, the more entropy that can be produced, but longer time scales require higher fidelity in predicting future states. Prediction in this case simply means that some of the temporal strategies the system possesses will be successful. Mismatches between prediction and the true state, due to perturbations, noise and uncertainties, ultimately limit the time scale interval for entropy production.

Our receding horizon optimal control implementation of the MaxEP problem shows that when time scales are short, biological structure should be invested for immediate entropy production, which leads to methanol production following the introduction of methane (Fig. 18.5a, dashed lines). This is an R-selection strategy [38], which is a possible driving mechanism for cross feeding [40], because partial substrate oxidation can increase growth rate [37]. When time scale is increased, the system allocates resources to $\mathfrak{S}_3$ (the equivalent of grazers) as well as the later production of methanol that acts as a storage compound (Fig. 18.5a, solid lines). Systems oriented analyses of natural ecosystems indicate that the presence of grazers increases nutrient recycling and ecosystem productivity [31, 41, 42]. Predators, and trophic structures in general, increase the characteristic time scale of an ecosystem. It appears reasonable that organisms with long development times, or life histories, impart the long characteristic time scales observed in mature ecosystems, such as forests. Under this conjecture, bacterial systems may be closer to fire than an ecosystem composed of macroscopic organisms that provide the long characteristic time scale with respect to entropy production.

Experimentally, we expected more effective use of CH_4 in the gas-cycled treatment; that is, we expected entropy production to be similar between the control and gas-cycled MCs. Interestingly, the MaxEP-BGC model also has difficulties in producing entropy in the gas-cycled MCs (Table 18.2), but matches the experimental data well (Figs. 18.3 and 18.4). Because of methanol washout from the chemostat, the model only uses methanol as a storage compound near the end of the CH_4-on cycle, which limits the system's ability to store chemical potential. Storage of free energy in biological structure is also limited due to N requirements for $\mathfrak{S}$. Perhaps the experiment and model are lacking higher trophic levels (i.e., macrofauna) that would provide a time scale relevant to the 20 day gas-cycle period. Currently, the model uses cannibalism of $\mathfrak{S}_3$ as a means of trophic closure [33], so adding additional trophic levels may be one means of increasing the characteristic time scale in the model. As for the experiment, we are currently characterizing the eukaryotic community structure via cell counts.

Our MaxEP-BGC model currently focuses on microbes as reaction catalysts that dissipate chemical potential, but the MaxEP concept can be extended to macro-fauna and-flora as well. Metabolically, macroorganisms are rather prosaic; however, in addition to their longer characteristic time scales, they provide physical structure, increase the surface area of particulate matter via mastication and greatly enhance transport processes that often limit reaction rates [11, 18, 22]. Application of MaxEP to natural ecosystems will require understanding the functional contributions of macroorganisms, in particular with respect to transport processes, which is not a typical focus in ecology. More research needs to be done in this area.

18.6 Conclusions

We have been able to use the MaxEP conjecture to develop a microbial biogeochemistry model that reproduces reasonably well experimental data obtained from a methanotrophic microcosm experiment. By assuming that genomic information allows living systems to maximize entropy production over a characteristic time scale, we have been able to formulate the model as a receding horizon optimal control problem. Most of the model's degrees of freedom have been captured by the optimal control variables whose values are determined by maximizing entropy production over successive intervals of time. This approach greatly reduces the number of adjustable parameters whose values are often unknown, poorly constrained and seldom constant. Our results indicate that temporal strategies that are successful over greater durations of time will result in greater entropy production. From this hypothesis, we have developed a methanotrophic microcosm experiment to study how microbial communities respond, adapt and evolve to time varying inputs of energy. Based on experimental data to date, there appears to be good agreement between the MaxEP-BGC model results and experimental data.

All organisms possess genomic and acquired information that dictates survival strategies and life cycles that operate over defined characteristic time scales. These time scales can be as short as minutes or hours (i.e., for some bacteria) to as long as centuries or more (i.e., some tree species). Our approach has illustrated the importance that temporal strategies have on ecosystem dynamics, but our choice of time scale (for both Δt and Δt^*) has been somewhat arbitrary based on our intuitive understanding of bacterial growth and the reduced complexity of our experimental microcosms. Natural ecosystems are comprised of populations of different organisms that operate over a multitude of time scales. However, we hypothesize that organisms with long time scales can access more free energy (and ultimately producing more entropy) than those operating on short time scales provided the system is stable enough for long term predictions. Viewing ecosystems as a collection of free energy dissipating machines adaptively operating over a spectrum of time scales may help us understand how these systems assemble, operate and respond to disturbances of differing magnitude and frequency. Further research is needed relating the ecological concepts of temporal strategies and succession to quantitative measures and representations of time scales for the dissipation of free energy.

Acknowledgments This research was supported by NSF grant EF-0928742 (Huber, Fernández González, and Vallino), NSF grant OCE-1058747 (Algar, Vallino) and NSF grants CBET-0756562, OCE-1058747 (Vallino). We thank Stefanie Strebel for sample analyses and assistance in operation of the microcosms.

Appendix

Tables A.1 and A.2.

Table A.1 Mass balance equations for the rates of change of chemical species concentrations in the microcosm model used for constraints in Eq. (18.6b)*

$$\dot{c}_{CH_4}(t) = -r_{1,1} - r_{2,1} + \left(F_L(c^f_{CH_4} - c_{CH_4}) + k_L a(p_{CH_4}/k_{CH_4}(T) - c_{CH_4})\right)/V_L$$

$$\dot{c}_{CH_3OH}(t) = (1-\varepsilon_1)(r_{1,1} + r_{2,1}) - r_{1,2} - r_{2,2} + \left(F_L(c^f_{CH_3OH} - c_{CH_3OH}) + k_L a(p_{CH_3OH}/k_{CH_3OH}(T) - c_{CH_3OH})\right)/V_L$$

$$\dot{c}_{H_2CO_3}(t) = (1-\varepsilon_2)(r_{1,2} + r_{2,2}) + \varepsilon_3(1-\varepsilon_3)\sum_{i=1}^{4} r_{i,3} + (1-\varepsilon_4)r_{1,4} + \left(F_L(c^f_{H_2CO_3} - c_{H_2CO_3}) + k_L a(p_{CO_2}/k_{H_2CO_3}(T) - c_{H_2CO_3})\right)/V_L$$

$$\dot{c}_{dC}(t) = (1-\varepsilon_3)^2 \sum_{i=1}^{4} r_{i,3} - r_{1,4} + F_L(c^f_{dC} - f_{PL}c_{dC})/V_L$$

$$\dot{c}_{HNO_3}(t) = -\varepsilon_1\gamma_1 r_{1,1} - \varepsilon_2\gamma_2 r_{1,2} + F_L(c^f_{HNO_3} - c_{HNO_3})/V_L$$

$$\dot{c}_{NH_3}(t) = -\varepsilon_1\gamma_1 r_{2,1} - \varepsilon_2\gamma_2 r_{2,2} + \varepsilon_3 \sum_{i=1}^{4}((2-\varepsilon_3)\gamma_i - \gamma_3)\, r_{i,3} + d_{1,4}r_{1,4} + F_L(c^f_{NH_3} - c_{NH_3})/V_L$$

$$\dot{c}_{dN}(t) = (1-\varepsilon_3)^2 \sum_{i=1}^{4} \gamma_i r_{i,3} - \gamma_{dCN} r_{1,4} + F_L(c^f_{dN} - f_{PL}c_{dN})/V_L$$

$$\dot{c}_{O_2}(t) = -\sum_{i=1}^{2}\sum_{j=1}^{2} a_{i,j}r_{i,j} - \sum_{i=1}^{4} a_{i,3}r_{i,3} - a_{1,4}r_{1,4} + \left(F_L(c^f_{O_2} - c_{O_2}) + k_L a(p_{O_2}/k_{O_2}(T) - c_{O_2})\right)/V_L$$

$$\dot{p}_{CH_4}(t) = \left(F_G(p^f_{CH_4} - p_{CH_4}) + k_L a\, RT(c_{CH_4} - p_{CH_4}/k_{CH_4}(T))\right)/V_G$$

$$\dot{p}_{CH_3OH}(t) = \left(F_G(p^f_{CH_3OH} - p_{CH_3OH}) + k_L a\, RT(c_{CH_3OH} - p_{CH_3OH}/k_{CH_3OH}(T))\right)/V_G$$

$$\dot{p}_{CO_2}(t) = \left(F_G(p^f_{CO_2} - p_{CO_2}) + k_L a\, RT(c_{H_2CO_3} - p_{CO_2}/k_{H_2CO_3}(T))\right)/V_G$$

$$\dot{p}_{O_2}(t) = \left(F_G(p^f_{O_2} - p_{O_2}) + k_L a\, RT(c_{O_2} - p_{O_2}/k_{O_2}(T))\right)/V_G$$

$$\dot{c}_{S_j}(t) = \varepsilon_j \sum_{i=1}^{ns_j} r_{i,j} - r_{j,3} + F_L(c^f_{S_j} - f_{PL}c_{S_j})/V_L \quad \text{for } j = 1,\ldots,4$$

*The superscript f refers to concentration of variables in the feed stream, F_L and F_G are the liquid and gas volumetric feed rates, respectively, k_La is the liquid-side mass transfer coefficient, $k_h(T)$ is a Henry's law coefficient for solute h, V_G is the gas headspace volume, and f_{PL} is the fraction of particulate matter loss due to dilution; that is, not associated with the biofilm

Table A.2 Nomenclature

Variable	Definition	Units
$a_{i,j}$	Oxygen stoichiometric coefficient for reaction $r_{i,j}$ (see Table 18.1)	
$b_{i,j}$	Water stoichiometric coefficient for reaction $r_{i,j}$ (see Table 18.1)	
c_i	Concentration of species i ($\mathbf{c}$ in vector form)	mmol m^{-3}
c_i^f	Concentration of species i in microcosm feed	mmol m^{-3}
c_{S_j}	Concentration of biological structure j	mmol m^{-3}
$d_{i,j}$	Ammonia stoichiometric coefficient for reaction $r_{i,j}$ (see Table 18.1)	
dC	Detrital organic carbon	
dN	Detrital organic nitrogen	
f_{PL}	Fraction of particulate matter loss due to dilution	
$\mathbf{f}$	Vector function of state equations (see Table A1)	
$k_i(T)$	Henry's law coefficient for solute i	Pa m^3 $mmol^{-1}$
$k_L a$	Air–water gas transfer coefficient, liquid side	m^3 d^{-1}
k_w	Optimization discounting parameter	d^{-1}
n_c	Number of chemical species	
n_{knots}	Number of grid points for discretizing control variables over an optimization interval (see Table 18.2)	
n_S	Number of biological structures, $\mathfrak{S}_j$	
n_{S_j}	Number of sub-reactions associated with $\mathfrak{S}_j$	
p_i	Partial pressure of gas species i	Pa
p_i^f	Partial pressure of gas species i in feed gas	Pa
$r_{i,j}$	Reaction rate	mmol m^{-3} d^{-1}
Δt	Optimization update interval	d
Δt^*	Optimization interval	d
t	Time	d
$\mathbf{u}$	Vector of control variables (ε, ω)	
$\mathbf{x}$	Vector of state variables ($\mathbf{c}$, $\mathbf{p}$, $\mathbf{c}_S$)	
F_G	Gas flow rate to microcosms	m^3 d^{-1}
F_L	Liquid flow rate to microcosms	m^3 d^{-1}
$\Delta_r G_{r_{i,j}}$	Gibbs free energy of reaction for reaction $r_{i,j}$	kJ $mmol^{-1}$
R	Gas constant (units depend on equation)	
S	System entropy	kJ K^{-1}
$\mathfrak{S}_j$	Biological structure j that catalyzes reaction $r_{i,j}$	
T	Temperature	K
V_G	Gas volume of microcosm	m^{-3}
V_L	Liquid volume of microcosm	m^{-3}
α_i	Hydrogen atoms in unit carbon formula for biological structure i	
β_i	Oxygen atoms in unit carbon formula for biological structure i	
γ_i	Nitrogen atoms in unit carbon formula for biological structure i	
ε_j	Growth efficiency for biological structure j. (Optimal control variable)	
κ_j	Substrate affinity parameter in reaction $r_{i,j}$	mmol m^{-3}
ν_j	Maximum specific reaction rate for reaction $r_{i,j}$	d^{-1}
σ	Entropy produced from irreversible processes within system	kJ K^{-1}
$\dot{\sigma}$	Rate of internal entropy production	kJ K^{-1} d^{-1}
χ_G	Parameter in free energy weighting function, f_G	mmol kJ^{-1}

(continued)

Table A.2 (continued)

Variable	Definition	Units
$\omega_{i,j}$	Partition coef. of biological structure j to sub-reaction $r_{i,j}$. (Optimal control variable)	
$\Lambda_{i,j,k}$	Stoichiometric exponent for reaction $r_{i,j}$ for c_k	
$\langle\rangle$	Expectation operator	

References

1. Adami, C.: Information theory in molecular biology. Phys. Life Rev. **1**(1), 3–22 (2004). doi:10.1016/j.plrev.2004.01.002
2. Alberty, R.A.: Thermodynamics of biochemical reactions. Wiley, Hoboken (2003)
3. Battley, E.: Absorbed heat and heat of formation of dried microbial biomass: studies on the thermodynamics of microbial growth. J. Therm. Anal. Calorim. **74**(3), 709–721 (2003). doi:10.1023/B:JTAN.0000011003.43875.0d
4. Battley, E.H.: The development of direct and indirect methods for the study of the thermodynamics of microbial growth. Thermochim. Acta **309**(1–2), 17–37 (1998)
5. Berut, A., Arakelyan, A., Petrosyan, A., Ciliberto, S., Dillenschneider, R., Lutz, E.: Experimental verification of Landauer/'s principle linking information and thermodynamics. Nature **483**(7388), 187–189 (2012). doi:10.1038/nature10872
6. Brugnano, L., Magherini, C.: The BiM code for the numerical solution of ODEs. J. Comput. Appl. Math. **164–165**, 145–158 (2004). doi:10.1016/j.cam.2003.09.004
7. Chen, H., Allgower, F.: A quasi-infinite horizon nonlinear model predictive control scheme with guaranteed stability. Automatica **34**(10), 1205–1217 (1998). doi:10.1016/S0005-1098(98)00073-9
8. Dewar, R.: Information theory explanation of the fluctuation theorem, maximum entropy production and self-organized criticality in non-equilibrium stationary states. J. Phys. A: Math. Gen. **36**, 631–641 (2003)
9. Dewar, R.: Maximum entropy production as an inference algorithm that translates physical assumptions into macroscopic predictions: don't shoot the messenger. Entropy **11**(4), 931–944 (2009). doi:10.3390/e11040931
10. Dewar, R.C.: Maximum entropy production and the fluctuation theorem. J. Phys. A: Math. Gen. **38**(21), L371–L381 (2005). doi:10.1088/0305-4470/38/21/L01
11. Erwin, D.H.: Macroevolution of ecosystem engineering, niche construction and diversity. Trends Ecol. Evol. **23**(6), 304–310 (2008). doi:10.1016/j.tree.2008.01.013
12. Falkowski, P.G., Fenchel, T., DeLong, E.F.: The microbial engines that drive earth's biogeochemical cycles. Science **320**(5879), 1034–1039 (2008). doi:10.1126/science.1153213
13. Friedrichs, M.A.M., Dusenberry, J.A., Anderson, L.A., Armstrong, R.A., Chai, F., Christian, J.R., Doney, S.C., Dunne, J., Fujii, M., Hood, R., McGillicuddy, D.J., Jr., Moore, J.K., Schartau, M., Spitz, Y.H., Wiggert, J.D (2007) Assessment of skill and portability in regional marine biogeochemical models: Role of multiple planktonic groups. J.Geophys.Res. **112**(C8), C08001
14. Horner, J.D., Gosz, J.R., Cates, R.G.: The role of carbon-based plant secondary metabolites in decomposition in terrestrial ecosystems. Am. Nat. **132**(6), 869–883 (1988)
15. Huber, J.A., Cantin, H.V., Huse, S.M., Mark Welch, D.B., Sogin, M.L., Butterfield, D.A (2010) Isolated communities of Epsilonproteobacteria in hydrothermal vent fluids of the Mariana Arc seamounts. FEMS Microbiology Ecology **73**(3), 538–549. doi:10.1111/j.1574-6941.2010.00910.x
16. Huber, J.A., Welch, D.B.M., Morrison, H.G., Huse, S.M., Neal, P.R., Butterfield, D.A., Sogin, M.L.: Microbial population structures in the deep marine biosphere. Science **318**(5847), 97–100 (2007). doi:10.1126/science.1146689

17. Jaynes, E.T.: The Gibbs paradox. In: Smith, C.R., Erickson, G.J., Neudorfer, P.O. (eds.) Maximum entropy and Bayesian methods, pp. 1–22. Kluwer Academic Publisheres, Dordrecht (1992)
18. Jones, C.G., Lawton, J.H., Shachak, M.: Organisms as ecosystem engineers. Oikos **69**(3), 373–386 (1994)
19. Kleidon, A., Fraedrich, K., Kunz, T., Lunkeit, F.: The atmospheric circulation and states of maximum entropy production. Geophys. Res. Lett. **30**(23), 1–4 (2003). doi:10.1029/2003GL018363
20. Kondepudi, D., Prigogine, I.: Modern thermodynamics: from heat engines to dissipative structures. Wiley, New York (1998)
21. Kuenen, J.G.: Anammox bacteria: from discovery to application. Nat. Rev. Micro. **6**(4), 320–326 (2008). doi:10.1038/nrmicro1857
22. Lawton, J.H.: What do species do in ecosystems? Oikos **71**(3), 367–374 (1994)
23. Lennon, J.T., Jones, S.E.: Microbial seed banks: the ecological and evolutionary implications of dormancy. Nat. Rev. Micro. **9**(2), 119–130 (2011). doi:10.1038/nrmicro2504
24. Lewis, K.: Persister cells. Annu. Rev. Microbiol. **64**(1), 357–372 (2010). doi:10.1146/annurev.micro.112408.134306
25. Lineweaver, C.H., Egan, C.A.: Life, gravity and the second law of thermodynamics. Physics of Life Reviews **5**(4), 225–242 (2008). doi:10.1016/j.plrev.2008.08.002
26. Lorenz, R.: Computational mathematics: full steam ahead-probably. Science **299**(5608), 837–838 (2003)
27. Lorenz, R.D., Lunine, J.I., Withers, P.G.: Titan, mars and earth: entropy production by latitudinal heat transport. Geophys. Res. Lett. **28**(3), 415–418 (2001)
28. Makela, T., Annila, A.: Natural patterns of energy dispersal. Phys. Life Rev. **7**(4), 477–498 (2010). doi:10.1016/j.plrev.2010.10.001
29. Martyushev, L.: Some interesting consequences of the maximum entropy production principle. J. Exp. Theor. Phys. **104**(4), 651–654 (2007). doi:10.1134/S1063776107040152
30. Mayne, D.Q., Rawlings, J.B., Rao, C.V., Scokaert, P.O.M.: Constrained model predictive control: Stability and optimality. Automatica **36**(6), 789–814 (2000). doi:10.1016/S0005-1098(99)00214-9
31. Mazancourt, D.C., Loreau, M.: Grazing optimization, nutrient cycling, and spatial heterogeneity of plant-herbivore interactions: should a palatable plant evolve? Evolution **54**(1), 81–92 (2000)
32. Meysman, F.J.R., Bruers, S.: A thermodynamic perspective on food webs: quantifying entropy production within detrital-based ecosystems. J. Theor. Biol. **249**(1), 124–139 (2007). doi:10.1016/j.jtbi.2007.07.015
33. Mitra, A.: Are closure terms appropriate or necessary descriptors of zooplankton loss in nutrient-phytoplankton-zooplankton type models? Ecol. Model. **220**(5), 611–620 (2009). doi:10.1016/j.ecolmodel.2008.12.008
34. Morrison, P.: A thermodynamic characterization of self-reproduction. Rev. Mod. Phys. **36**(2), 517 (1964)
35. Niven, R.K.: Steady state of a dissipative flow-controlled system and the maximum entropy production principle. Physical review E (Statistical, Nonlinear, and Soft Matter Physics) **80**(2), 021113–021115. (2009) doi:10.1103/PhysRevE.80.021113
36. Paltridge, G.W.: Global dynamics and climate-a system of minimum entropy exchange. Q. J. Roy. Met. Soc. **104**, 927–945 (1975)
37. Pfeiffer, T., Bonhoeffer, S.: Evolution of cross-feeding in microbial populations. Am. Nat. **163**(6), E126–E135 (2004)
38. Pianka, E.R.: R-selection and K-selection. Am. Nat. **104**(940), 592–597 (1970)
39. Powell, M.J.D.: The BOBYQA algorithm for bound constrained optimization without derivatives. Optim. Online **5**, 1–39 (2010)
40. Rosenzweig, R.F., Sharp, R.R., Treves, D.S., Adams, J.: Microbial evolution in a simple unstructured environment: genetic differentiation in escherichia coli. Genetics **137**(4), 903–917 (1994)

41. Schmitz, O.J., Hawlena, D., Trussell, G.C.: Predator control of ecosystem nutrient dynamics. Ecol. Lett. **13**(10), 1199–1209 (2010). doi:10.1111/j.1461-0248.2010.01511.x
42. Urabe, J., Elser, J.J., Kyle, M., Yoshida, T., Sekino, T., Kawabata, Z.: Herbivorous animals can mitigate unfavourable ratios of energy and material supplies by enhancing nutrient recycling. Ecol. Lett. **5**(2), 177–185 (2002). doi:10.1046/j.1461-0248.2002.00303.x
43. Vallino, J.J.: Modeling microbial consortiums as distributed metabolic networks. Biol. Bull. **204**(2), 174–179 (2003)
44. Vallino, J.J.: Ecosystem biogeochemistry considered as a distributed metabolic network ordered by maximum entropy production. Philos. Trans. Royal Soc. B: Biol. Sci. **365**(1545), 1417–1427 (2010). doi:10.1098/rstb.2009.0272
45. Vallino, J.J.: Differences and implications in biogeochemistry from maximizing entropy production locally versus globally. Earth Syst. Dynam. **2**(1), 69–85 (2011). doi:10.5194/esd-2-69-2011
46. Wang, J., Bras, R.L.: A model of evapotranspiration based on the theory of maximum entropy production. Water Resour. Res. **47**(3 (W03521)), 1–10 (2011) doi:10.1029/2010WR009392

Chapter 19
Maximum Entropy Production and Maximum Shannon Entropy as Germane Principles for the Evolution of Enzyme Kinetics

Andrej Dobovišek, Paško Županović, Milan Brumen and Davor Juretić

Abstract There have been many attempts to use optimization approaches to study the biological evolution of enzyme kinetics. Our basic assumption here is that the biological evolution of catalytic cycle fluxes between enzyme internal functional states is accompanied by increased entropy production of the fluxes and increased Shannon information entropy of the states. We use simplified models of enzyme catalytic cycles and bioenergetically important free-energy transduction cycles to examine the extent to which this assumption agrees with experimental data. We also discuss the relevance of Prigogine's minimal entropy production theorem to biological evolution.

19.1 Introduction

During biological evolution, the earliest cells already contained complex macromolecules with a remarkable capacity to speed up chemical reactions. These macromolecules are proteins, called enzymes. The kinetic and structural properties of enzymes are outcomes of evolution. Are present-day enzymes optimized by evolution, and if so, what physical or statistical principles govern their optimization?

A. Dobovišek (✉) · M. Brumen
Natural Sciences and Mathematics, Medicine, and Health Sciences, University of Maribor, Slomškov trg 15, SI-2000 Maribor, Slovenia
e-mail: andrej.dobovisek@gmail.com

P. Županović · D. Juretić
Department of Physics, Faculty of Science, University of Split, Teslina 12, 21000 Split, Croatia

M. Brumen
Jožef Stefan Institute, Jamova 39, SI-1000 Ljubljana, Slovenia

R. C. Dewar et al. (eds.), *Beyond the Second Law*,
Understanding Complex Systems, DOI: 10.1007/978-3-642-40154-1_19,

These are the questions that will be considered in this chapter, within the broader picture connecting enzymes to the energy conversions that are crucial to life.

As a rule, biological processes are non-linear and take place far from the equilibrium. They require an input of free energy to operate and hence do not support blockages in the hierarchy of free-energy conversion. Energy conversion (also called transduction) is thus the central concept in bioenergetics. Many free-energy transducers are integral membrane proteins, i.e. proteins embedded in biological membranes. For example, inner mitochondrial membranes and chloroplast thylakoid membranes convert redox and light energy, respectively, into a trans-membrane electrochemical potential difference. In this way, free-energy input from photons or from carbohydrates is converted into a "user-friendly" form, as gradients in ion concentration and potential, which cells or organelles can then use at their convenience. This primary step in energy conversion is performed by specialized membrane proteins that act as ion pumps.

The next steps in the hierarchy of free-energy transduction are also performed by integral membrane proteins, which use electrochemical potential gradients to drive the synthesis of ATP from ADP. ATP molecules are converted back to ADP in living cells as soon as they are created, the free energy thus generated being used to drive all the subsequent steps of free-energy transduction involved in enzyme kinetics and biochemistry. A dynamic equilibrium is established such that sustained high ATP and low ADP concentrations maintain cells far from thermochemical equilibrium.

Energy transduction is also a central concept in physics, from the energy conservation principle to thermodynamics. One of the better known results from non-equilibrium thermodynamics is Prigogine's theorem of minimal entropy production [1]. It is a simple consequence [2] of Onsager's linear relationships between fluxes and forces [3, 4] valid close to thermodynamic equilibrium. The theorem defines a non-equilibrium stationary state, called the static head state. Non-equilibrium stationary states are the main interest to us here. Free-energy transduction and efficiency are zero in the static head stationary state, which can be considered as the closest non-equilibrium relative of the equilibrium state. Coupling downhill and uphill free energy changes is essential for all life, but this is impossible in the static head state. Life must look to other non-equilibrium steady states with a non-vanishing efficiency of free-energy transduction.

Elevating Prigogine's theorem to a general principle [5], or fundamental law of nature relevant for biochemical processes [6], has been criticized from a mathematical stance [7]. Technically, the differentiation of the entropy production expression in its most general form, and with all constraints taken into account, should be done first, before introducing the assumption of being close to equilibrium. This procedure does not lead to Prigogine's principle.

An excellent test case is photosynthesis. Bacterial and chloroplast photosynthesis is responsible for free-energy transduction on Earth that is much more intensive than that of a comparable Sun volume [8]. High free-energy-transduction in macroscopic systems is closely associated with high entropy production, a convenient measure of how far such systems are from thermodynamic equilibrium.

By assuming or requiring minimum entropy production in photosynthesis (i.e. by assuming that photosynthesis operates close to equilibrium), Andriesse and Hollestelle [9] introduced the mathematical error noted by Ross and Vlad [7]. Closer inspection of the mathematical restrictions used by Andriesse [10] revealed that the close-to-equilibrium state assumed by Andriesse and Hollestelle [9] cannot be different from thermodynamic equilibrium, where entropy production is zero in accordance with the second law of thermodynamics. The claim that photosynthesis is associated with negative entropy production, and that photosynthesis violates the second law [11], has been amply criticized elsewhere [12, 13] and will not be discussed here.

The application of other optimization principles to enzyme kinetics, such as the maximal metabolic flux principle [14], has two problems besides lacking a basis in physics. First, if one tries to apply it to branched biochemical networks, it is not clear which flux should be maximized. Therefore, its application is restricted to the cyclic processes [14–17]. Secondly, the maximum flux principle involves no trade-off between fluxes and forces. As a result, an infinite flux could be predicted [18]. To exclude such a non-physical situation, one is then forced to introduce various ad hoc flux limitations [14].

The maximal metabolic flux principle is also challenged by the observation that most metabolic enzymes operate far below their maximal capacities in vivo [19] (where capacities are defined as the upper limits of flux rates). In principle, capacities should match maximum loads, so that an optimality hypothesis can be proposed [20] according to which natural selection eliminates excess capacities. This ad hoc and non-physical optimization principle is also challenged by the observation that excess capacities are also ubiquitous in vertebrate bodies [21]. The entrance of system biology into this field contributed to the realization that distributed control and conservation of optimal metabolic flux within given external constraints are much more important objectives for natural selection than maximal metabolic flux [22, 23] or maximal catalytic efficiency (the "perfect enzyme theory" proposed by Albery and Knowles in 1976 [15]). However, this development still leaves open the possibility of a role for general physical principles governing inorganic and organic matter alike.

In this chapter we analyze kinetic schemes common to biochemistry (Michaelis-Menten kinetics) and bioenergetics (photosynthesis, ATP synthesis). Our aim is to compare the predictions of the maximum entropy production (MaxEP) and maximum Shannon entropy (MaxEnt) principles with experimental data. MaxEnt is a powerful statistical principle that, according to Jaynes [24], predicts the best unbiased distribution of probabilities (see also Dewar and Maritan Chap. 3 [25]). For enzyme kinetics and bioenergetics, MaxEnt pertains to the probability of occupation of molecular states, while MaxEP pertains to the fluxes and corresponding thermodynamic forces. Taking all constraints into account (such as the generalized Kirchhoff's laws for networks of enzyme functional states) MaxEnt predicts a maximally-diverse occupancy distribution in a stationary non-equilibrium system, which must be based on a complex and flexible enzyme structure. It has been proposed that MaxEnt provides a theoretical basis for MaxEP

[25], in which case MaxEP, like MaxEnt, is also a statistical principle which can serve as the selection criterion for the most probable states and transition fluxes between states under given constraints. In the majority of stationary states with known constraints considered here, we find qualitative agreement (within an order of magnitude or better) between MaxEP and MaxEnt predictions for forward kinetic constants and their measured values.

On the basis of this agreement, we suggest that physical and biological evolution can be considered from a common perspective. They are coupled such that the increase in entropy export associated with the evolution of biological macromolecules and bioenergetic processes also serves to increase the probability of functionally important states. In other words, biological evolution serves to accelerate the thermodynamic evolution of the system and its environment.

The remainder of this chapter is structured as follows. Section 19.2 extends a recent study [26] by presenting new results for the application of MaxEnt to generalised Michaelis-Menten kinetics. Section 19.3 focuses on the light-driven bacteriorhodopsin photocycle, from the viewpoint of MaxEnt and MaxEP. Section 19.4 applies MaxEnt and MaxEP to a kinetic model for chloroplast ATP-synthase, to predict a key transition state parameter. The Discussion brings these results together and offers a new statistical interpretation for the evolution of enzyme kinetics and bioenergetics.

19.2 Entropy Production and Shannon Entropy for Reversible Michaelis-Menten Kinetics

In this section we use a simple single-substrate model to suggest that biological evolution is accompanied by an increase in the entropy production of the internal transition of an enzyme reaction, as well as an increase in the Shannon information entropy of the entire enzyme reaction scheme.

19.2.1 General Considerations and Basic Assumptions

The simplest generally accepted model for basic one-substrate enzyme reactions is the Michaelis-Menten mechanism. Following its introduction 100 years ago, it is now a standard feature in all textbooks on biochemistry and enzyme kinetics. Here we consider the generalized three-state reversible enzyme reaction scheme shown in Fig. 19.1. Forward and backward transitions occur between the free enzyme molecule (E), the enzyme molecule in complex with a substrate molecule (ES), and the enzyme molecule in complex with a product molecule (EP).

Reactions $E + S \rightarrow ES$ and $E + P \rightarrow EP$ involve binding of a substrate (S) or product (P) molecule to the active site of the enzyme; consequently, their

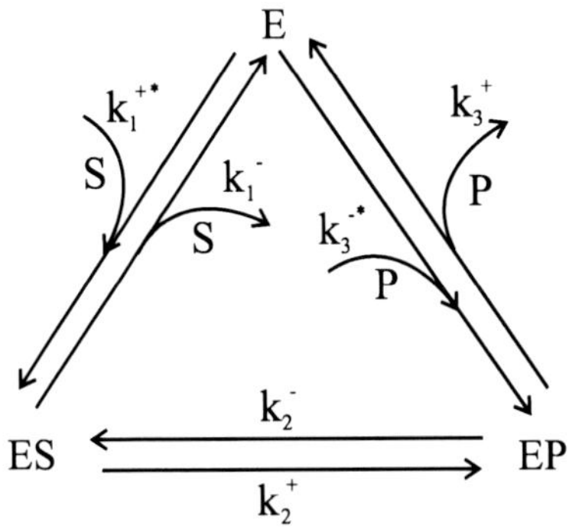

Fig. 19.1 Reversible three-state kinetic scheme for a single-substrate enzyme reaction. State 1 = free enzyme (E), state 2 = enzyme—substrate complex (ES), state 3 = enzyme—product complex (EP), substrate (S), product (P). k_1^{+*} and k_3^{-*} are second order rate constants and k_1^-, k_2^+, k_2^-, k_3^+ are first order rate constants

metabolic fluxes depend on the substrate (S) and product (P) concentrations. Both reaction steps are thus parameterised by second order rate constants k_1^{+*} and k_3^{-*} measured in units $s^{-1}\mu M^{-1}$ (see Fig. 19.1). Quasi first order rate constants can then be defined for these reactions by $k_1^+ = k_1^{+*}$ $[S]$ and $k_3^- = k_3^{-*}$ $[P]$ (s^{-1}). Metabolic fluxes of the other reactions in Fig. 19.1 are independent of $[S]$ and $[P]$ and so can be described by first order rate constants (k_1^-, k_2^+, k_2^- and k_3^+, s^{-1}). Equilibrium constants K_i for individual transitions are defined as the ratios of the forward and backward rate constants. Thus the equilibrium constants for transitions ES↔EP, E + S↔ES and E + P↔EP are $K_2 = k_2^+/k_2^-$, $K_1^* = k_1^{+*}/k_1^-$ and $K_3^* = k_3^+/k_3^{-*}$, respectively. In the latter two cases, the equilibrium constants are denoted with an asterisk, because they involve the second order rate constants k_1^{+*} and k_3^{-*}. For any steady state, the product of the equilibrium constants defines the overall thermodynamic force (or affinity) X [27]. For the equilibrium state, the overall equilibrium constant is the ratio of the equilibrium product and substrate concentrations ($[P]_e$ and $[S]_e$ respectively) and can also be expressed as the product of transition equilibrium constants: $K_{\text{overall}}^{\text{eq}} = [P]_e/[S]_e = K_1^* K_2\, K_3{}^*$. In Fig. 19.1, the transition ES↔EP is the internal enzyme transition.

Enzyme reactions involve metabolic fluxes (J) and thermodynamic forces (X) that govern these fluxes. The associated entropy production rate is defined as the product of the metabolic flux and the corresponding thermodynamic force, divided by absolute temperature (T)

$$\sigma = \frac{JX}{T}. \qquad (19.2.1)$$

For the three-state model (Fig. 19.1), J is the net flux of any given transition, because there is only one cycle and only one flux (which must be the same for all transitions in accordance with Kirchhoff's junction rule). For given substrate and product concentrations, the total thermodynamic force of the overall reaction is a constant. The sum of the affinities (i.e. thermodynamic forces) associated with chosen transitions equals the total thermodynamic force X (in accordance with Kirchhoff's loop rule). One of our basic results from previous research [26] shows that there is a unique maximum for the entropy production of any given transition with respect to variation in its forward rate constant. This is because the associated transition flux and affinity are, respectively, monotonically increasing and decreasing functions of the forward rate constant. In other words, there is a simple trade-off between thermodynamic flux and force.

The Shannon information entropy of the enzyme model in Fig. 19.1 is defined as

$$H = -\sum_{i=1}^{3} p_{\mathrm{i}} \ln(p_{\mathrm{i}}), \tag{19.2.2}$$

where p_{i} (i = 1…3) are probabilities that the enzyme is found in one of its functional states (E, ES or EP, respectively). These probabilities are given by:

$$p_i = \frac{[X_{\mathrm{i}}]}{[E]_{\mathrm{t}}} \tag{19.2.3}$$

where $[E]_{\mathrm{t}}$ is the total enzyme concentration and $[X_{\mathrm{i}}]$ are the concentrations of the enzyme species E, ES or EP (for i = 1, 2, 3, respectively). It follows that $\Sigma p_{\mathrm{i}} = 1$. The p_{i} are thus functions of the rate constants and of the concentrations $[S]$ and $[P]$. Pettersson [16] showed that, at the early stages of evolution, rate constant of the reaction ES → E + S was very high, suggesting that early enzymes were most probably free molecules. In terms of the probabilities p_{i} (i = 1…3) this means that $p_1 \cong 1$ and $p_2 = p_3 \cong 0$. According to Eq. (19.2.2) the associated Shannon information entropy of the entire reaction is close to zero. Pettersson [16] also claims that in early evolution, the metabolic flux of the internal transition ES → EP was very low due to the low value of its forward reaction constant, k_2^+. In view of the above trade-off between flux and force, this means that the low metabolic flux of this transition was associated with a high thermodynamic force (i.e. high affinity).

This situation would undoubtedly have been very unfavourable for enzyme catalysis. In such a situation, progress in enzyme evolution would have occurred through the spontaneous evolutionary increase in the value of the forward rate constant k_2^+. This would have caused an increase in the transition flux of reaction ES → EP (Eq. 19.2.4) and an increase in the associated entropy production, (until the associated affinity had decreased too much). Because the Shannon information entropy was close to zero at the early stages of evolution, and because the p_{i} are functions of k_2^+, an evolutionary increase in k_2^+ would have increased the Shannon information entropy. In other words, due to the increase in k_2^+ other functional enzyme states could now also be occupied with higher probability.

This suggests that the reaction ES → EP could have been an important link in the entire enzyme reaction, as one of the main targets for evolutionary change.

Another reason for the evolutionary importance of the ES → EP transition is that, in contrast to the other two transitions, ES → EP does not depend on the substrate or product concentrations. Since this transition involves the enzyme-mediated decrease in activation energy for the formation of ES and for the emergence of EP, it is expected to be the most sensitive to evolutionary pressure. Therefore we have applied optimization principles to the ES → EP transition, focusing on optimization of the forward rate constant k_2^+ (although other rate constants are probably appropriate candidates for optimization too).

It is important to note that enzyme reaction rate constants are not independent. An increase in k_2^+, as demanded by evolutionary pressure, must be followed by changes in other rate constants. However, those changes are subject to certain constraints. Based on previous work, we assumed the following key constraints:

Assumption 1 As mentioned above, $K_{\text{overall}}^{\text{eq}} = [P]_e/[S]_e = K_1^* K_2 K_3^*$. Because the transition equilibrium constants can be further expressed by forward and backward rate constants, a given set of kinetic parameters has to meet the same equilibrium condition. Thus, in any steady state with constant non-equilibrium concentrations of product [P] and substrate [S], $K_{\text{overall}}^{\text{eq}} = K_1^* K_2 K_3^*$ is a fixed parameter, independent of the catalytic properties of the enzyme [14].

Assumption 2 The second order rate constants k_1^{+*} and k_3^{-*} are not under evolutionary pressure since their values are limited by the diffusion of substrate and product molecules to the active site of the enzyme. The maximal values of these two rate constants are usually taken as equal, i.e. $k_3^{-*} = k_1^{+*} = \text{constant}$ [14].

Assumption 3 From assumption 2, only first order rate constants are subject to evolution. Here we focused on the optimization of the internal enzyme transition, taking k_1^- and k_3^+ as fixed constants [26].

Assumptions 2 and 3 imply that the equilibrium constants K_1^* and K_3^* are fixed parameters. It then follows from assumption 1 that the equilibrium constant of the internal transition K_2 is also fixed. Overall, these assumptions imply that only the rate constants k_2^+ or k_2^- can be targets of evolution. We chose k_2^+ as the variable to be optimized, with k_2^- expressed as $k_2^- = k_2^+/K_2$. The entropy production of the internal enzyme transition and the Shannon information entropy of the entire reaction can then be calculated as functions of the forward rate constant k_2^+ (see Sect. 19.2.2).

We did this for the three-state reversible enzyme reaction scheme of β-Lactamase enzymes. β-Lactamase enzymes are interesting for three reasons. Firstly, experimental kinetic studies suggest that these enzymes are almost fully evolved. Secondly, the kinetics of these enzymes is well known. Thirdly, β-Lactamase enzymes play an important role in the resistance of pathogens to β-lactam antibiotics [17]. Both MaxEP and MaxEnt principles were used to assess whether the enzyme has approached a fully evolved state. However, fully evolved enzymes are rare in nature, if they exist at all, and we do not expect to find more than an order of magnitude agreement between predicted optimal rate constants and measured rate constants.

19.2.2 Entropy Production and Shannon Information Entropy in the Three-State Enzyme Kinetic Scheme

Here we calculate the entropy production of the internal enzyme transition ES↔EP, and the Shannon entropy of the entire reaction, as functions of the forward rate constant k_2^+. The net metabolic flux for ES↔EP is

$$J \equiv \frac{d[P]}{dt} = k_2^+[ES] - k_2^-[EP] \tag{19.2.4}$$

and the corresponding thermodynamic force (or affinity) is the difference in chemical potentials between states ES and EP,

$$A_{23} = RT \ln\left(K_2 \frac{[ES]}{[EP]}\right) \tag{19.2.5}$$

where R is the gas constant and K_2 is the equilibrium constant for ES↔EP. While the overall thermodynamic force X is assumed to be constant, the affinity (19.2.5) is a function of the concentrations [ES] and [EP].

The concentrations $[E]$, $[ES]$ and $[EP]$ are calculated from the enzyme kinetic equations

$$\frac{d[E]}{dt} = -k_1^{+*}[S][E] + k_1^-[ES] - k_3^{-*}[P][E] + k_3^+[EP] \tag{19.2.6}$$

$$\frac{d[ES]}{dt} = k_1^{+*}[S][E] - k_1^-[ES] - k_2^+[ES] + k_2^-[EP] \tag{19.2.7}$$

$$\frac{d[EP]}{dt} = k_2^+[ES] - k_2^-[EP] - k_3^+[EP] + k_3^{-*}[P][E]. \tag{19.2.8}$$

We assume that the enzyme reaction takes place in a non-equilibrium stationary state with fixed, known concentrations of $[S]$ and $[P]$. Summing Eqs. (19.2.6)–(19.2.8) implies that the total enzyme concentration $[E]_t$ is constant, i.e. $d[E]_t/dt = 0$ where

$$[E]_t = [E] + [ES] + [EP]. \tag{19.2.9}$$

In this stationary state, the time derivatives on the left-hand sides of Eqs. (19.2.6)–(19.2.8) are then zero. Equations (19.2.6)–(19.2.8) then form a system of three equations in the three unknown stationary concentrations $[E]$, $[ES]$ and $[EP]$, whose solutions can be expressed in terms of $[S]$, $[P]$, $[E]_t$ and all the rate constants. In these equations all parameters except k_2^+ and k_2^- are fixed. Since the equilibrium constant K_2 is also a fixed parameter, the rate constant k_2^- is given by $k_2^- = K_2/k_2^+$. Thus, the solutions depend only on the forward rate constant k_2^+, and are given by

$$[E] = [E]_t \frac{Ak_2^+ + B}{Gk_2^+ + H} \tag{19.2.10}$$

$$[ES] = [E]_t \frac{Ck_2^+ + D}{Gk_2^+ + H} \tag{19.2.11}$$

$$[EP] = [E]_t \frac{Ek_2^+ + F}{Gk_2^+ + H} \tag{19.2.12}$$

where $A = k_3^+ + k_1^-/K_2$, $B = k_3^+ k_1^-$, $C = (k_1^{+*}[\mathrm{S}] + k_3^{-*}[\mathrm{P}])/K_2$, $\mathrm{D} = k_3^+ k_1^{+*}[\mathrm{S}]$, $E = CK_2$, $F = k_1^- k_3^{-*}[\mathrm{P}]$, $G = A + C + E$, $H = B + D + F$.

The transition flux (J) and affinity (A_{23}) as functions of k_2^+ are then obtained by substituting (19.2.11) and (19.2.12) into (19.2.4) and (19.2.5), and the associated entropy production is then given from (19.2.1) by

$$\sigma(k_2^+) = R[E]_t \frac{k_2^+(DK_2 - F)}{K_2(Gk_2^+ + H)} \ln\left(K_2 \frac{Ck_2^+ + D}{Ek_2^+ + F}\right). \tag{19.2.13}$$

Likewise, the Shannon information entropy as a function of k_2^+ is obtained from Eqs. (19.2.2), (19.2.3) and (19.2.10)–(19.2.12), giving

$$H(k_2^+) = -\left(\frac{1}{k_2^+ G + H}\right)\left((k_2^+ A + B)\ln\left(\frac{k_2^+ A + B}{k_2^+ G + H}\right) + (k_2^+ C + D)\ln\left(\frac{k_2^+ C + D}{k_2^+ G + H}\right) + (k_2^+ E + F)\ln\left(\frac{k_2^+ E + F}{k_2^+ G + H}\right)\right). \tag{19.2.14}$$

The optimal values of the forward rate constant k_2^+ predicted by MaxEP and MaxEnt are then obtained from the conditions

$$\frac{d\sigma}{dk_2^+} = 0 \tag{19.2.15}$$

$$\frac{dH}{dk_2^+} = 0, \tag{19.2.16}$$

respectively.

19.2.3 Comparison with Experimental Results

In this section we compare experimental values of the forward rate constants k_2^+ for three types of β-Lactamase enzymes [17] with the optimal values predicted by MaxEP and MaxEnt. Here we assume $[E]_t = 1\ \mu\mathrm{M}$ and $[S] = 1{,}500\ \mu\mathrm{M}$ for all three types of β-Lactamase enzymes. Other parameters are given in Table 19.1.

Table 19.1 Model parameters values used for three types of β-Lactamase enzymes

Enzyme	$k_1^{+*} = k_3^{-*}$ [μ $M^{-1}s^{-1}$]	k_1^- [s^{-1}]	k_2^+ [s^{-1}]	k_2^- [s^{-1}]	k_3^+ [s^{-1}]	K_1^* [μ M^{-1}]	K_2	K_3^* [μ M]	$[P]$ [μ M]
PC1 β-Lactamase	22	196	173	1.3	96	0.11	133	4.4	7.9
RTEM β-Lactamase	123	11,800	2,800	47	1,500	0.01	59	12	23
β-Lactamase I	41	2,320	4,090	141	3,610	0.018	29	88	95

Table 19.2 The comparison of experimental and predicted values of the forward rate constants k_2^+ for three types of β-Lactamase enzymes

Enzyme	k_2^+ [s^{-1}] (MaxEP)	k_2^+ [s^{-1}] (MaxEnt)	k_2^+ [s^{-1}] (Observed)
PC1—β-Lactamase	281	94.5	173
RTEM β-Lactamase	4,034	1,091	2,800
β-Lactamase I	6,669	3,548	4,090

Previous analysis indicated that all β-Lactamase reactions take place far from equilibrium [26].

The comparison of observed and predicted values of k_2^+ is presented in Table 19.2.

The values of k_2^+ predicted by MaxEP and MaxEnt are of the same order of magnitude as the observed values. In all cases, observed values of k_2^+ are higher than those predicted by MaxEnt and lower than those predicted by MaxEP. These results are reasonable given that the model neglects interactions between molecules, while the Michaelis-Menten mechanism assumes very dilute solutions and neglects macromolecular crowding and the gel-like state of water near hydrophilic surfaces in living cells. These effects can significantly alter enzyme reaction rates, as well as association and dissociation rates.

Our results using the parameter values in Table 19.2 also show that an evolutionary tendency towards maximal entropy production (σ), simulated by an increase in the forward rate constant k_2^+, would result in state probabilities $p_2 = p_3 \cong 0.5$ and $p_1 \cong 0$, consistent with theoretical results [15] showing that, for nearly fully evolved enzymes, the condition $k_2^+ = k_3^+$ holds. Based on this result, Christensen [17] also concluded that β-Lactamase enzymes are nearly fully evolved enzymes.

Thus our theoretical predictions from MaxEnt are consistent with existing theoretical and experimental studies. In a fully evolved enzyme, one might argue that all functional states should be occupied with the same probability. There is no reason why evolution would favour one functional state over another, because this would limit the reaction rate. All reaction steps should be equally fast and this would be achieved by a uniform probability distribution in a fully evolved state. The Shannon information entropy H attains its maximum when probability values are equal ($p_1 = p_2 = p_3 = 1/3$) for the condition $k_1^*[S] = k_2^+ = k_3^+$. When constraints are introduced (such as experimentally measured rate or equilibrium constants) the optimal rate constants lead to lower maximal H or σ values.

19.3 The Five-State Model for Bacteriorhodopsin Photosynthesis

In this section we explore whether MaxEnt and MaxEP can be applied to bioenergetics at the next level of complexity beyond enzymes in solution. Most enzymes important in free-energy transduction are integral membrane proteins embedded in topologically closed membranes. Such enzymes can be conveniently classified into primary and secondary proton pumps. The former can be involved in respiration or photosynthesis, while the latter (e.g. ATP synthase) use the proton motive force created by the primary pumps to drive ATP synthesis. In this section we consider a proton pump involved in photosynthesis, while ATP synthase will be considered in Sect. 19.4.

Firstly, we address the question of whether photosynthesis is open to thermodynamic analysis. Some scientists have answered in the negative [27] and have even claimed that photosynthesis violates the second law of thermodynamics [11]. In contrast, Meszena and Westerhoff [28] developed a thermodynamic description of the light absorption transition during photosynthesis. We have also pointed out that the separation of the light and dark reactions is an essential step [29] which allows the concepts from enzyme kinetics and irreversible thermodynamics of the previous section to be applied to thermodynamic analysis of the photocycle here.

Bacteriorhodopsin (bR) from *Halobacterium salinarium* is probably the simplest photosynthetic system where free-energy transduction theory and experiments may meet in the future. However, there is still a lack of consensus as to the best kinetic model for the bR-photocycle [30, 31]. Here we used the five-state kinetic model [29] also used by Juretić and Westerhoff [32].

There are seven possible transitions and six cycles in this model, but only four internal transitions (with forward rate constants k_1–k_4) are directly connected with the proton uptake and export responsible for generating proton motive force (Fig. 19.2, bold arrows). Stationary state probabilities, state transition fluxes J and corresponding affinities A are all functions of the rate constants and may be calculated using Hill's diagram method [27]. Only the forward rate constants k_1–k_4 are optimized. The application of the generalized Kirchhoff's laws for enzyme kinetic schemes (junction and loop rules) gives relationships between fluxes and affinities. The latter are related to equilibrium constants. Among the affinities, we may single out the primary and secondary force. Kirchhoff's loop law applied to the light-activated cycle of excitation and direct relaxation gives the primary force

$$X_{\text{prim}} = RT \ln\left(\frac{\alpha_{01} k_d}{\alpha_{10} k_{-d}}\right) \tag{19.3.1}$$

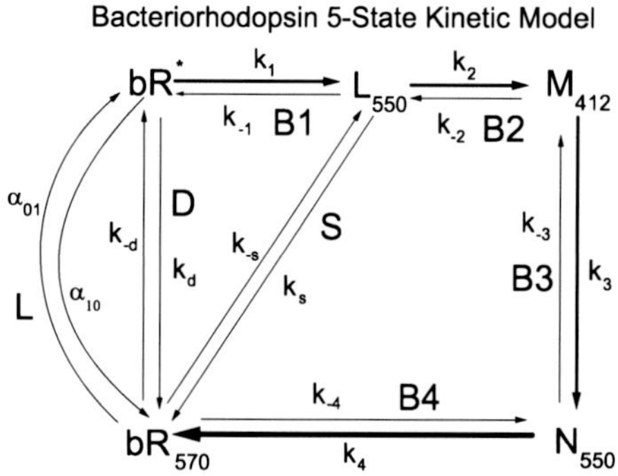

Fig. 19.2 The five-state kinetic model for the bacteriorhodopsin photocycle. The five states are spectroscopic states bR570 (the *ground state*), bR* (*excited state*), L550, M412 and N550. Proton transfer is assumed to take place through the B2 transition (proton release to the extracellular space) and B4 transition (proton absorption from the cytoplasmic space). The slip transition S is a non-productive transition

where α and k are the relevant rate constants as shown in Fig. 19.2. Bacteriorhodopsin is a light-driven proton pump, which transfers protons from the space of low proton concentration (cytoplasmic side of membrane) to the space of high proton concentration (extracellular side), thereby creating the proton motive (secondary) force X_{sec}. Kirchhoff's loop rule applied to the cycle containing proton release and proton uptake transitions gives this as

$$X_{sec} = \Delta\mu_{H^+} = F\Delta\Psi + RT\ln\left(\left[H^+_{in}\right]/\left[H^+_{out}\right]\right), \tag{19.3.2}$$

where F is the Faraday constant, while $\Delta\Psi$ is the transmembrane potential. We fix the ratio between photon energy and external temperature at $h\nu/k_BT = 84.67$ and the secondary force at either the commonly measured value $X_{sec} = -18.84$ kJ mol^{-1} (-195 mV) or the maximal observed [33] value $X_{sec} = -26.82$ kJ mol^{-1} (-278 mV). Equilibrium constants in productive transitions B1 to B4 (Fig. 19.2) are assumed to be high so that a forward net flux for these transitions is preferred.

The optimization procedure consisted in searching for models that have free-energy transduction efficiency greater than 10 % and maximal entropy production with respect to each of the four forward rate constants k_1–k_4. The optimal values of these rate constants are found iteratively. When the lower value of proton motive force was used ($X_{sec} = -195$ mV), the optimal rate constants from seven optimized kinetic models (Table 2 in the [29]) fell within the following ranges: k_1 from 10^9 to 10^{10} s^{-1}, k_2 from 10^3 to 10^7 s^{-1}, k_3 from 10^2 to 10^3 s^{-1} and k_4 from 10^2 to 10^3 s^{-1}. The light absorption rate in these models varied from $\alpha_{01} = 20$ to 100 s^{-1}, while the slip rate constant (for unproductive return to the ground state) varied from 10^2 to 10^6 s^{-1}. Due to the non-linear flux-force relationships (Eqs. 19.3.1 and 19.3.2), and in spite of the slip transition presence, the affinity transfer efficiency and quantum yield had high optimal values of >95 and >89 %, respectively, in all of the seven models. This result implies that most (about 90 %) of the thermodynamic force and flux is channelled through the charge separation pathway, representing an enormous advantage for the non-linear mode of operation, which overcomes the impedance matching requirement derived from the maximum power transfer theorem [34]. The latter theorem restricts the affinity (power) transfer efficiency to 50 % in the linear mode of operation.

In order to calculate the affinity transfer efficiency and quantum yield, we used the chemical analogue of Kirchhoff's laws for the enzyme "circuit" of functionally important states. This was done first for a three-state toy model for photosynthesis [29] consisting of ground state (P), excited state (P*), and charge separated state (P^+) connected with light-activated transition (L) between states P and P*, non-productive relaxation (D) between states P* and P, and charge separation transitions B1 and B2 between states P*–P^+ and states P^+–P respectively. From [29], the corrected Eq. (22) for total entropy production σ of the three-state model reads:

$$\sigma T = A_{PP^*}(L)J(L) + A_{P^*P}(D)J(D) + A_{P^*P^+}(B1)J(B1) + A_{P^+P}(B2)J(B2)$$

The corrected[1] Eq. (24) from [29] are applications of the Kirchhoff's loop rule to:

a. the "light-cycle" with "light-force" X_L [28], and the maximal free-energy A_{oc} that the system can absorb while in chemical equilibrium with radiation at higher effective temperature:

$$A_{P^*P}(D) = A_{oc} - X_L$$

b. the charge separation (productive) pathway where the secondary force X_{sec} is generated:

$$A_{P^*P^+}(B1) + A_{P^+P}(B2) - A_{P^*P}(D) = X_{sec}$$

The affinity transfer efficiency is defined as:

$$A_{P^*P}(D)/A_{oc} = 1 - X_L/A_{oc}$$

in which $A_{P*P}(D)$ is the sum of the affinities in the charge separation pathway. The quantum yield is calculated as J(B2)/J(L) in the three-state model, and as J(B4)/J(L) in the five-state model.

Kirchhoff's rules may be applied similarly to the five-state model (Fig. 19.2). Our five-state model for the bacteriorhodopsin photosynthesis can be compared with the similar kinetic model bR* → L→M → O→bR that also has four productive transitions in the fast decay route of the bR-photocycle [35]. Experimental values for the forward rate constants associated with these four internal transitions depend on the experiments and the models used to interpret them:

[1] Expressions given in [29] involve printing errors.

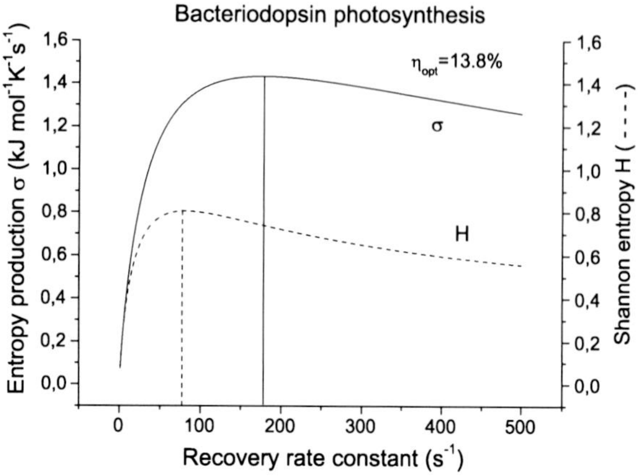

Fig. 19.3 The dependence of the entropy production σ and Shannon's information entropy H on the recovery rate constant k_4. Maximal values for σ and H correspond to optimal k_4 values of 177 and 81 s^{-1} respectively. Other optimal forward constants in this case are $k_1 = 3.32 \times 10^9\ s^{-1}$, $k_2 = 9.2 \times 10^3\ s^{-1}$ and $k_3 = 1{,}570\ s^{-1}$

$(k_1)_{exp} = 1.67 \times 10^5\ s^{-1}$, $(k_2)_{exp} = 3.03 \times 10^4\ s^{-1}$, $(k_3)_{exp} = 556\ s^{-1}$, $(k_4)_{exp} = 233\ s^{-1}$.

When the lower value for the proton motive force is used ($X_{sec} = -195$ mV), the optimal rate constants from the optimized kinetic models (Table 2 in the Ref. [29]) span the range of experimental values. The exception is the much shorter time constant for the first forward transition, for which the optimal k_1 lies in the range from 10^9 to $10^{10}\ s^{-1}$ in all such models. However, the relaxation of the excited state to the first spectroscopically detectable subsequent state (K_{590}) indeed happens in nanoseconds or picoseconds [36] rather than microseconds.

A maximal proton motive force of about 280 mV has been achieved in experiments [33]. When this value is used ($X_{sec} = -278$ mV) in the optimized model of bacteriorhodopsin light cycle (Fig. 19.3), we obtain an optimal overall free-energy transduction efficiency that exceeds 13 %. As in most models of the bacteriorhodopsin photocycle [37], here we have assumed that the first and the last relaxation steps are essentially irreversible, with high equilibrium constants.

The optimal free-energy transduction efficiency was calculated as:

$$\eta_{opt} = -\frac{X_{sec}J(B4)}{A_{oc}J(L)} \tag{19.3.3}$$

in which the optimal values for the fluxes were used.

The optimal values for the recovery rate constant k_4 predicted by MaxEP and MaxEnt differ by a factor of about two. However, it should be noted that more up-to-date kinetic schemes for bacteriorhodopsin light-activated cycle could also be used (e.g. including the K_{590} spectroscopic state). More important than the comparison of measured and predicted kinetic constants, however, is our MaxEP prediction (see also [29] and the chapter by Juretić and Županović in [38]) of a wide range of forward kinetic constants values in the descending order from $10^9\ s^{-1}$ to about $10^2\ s^{-1}$. This result agrees with experimental observations of the slowing down of the bacteriorhodopsin light cycle in all subsequent steps. We suspect that this MaxEP-predicted hierarchy of optimal kinetic constants will emerge regardless of the kinetic scheme used for bacteriorhodopsin photocycle, although this conjecture remains to be tested.

So far we have applied MaxEP and MaxEnt independently. The question of how to apply these principles simultaneously in bioenergetics is the subject of the next section.

19.4 MaxEnt, MaxEP and the Functional Design of the Rotary Enzyme ATP Synthase

ATP synthase is an important biomolecular nanomotor. From an evolutionary viewpoint it is a very ancient secondary proton pump, which exploits the proton motive force created by respiration or photosynthesis to drive the synthesis of adenosine triphosphate (ATP), the most commonly used "energy currency" in living cells. ATP synthase is embedded in the inner membrane of mitochondria or in the thylakoid membrane of chloroplasts. ATP is formed from adenosine diphosphate (ADP) and inorganic phosphate (P), assuming that activation energy is available. This activation energy is stored and released as elastic energy in the stalk-like axle of the ATP synthase nanomotor. The rotary mechanism is well understood [39]. The stator is an ensemble of three structural subunits. Translocation of protons through this protein, driven by the transmembrane electrochemical proton gradient, is accompanied by a stepped rotation of the stalk-like axle. Each 120° clockwise (or counter-clockwise) rotation is accompanied by the synthesis (or hydrolysis) of ATP. Here we will consider only the ATP synthase of chloroplast thylakoid membranes.

The number of protons translocated through the thylakoid membrane that is necessary for the synthesis of one ATP molecule is called the gearing ratio, $g \equiv H^+/ATP$. The gearing ratio g is related to the free energy E input per revolution,

$$E = 3g\Delta\mu_{\mathrm{H}^+}, \tag{19.4.1}$$

where

$$\Delta\mu_{\mathrm{H}^+} = 2.3RT\Delta pH - F\Delta\Psi \tag{19.4.2}$$

is the transthylakoid proton motive force. F is the Faraday constant, while ΔpH and $\Delta\Psi$ are the transmembrane differences in pH and electric potentials, respectively. The 120° stalk rotation has a short ($\approx$ 2ms) pause, called the catalytic dwell, at a certain relative angular position of stalk, denoted by κ (with $0 \leq \kappa \leq 1$). The catalytic dwell is so-called because it is associated with the internal transition (synthesis or hydrolysis of ATP) of ATP synthase. In accordance with our assumptions (see Sect. 19.2), this internal transition is most sensitive to evolution. Therefore we take κ as the variable that is optimised during evolution. The free energy (E) partly depends on external conditions (the difference between pH

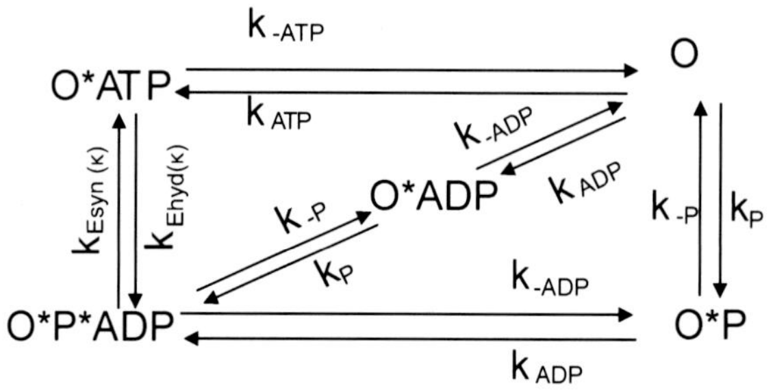

Fig. 19.4 Kinetic model of ATP synthase cycle for transitions between enzyme open (O) states. O*P, O*ADP, O*ATP and O*P*ADP are states which bind P, ADP, ATP, and ADP with P, respectively. Rate constants are expressed in same units when second-order rate constants are multiplied by substrate concentrations: $k_{ATP} = 20.8\ s^{-1}$, $k_{-ATP} = 270\ s^{-1}$, $k_{ADP} = 8{,}900\ s^{-1}$, $k_{-ADP} = 490\ s^{-1}$, $k_P = 810\ s^{-1}$ and $k_{-P} = 2{,}030\ s^{-1}$

factors outside and inside of the membrane), and we take this to be an adjustable parameter as explained below.

We describe the synthesis and hydrolysis of ATP using the five-state kinetic model shown in Fig. 19.4. The problem can be solved analytically, either by solving the steady-state rate equations directly [40] or by using Hill's diagram method [27]. Using experimental data obtained by Pänke and Rumberg [40, 41], we calculated the state probabilities $p_E(i|\kappa)$, the forward fluxes,

$$J_{E+}(\kappa) = k_{Esyn}(\kappa) p(O^* P^* ADP|\kappa)_E \tag{19.4.3}$$

and the backward fluxes

$$J_{E-}(\kappa) = k_{Ehyd}(\kappa) p(O^* ATP|\kappa)_E \tag{19.4.4}$$

Rate coefficients $k_{Esyn}(\kappa)$ and $k_{Ehyd}(\kappa)$ are calculated within the transition state theory [42] and are given by

$$k_{Esyn}(\kappa) = k^0_{syn} \exp(\kappa E/3RT), \tag{19.4.5}$$

$$k_{Ehyd}(\kappa)_g = k^0_{hyd} \exp(-(1-\kappa)E/3RT). \tag{19.4.6}$$

The values of specific binding change constants $k^0_{syn} = 1.15 \times 10^{-3} s^{-1}$ and $k^0_{hyd} = 4.5 \times 10^5 s^{-1}$ are taken from [40, 41]. Under controlled experimental conditions, the enzyme was illuminated in the presence of 1 mM ADP, 1 mM P and 10 μM ATP at $T = 300$ K [40, 41]. Fixed kinetic rate constants are given in the legend of Fig. 19.4.

The number of ATP molecules produced per enzyme per second is then

$$J_E(\kappa) = J_{E+}(\kappa) - J_{E-}(\kappa). \tag{19.4.7}$$

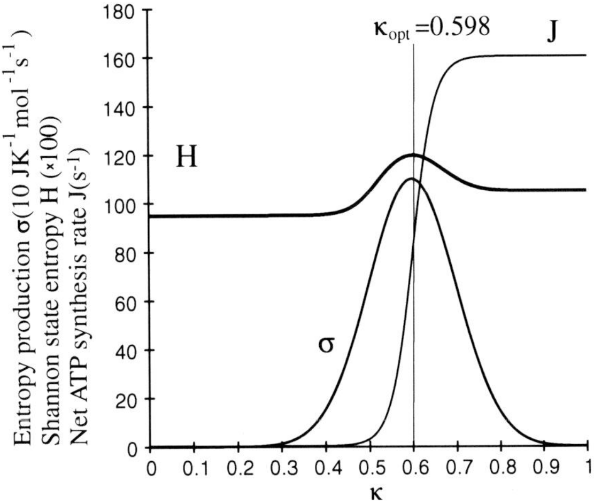

Fig. 19.5 Information entropy of state probabilities, entropy production and net ATP synthesis rate as a function of relative angular position of catalytic dwell at optimal input free energy $E_{opt} = 161.4\ \text{kJmol}^{-1}$

The Shannon information entropy of state probabilities, and entropy production of the internal enzymatic transitions, are

$$H_E(\kappa) = -\sum_{i=1}^{5} p_E(i|\kappa)\log p_E(i|\kappa), \tag{19.4.8}$$

$$\sigma_E(\kappa) = RJ_E(\kappa)\log\frac{J_{E+}(\kappa)}{J_{E-}(\kappa)}, \tag{19.4.9}$$

respectively.

Our hypothesis is that the information entropy $H(\kappa)$ and entropy production $\sigma(\kappa)$ of a fully evolved enzyme are maximized at a common value of κ, the relative angular position of the catalytic dwell. That is,

$$\frac{\partial H_E(\kappa)}{\partial\kappa} = 0, \tag{19.4.10}$$

$$\frac{\partial\sigma_E(\kappa)}{\partial\kappa} = 0. \tag{19.4.11}$$

In order to obtain a solution, we adjust the free energy input E until there is a common value of κ that satisfies both Eqs. (19.4.10) and (19.4.11). In other words, we are simultaneously optimizing κ and E.

The numerical calculations are shown in Fig. 19.5. The solution yields optimal values $\kappa_{opt} = 0.598$ and $E_{opt} = 161.4\,\text{kJ/mol}$. The former value is very close to the empirical estimate $\kappa_{opt} = 0.6$ [40]. From Eq. (19.4.1), $E_{opt} = 161.4\,\text{kJ/mol}$ corresponds to an optimal gearing ratio equal to the observed value $g = 4$ in chloroplasts [43] and to optimal proton motive force of $\Delta\mu_{H^+} = 13.4\,\text{kJ/mol}$ (2.43 pH difference equivalent) when information entropy is maximal with respect to $\Delta\mu_{H^+}$ too [42].

In summary, our calculations show that ATP synthase is a fully evolved enzyme in the sense of MaxEnt and MaxEP. It is also interesting that the optimal solution of MaxEnt and MaxEP coincides with an inflection point of the curve of ATP synthesis rate (J_E) versus both proton motive force and κ (Fig. 19.5 and [42]); this feature allows fast metabolic control with respect to short-term changes in proton motive force, as well as a high optimal output/input free energy ratio of 69 % [42].

MaxEnt is a powerful inference algorithm for solving problems with incomplete available information. In physics, the whole of equilibrium statistical physics can be derived from this principle [44, 45]. At first sight, it might seem that biological evolution, by building ever more structurally complex macromolecules (i.e. of low configurational entropy), has proceeded in the direction of entropy decrease rather than entropy increase. But when we look at the functional design of ATP synthase, as we have done here, we find that biological evolution is consistent with MaxEnt. There is no contradiction with the second law. The evolutionary optimization of ATP synthase can be interpreted as selection of the most probable functional design within the constraints considered here.

19.5 Discussion and Conclusions

In this chapter we asked: which optimality principles (if any) govern biological evolution, and in particular the evolution of enzyme kinetics? Might such a selection principle be found in physics? One of the first studies in this direction, by Prigogine and Wiame [46], correctly noted that biological processes are irreversible, and as such should be describable by irreversible thermodynamics. Since irreversible processes are characterised by entropy increase, entropy production was identified as a fundamental quantity in the description of biological processes, leading to the introduction of the concept of dissipative structures [1, 2, 47].

There were many attempts to extend Prigogine's theorem of minimum entropy production (MinEP) [1] from linear non-equilibrium thermodynamics to the realm of biological processes which, as a rule, are nonlinear and take place far from equilibrium [2]. However, Ross and Vlad [7] showed that Prigogine's MinEP principle did not apply far from equilibrium. Nevertheless, the question remains whether some other variational principle involving entropy production applies far from equilibrium. In the context of enzyme catalyzed reactions, we have shown that biological systems conform more closely to MaxEP than MinEP. From a physical stand point, MaxEP also seems more appropriate to the evolution of enzyme kinetics and bioenergetics than other principles such as maximum metabolic flux, because MaxEP takes into account the fact that evolution acts on thermodynamic forces (affinities) as well as fluxes.

One of our key results is that a unique maximum exists in the entropy production of an enzyme transition *ij* such that the associated flux and thermodynamic force are, respectively, increasing and decreasing functions of the forward rate constant k_{ij}^{+} resulting in a trade-off between flux and force.

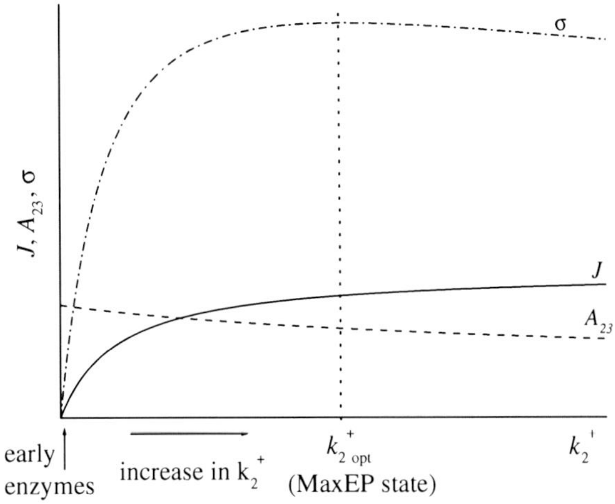

Fig. 19.6 Qualitative representation of metabolic flux J (*solid line*), transition affinity A_{23} (*dashed line*) and entropy production σ (*dash-dot line*) as functions of forward rate constant k_2^+

The fact, that the optimization of entropy production gives a more complete understanding of enzyme evolution than the maximization of metabolic flux, is also reflected mathematically. Specifically, the optimization of entropy production could be performed with a lower number of starting assumptions than the maximization of metabolic flux and thus gives more complete physical information.

The MaxEP principle also provides a specific prediction for the metabolic flux (substrate conversion into product) and transition affinity. This prediction can be understood from Fig. 19.6, where metabolic flux, affinity and entropy production are qualitatively represented as functions of the forward rate constant k_2^+. As discussed in Sect. 19.2.1, in early enzyme reactions low metabolic fluxes were governed by high thermodynamic forces, which does not seem to be beneficial for enzyme kinetics. As Fig. 19.6 also shows, early enzymes were operating in the regime of low entropy production. Spontaneous evolutionary increase in the forward rate constant k_2^+ leads to higher metabolic flux, lower affinity and higher entropy production. According to our hypothesis, evolutionary pressure on enzymes is directed towards the state of maximum entropy production, which is achieved when $k_2^+ = k_{2\text{opt}}^+$ (Fig. 19.6). Our model results presented in Sect. 19.2.3 (Table 19.2) show that β-Lactamase enzymes are indeed very close to the optimal state with maximal entropy production. However, since the metabolic flux (J) and transition affinity (A_{23}) are, respectively, monotonically increasing and decreasing functions of k_2^+ we see that, close to the optimal state of maximum entropy production, J and A_{23} are already close to their maximal and minimal values, respectively, which are achievable during evolution. Higher or lower values of flux and affinity can also be achieved, but only with further substantial increase in k_2^+ (several orders of magnitude) above $k_{2\text{opt}}^+$.

All this leads us to the following hypothesis: "A fully evolved kinetic scheme of an enzyme is characterized by maximal entropy production (MaxEP prediction) for functionally important internal enzyme transitions. A broad distribution of functionally important enzyme states is then established such that Shannon information entropy achieves the maximal value possible within external constraints (MaxEnt prediction)."

The results described in this chapter point to MaxEP as a relevant principle for the evolution of enzyme kinetics, and for biological evolution more generally. The intensive free-energy transduction that occurs during bacterial photosynthesis and respiration is more consistent with MaxEP than MinEP, as discussed in [29]. This has led us to postulate the Evolution Coupling Hypothesis [38] that biological evolution accelerates the thermodynamic evolution of life's environment, and that this holds at all levels of biological organization, from enzymes to biosphere.

The values of the forward rate constant k_2^+ predicted by the MaxEP and MaxEnt principles for β-Lactamase enzymes are approximately equal. For the bacteriorhodopsin kinetic scheme, the MaxEnt- and MaxEP-predicted values for the relaxation constant to the ground state differ by a factor of two (Fig. 19.2) for a proton motive force $X_{\mathrm{sec}} = -278$ mV but by only 5 % for $X_{\mathrm{sec}} = -195$ mV (not shown). Is this mere coincidence or does it reflect a deeper cause? For the case of ATP-synthase, MaxEnt and MaxEP together predict the optimal relative angular position of the catalytic dwell and gearing ratio in remarkable agreement with experiments [42]. It thus seems desirable to extend the application of these two principles to other bioenergetic systems.

Reaction cycles abound in biology and bioenergetics. For example, many types of molecular nanomotors, essential for cell functioning, operate in a cyclic manner, which should be amenable to analysis similar to that presented here; other schemes such as the Krebs cycle may require modification of these methods. In any case, we anticipate that future research will involve the profound theoretical breakthroughs that have been achieved in system biology during the last 20 years [23] in some combination with MaxEP and MaxEnt. In this way we may hope to further elucidate the connection between statistical physics and biological evolution.

References

1. Prigogine, I.: Introduction to Thermodynamics of Irreversible Processes. Wiley, New York (1967)
2. Martyushev, L.M., Seleznev, V.D.: Maximum entropy production principle in physics, chemistry and biology. Phys. Rev. **426**, 1–45 (2006)
3. Onsager, L.: Reciprocal relations in irreversible processes I. Phys. Rev. **37**, 405–426 (1931)
4. Onsager, L.: Reciprocal relations in irreversible processes II. Phys. Rev. **38**, 2265–2279 (1931)
5. Glansdorff, P., Prigogine, I.: Thermodynamic Theory of Structure, Stability and Fluctuations. Wiley, New York (1971)
6. Voet, D., Voet, J.G.: Biochemistry, 3rd edn. Wiley, New York (2004)
7. Ross, J., Vlad, M.O.: Exact solutions for the entropy production rate of several irreversible processes. J. Phys. Chem. A **109**, 10607–10612 (2005)
8. Metzner, H.: Bioelectrochemistry of photosynthesis: a theoretical approach. Bioelectrochem. Bioenerg. **13**, 183–190 (1984)

9. Andriesse, C.D., Hollestelle, M.J.: Minimum entropy production in photosynthesis. Biophys. Chem. **90**, 249–253 (2001)
10. Andriesse, C.D.: On the relation between stellar mass and luminosity. Astrophys. J. **539**, 364–365 (2000)
11. Jennings, R.C., Engelmann, E., Garlaschi, F., Casazza, A.P., Zucchelli, G.: Photosynthesis and negative entropy production. Biochim. Biophys. Acta Bioenerg. **1709**, 251–255 (2005)
12. Lavergne, L.: Commentary on photosynthesis and negative entropy production by Jennings and coworkers. Biochim. Biophys. Acta **1757**, 1453–1459 (2006)
13. Knox, R.S., Parson, W.W.: Entropy production and the second law in photosynthesis. Biochim. Biophys. Acta **1767**, 1189–1193 (2007)
14. Heinrich, R., Schuster, S., Holzhütter, H.G.: Mathematical analysis of enzymic reaction systems using optimization principles. Eur. J. Biochem. **201**, 1–21 (1991)
15. Albery, W.J., Knowles, J.R.: Evolution of enzyme function and the development of catalytic efficiency. Biochemistry **15**, 5631–5640 (1976)
16. Pettersson, G.: Effect of evolution on the kinetic properties of enzymes. Eur. J. Biochem. **184**, 561–566 (1989)
17. Christensen, H., Martin, M.T., Waley, G.: β-lactamases as fully efficient enzymes. Determination of all the rate constants in the acyl-enzyme mechanism. Biochem. J. **266**, 853–861 (1990)
18. Pettersson, G.: Evolutionary optimization of the catalytic efficiency of enzymes. Eur. J. Biochem. **206**, 295–298 (1992)
19. Suarez, R.K., Staples, J.F., Lighton, J.R.B., West, T.G.: Relationships between enzymatic flux capacities and metabolic flux rates in muscles: nonequilibrium reactions in muscle glycolysis. Proc. Natl. Acad. Sci. **94**, 7065–7069 (1997)
20. Taylor, C.R., Weibel, E.R.: Design of the mammalian respiratory system. I. Problem and strategy. Respir. Physiol. **44**, 1–10 (1981)
21. Diamond, J.M.: Evolution of biological safety factors: a cost/benefit analysis. In: Weibel, E.R., Taylor, C.R., Bolis, L.C. (eds.) Principles of Animal Design: The Optimization and Symmorphosis Debate, pp. 21–27. Cambridge University Press, Cambridge (1998)
22. Suarez, R.K., Darveau, C.A.: Multi-level regulation and metabolic scaling. J. Exp. Biol. **208**, 1627–1634 (2005)
23. Reaves, M.L., Rabinowitz, J.L.: Metabolomics in systems microbiology. Curr. Opin. Biotechnol. **22**, 17–25 (2011)
24. Jaynes, E.T.: Probability Theory: The Logic of Science. Cambridge University Press, Cambridge (2003)
25. Dewar, R.C., Maritan, A.: The theoretical basis of maximum entropy production. In: Dewar, R.C., Lineweaver, C.H., Niven, R.K., Regenauer-Lieb, K. (eds.) Beyond the Second Law: Entropy Production and Non-Equilibrium Systems. Springer, Berlin (2013)
26. Dobovišek, A., Županović, P., Brumen, M., Bonačić-Lošić, Ž., Kuić, D., Juretić, D.: Enzyme kinetics and the maximum entropy production principle. Biophys. Chem. **154**, 49–55 (2011)
27. Hill, T.L.: Free Energy Transduction in Biology. The Steady State Kinetic and Thermodynamic Formalism. Academic Press, New York (1977)
28. Meszena, G., Westerhoff, H.V.: Non-equilibrium thermodynamics of light absorption. J. Phys. A: Math. Gen. **32**, 301–311 (1999)
29. Juretić, D., Županović, P.: Photosynthetic models with maximum entropy production in irreversible charge transfer steps. J. Comp. Biol. Chem. **27**, 541–553 (2003)
30. Lukács, A., Papp, E.: Bacteriorhodopsin photocycle kinetics analyzed by the maximum entropy method. J. Photochem. Photobiol., B **77**, 1–16 (2004)
31. Kakareka, J.W., Smith, P.D., Pohida, T.J., Hendler, R.W.: Simultaneous measurements of fast optical and proton current kinetics in the bacteriorhodopsin photocycle using an enhanced spectrophotometer. J. Biochem. Biophys. Methods **70**, 1116–1123 (2008)
32. Juretić, D., Westerhoff, H.V.: Variation of efficiency with free-energy dissipation in models of biological energy transduction. Biophys. Chem. **28**, 21–34 (1987)

33. Michel, H., Oesterhelt, D.: Electrochemical proton gradient across the cell membrane of Halobacterium halobium: Effect of N, N'-dicyclohexylcarbodiimide, relation to intracellular adenosine triphosphate, adenosine diphosphate, and phosphate concentration, and influence of the potassium gradient. Biochemistry **19**, 4607–4619 (1980)
34. Boylestad, R.: Introductory Circuit Analysis. Prentice-Hall, Upper Saddle River (1999)
35. Hendler, R.W., Shrager, R.I., Bose, S.: Theory and practice for finding a correct kinetic model for the bacteriorhodopsin photocycle. J. Phys. Chem. B **105**, 3319–3328 (2001)
36. Chizhov, I., Chernavskii, D.S., Engelhard, M., Mueller, K.H., Zubov, B.V., Hess, B.: Spectrally silent transitions in the bacteriorhodopsin photocycle. Biophys. J. **71**, 2329–2345 (1996)
37. Lanyi, J.K.: Proton transfers in the bacteriorhodopsin photocycle. Biochim. Biophys. Acta **1757**, 1012–1018 (2006)
38. Juretić, D., Županović, P.: The free-energy transduction and entropy production in initial photosynthetic reactions. In: Kleidon, A., Lorenz, R.D. (eds.) Non-Equilibrium Thermodynamics and the Production of Entropy, pp. 161–171. Springer, Berlin (2005)
39. Boyer, P.D.: The ATP synthase –a splendid molecular machine. Ann. Rev. Biochem. **66**, 717–749 (1997)
40. Pänke, O., Rumberg, B.: Kinetic modeling of rotary CF0F1-ATP synthase: storage of elastic energy during energy transduction. Biochim. Biophys. Acta **1412**, 118–128 (1999)
41. Pänke, O., Rumberg, B.: Kinetic modelling of the proton translocating CF0CF1-ATP synthase from spinach. FEBS Lett. **383**, 196–200 (1996)
42. Dewar, R.C., Juretić, D., Županović, P.: The functional design of the rotary enzyme ATP synthase is consistent with maximum entropy production. Chem. Phys. Lett. **430**, 177–182 (2006)
43. Turina, P., Samoray, D., Graeber, P.: H +/ATP ratio of proton transport-coupled ATP synthesis and hydrolysis catalysed by CF0F1-liposomes. Eur. Mol. Biol. Org. J. **22**, 418–426 (2003)
44. Jaynes, E.T.: Information theory and statistical mechanics. Phys. Rev. **106**, 620–630 (1957)
45. Jaynes, E.T.: Information theory and statistical mechanics II. Phys. Rev. **108**, 171–190 (1957)
46. Prigogine, I., Wiame, J.M.: Biologie et thermodynamique des phénomènes irréversibles. Experientia **2**, 451–453 (1946)
47. Kondepudi, D., Prigogine, I.: Modern Thermodynamics: From Heat Engines to Dissipative Structures. Wiley, Chichester (1998)

Chapter 20
Entropy Production and Morphological Selection in Crystal Growth

Leonid M. Martyushev

Abstract This chapter discusses morphological transitions during non-equilibrium crystallization and coexistence of crystals of different shapes from the viewpoint of the maximum entropy production principle (MEPP).

Notation
Symbol Meaning

Roman Symbols

D	Diffusion coefficient
V	Local velocity of the crystal
k, l	Wave-numbers of perturbing modes
u	Solute concentration
u_1	Equilibrium solute concentrations near the crystal surface
u_2	Density of the crystal
R	Size of a crystal
R^*	Critical radii of nucleation of a crystal
R_{crit}	Critical size for instability of a crystal
R^{max}	Maximal size for instability of a crystal (spinodal)
R^{min}	Minimal size for instability of a crystal (binodal)
R_{EP}^{min}	Minimum critical size, which was calculated using entropy production

Greek Symbols

α	Dimensionless parameters characterizing the growth regime
β	Coefficient of attachment kinetics

L. M. Martyushev (✉)
Institute of Industrial Ecology, Russian Academy of Sciences, 20A Sophy Kovalevskaya Street, Ekaterinburg, Russia 620219,
e-mail: Leonidmartyushev@gmail.com

L. M. Martyushev
Ural Federal University, 19 Mira Street, Ekaterinburg, Russia 620002,

R. C. Dewar et al. (eds.), *Beyond the Second Law*,
Understanding Complex Systems, DOI: 10.1007/978-3-642-40154-1_20,

δ Initial amplitude of perturbation
Σ Local entropy production
Ω The equation of the crystal surface

Subscripts
C Cylindrical
S Spherical

This chapter discusses morphological transitions during non-equilibrium crystallization and coexistence of crystals of different shapes from the viewpoint of the maximum entropy production principle (MEPP). We advance the following hypothesis: a necessary condition for a morphological transition is a larger entropy production in the final non-equilibrium phase, and equality of the entropy productions of two non-equilibrium phases is a condition for their coexistence. The basic results are as follows. (1) We explain the experimentally observed phenomenon of coexistence of morphological phases during non-equilibrium crystallization by metastability of crystallization growth regimes. (2) We prove by analytical and numerical methods that metastable regions limited by a minimum size (binodal) and a maximum size (spinodal) exist for simple morphological transitions during non-equilibrium crystallization. (3) Considering an agreement between analytical and numerical calculations, we infer that the maximum entropy production principle allows finding the binodal of a morphological transition during non-equilibrium crystallization.

20.1 Morphological Stability and the Coexistence of Morphological Phases

In the non-equilibrium phenomenon of crystal growth from a supersaturated/supercooled solution/vapor/melt, a regular-shaped nucleus starts distorting and transforming into, for example, dendrite structures upon reaching a certain size. The formation of snowflakes (ice crystals) from supersaturated water vapour (Fig. 20.1) is the most common example of non-equilibrium crystal growth. Such changes from one crystal shape to another during the growth process signal a loss of morphological stability of the initial shape.

This loss of morphological stability arises as follows. In the case of growth under condition very close to equilibrium (as it is called equilibrium growth), the external shape (morphology) of crystals is determined by a minimum of the surface energy, and as a result crystals have a cylindrical (spherical) or a regular polyhedral shape. Such a perfect shape rarely persists in the process of growth under conditions far from equilibrium (as it is called non-equilibrium growth).

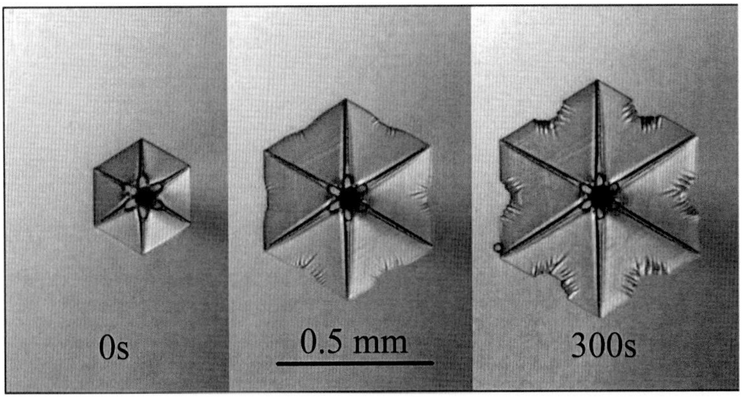

Fig. 20.1 Timeseries of a growing snow crystal. Reproduced with permission from www.its.caltech.edu/~atomic/snowcrystals (Kenneth G. Libbrecht) where details of the experiment and other splendid photos of snow crystals can be found

Indeed, the solute concentration becomes highly inhomogeneous around the crystal, which grows relatively rapidly (the concentration rapidly increases with distance from the crystal surface). A distortion appearing for one reason or another on the surface (e.g., due to an external mechanical disturbance) will find itself in a region of greater solute concentration and can therefore grow at a faster rate. On the other hand, such a distortion has greater curvature, which results in a decrease of the local supersaturation near this irregularity (because the equilibrium concentration of solute is dependent on surface energy and curvature). Thus, two competing effects are at play. Diffusion processes in the solution promote the growth of the irregularity and so contribute to instability of the crystal shape, whereas surface phenomena contribute to its stability. The stabilising surface phenomena overcome the destabilising diffusion processes for small crystal sizes, but not for large crystal sizes. There is therefore a critical size (R_{crit}) beyond which a crystal morphology becomes unstable.

The results of numerous experiments and computer simulation [1–12] have shown that in some case of non-equilibrium crystallization the coexistence of different crystal growth morphologies is possible. Figure 20.2 shows two examples of the coexistence observed in experiments on the crystallization of ammonium chloride from an aqueous solution. The studies [12–14] indicate that the transition from one morphology to another under changing conditions (e.g., the extent of supercooling/supersaturation) can involve discontinuities in both the growth rate and its slope. On this basis, an analogy between equilibrium phase diagrams and morphological diagrams can be drawn, and the notions of morphological transitions of the first and the second order can be introduced [14]. For example, Shibkov and co-authors [12] have investigated the non-equilibrium growth of ice in a film of supercooled water. In the case of supercooling by 7.5 °C, these authors

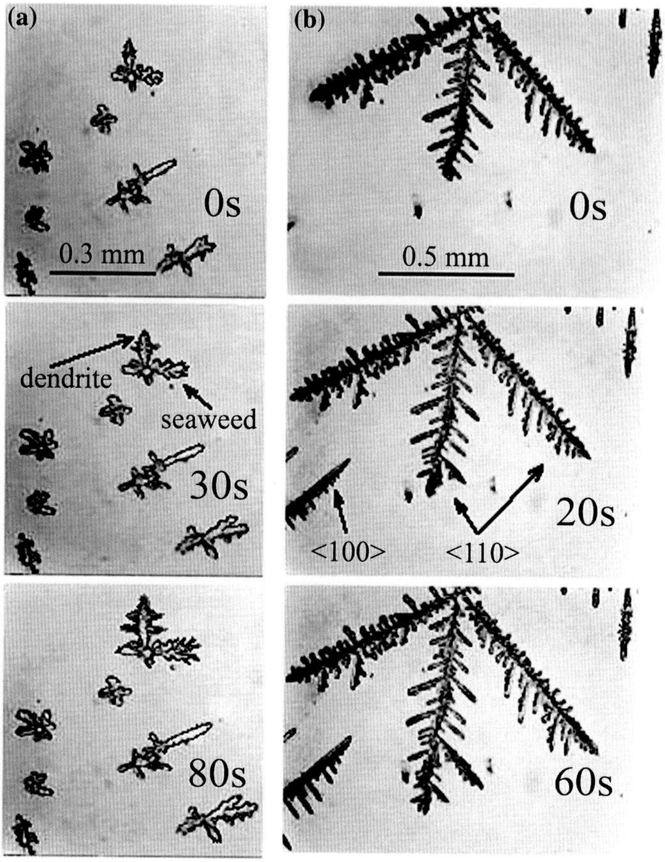

Fig. 20.2 Structures observed during crystallization of ammonium chloride from an aqueous solution initially saturated at 40 °C. **a** Coexistence of dendrites and seaweeds; **b** Coexistence of dendrites of ⟨110⟩ and ⟨100⟩ type

have found a morphological transition of the first order from needle-shaped ice crystals to the so-called plate ice crystal.

The important question arising in theory of morphology transitions is as follows: what can be considered as a principle for a morphology selection? The hypothesis that maximum entropy production can determine this selection far from equilibrium is proposed [13, 14]. A considerable development of this idea has been recently accomplished. Brief and accessible presentation of these results is the objective of this chapter.

20.2 Classical Stability Analysis and Numerical Solution of a Simple Model

Here we consider the simplest (but, equally, the most mathematically developed) model used for studying the above phenomena for almost half a century [15, 16].[1] The following approximations are used: quasi-stationarity of the growth process (valid at low supersaturation); isotropy of surface attachment kinetics and surface tension, and an initially cylindrical (C) or spherical (S) morphology. The mathematical representation of the problem is as follows:

$$\nabla^2 u = 0, \tag{20.1}$$

$$D\nabla_n u|_{\Omega} = \beta(u - u_1)|_{\Omega}, \tag{20.2}$$

$$u(r_\infty) = u_\infty, \tag{20.3}$$

$$V = \frac{D}{u_2 - u}\nabla_n u\bigg|_{\Omega}, \tag{20.4}$$

The first equation is the Laplace equation for a solute concentration u (u is a function of cylindrical or spherical coordinates). The second equation defines the solution concentration at the crystal boundary under arbitrary attachment kinetics. Here D is the diffusion coefficient; β is the coefficient of attachment kinetics; u_1 is the equilibrium solute concentrations near the crystal surface; Ω is the equation of the crystal surface; ∇_n is the component of the nabla operator normal to the surface. The third equation is the boundary condition for the solute concentration at a distance r_∞ from a crystal (it is constant and is equal to u_∞); r_∞ is finite for cylindrical and infinite for spherical crystal morphology. The Eq. (20.4) defines the local velocity of the crystal V. Here u_2 is the density of the crystal.

Stability analysis are traditionally performed using perturbations of the crystal surface by single harmonics of initial amplitude δ, with equations (20.1)–(20.4) being used to determine the critical size of the crystal at which the perturbation amplitude begins to grow. Thus, the critical size of a crystal (R_crit) is the main characteristic of a morphological transition. If we assume that δ is infinitely small and limit ourselves to the linear perturbation theory, the critical sizes for instability are known for the growth of a spherical $R_\mathrm{S}^\mathrm{max}$ and a cylindrical $R_\mathrm{C}^\mathrm{max}$ crystal[2] [15–20]. These critical sizes (normalized by the critical radius (size) of homogeneous nucleation[3]), are as follows:

---

[1] It is only in this approximation that the analytical solutions can be advanced sufficiently far.

[2] Superscript "Max" refers the fact that this is the largest possible critical size; the actual critical size will be smaller under finite-amplitude perturbation.

[3] Above this size the crystal itself growth and below this size it shrinks.

$$R_S^{\max} = \frac{(1+0.5(l+1)(l+2))}{2}\left[1+\sqrt{1+2\alpha_S\frac{(l+1)(l+2)}{(1+0.5(l+1)(l+2))^2}}\right], \quad (20.5)$$

$$R_C^{\max} = \frac{1+A_\lambda k(k+1)+\sqrt{(1+A_\lambda k(k+1))^2+4\alpha_C k(k+1)}}{2}, \quad (20.6)$$

where $\alpha_S = D/(\beta R_S^*)$ and $\alpha_C = D/(\beta R_C^*)$ are dimensionless parameters characterizing the growth regime (diffusion-limited at small values, and surface-limited phenomena at large values); A_λ is a dimensionless parameter related to r_∞; l and k are wave-numbers of perturbing modes; R_S^* and R_C^* are critical radii of nucleation of spherical and cylindrical crystals.

Formulas (20.5)–(20.6) fully determine stability of growing spherical and cylindrical particles with respect to infinitely small perturbations. Using the terminology available in physics of equilibrium phase transitions, these sizes can be thought of as spinodals of morphological transitions.

The above linear analysis says nothing about what will happen if the perturbation amplitude δ is finite. To address this shortcoming, a weakly non-linear analysis (to third order in δ) of Eqs. (20.1)–(20.4) for a cylindrical crystal has been made [21–25]. It was found that the critical size for instability (R_{crit}) decreases as the initial perturbation amplitude δ is increased (and, in some cases, reaches a constant nonzero value). The results of this weakly non-linear analysis hold for relatively small perturbation amplitudes. The behavior of R_{crit} for arbitrary-amplitude perturbations can only be understood from a numerical solution of Eqs. (20.1)–(20.4).

This numerical approach to the study of morphological stability under arbitrary-amplitude perturbations has been investigated in work by the present author and colleagues [26–28]. Our calculations were performed using a finite element method. Equations (20.1)–(20.4) were solved numerically for growing cylindrical and spherical crystals. Perturbations were imposed as cosine functions in the first case and axial-symmetric spherical functions in the second case. The results of the above linear and third-order non-linear analyses were used to check the algorithm for low-amplitude perturbations. The computed dependence of the critical size R_{crit} on the perturbation amplitudes δ for the cylindrical (Fig. 20.3) and spherical problems were qualitatively similar [27, 28]. We found that for any growth regime (from diffusion-limited to surface-limited) and for any modes of the initial harmonic perturbations, the dependences of R_{crit} on δ are similar and have two characteristic specific points: at $\delta \to 0$ (where $R_{crit} = R^{\max}$) and a minimum point (where $R_{crit} = R^{\min}$). While the first point ($R_C^{\max}$—in Fig. 20.3) has been well-studied in analytical terms by methods of classical linear analysis for stability [Eqs. (20.5) and 20.6], the presence of the minimum point ($R_C^{\min}$ in Fig. 20.3) is an interesting and non-trivial result.

The signification of this minimum for the morphological stability of crystal may be understood as follows. Consider a growing crystal of cylindrical shape. Assume

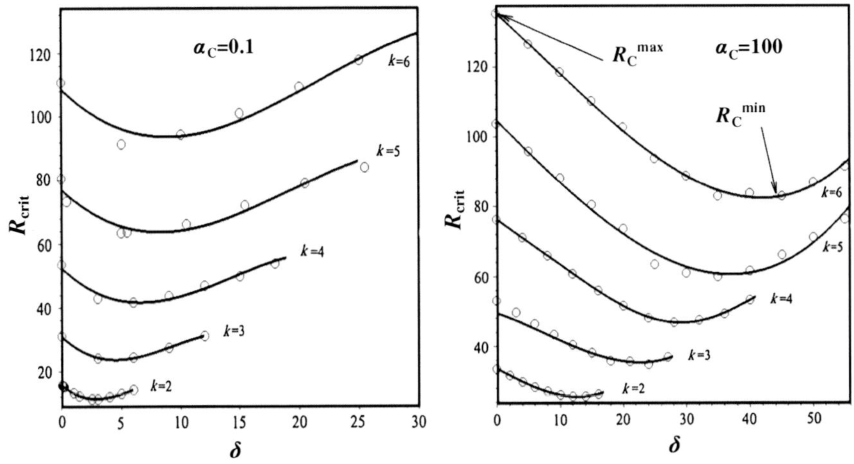

Fig. 20.3 For cylindrical crystals, the dependence of the critical size for morphological instability (R_{crit}) on the perturbation amplitude (δ) in two growth regimes (α_C) and perturbation modes (k). The points are numerical solution of Eqs. (20.1)–(20.4); the solid lines are cubic polynomial interpolations

that it is growing in a solution subject to perturbations of a certain wave-number k and an *arbitrary* amplitude δ. Then, in accordance with the results shown in Fig. 20.3, the loss of stability and the transition to a new growth morphology first takes place at some critical size R_C^{min} corresponding to the minimum in the dependence of R_{crit} on δ which occurs at some value $\delta = \delta_{min}$. If the amplitude of the perturbations is reduced to a value $\delta < \delta_{min}$, the critical size for instability (R_{crit}) will increase (corresponding to points in Fig. 20.3 lying to the left of the minimum).[4] Thus, the transition to a new growth morphology can be observed in the range $R_C^{min} < R < R_C^{max}$ depending on the amplitude of perturbations. This range will be referred to as a metastable region by analogy with the theory of equilibrium phase transitions.

Figure 20.4 shows how the extent of the metastable region depends on the growth regime at different harmonics (i.e. different perturbation wave-number k). In the diffusion-limited regime $\alpha < 1$, the metastable regions corresponding to different harmonics do not intersect. In the intermediate regime ($1 < \alpha < 10$) the minimum critical size (R_C^{min}) of the harmonic $k + 1$ and the maximum critical size (R_C^{max}) of the harmonic k approach each other, and at $\alpha > 10$ they intersect. As a result, the metastable regions overlap at adjacent harmonics. If $\alpha > 100$ (in the surface-limited regime), three or more metastable regions can intersect. Therefore, in the intermediate and surface-limited regimes of growth in a medium with

[4] In line with the terminology accepted in the theory of equilibrium phase transitions, it is reasonable if the calculated minimum critical size R_C^{min} is called, by analogy, a *binodal*, and the stability size R_C^{max}, which was observed when the perturbation amplitude was almost zero, is termed a *spinodal* of a non-equilibrium transition.

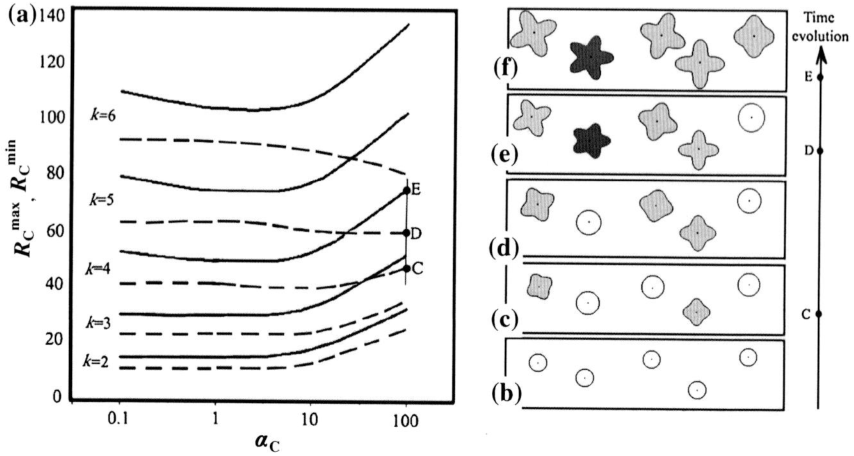

Fig. 20.4 **a** For cylindrical crystals, the dependence of the maximal critical size (R_C^{max}, *solid lines*) and the minimal critical size (R_C^{min}, *dashed lines*) on the growth regime (α_C) for different perturbation modes (k). **b–f** The possible time evolution of crystals of different morphologies for perturbation modes $k \geq 4$ (the growth conditions correspond to the trajectory CDE)

perturbations of different amplitudes (δ) and wave-numbers (k), a large number of crystals of different shapes—different morphological phases—can coexist and develop from nuclei.

To take a particular example, suppose we have a number of cylindrical crystals growing in a medium with fluctuations at $k \geq 4$ of arbitrary amplitude δ, leading to distortions of the crystal surface. We shall assume that large amplitude perturbations are rarer than small amplitude perturbations. Let the physico-chemical properties of the solution and growing crystal correspond to the growth regime $\alpha_C = 100$. Then the growth of the crystals follows a straight line CE (Fig. 20.4a). Up to the point C, all crystals have a cylindrical shape (Fig. 20.4b). In the interval CD, two different morphologies can be observed simultaneously (Fig. 20.4c–d): cylindrical crystals (which are more probable in view of the above assumptions of the perturbation statistics) and crystals that have lost stability under perturbations with $k = 4$. Beyond the point D (corresponding to the minimal critical size R_C^{min} for instability under perturbations with $k = 5$), a third morphology can appear on account of the loss of stability by cylindrical particles under perturbations with $k = 5$. Thus, three morphological phases coexist within the interval DE (Fig. 20.4e). Beyond the point E (corresponding to the maximum critical size R_C^{max} for instability under perturbations with $k = 4$) all the remaining cylindrical particles become unstable with respect to perturbations with $k = 4$ (Fig. 20.4f).

In summary, the numerical calculations of the morphological stability of growing cylindrical and spherical particle [26–28] have revealed the presence of metastable regions, leading to the possibility that crystals of different morphologies can appear and grow simultaneously.

20.3 Maximum Entropy Production Principle and Morphology Selection

The previous section we saw how metastable regions, delimited by minimum (binodal, R^{min}) and maximum (spinodal, R^{max}) critical size for instability, can exist for simple morphological transitions during non-equilibrium crystallization. While an analytical solution for R^{max} can be found [Eqs. (20.5), (20.6)]. But is there an analytical method for calculating the minimum critical size R^{min}?

As is seen from above calculations, the coexistence of crystals of different morphologies is possible for $R > R^{min}$. These crystals have different velocities and, as result, different energy dissipations per unit of time (entropy productions). Maximum entropy production principle[5] (MEPP) has been theoretically developed and successfully applied in various domains of physics and chemistry (see overview works [31–33]) for more than fifty years. According to this principle, the process with maximum entropy production is most preferred among the possible non-equilibrium processes. As a consequence, a hypothesis on the use of the entropy production for selection of crystal pattern formations has been advanced in a number of works [13, 14, 17, 29, 34–40]. This hypothesis may be stated as follows [29, 30]: a necessary condition for a transition from one crystal morphology to another is a larger entropy production in the final morphological phase, and equality of the entropy productions of two non-equilibrium phases determines the condition (curve, point) at which two distinct phases may coexist.[6]

As is shown in [17–20, 40, 41], the local entropy production Σ in an element of solution volume $d\upsilon$ near the crystal surface can be written as

$$\Sigma \propto V^2 d\upsilon, \tag{20.7}$$

where V is the local velocity of the crystal surface. According to the above MEPP hypothesis, if $\Delta\Sigma$ denotes the difference between the local entropy productions near the surface of two morphological phases,[7] then the solution of $\Delta\Sigma = 0$ for the crystal size R determines the minimum critical size R^{min} for the transition between them. The solution of this equation for spherical and cylindrical geometries in the case of the mathematical model (20.1)–(20.4) gave expressions for the minimum critical sizes of spherical and cylindrical crystals of the form[8] [17–20]:

---

[5] This principle can be most generally formulated as follows [29–31]: at each level of description, with preset external constraints, the relationship between the cause and the response of a non-equilibrium system is established such as to maximize the entropy production.

[6] In line with the terminology used in the theory of equilibrium phase transitions, this curve is named as binodal (it separates the region, in which the phase is stable, from the region, in which it is metastable and unstable).

[7] For example, one morphological phase implies the initial (spherical or cylindrical) form of growth, and the other phase means the initial form with some added harmonic.

[8] These radii were rendered dimensionless to the critical radius of nucleation.

Table 20.1 For a cylindrical crystal, the minimal critical size for instability determined numerically (R_C^{min}, Fig. 20.4a) and analytically from entropy production ($R_{C,EP}^{min}$, Eq. 20.9)

α_C	k	R_C^{min}	$R_{C,EP}^{min}$	α_C	k	R_C^{min}	$R_{C,EP}^{min}$	α_C	k	R_C^{min}	$R_{C,EP}^{min}$
0.1	2	11.3	10.8	1	2	11.8	10.6	10	2	14.2	9.2
	3	23.6	24.6		3	23.9	24.4		3	25.1	23.0
	4	41.7	43.2		4	42.0	43.0		4	40.9	41.7
	5	63.8	66.6		5	63.8	66.5		5	60.8	65.2
	6	93.7	94.9		6	92.5	94.5		6	89.3	93.6

$$R_{S,EP}^{min} = \frac{l^3 + 2l^2 + l - 2 - 2\alpha_S(l+1) + \sqrt{(l^3 + 2l^2 + l - 2)^2 + 4\alpha_S^2(l+1)^2 + 4\alpha_S(l^4 + l^3 - l^2 + l + 2)}}{4l}, \quad (20.8)$$

$$R_{C,EP}^{min} = \frac{1}{2}\left\{1 - \frac{\alpha_C k}{2k-1} + \frac{2A_\lambda k(k^2-1)}{2k-1} + \sqrt{\left[1 - \frac{\alpha_C k}{2k-1} + \frac{2A_\lambda k(k^2-1)}{2k-1}\right]^2 + 4\alpha_C \frac{k(2k^2-1)}{2k-1}}\right\}. \quad (20.9)$$

Using (20.5), (20.6), (20.8), and (20.9), full morphological phase diagrams of the regions of stable ($R < R_{EP}^{min}$), metastable ($R_{EP}^{min} < R < R^{max}$), and unstable ($R > R^{max}$) growth of spherical and cylindrical crystals can be easily analyzed as a function of the model parameters [17–20]. In a wide interval of α_S and α_C the metastable regions corresponding to different perturbing harmonics overlap (similar to those determined numerically, see, e.g., Fig. 20.4), pointing to the possibility that a great number of morphological phases can coexist. It was found theoretically [17–20] that there is a discontinuous increase in the mass of a crystal at the morphological transitions. The value of this stepwise change decreases as the kinetic crystallization coefficient diminishes; the relative supersaturation decreases; the surface tension coefficient increases; and the numbers of perturbing harmonics grow.

The most important question is to what extent the minimum critical radius, which was calculated using the entropy production (R_{EP}^{min}), coincides with its numerically predicted counterpart (R^{min}). A quantitative comparison of the results obtained for a cylindrical crystal[9] [26, 27] is given in Table 20.1. Whatever the perturbation mode, the accuracy of prediction based on (20.9) for the diffusion and intermediate regimes of growth was very high, with the discrepancy being just 2–10 %. However, for $\alpha_C = 10$, the coincidence was much worse at perturbations with small k (a maximum discrepancy as large as 35 % at $k = 2$).

This discrepancy may reflect insufficient accuracy of the analytical calculations performed in this range of the parameters. The analytical calculations are based on a comparison of the entropy production in a solution at the surface of unperturbed and perturbed growing crystals. The entropy production is essentially a measure of non-equilibrium, but non-equilibrium is extremely small for this parameter range. Indeed, the farther we are from the diffusion growth regime (α_C increases), the more homogeneous is the diffusion field at the surface of a particle. Also, the

[9] The results were similar for a spherical crystal [28].

longer the wavelength of the surface perturbation (i.e. the smaller the value of k), the closer is the curvature (and, hence, the equilibrium concentration) of such a perturbed surface to its unperturbed value. For these reasons, both the absolute value of the entropy production and the difference of the entropy productions for perturbed and unperturbed forms of the crystal growth are very small, and this can lead to inaccurate analytical result for the large α_C.

In summary, good quantitative agreement has been shown between the minimum critical sizes for instability predicted analytically using the entropy production and those calculated numerically, especially when there is a relatively large concentration gradient near the crystal surface.

20.4 Conclusion

This brief review summarizes the results of recent studies of the morphological stability of growing crystals from the viewpoint of the maximum entropy production principle. The key findings are as follows:

1. As the perturbation amplitude increases from zero, the critical size for morphological instability decreases from a maximum value (spinodal) R^{max} (the boundary of stability under infinitely small perturbations) to a minimum value (binodal) R^{min}. The morphological transition occurs in the metastable region $R^{min} \leq R \leq R^{max}$. The notion of the metastable region and its dependence on the amplitude of perturbations allows us to understand the experimentally observed coexistence of different growth morphologies.
2. The hypothesis that a necessary condition for the occurrence of a morphological transition is a larger entropy production in the final phase is verified. So, the minimum critical size (binodal) R^{min} can be found from the condition that the difference of the entropy productions in the first (unperturbed) and second (perturbed) growth regimes is zero.

The thermodynamic approach applied here to morphological transitions during crystallization can be used to study pattern formation in other non-equilibrium systems. Examples include hydrodynamic and thermal instabilities in liquids or plasmas, and the fracture of solids under deformations. Work in this direction is now underway [42–44].

Glossary

Binodal (binodal curve or coexistence curve) denotes the condition at which two distinct equilibrium or non-equilibrium phases may coexist. Beyond the binodal, the perturbations (or fluctuations) of phase will lead to phase transition. Before the binodal, the phase will be stable with respect to any perturbations (or fluctuations).

For equilibrium phase transition, the binodal is defined by the condition at which the chemical potential is equal in each equilibrium phase. There is hypothesis that for non-equilibrium phase transition, the binodal is defined by the condition at which the entropy production is equal in each non-equilibrium phase.

Spinodal (spinodal curve) denotes the boundary of absolute instability of equilibrium or non-equilibrium phases. Beyond the spinodal, infinitesimally small perturbations (or fluctuations) of phase will lead to phase transition. Before the spinodal, the phase will be at least stable or metastable with respect to perturbations (or fluctuations).

Reference[10]

1. Shochet, O., Ben-Jacob, E.: Coexistence of morphologies in diffusive patterning. Phys. Rev. E **48**(6), R4168–R4171 (1993)
2. Chan, S.K., Reimer, H.H., Kahlweit, M.J.: On the stationary growth shape of NH_4Cl dendrities. J. Cryst. Growth **32**, 303–315 (1976)
3. Sawada, Y., Dougherty, A., Gollub, J.P.: Dendritic and fractal patterns in electrolytic metal deposits. Phys. Rev. Lett. **56**(12), 1260–1263 (1986)
4. Shochet, O., Kassner, K., Ben-Jacob, E., et al.: Morphology transitions during non-equilibrium growth. II. Morphology diagram and characterization of the transition. Phys. A **187**, 87–111 (1992)
5. Ihle, T., Müller-Krumbhaar, H.: Fractal and compact growth morphologies in phase transitions with diffusion transport. Phys. Rev. E **49**(4), 2972–2991 (1994)
6. Honjo, H., Ohta, S., Matsushita, M.: Phase diagram of a growing succionitrile crystal in supercooling-anisotropy phase space. Phys. Rev. A **36**(9), 4555–4558 (1987)
7. Sawada, Y., Perrin, B., Tabeling, P., Bouissou, P.: Oscillatory growth of dendritic tips in a three-dimensional system. Phys. Rev. A **43**(10), 5537–5540 (1991)
8. Flores, A., Corvera-Poir, E., Garza, C., Castillo, R.: Growth and morphology in Langmuir monolayers. Europhys. Lett. **74**(5), 799–805 (2006)
9. Harkeand, M., Motschmann, H.: On the transition state between the oil water and air water interface. Langmuir **14**(2), 313–318 (1998)
10. Akamatsu, S., Faivre, G., Ihle, T.: Symmetry—broken double fingers and seaweed patterns in thin-film directional solidification of a non-faceted cubic crystal. Phys. Rev. E **51**(5), 4751–4773 (1995)
11. Lamelas, F.J., Seader, S., Zunic, M., Sloane, C.V., Xiong, M.: Morphology transitions during the growth of alkali halides from solution. Phys.Rev. B. **67**, 045414(11) (2003)
12. Shibkov, A.A., Golovin, YuI, Zheltov, M.A., et al.: Morphology diagram of non-equilibrium patterns of ice crystals growing in supercooled water. Phys. A **319**, 65–72 (2003)
13. Ben-Jacob, E., Garik, P., Mueller, T., Grier, D.: Characterization of morphology transitions in diffusion-controlled systems. Phys. Rev. A **38**(3), 1370–1380 (1989)
14. Ben-Jacob, E., Garik, P.: The formation of patterns in non-equilibrium growth. Nature **343**, 523–530 (1990)
15. Mullins, W.W., Sekerka, R.F.: Morphological stability of a particle when growth is controlled by diffusion or heat flow. J. Appl. Phys. **34**, 323–340 (1963)

[10] Martyushev L. M. has also published under the alternate (French) spelling Martiouchev L.M.

16. Coriell, S.R., Parker, R.L.: Stability of the shape of a solid cylinder growing in a diffusion field. J. Appl. Phys. **36**(2), 632–637 (1965)
17. Martiouchev, L.M., Seleznev, V.D., Kuznetsova, I.E.: Application of the principle of maximum entropy production to the analysis of the morphological stability of a growing crystal. J. Exper. Theor. Phys. **91**(1), 132–143 (2000)
18. Martyushev, L.M., Kuznetsova, I.E., Seleznev, V.D.: Calculation of the complete morphological phase diagram for non-equilibrium growth of a spherical crystal under arbitrary surface kinetics. J. Exper. Theor. Phys. **94**(2), 307–314 (2002)
19. Martiouchev, L.M., Sal'nicova, E.M.: An analysis of the morphological transitions during non-equilibrium growth of a cylindrical crystal from solution. Tech. Phys. Lett. **28**(3), 242–245 (2002)
20. Martyushev, L.M., Sal'nicova, E.M.: Morphological transition in the development of a cylindrical crystal. J. Phys.: Cond. Matter. **15**, 1137–1146 (2003)
21. Brush, L.N., Sekerka, R.F., McFadden, G.B.: A numerical and analytical study of nonlinear bifurcations associated with the morphological stability of two-dimensional single crystal. J. Cryst. Growth **100**, 89–108 (1990)
22. Debroy, P.P., Sekerka, R.F.: Weakly nonlinear morphological instability of a cylindrical crystal growing from a pure undercooled melt. Phys. Rev. E **53**(6), 6244–6252 (1996)
23. Debroy, P.P., Sekerka, R.F.: Weakly nonlinear morphological instability of a spherical crystal growing from a pure undercooled melt. Phys. Rev. E **51**, 4608–4651 (1995)
24. Martyushev, L.M., Sal'nicova, E.M., Chervontseva, E.A.: Weakly nonlinear analysis of the morphological stability of a two-dimensional cylindrical crystal. J. Exper. Theor. Phys. **98**(5), 986–996 (2004)
25. Martyushev, L.M., Chervontseva, E.A.: Morphological stability of a two-dimensional cylindrical crystal with a square-law supersaturation dependence of the growth rate. J. Phys.: Cond. Matter. **17**, 2889–2902 (2005)
26. Martyushev, L.M., Serebrennikov, S.V.: Morphological stability of a crystal with respect to arbitrary boundary perturbation. Tech. Phys. Lett. **32**(7), 614–617 (2006)
27. Martyushev, L.M., Chervontseva, E.A.: On the problem of the metastable region at morphological instability. Phys. Lett. A. **373**, 4206–4213 (2009)
28. Martyushev, L.M., Chervontseva, E.A.: Coexistence of axially disturbed spherical particle during their nonequilibrium growth. EPL (Europhys. Lett.) **90**, 10012(6 pages) (2010)
29. Martyushev, L.M., Konovalov, M.S.: Thermodynamic model of nonequilibrium phase transitions. Phys. Rev. E. **84**(1), 011113(7 pages) (2011)
30. Martyushev, L.M.: Entropy production and morphological transitions in non-equilibrium processes. arXiv:1011.4137v1
31. Martyushev, L.M., Seleznev, V.D.: Maximum entropy production principle in physics, chemistry and biology. Phys. Rep. **426**, 1–45 (2006)
32. Kleidon, A., Lorenz, R.D. (eds.): Non-equilibrium thermodynamics and the production of entropy in life, Earth, and beyond. Springer, Heidelberg (2004)
33. Ozawa, H., Ohmura, A., Lorenz, R.D., Pujol, T.: The second law of thermodynamics and the global climate systems—a rewiew of the maximum entropy production principle. Rev. Geophys. **41**(4), 1018–1042 (2003)
34. Sawada, Y.: A thermodynamic variational principle in nonlinear systems far from equilibrium. J. Stat. Phys. **34**, 1039–1045 (1984)
35. Kirkaldy, J.S.: Entropy criteria applied to pattern selection in systems with free boundaries. Metall. Trans. **16A**, 1781–1797 (1985)
36. Kirkaldy, J.S.: Spontaneous evolution of spatiotemporal patterns in materials. Rep. Prog. Phys. **55**, 723–795 (1992)
37. Hill, A.: Entropy production as the selection rule between different growth morphologies. Nature **348**, 426–428 (1990)
38. Hill, A.: Reply to Morphologies of growth, written by Lavenda B.H. Nature **351**, 529–530 (1991)

39. Wang, Mu: Nai-ben Ming.: Alternating morphology transitions in electro chemical deposition. Phys. Rev. Lett. **71**(1), 113–116 (1993)
40. Martiouchev, L.M., Seleznev, V.D.: Maximum-Entropy production principle as a criterion for the morphological-phase selection in the crystallization process. Dokl. Phys. **45**(4), 129–131 (2000)
41. Martyushev, L.M., Kuznetsova, I.E., Nazarova, A.S.: Morphological phase diagram of a spherical crystal growing under non-equilibrium conditions at the growth rate as a quadratic function of supersaturation. Phys. Solid State **46**(11), 2115–2120 (2004)
42. Martyushev, L.M.: Some interesting consequences of the maximum entropy production principle. J. Exper. Theor. Phys. **104**(4), 651–654 (2007)
43. Niven, R.K.: Simultaneous extrema in the entropy production for steady-state fluid flow in parallel pipes. J. Non-Equilib. Thermod. **35**, 347–378 (2010)
44. Martyushev, L.M., Birzina, A.I., Konovalov, M.S., Sergeev, A.P.: Experimental investigation of the onset of instability in a radial Hele-Shaw cell. Phys. Rev. E. **80**(6), 066306(9 pages) (2009)

Chapter 21
Maximum Entropy Production by Technology

Peter K. Haff

Abstract The dominant mode of entropy production enabled by the large-scale technological systems that power the world economy is the degradation of chemical energy in fossil fuels. One key parameter determining the rates of fossil fuel consumption and entropy production is the price of energy. The Rayleigh-Benard cell provides a laboratory analog in which, for a given driving force, the rate of entropy production is determined by the value of the thermal boundary layer thickness, whose inverse plays a role similar to that of price in large fossil fuel systems. In steady, serial systems like the diffusion-advection-diffusion Rayleigh-Benard cell or the oilfield-pipeline-city "technology cell", an auto-control parameter like price or boundary layer thickness is required to coordinate spatially separated energy source and sink dynamics. For complex fossil fuel technologies the implicit and often unknown dependence of such control parameters on intrinsic system variables can hide internal constraints. If applied in the absence of knowledge of such constraints, the principle of Maximum Entropy Production (MaxEP) would yield, for sufficiently complex systems, an upper limit to rather than the actual value of the entropy production rate. Internal constraints on technology-enabled energy consumption, however, may represent only temporary hangups on the road toward a larger entropy production rate.

List of Symbols

Symbol Meaning (SI units)

Roman Symbols

C_P	Specific heat capacity at constant pressure ($m^2\ s^{-2}\ K^{-1}$)
D	Separation of Rayleigh–Benard (RB) plates (henceforth "plates") (m)
g	Acceleration of gravity ($m\ s^{-2}$)
k	Thermal conductivity ($kg\ m\ s^{-3}\ K^{-1}$)

P. K. Haff (✉)
Nicholas School of the Environment, Duke University, Durham, NC 27708, USA
e-mail: haff@duke.edu

R. C. Dewar et al. (eds.), *Beyond the Second Law*,
Understanding Complex Systems, DOI: 10.1007/978-3-642-40154-1_21,

i	Index identifying plate
q	Heat flux between plates (kg s^{-3})
q_i	Heat flux through boundary layer at plate i (kg s^{-3})
P	Price per unit energy (\$ kg^{-1} m^{-2} s^2)
P_e	Equilibrium price per unit energy (\$ kg^{-1} m^{-2} s^2)
Ra	Rayleigh number
Ra*	Critical Rayleigh number
T_i	Temperature of plate i (K)
ΔT	Temperature difference between plates (K)
ΔT_i	Temperature difference across thermal boundary layer at plate i (K)
Δx	Distance increment (m)

Greek Symbols

α	Thermal volumetric expansion coefficient (K^{-1})
δ	Thermal boundary layer thickness (m)
δ_i	Thermal boundary layer thickness at plate i (m)
$\delta_{\min}$	Constant (m)
ξ	Fluid constant (m^{-3} K^{-1})
γ	Transport control parameter (m^{-1})
γ_e	Equilibrium transport control parameter (m^{-1})
κ	Thermal diffusivity (m^2 s^{-1})
ρ	Mass density (kg m^{-3})
σ	Rate of entropy production (kg s^{-3} K^{-1})
ν	Kinematic viscosity (m^2 s^{-1})

21.1 Technology, The Economy, and Entropy

By the term "technology" we refer collectively to the interlinked regional and global systems associated with processes such as communication, transportation, power generation and transmission, food production, the manufacture of goods and their distribution, and so on. Technology as defined here is similar to the concept of technology used by Arthur [1], and includes not just artifacts like shovels, transmission lines, and computers, but also the people, processes, shared-knowledge, rules, protocols, and social organizations without which technology would be only a large and inert collection of stuff. In this view, technology is more than just a form of material capital subject to human use and deployment. As is evident from a consideration of the surprising (i.e., unpredicted and unanticipated) emergence and developmental trajectory of the internet, cellular telephony, nuclear weapons systems, and other large-scale expressions of technological change, technology at large-enough scale is in essence an autonomous phenomenon beyond the control or detailed understanding of any person or organization. Technology is an emergent phenomenon representing a new phase or paradigm in earth evolution. Humans are essential components of technology and "the economy" is the part of technology that is directly connected with decision making, entrepreneurship, money, finance, consumption, markets, and other phenomena whose use reflects human behavior. In a crude way one could say that the rules of economics are elements of the software on which the hardware or material part of technology runs.

The attempt to connect entropy considerations to technology (or economics) is motivated by the example of thermodynamics, especially its second law, as a branch of physics that can be usefully applied to analyses of many complex systems whose dynamics is incompletely known, such as the earth's atmosphere. It has also been motivated by the example of statistical mechanics which treats in a probabilistic way the relation between system micro and macro variables, such as molecular velocities and pressure values in a gas. Treated as a method of inference, relations originating in statistical mechanics have application to non-thermodynamic systems, perhaps even economic systems. An interesting question is, is it possible to replace with a cleaner physical explanation at least some of the more opaque (at least to physical scientists) relations of apparently messy social sciences like economics? Below we argue that although it is not likely that a social science understanding of either economics or technology will be replaced by thermodynamics or statistical mechanics, the maximum entropy production framework might still provide a useful perspective on technological change.

Discussion of entropy, maximum entropy, maximum entropy production, and related concepts such as maximum power production in relation to systems outside basic physics, including biological and economic systems, began to appear in the last century. In the early and mid-twentieth century Schrodinger [2], Bertalanffy [3], Odum [4], and others and more recently authors such as Harte [5] and Dewar [6] discussed entropy and entropic processes in biological organisms and ecosystems. The connection between entropy and economics was the main focus of the treatise by Georgescu-Roegen [7] and more recently has been discussed by Ruth [8], Annila and Salthe [9], and others. In the present work we compare entropy production in a simple (to define) physical system, the Rayleigh-Benard cell, with that in an idealized, energy consuming, technological system. Our aim is to identify more clearly the limitations and the potential applications of the principle of Maximum Entropy Production (MaxEP) in technological systems.

21.2 Technology as a Geologic Phenomenon

The principle of Maximum Entropy Production suggests that sufficiently complex dynamic systems will configure themselves, when driven hard enough by an external force, to produce entropy at the maximum rate allowed by existing constraints [10]. Where more than one dynamical configuration is allowed by these constraints and the equations of motion, then MaxEP may provide a closure criterion that allows determination of the rate of system entropy production or the rate of energy dissipation.

A role for MaxEP has been suggested in the evolution of earth systems over geologic time [11]. For example, plants capture a fraction of the low entropy sunlight that impinges on the surface of the earth and support entropy production at a higher rate per unit area of the earth's surface than was possible with the prebiotic mineral surface. A strong interpretation of MaxEP would suggest that the biotic surface, evidently one solution to the earth's equations of motion, was selected by entropy maximization. The fact that the earth's land surface was nearly abiotic over most of its history (vascular land plants emerged about 450 million years ago during the Silurian period) would be explained by the fact that the transformation of available free energy to unusable low temperature heat is often subject to delays. The degradation of usable energy to heat from the energy available in the gravitational potential energy of the universe (the ultimate source of free energy) is subject to many internal constraints or hangups [12]. For example our galaxy has been preserved for billions of years by a spin hangup that prevents a rotating extended object from immediately collapsing under its own gravitational attraction. At a smaller scale a mechanical hangup delays for a short period of time the transformation of food calories into heat whenever we exert ourselves to set a mouse trap.

An energy hangup is a temporary condition, subject to change. A radical change in consumption rate of free energy and rate of entropy production by earth surface systems has occurred in recent geologic history with the emergence of technology. The strata of coal, pools of oil, and deposits of natural gas that power the lion's share (80 %) of modern technology [13] represent metastable accumulations of free energy that Nature has not hitherto been able to convert to heat. These stranded energy resources have languished for millions of years in the earth's crust. Nature has occasionally been able to find ingenious ways to access otherwise inaccessible energy resources. One spectacular example is the Oklo natural nuclear reactor in Gabon [14] in which uranium atoms became sufficiently concentrated by natural hydrogeological processes to support a fission chain reaction. Technology, an even more startling innovation of Nature, has emerged as the principal mechanism by which the earth degrades deeply buried chemical energy, and, far from being a curiosity like Oklo, has come to define the latest epoch of the evolution of the earth. (This epoch has been dubbed the *Anthropocene* [15], although, given the insignificance of humans as agents of geological change prior to the emergence of technology, I am tempted to call it the *Technocene*).

The emergence of large-scale fossil fuel technology is intimately connected with the emergence of transport mechanisms able to advect large quantities of chemical energy across the surface of the earth. For quasi-steady-state systems, on which we focus here, rapid consumption of chemical energy on a large scale generally requires a fast transport mechanism that can carry energy originating in a low entropy resource to points of consumption and entropy generation. There are two principal ways to transport the energy-rich mass—by diffusion and by flow or advection. In the present context the word "diffusion" is shorthand for "diffusion-like", and refers to a mode of mass transport which, in comparison to advection in the same system, is characterized by short displacements with frequent changes in direction of motion [16, 17]. Under otherwise similar conditions, advection of a resource, when it occurs, is generally faster than diffusion of the same resource. The short randomly directed mean free paths characteristic of diffusion are not well suited to compete with unidirectional displacements of the same quantity, unless the system is only weakly driven. If an energy resource is momentarily being consumed at a high rate, as wood is consumed in a fire, the local rate of energy consumption will diminish as the resource becomes depleted unless additional energy is supplied from another location. In the absence of an advected supply the subsequent rate of energy consumption would be limited by the usually slow rate of diffusion of the resource into the consumption zone. For this reason, on the earth's surface high steady-state rates of entropy production are generally associated with systems like highways and rivers that can support advective transport of energy from more distant sources to a location where it is consumed.

Entropy generation by the earth's atmosphere through global advective transport of energy originating in solar radiation is a well-discussed example of an apparently successful application of MaxEP [10, 18]. The solid earth on the other hand, until recently, had offered no advective mechanism of comparable (global) scale able to support rapid sustained consumption of energy like that stored in fossil fuels. After hundreds of millions of years from the time of formation of many of today's fossil fuel deposits, one might have thought that such a mechanism did not exist. Alternatively one might have viewed this situation as a temporary state of affairs dictated by the existence of an energy hangup, and that eventually the barrier to consumption would be surmounted. This turned out to be the case. The relevant mechanism(s) for effecting rapid fossil energy consumption became available with the emergence of technology. Its global reach, large rate of energy consumption, massive cooption of planetary resources, and independence (at large scale) from human control lead us to consider technology as the most recent paradigm of earth evolution. Other geologic paradigms include systems such as the hydrosphere, atmosphere, and biosphere.

21.3 Advective Pathways

In discussing MaxEP it is useful to think about advection in the following way. From the point of view of statistical mechanics advection is, under a suitable driving force, always a potentially available, and indeed statistically likely, transport mechanism in the following sense. Consider a blob of material that forms part of a larger system. This might be a blob of air in the atmosphere, a blob of water in a reservoir, a blob of dirt on a hillside, and so on. Under differences in temperature or of elevation these blobs may be transported to lower temperature regions, or points of lower elevation, generating entropy in the process. The macroscopic state of such a blob is realized by many distinct microstates, defined for example by different arrangements of accessible molecular coordinates and velocities. If the blob macrostate is one of thermal equilibrium, then that macrostate can be realized by more microstates than can any other blob macrostate, i.e., it is the most probably macrostate.

If we imagine ourselves co-moving with a hypothetical uniformly translating blob, then, under the principle of Galilean invariance, the microscopic probability distribution underlying the macrostate of the subsystem as observed in the co-moving frame would be identical to that of the stationary blob. The co-moving system, seen from the original "stationary" frame, corresponds to an advective state and in this sense the advecting blob is as statistically likely as the thermal equilibrium state. That is, in reference to the usual (approximate) classical picture of stationary system microstates [19], there exist the same number of distinct microscopic pathways (we are considering configuration space pathways not phase space pathways) corresponding to advection at a given speed and in a given direction as there are microstates that realize the macrostate of the stationary blob, i.e., a maximal number. If one imagined each molecule to be tagged with fluorescent paint then each microstate of the blob would be the starting configuration for a bundle of fluorescent streaks that traced out the advective evolution of that microstate. The potential number of advective pathway bundles would equal the number of stationary system microstates. Advection does not occur unless a force is available that can enable access to these pathways, but, nonetheless, pathways for fast transport and thus fast entropy production are always potentially available. The unanswered question is whether system dynamics can enable access to them.

21.4 Hidden Constraints

MaxEP implies that under broad conditions system dynamics of sufficiently complex dynamical systems should converge to that of a maximum entropy production state [20, 21]. Theories of MaxEP are not based on details of system dynamics but on general considerations of statistical mechanics. A necessary assumption is that there are no hidden (to us) internal dynamical constraints that would reduce entropy production from what we would otherwise infer on the basis of a given global driving force such as the temperature drop across a Rayleigh-Benard cell or a chemical affinity like that which underlies the flux of fossil fuels from mines and wells to homes, cars, and factories. In these formulations of MaxEP it is assumed that a complex system will have enough internal flexibility to "find" one of the maximum-entropy-producing microstates that are potentially available to it. The present chapter emphasizes that we should not be surprised to see this assumption fail in certain kinds of complex systems.

The earth's atmosphere, evidently, does fulfill the conditions required of MaxEP. However, what is true for a relatively free-flowing fluid like the earth's atmosphere may not be true for systems with more enduring structural elements.

For example, in the fossil-fuel-based technological systems of interest here, the discreteness of parts, the high friction that exists between moving solid components, the frozen design of transport systems (e.g., railroads) and other components, and the dependence of function on poorly understood or unrecognized political, cultural and economic factors, whose physical manifestation is unclear, may conceal obstacles to advective flow and thus to entropy production. Even if the system is eventually able to breach such choke points, we know from the history of the earth—from the long time delay between the formation of the earth's mineral surface and the emergence of its biological surface for example, and the additional delay prior to the emergence of oil and coal burning technology—that the transition from potential to realized transport pathways even in the presence of exceptional complexity can be difficult for a system to discover.

As a general principle, MaxEP may have a wider applicability as a limiting or asymptotic (in time) condition than as a statement of what to expect from the behavior of a system at a particular instant. MaxEP in the event could still provide clues about what might be expected of the future behavior of a complex system of interest, in which case it would become more a tool of anticipation and expectation than of specific prediction. The probabilistic basis for our understanding of MaxEP also has another implication. If a condition of maximum entropy production obtains for the reason that there is an overwhelming number of ways in which it could be realized, then we should expect, if a system subject to an internal constraint finds a way around the constraint, that the constrained state, which would *ex post facto* be a statistically rare configuration, will not be likely to reappear. In other words, under MaxEP a driven complex system should tend to evolve ratchet-like over time toward a maximum entropy production state.

21.5 Jevons' Paradox

A possible example of such a ratchet effect and an illustration of how MaxEP might eventually play a role in analysis of economic processes is provided by what in economics is called Jevons' Paradox [22]. This is the proposition that increasing the efficiency of, or the efficient use of, technological systems may offer more rather than less opportunity to satisfy the incentives that drive human energy consumption, leading to an increase rather than a decrease in energy consumption. Thus increasing efficiency might decrease energy consumption but by less than the amount of energy nominally saved by the efficiency measure (rebound effect), or might cause consumption to actually increase (Jevons' Paradox or backfire effect). Jevons' Paradox is controversial and arguments have been adduced both supporting and opposing its validity [23, 24]. The rebound effect has been documented in a number of specific cases, but direct evidence for the backfire effect is sparse and arguments in its support are mainly theoretical. Historically, increases in technological efficiency have been accompanied by increases in energy use, but the causal relationship is unclear [24].

Difficulty in documenting Jevons' Paradox might be expected if the backfire effect is mainly a macroeconomic phenomenon, with the energetic consequences of increases in efficiency in a specific technology diffusing out along complex and difficult to trace causal chains from the micro to the macro realm before the aggregated effect of increased energy use finally becomes measurable (but difficult to attribute). This is reminiscent of the conditions under which MaxEP is supposed to apply—the existence of a complex system powered by a large supply of free energy in the presence of multiple pathways by which that energy can be dissipated. Jevons' Paradox is usually explained as being a consequence of the introduction of technological efficiency "causing" an increase in energy consumption. The problem with such a causal explanation in which one tries to trace through a series of specific cause-and-effect relations is the same as the problem of causal explanation in any system where detailed dynamics is drowned in complexity—namely, it is not in practice possible to anticipate complex emergent (macro) behavior by simply following micro-variables [25]. MaxEP attempts to solve this problem by discarding micro-variables and appealing to statistics. The effects of system dynamics and structure are retained in a generalized way in the form of constraints which the statistics must respect. For example in the application of MaxEP to the transport of heat by the earth's atmosphere, energy loss to space is constrained to occur according to the requirements of the Stefan-Boltzmann radiation law [10]. In the context of Jevons' Paradox, adoption of energy-efficient technology can be considered in essence the removal of a constraint. Processes that could not previously be executed and gizmos that could not previously be fabricated can now be performed and manufactured. If for the sake of illustration we took the total pool of energy potentially available for consumption to be that contained in the world's recoverable fossil fuel deposits, then MaxEP would imply that the removal of constraints would result in an increased rate of energy drawdown from that pool, because there are more ways (pathways) by which this could occur than there are ways in which it could be avoided. Roughly speaking it is easier to squander a new resource than to use it wisely because there are more opportunities to do the former than the latter. The alleged "paradoxical" nature of backfire can thus be given a natural if qualitative explanation in the context of MaxEP, and at the end of this chapter we make some comments based on MaxEP about possible future rates of consumption of fossil fuel energy. However, MaxEP, like Jevons' Paradox, remains a hypothesis or conjecture rather than an established physical principle, and thus our discussion of energy efficiency and, more broadly, of energy consumption by the technosphere represents only a suggestive framework for thinking about the role of maximum entropy production in a technological world.

21.6 Entropy Production in the Rayleigh-Benard Cell

Because internal constraints affect system function, it is useful to consider in some detail a constrained entropy producing system about which we have some physical understanding before looking at entropy production in more complex technological

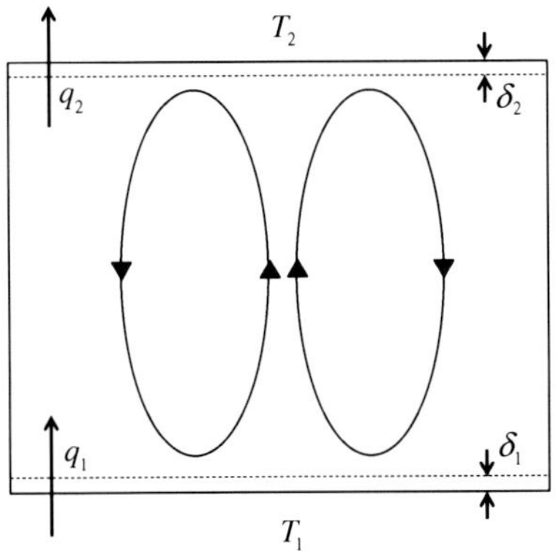

Fig. 21.1 Schematic drawing of Rayleigh-Benard apparatus showing convection cells and thermal boundary layers $\delta_{1,2}$. The temperature of the bottom plate T_1 is higher than that T_2 of the upper plate. In steady-state the average heat flux through the bottom plate q_1 equals the average heat flux q_2 through the upper plate, and under steady conditions the average boundary layer thicknesses $\delta_{1,2}$ are equal

systems. For this purpose we consider the Rayleigh-Benard (RB) cell, a constrained entropy producing system that has been offered [26] as an example of a system that generates entropy at the maximum possible rate. Here we reinterpret entropy production in this system and use the results of this reinterpretation to suggest how we might think about maximum entropy production in large-scale fossil-fuel-consuming technological systems, where, we argue, analogous constraints operate.

In an RB cell, Fig. 21.1, the space between two horizontal parallel plates is filled with a fluid having a positive coefficient of thermal expansion. The bottom plate is maintained at a temperature T_1 (K) and the top plate at a temperature T_2, with $T_1 > T_2$. At low enough temperature difference, $\Delta T = T_1 - T_2$, heat simply diffuses from bottom to top, with the heat flux given by $q = -k\Delta T/\Delta x$ (kg s^{-3}) where k (kg m s^{-3} K^{-1}) is the thermal conductivity of the fluid and $\Delta x = D$ (m) is the separation of the plates. The rate of entropy production and export by the cell is $\sigma = q(1/T_2 - 1/T_1)$ (kg s^{-3} K^{-1}). As ΔT increases, buoyancy forces become important and advective pathways become accessible. Thus at some value of ΔT buoyancy is strong enough that a hot blob of fluid is propelled upward toward the cold plate. If its velocity is high enough the advected heat flux will exceed the diffusive heat flux calculated above, and the rate of entropy production will increase. Advective motion begins when the dimensionless Rayleigh number $\mathrm{Ra} = D^3 g\alpha\Delta T/\nu\kappa$ exceeds a certain critical value Ra*. Here, α (K^{-1}) is the thermal volumetric expansion coefficient, ν (m^2 s^{-1}) is the kinematic viscosity of the fluid, κ (m^2 s^{-1}) is its thermal diffusivity, g(m s^{-2}) is the acceleration of gravity, and ρ is fluid density (kg m^{-3}). The variables k and κ are related by $C_P = k/\rho\kappa$, where C_P (m^2 s^{-2} K^{-1}) is the specific heat capacity of the fluid at constant pressure.

For $\mathrm{Ra} > \mathrm{Ra}^*$, most of the space between the plates is occupied by advecting fluid at relatively uniform temperature. Immediately adjacent to each boundary, however, there exists a thin layer, the thermal boundary layer, of thickness δ (m), where diffusion dominates advection. Diffusion dominates here because the effective Rayleigh number of the layer is smaller than Ra*. Physically, the boundary layer remains diffusive even for large ΔT because there is always some short-enough distance near the plates for which the diffusive transport time scale is shorter than the corresponding advective transport time scale The boundary layer thickness is determined [26] by the critical Rayleigh number, $\mathrm{Ra}^* = \delta_i^3 g\alpha\Delta T_i/\nu\kappa = \xi\delta_i^3\Delta T_i$, where $\xi = g\alpha/\nu\kappa$ (m^{-3} K^{-1}) is a constant, characteristic of the fluid, and δ_i and ΔT_i refer respectively to the boundary layer thickness and to the temperature drop that occurs across the boundary layer at plate i, with $i = 1$ for the hot lower plate and $i = 2$ for the cold upper plate.

Diffusion and advection operate in series in the RB cell, with advection dominating long distance transport through most of the interplate volume, and diffusion dominating transport across the thin boundary layer at each plate. Most of the temperature drop in the cell occurs in the thermal boundary layers, whereas the flow across the main body of the cell generates relatively little entropy. The thermal boundary layer, though thin, is critical to the dynamics of the cell. It can be thought of as a localized auto-control subsystem that determines the overall heat flux through the cell. A diffusive boundary layer is necessary in a standard RB cell in order to communicate heat from the plate to the flow (and vice versa). A flowing fluid is essentially a mechanical part whose macroscopic (flow) degrees of freedom cannot directly absorb or emit heat or entropy. Entropy (and heat) is transferrable to and from the flow only by microscopic diffusion.

Since presumably the steady-state value of δ cannot be not less than the value determined above in terms of the critical Rayleigh number, Ra*, because that relationship is what defines the range of diffusive dominance, the rates of heat transport and thus entropy production in the RB cell appear to be as large as they can be, i.e., maximal. However, we have inferred this fact independently of MaxEP by considering the physical meaning of the critical Rayleigh number. Moreover, we might just as well argue that the rate of entropy production is minimal, since δ cannot assume any larger value on average than that determined by Ra*, else buoyancy forces would dominate transport in the boundary layer. That is, a straight forward application of MaxEP does not apply to the RB cell because there is nothing to vary—there is no variational parameter that would allow us to find a maximum entropy production state. If we knew nothing about the internal dynamics of the cell and blindly applied MaxEP using say the flux q as a variational parameter, then there would be nothing to limit the flux of energy through the cell, and entropy production predicted on the basis of MaxEP would be infinite, or, rather, a value limited by the capacity of the laboratory power supply.

Nonetheless, the RB cell example can provide insight into entropy production and the role of MaxEP in more complex systems for which one does not have as clear a picture of internal dynamics. In the RB cell the time average heat flux through the hot boundary layer is $q_1 = k\Delta T_1/\delta_1$, where ΔT_1 is the temperature drop across δ_1. Under steady-state conditions, $q_1 = q_2$, where q_2 is the flux

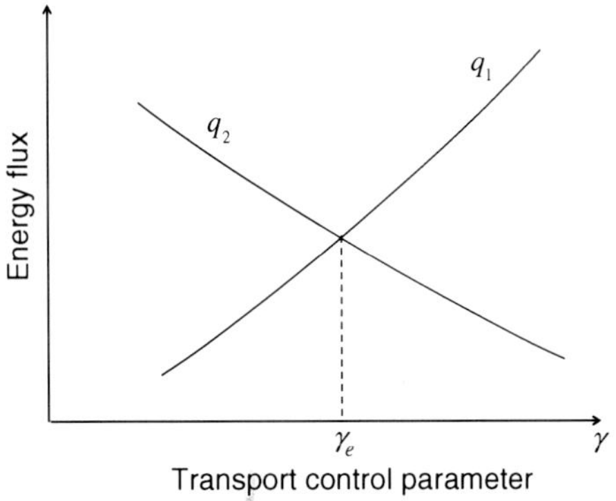

Fig. 21.2 Energy fluxes at hot plate, q_1, and cold plate, q_2, in an RB cell, versus transport parameter $\gamma = \delta_1^{-1}$. The steady-state condition is at $\gamma = \gamma_e$

through the cold boundary layer. The average temperature drops at the two plates, as well as the average boundary layer thicknesses, are also equal, $\Delta T_1 = \Delta T_2 = \Delta T/2$, and $\delta_1 = \delta_2$, respectively. With the plate temperatures held constant, we now imagine a small fluctuation in the temperature drop ΔT_1 across the hot layer, so that ΔT_1 and ΔT_2 become momentarily unequal. From the definition of the critical Rayleigh number $\delta_{1,2} = \left(\mathrm{Ra}^*/\xi\Delta T_{1,2}\right)^{1/3}$ or $\Delta T_{1,2} = \mathrm{Ra}^*/\xi\delta_{1,2}^3$, so that $1/\delta_2^3 = 1/\delta_{\min}^3 - 1/\delta_1^3$, where $\delta_{\min} = (\mathrm{Ra}^*/\xi\Delta T)^{1/3}$ is a constant. Thus, at fixed ΔT, an increase in ΔT_1 is accompanied by a decrease in δ_1 and a corresponding increase in δ_2, and vice versa. The dependence of the virtual fluxes $q_{1,2} = (k\mathrm{Ra}^*/\xi)\delta_{1,2}^{-4}$ on the transport control parameter $\gamma = 1/\delta_1$ (m^{-1}) is shown qualitatively in Fig. 21.2.

The steady state or "equilibrium" condition of the cell corresponds to the value $\gamma = \gamma_e$ where the two curves cross. If, due to a fluctuation in ΔT_1, the hot boundary layer thickness δ_1 momentarily decreases in value, so that γ exceeds γ_e, then the hot plate would begin to supply an increased heat flux to the main body of the cell. The cold boundary layer system under this condition, however, can absorb only a decreased heat flux in virtue of the corresponding increase in δ_2. The increased heat flow at the hot plate tends to reduce the temperature fluctuation, and the energy flux through the hot boundary layer subsequently decreases (relaxes toward equilibrium). At the same time, the heat flux to the cold plate increases as the system moves back toward steady-state at $\gamma = \gamma_e$. The boundary layer provides the mechanism by which the physically separate energy source and sink coordinate their behavior. Such a mechanism is needed in any system in which two remote regions must agree on the serial production and consumption of the same packet of energy. It is worth noting that the present analysis at or near steady-state says nothing about the dynamical route that a system may take in arriving at a steady condition, whose resolution presents a much harder problem.

21.7 Boundary Layers in the Technology Cell: The Role of Price

The problem of coordinating the rate of energy production and consumption in the RB cell is similar to that faced by technological systems that rapidly advect energy from a localized resource to a distant consumption zone. For simplicity we consider an idealized "technology cell" in which energy consumption is located at B, say a major city or industrialized zone, and the source of energy is located at some distant point A, such as an oil field or coal mining region. Region A corresponds to the hot plate in the RB cell, and region B to the cold plate. For simplicity we only consider transport of chemical energy. Energy-rich mass must be transported from A to B, where we assume each unit of chemical energy is soon dissipated (burned) to produce heat at ambient temperature T together with a corresponding quantity of entropy. The technology cell, as the concept is used here, includes the economic mechanisms associated with energy use as well as the physical operations of energy production, transport, and consumption.

The dynamics of the entropy production process in the technology cell is similar to that in the RB system in that a low-entropy-producing advective transport corridor (shipping lanes, highways, railroads, etc.) is bounded on one end by a non-advective consumptive zone (population center) and at the other end by a non-advective production zone (the location of the physical oil or coal operation). The source and sink zones of the technology cell are taken to be "diffusive" in the sense that the slower speed of local transport and the spatial complexity of extraction and preparation for long distance transport in the production zone, and the low speed and finely reticulated delivery pathways to individual customers in the populated consumption zone, are easily distinguished from the long pathways and high-speed transport characterized by directed conveyance through the intervening advective zone. As in the RB cell, a diffusive boundary layer is needed at the consumptive end of the technology cell to handle what would otherwise be a mismatch between the amount of energy each consumer can absorb per unit time and the quantity of energy delivered to the consumption zone per unit time by rail, highway or other advective mode. A similar boundary layer is needed in the production zone to match the size of the units of energy extracted from the ground and their rate of extraction, to the bulk advective transport mechanisms by which the energy is moved to distant locations.

The differences between the RB cell and the technology cell are many of course, including that fossil-energy flux is powered by a chemical affinity rather than a temperature gradient, that the length scale of the technological system may be thousands of kilometers rather than a few centimeters, that truck, rail, and/or ship transport typically substitute for an advecting fluid, that source and sink zones in the technology cell have distinct dynamics, such as pumping or mining on one end and factory operation and home heating on the other unlike in the RB cell where each zone operates by the same mechanism, and that the function of the technology cell depends on economic as well as physical variables. Despite these

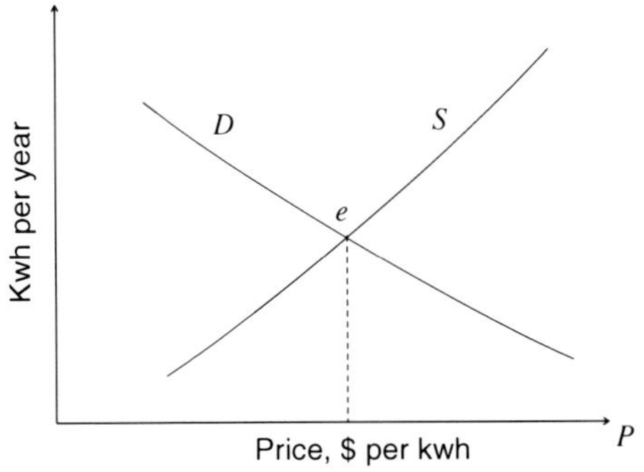

Fig. 21.3 Energy supply (*S*) and demand (*D*) curves for the technology cell. Curve *D* is the putative energy consumption rate in the "cold" zone of the technology cell and curve *S* is the putative energy production rate in the "hot" zone. Point *e* is the equilibrium point that defines the realized energy flow between the two ends of the cell. The axis orientations on this "supply/demand" graph are reversed from the usual convention in economics (where the vertical axis usually represents the price per unit good and the horizontal access the quantity of goods per unit time)

differences the similarities are strong enough that we are led to analyze entropy production in the technology cell using what we have learned from the RB cell. Thus we identify a variable that plays a role similar to γ in the RB cell, i.e., that coordinates the rate of energy transport through two widely separated diffusive "chemical" boundary layers. That variable is price per unit energy, P (\$ kg^{-1} m^{-2} s^{2}), where an equilibrium price per unit energy P_e determines the average energy flux through the cell, Fig. 21.3. Physical properties, such as system-wide advection for fast long-distance transport, localized diffusion zones as the sources of usable energy and as sinks for degraded energy, and the necessity in a steady-state system of a control parameter jointly shared by both source and sink, survive the transformation from the RB cell to large-scale technological systems because they represent functionality essential to a high-metabolism bipolar system.

There may be other determinants of the rate of energy consumption besides price—under some circumstances gasoline can be rationed, for example. One might also argue that market mechanisms that enable meeting of supply and demand are better represented by transaction costs than price, which costs might thus be a preferred mirror to the role of the thermal boundary layer. However, in the present exploratory study we stick to price to illustrate the main concepts.

It is worth noting that the condition of equilibrium or steady-state is defined here for the relatively fast time scales in which the dynamics of interest plays out, and not for much slower time scales over which otherwise effectively constant parameters that define the background conditions of the dynamic state may change. For example, a drift in the temperature difference between the plates in the RB cell can be ignored (up to a point) if the drift time is slow relative to time scales such as the mean recycle time of a parcel of water as it makes the rounds from hot to cold plate and back to hot again. Similarly, in the technology cell, equilibrium is defined for short enough periods of time that factors such as the regulatory climate or availability of new technical means can be considered fixed. Such assumptions are similar to those that are routinely made in thermodynamics, which is seldom applied to a body in true thermal equilibrium, but is assumed to be valid when applied to bodies where rates, such as the rate of change of temperature of a heat reservoir, whose nonzero value would violate strict equilibrium, can nonetheless be taken effectively to vanish.

A steady-state analysis of the technology cell, as in the case in the RB cell, is limited by its inability to specify the approach to "equilibrium"; for example, the above analysis does not address the role that profits and entrepreneurship play in reaching the condition of equilibrium, nor in general answer the question of *how* the behavior of many individual microscale entities in the pole regions (buyers and sellers) results in macroscale phenomena such as the bulk transfer of goods. The response of the macro regime reflects the specific conditions of the system in question—for example the (macro) advective, buoyant response to the (micro) diffusive heating of the fluid in the RB cell is a mechanism specific to that system. The model offered here treats only the requirement of a certain relation between two *types* of dynamics (diffusive or advective) in a high-speed production-advection-consumption system.

21.8 Entropy Production in the Technology Cell

In the technology cell there is an intermediate value of price that can accommodate a flow rate of energy that will simultaneously satisfy the requirements for transport through both the production and consumption zones. In an economic analysis this "equilibrium" price is determined by the crossing of supply and demand curves, Fig. 21.3. Here the supply curve represents the amount of free energy per year or other unit of time that the supplier (producer) wants to provide at a given price, and the demand curve is the amount of free energy that a potential buyer is willing to buy (consume) at the given price, all other factors remaining constant.

The interpretation of the supply and demand curves in the technology cell is similar to the interpretation of the flux curves in the RB cell. If the price of oil or coal in the technology cell should fluctuate upward from the equilibrium price, then in the consumption zone the diffusive throughput of fuel, that is, the quantity of fuel demanded per unit time for burning in houses, offices, and factories, as shown by the demand curve, would decrease, even though at the higher price the production zone could supply a flux of energy in excess of demand. A higher price in the technology cell would generate the analog of a larger (than equilibrium value) thermal boundary layer thickness at the cold plate in the RB cell. If instead the price should fluctuate downward, in analogy to an increase in boundary layer thickness in the hot layer in the RB cell, the transport rate of free energy through the production zone, i.e., the supply rate, would decrease, even though under these conditions demand, reflected in the willingness of the consumptive layer to absorb additional energy if it were supplied at such a price, would increase. As in the RB cell, in the technology cell average energy consumption is set by the equilibrium value of the control parameter, the equilibrium price, P_e. If the price drifts away from the equilibrium price, "market forces" work to push it back toward the equilibrium value. The idea of an equilibrium price for a good is conditioned on other variables that might potentially affect price remaining constant. These include variables that describe the effects of costs, tax policy, government regulations, the state of extraction technology and of technology-enabled efficiency, degree of competitiveness in the market, and so on. Fixing these variables is equivalent to imposing constraints on energy consumption by the technology cell.

Some factors that appear in the short term as constraints on energy consumption are probably more realistically identified as slower modes of system dynamics that are potentially subject to relaxation, with consequent increase in rate of entropy production. For example, a fixed number of consumers represents a constraint that limits the ways in which the entropy production rate could increase. But over longer time scales products of technology, including increased availability of food, sanitation and other goods and services conducive to health and longevity, may lead to an increase in population and a consequent increase in rate of energy use. An increase in technological efficiency may also increase the rate of energy usage (Jevons' Paradox). If MaxEP is applicable to such systems it would be a conditional applicability. At any point in time the technology cell would be in a conditional state of maximum entropy production as determined by the equilibrium price, with the conditioning provided by constraints many of which are only nominal. In that case one expects that over time the technology cell would discover ways to change the supply–demand relation in a way that increases energy flow.

An argument against the idea that equilibrium price defines a state of maximum entropy production is the observation that someone could simply set fire directly to the oil production fields, burning the oil at a high rate that has nothing to do with supply and demand, as happened in the torching of the Kuwaiti oil fields in 1991 [27]. It may seem that if oil were pumped at the maximum possible rate (according to the capacity of oil field extraction technology) and immediately burned then entropy production would increase above the supply–demand crossing point determined in the absence of burning. Such a condition, however, would be only a fluctuation, and could not be maintained in the long run. For steady entropy production, with which we are concerned here, a steady driving force is required. In the case of the burning oil field there is no force to drive a sustained flow of oil because there is no demand for what under these circumstances would be a useless product. The price to burn oil on the spot would be too low to sustain a significant energy flux.

Relaxation of constraints can be illustrated by considering a gedanken experiment in which an RB cell is able to change its own rate of entropy production. In the RB cell the boundary layer thickness $\delta = (2\nu\kappa \mathrm{Ra}^*/g\alpha\Delta T)^{1/3}$ and the steady equilibrium energy flux, $q = k\Delta T/2\delta = (k^2\rho C_P g\alpha/16\nu \mathrm{Ra}^*)^{1/3}\Delta T^{4/3}$ reflect the combined effects of system properties such as viscosity, thermal diffusivity, density, and volumetric expansion coefficient. To the extent that these parameters are fixed the above relations describe a constraint on cell dynamics. However, if the RB cell had the capability of spontaneously changing one or more system parameters, then it would be able to increase its rate of entropy production. For example, if, with fixed driving force ΔT, the system were able to spontaneously reduce fluid kinematic viscosity v, then energy flux through the cell as well as the rate of entropy production would increase. What is in essence a mechanical constraint (due to friction) would have been relaxed, δ would decrease, and the same driving force would become capable of moving more energy per unit time through the cell. MaxEP implies that if the RB cell were able to access a dynamical state of lower viscosity, it would switch to that state, increasing its entropy production as a result.

If MaxEP, in the conditional form described above, applies to large-scale, high-metabolism technological systems like those fossil fuel systems that power the economy, then we might expect these systems to behave *as if* they were trying to circumvent the effects of any constraints on energy consumption. Oil companies display their implicit belief in an intuitive version of this prediction of MaxEP when they invest billions of dollars in cutting edge technology (like horizontal drilling) with no real assurance that new policies or political constraints will not make those investments a failure. They believe that if known physical barriers to production (e.g., known external constraints due to difficulty of accessing deposits) are reduced, then one way or another political or other internal constraints will eventually give way and the newly accessible energy resources will be consumed (and entropy production increased).

The interpretation of MaxEP presented here may also, and for the same reason, raise our own skepticism about the likely long-term effectiveness on constraining fossil fuel use of instruments like policy statements, regulations, treaties, and political decisions, which, from the point of view of MaxEP, would appear as rules made to be broken. Even what appears as a straightforward and foolproof strategy for decreasing the rate of energy use, namely, increasing the energy efficiency of technology, may backfire to result in increased energy consumption.

The line of argument presented above suggests that the principle of Maximum Entropy Production when applied to complex technological systems like those that (literally) power the economy may find greater utility in indicating where world energy consumption is headed than as an explanation of its present state of energy use. The latter would appear to lie in the details of system dynamics, i.e., in the bailiwick of economics. If MaxEP should finally be put on a sound theoretical footing, the resulting psychological impact of the knowledge that an increasing rate of energy use is not something humans can ultimately control might be a more important practical consequence of MaxEP than would be any purely physical application of the principle, because it would change the way that humans look at their place in the world.

Acknowledgments Comments, criticism, and advice by David Furbish, Adrian Down, Kenneth Ells, Evan Goldstein, Tim Johnson, Pat Limber, Dylan McNamara, Robert Lanfear, and Stacey Worman are appreciated. I also thank Robert Niven, Carsten Herrmann-Pillath and an anonymous referee for helpful remarks in revision.

References

1. Arthur, B.: The Nature of Technology: What It is and How It Evolves. Free Press, NY (2009)
2. Schrodinger, E.: What is Life?. Cambridge University Press, Cambridge (1944)
3. von Bertalanffy, L.: General System Theory. George Braziller, NY (1969). (revised edition)
4. Odum, H.T.: Ecological and General Systems. University of Colorado Press, Niwot, Colorado (1994). (revised edition)
5. Harte, J.: Maximum Entropy and Ecology. Oxford University Press, Oxford (2011)
6. Dewar, R.C.: Maximum entropy production and plant optimization theories. Phil. Trans. R. Soc. B **365**, 1429–1435 (2010). doi:10.1098/rstb.2009.0293
7. Georgescu-Roegen, N.: The Entropy Law and Economic Process. Oxford University Press, London (1971)
8. Ruth, M.: Insights from thermodynamics for the analysis of economic processes. In: Kleidon, A., Lorenz, R. (eds.) Non-equilibrium Thermodynamics and the Production of Entropy. Springer, Berlin (2005)
9. Annila, A., Salthe, S.: Economies evolve by energy dispersal. Entropy **11**, 606–633 (2009). doi:10.3390/e11040606
10. Kleidon, A., Lorenz, R.: Entropy production by earth system processes. In: Kleidon, A., Lorenz, R. (eds.) Non-equilibrium Thermodynamics and the Production of Entropy. Springer, Berlin (2005)
11. Kleidon, A.: Non-equilibrium thermodynamics, maximum entropy production and Earth-system evolution. Phil. Trans. R. Soc. A **368**, 181–196 (2010). doi:10.1098/rsta.2009.0188
12. Dyson, F.J.: Energy in the universe. Sci. Am. **225**, 51–59 (1971)
13. International Energy Agency: Key World Energy Statistics 2011. http://www.iea.org/publications/free_new_desc.asp?pubs_ID=1199 (2011). Accessed 1 Dec 2011
14. Brookins, D.G.: Radionuclide behavior at the Oklo nuclear reactor, Gabon. Waste Manage **10**, 285–296 (1990)
15. Crutzen, P.J., Stoermer, E.: The anthropocene. Glob. Change Newsl. **41**, 17–18 (2000)
16. Haff, P.K.: The landscape Reynolds number and other dimensionless measures of earth surface processes. Geomorphology **91**, 178–185 (2007)
17. Haff, P.K.: Hillslopes, rivers, plows, and trucks: mass transport on earth's surface by natural and technological processes. Earth Surf. Proc. Land. **35**, 1157–1166 (2010). doi:10.1002/esp.1902
18. Paltridge, G.W.: Global dynamics and climate—a system of minimum entropy exchange. Q. J. R Meteor. Soc **101**, 475–484 (1975)
19. Schrodinger, E.: Statistical Thermodynamics. Dover, New York (1989)
20. Niven, R.K.: Steady state of a dissipative flow-controlled system and the maximum entropy production principle. Phys. Rev. E **80**, 021113 (2009). doi: 10.1103/PhysRevE.80.021113
21. Dewar, R.C., Maritan, A.: The theoretical basis of maximum entropy production (2012) (this volume)
22. Jevons, W.S.: The Coal Question, 2nd edn. Macmillan, London (1866). http://www.pdrap.org/The_Coal_Question/The_Coal_Question.pdf. Accessed 6 Aug 2012
23. Polimeni, J.M., Polimeni, R.I.: Jevons' paradox and the myth of technological liberation. Ecol. Complex **3**, 344–353 (2006)
24. Sorrell, S.: Jevon's paradox revisited: the evidence for backfire from improved energy efficiency. Energ. Policy **37**, 1456–1469 (2009)
25. Werner, B.T.: Modeling landforms as self-organized, hierarchical dynamical systems. In: Wilcock, P.R., Iverson, R.M. (eds.) Prediction in Geomorphology, Geophysical Monograph 135. American Geophysical Union, Washington DC (2003). doi:doi.1029/135GM10

26. Ozawa, H., Shimokawa, S., Sakuma, H.: Thermodynamics of fluid turbulence: a unified approach to the maximum transport properties. Phys. Rev. E 026303 (2001). doi: 10.1103/PhysRevE.64.026303
27. Campbell, R.W.: Iraq and Kuwait: 1972, 1990, 1991, 1997. In: Campbell, R. W. (ed.) Earthshots: Satellite Images of Environmental Change. United States Geological Survey (1999). http://earthshots.usgs.gov Dec 2011. Accessed 24 July 2012

Chapter 22
The Entropy of the Universe and the Maximum Entropy Production Principle

Charles H. Lineweaver

Abstract If the universe had been born in a high entropy, equilibrium state, there would be no stars, no planets and no life. Thus, the initial low entropy of the universe is the fundamental reason why we are here. However, we have a poor understanding of why the initial entropy was low and of the relationship between gravity and entropy. We are also struggling with how to meaningfully define the maximum entropy of the universe. This is important because the entropy gap between the maximum entropy of the universe and the actual entropy of the universe is a measure of the free energy left in the universe to drive all processes. I review these entropic issues and the entropy budget of the universe. I argue that the low initial entropy of the universe could be the result of the inflationary origin of matter from unclumpable false vacuum energy. The entropy of massive black holes dominates the entropy budget of the universe. The entropy of a black hole is proportional to the square of its mass. Therefore, determining whether the Maximum Entropy Production Principle (MaxEP) applies to the entropy of the universe is equivalent to determining whether the accretion disks around black holes are maximally efficient at dumping mass onto the central black hole. In an attempt to make this question more precise, I review the magnetic angular momentum transport mechanisms of accretion disks that are responsible for increasing the masses of black holes

22.1 The Entropy of the Observable Universe

Stars are shining, supernovae are exploding, black holes are forming, winds on planetary surfaces are blowing dust around, and hot things like coffee mugs are cooling down. Thus, the entropy of the universe $S_{\rm uni}$, is increasing, and has been

C. H. Lineweaver (✉)
Planetary Science Institute, Research School of Astronomy and Astrophysics and the Research School of Earth Sciences, Australian National University, Canberra, ACT 0200, Australia
e-mail: charley.lineweaver@anu.edu.au

R. C. Dewar et al. (eds.), *Beyond the Second Law*,
Understanding Complex Systems, DOI: 10.1007/978-3-642-40154-1_22,

increasing since the hot big bang 13.8 billion years ago [1]. The universe obeys the second law of thermodynamics:

$$dS_{uni} \geq 0. \tag{22.1}$$

In the entropy literature there is often confusion about both the boundary of "the system" and the distinction between the rate of increase of the entropy of the system dS/dt, and the rate entropy is produced by the system σ [2]. For example, in the Earth system, many processes are producing entropy (therefore, naively, the entropy of the Earth should be increasing), but the entropy produced is being exported into the interstellar radiation field (therefore the entropy of the Earth could be constant). Assuming a steady state for the Earth means that the amount of entropy exported is equal to the amount of entropy produced, thus $dS/dt = 0$ but, $\sigma > 0$ [3]. Such an entropy-producing steady state can only happen when the system, or control volume, is different from (e.g. hotter than) the environment. This difference allows the system to export the entropy it produces, to the environment.

The entropy of the universe is more simple to deal with because the boundaries of the system are not an issue. We have much evidence that the universe is homogeneous on scales above $\sim$100 million light years [4]. This homogeneity makes the distinction between a very large control volume (100 million light years)^3 and its environment, meaningless. Volumes of the universe that are at least that big are essentially identical. That is, they are so large that their average density of black holes, supernovae, stars and planets, accurately represents the average density of these objects everywhere in the universe. Thus, the amount of entropy being produced by these structures in any large control volume, is the same as the entropy being produced in the neighbouring control volumes. Thus, in cosmology we can ignore the system boundary problem. Without an environment into which to dump entropy, we have,

$$dS_{uni}/dt = \sigma_{uni} > 0. \tag{22.2}$$

Thus, we can ignore the distinction between dS_{uni}/dt and σ_{uni}. We can consider a representative sample volume of the universe (say the current observable universe) without worrying about the net import or net export of heat or mass or entropy across any boundary, because there is no net import or export. For smaller, unrepresentative volumes of the universe, $V <$ (million light years)^3, this simplicity does not exist because there can be local inhomogeneities: an over-density of matter such as a galaxy cluster or a giant wall of galaxies, or an under-density of matter such as a cosmic void. In our analysis of cosmic entropy [5] the control volume is the observable universe—the sphere around us with a radius equal to the distance light has traveled since the big bang.

22.1.1 Expansion of the Universe is Isentropic

Since 1929, we have known that the universe is expanding. This expansion is isentropic [1, 6]. That is, the entropy of relativistic particles such as photons, gravitons and neutrinos does not increase or decrease with the expansion. This is because the entropy of a gas of relativistic particles is proportional to the number of particles N, which does not change as the universe expands. If we follow the entropy of a comoving volume of the universe, forward or backward in time, the number of photons in that volume does not change.

Another way to understand that the expansion of the universe is isentropic is to use the fact that the entropy density s of photons (or any relativistic particle) is proportional to the temperature cubed: $s \sim T^3$. Also, the temperature of relativistic particles is inversely proportional to the size of the universe (represented by the scale factor a): $T \sim 1/a$ (the particles lose energy since their wavelengths expand with the universe, $\lambda \sim a$). Also, the volume under consideration is proportional to the cube of the size of the universe: $V \sim a^3$. Combining these facts lets us derive that the entropy S of the photons in any volume expanding with the expansion of the universe, is $S = sV \sim a^{-3}a^3 = \text{constant}$. In addition, the expansion of the universe does not increase the rate at which mass accretes into black holes. Thus, expansion does not increase the entropy of the universe. The adiabatic expansion of an ideal gas into empty space is irreversible and thus the entropy, which is proportional to volume, increases. This is not the case in cosmology because the CMB photons are not expanding into empty space.

22.1.2 The Entropy Budget of the Universe

The entropy of a black hole of mass M_{BH} is proportional to the square of the mass [7–9]:

$$S_{\text{BH}} = k(4\pi G/c\hbar)\, M_{\text{BH}}^2 \tag{22.3}$$

where k is Boltzmann's constant, G is Newton's constant, c is the speed of light and $\hbar$ is Planck's constant divided by 2π. To obtain the entropy of black holes in the universe, we multiplied Eq. (22.3) by the mass function of black holes and then integrated over mass and volume [5]. The result is: $S_{\text{BHs}} \sim 3.1 \times 10^{104}\, k$. The M_{BH}^2-weighted black hole mass function peaks in the range $\sim 10^9 \sim 10^{10}$ solar masses. Therefore, such supermassive black holes at the cores of the most massive elliptical galaxies (which are in the central regions of the most massive clusters of galaxies), are the source of most of the entropy in the universe (Fig. 22.1).

The entropy density of non-relativistic particles can be computed from the Sakur-Tetrode equation [10] which gives the entropy per baryon, which we then multiplied by the density of baryons. The second largest contribution to the entropy comes from the photons of the cosmic microwave background and a close

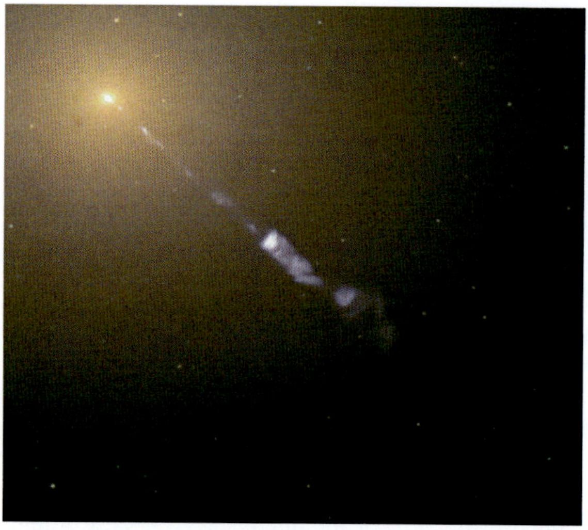

Fig. 22.1 M87 is the closest giant elliptical galaxy at the core of the Virgo Cluster of galaxies, of which our galaxy is an outlying member. The black hole at the center of M87 has a mass $\sim 7 \times 10^9\ M_{\mathrm{Sun}}$. Black holes of this mass are called supermassive black holes and dominate the entropy budget of the universe. The central black hole is larger than the radius of Pluto's orbit. The accretion disk which feeds the central black hole is $\sim$0.4 light years in diameter and is rotating at velocities of up to $\sim$1,000 km/s. The accretion rate onto the black hole is 0.1 M_{Sun}/year. Magnetic fields in the accretion disk collimate the ejected material forming the prominent relativistic jet coming out of the black hole in the upper left of the image. *Image* Hubble Space Telescope/STScI/AURA

third is from cosmic neutrinos. Both of these are a quadrillion ($=10^{15}$) times smaller than the entropic contribution from black holes. An important distinction to make is between the entropy content of various components of the universe (Table 22.1) and entropy production. The dominant sources of entropy production are the accretion disks around black holes (Sect. 22.3).

22.2 The Entropy Gap and the Initial Entropy of the Universe

The early universe was close to thermal equilibrium. Direct evidence for this comes from the high level of isotropy of the temperature maps of the cosmic microwave background (CMB) [11, 12]. CMB photons give us a direct view of the universe as it was $\sim$380,000 years after the big bang when the entire universe had a temperature of $\sim$3,000 K. Tiny temperature fluctuations in the CMB maps have a $\Delta T/T \sim 10^{-5}$. That is, the anisotropies seen in the maps (hot spots and cold spots) are deviations of amplitude $\Delta T \sim 30\ \mu$K around the current average temperature $T = 3$ K. If CMB photons were its only component, the universe would

Table 22.1 Entropy [k] of the various components of the observable universe

Black holes	$S_{BHs} \sim 3.1 \times 10^{104}$
Cosmic microwave background photons	$S_{photons} \sim 5.4 \times 10^{89}$
Cosmic neutrinos	$S_{neutrinos} \sim 5.2 \times 10^{89}$
Dark matter	$S_{DM} \sim 2 \times 10^{88}$
Cosmic graviton background	$S_{gravitons} \sim 6.2 \times 10^{87}$
Interstellar medium and intergalactic medium	$S_{ISMIGM} \sim 7.1 \times 10^{81}$
Stars	$S_{stars} \sim 9.5 \times 10^{80}$

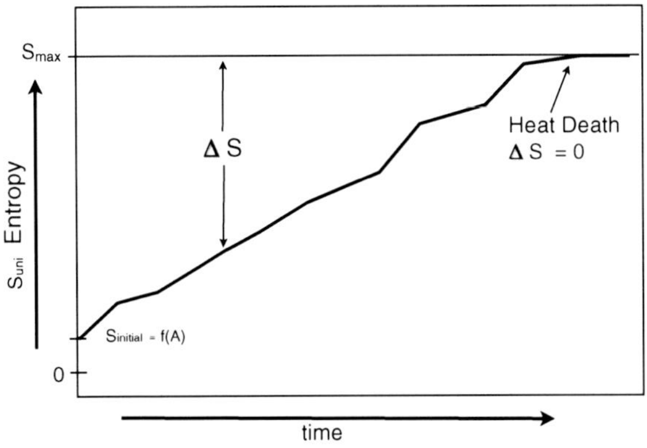

Fig. 22.2 The Entropy of the Universe as a Function of Time. $S_{uni}(t)$ monotonically increases. We define S_{max} as a constant equal to the largest entropy that the universe will ever have, $S_{uni}(t \to \infty) = S_{max}$. We define the entropy gap as $\Delta S(t) = S_{max} - S_{uni}(t)$. When $\Delta S = 0$, the universe reaches an equilibrium heat death [13]. The low initial entropy of the universe is due to the low gravitational entropy [1, 14, 16], which, one day, should be parametrized by the large scale structure normalization A (a parameter used by cosmologists to quantify the initial clumping of matter). If the universe were born with a high entropy, we would have $S_{initial} \sim S_{max}$, and $\Delta S \sim 0$, and a lifeless universe. Figure from [1]

have started out in equilibrium, at maximum entropy ($\Delta S = 0$) and would have stayed there. Nothing would have happened and no life would be possible. Such a universe is unobservable by life forms of any kind. The second law of thermodynamics (Eq. 22.1) tells us that as long as life or any other irreversible dissipative process exists in the universe, the entropy of the universe S_{uni} will increase. Thus the entropy of the very early universe had to have some initially low value $S_{initial}$, where "low" means low enough compared to the maximum possible entropy S_{max} so that the entropy gap ΔS ($=S_{max} - S_{uni}(t)$) was large and could produce and support irreversible processes, such as stars and life forms [1] (Fig. 22.2).

Trying to understand the low initial entropy of the universe is an important unresolved issue of cosmology [13–16]. Figure 22.3 summarizes a few hypotheses. The "uniform" distribution in Fig. 22.3 is just a toy model without physical justification. However, physically plausible arguments can be made for both the "Penrose" and the "smooth energy dump" distributions. In standard

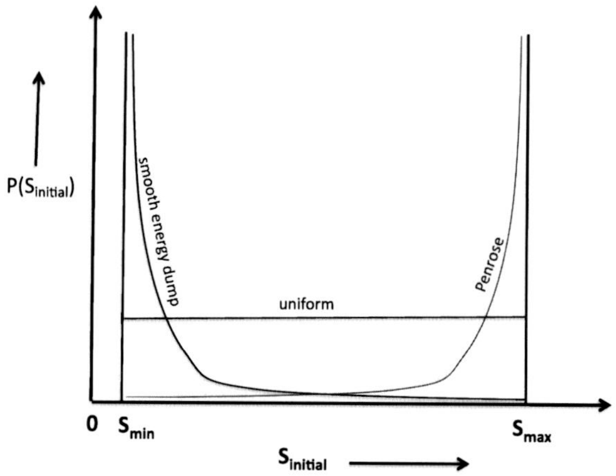

Fig. 22.3 Three conflicting expectations about the origin of the initial entropy of the universe. $P(S_{initial})$ is the probability distribution from which the initial entropy of our universe $S_{initial}$ (or of other universes) could have been drawn. One could imagine a uniform distribution in which all values between S_{min} and S_{max} are equally likely (*horizontal line*). Penrose's idea ([14], Chap. 27) is that there are many more ways to have high initial entropy than low initial entropy. In inflationary models, a "smooth energy dump" of the non-clumpable false vacuum energy constrains the resulting matter to a smooth homogeneous distribution with low gravitational entropy [15]

thermodynamics there are many more ways to be at high entropy than at low entropy. Motivated by this idea and applying it to the early universe, Penrose makes the assumption that there are many more ways for the universe to have had high initial entropy than low initial entropy. Thus he refers to "our extraordinarily special big bang" ([14], p 726, Chap 27 and Fig. 27.4) because contrary to his assumption and expectation, our universe started out at low entropy.

If there are many more ways to be at S_{max} (in the absence of other constraints) Penrose would be correct that it is much more likely that the universe should have been born at or near maximum entropy (and our expectations should be that $S_{initial} \sim S_{max}$). However, at the beginning, did the universe have access to all those ways? Or were there constraints associated with the origin of matter that restrict the universe to having a smooth matter distribution and therefore low gravitational entropy?

It is possible that there were physical constraints associated with the physics of inflation. Inflation starts from an initially smooth distribution of false vacuum energy (quantum fluctuations of false vacuum, this can also be understood as a higher zero-point energy than the current zero-point energy of the vacuum state of the universe). See [15]. Part of the definition of vacuum energy is that it does not, and cannot clump. This false vacuum energy is homogeneously distributed (subject to quantum fluctuations). When the false vacuum decays during reheating creating all the energy and matter in the universe, it may only be possible for this to happen as a smooth energy dump, resulting in a universe with a relatively smooth distribution of matter (and therefore low initial gravitational entropy). Thus inflation provides a natural initial condition that could explain why the initial entropy of our universe ($S_{initial}$ in Fig. 22.2) is so low. Homogeneously distributed matter (i.e. with low gravitational entropy) could well be an initial constraint (boundary condition) associated with the origin of matter from false vacuum energy.

The low gravitational entropy of the homogeneously distributed matter is what gives the universe its low initial entropy [1, 16]. Penrose ([14], p 706) explains:

> A uniformly spread system of gravitating bodies would represent relatively low entropy (unless the velocities of the bodies are enormously high and/or the bodies are very small and/or greatly spread out, so that the gravitational contributions become insignificant), whereas high entropy is achieved when the gravitating bodies clump together.

For an elaboration of this view see [17–19].

22.2.1 Anthropic Reasoning Cannot Rescue Penrose's Model

In Penrose's model, if the initial entropy is too close to S_{max}, the entropy gap ΔS will not be large enough to produce stars and life. Thus, in Penrose's model, an anthropic argument (in the context of a multiverse scenario in which the probability distribution of $S_{initial}$, P($S_{initial}$) is exhaustively sampled) has to be invoked to explain why $S_{initial} \ll S_{max}$ [20]. That is, although universes with $S_{initial} \sim S_{max}$ greatly outnumber universes with low initial entropy, life (and observers like us) are only possible in universes with low initial entropy.

Sagan [21] has poetically described the low entropy requirements for life: *"If you wish to make an apple pie from scratch, you must first invent the universe."* However, the entire universe did not have to be at low entropy in order for our part of the universe to have low entropy. Feynman [22] discussed the idea of whether our low entropy part of the universe could be a low entropy fluctuation, i.e. a low entropy sub-set of a larger universe that is much closer to maximum entropy:

> [F]rom the prediction that the world is a fluctuation, all of the predictions are that if we look at a part of the world we have never seen before, we will find it mixed up, and not like the piece we just looked at. If our order were due to a fluctuation, we would not expect order anywhere but where we have just noticed it...Every day [astronomers] turn their telescopes to other stars, and the new stars are doing the same thing as the other stars. We therefore conclude that the universe is not a fluctuation, and that the order is a memory of conditions when things started. This is not to say that we understand the logic of it. For some reason, the universe at one time had a very low entropy for its energy content, and since then the entropy has increased.

Feynman's argument, based on new stars coming into view, can be made more rigorous by basing it on the increasing particle horizon. If we are living in a rare low entropy fluctuation that has enabled us to be here, then when we view previously unobserved parts of the universe (more specifically when we observe parts of the universe that we had not been in causal contact with), we should find them to be close to maximum entropy. The entropy fluctuation that made us should be of minimal extent. As the size of the observable universe increases, new parts of the universe that were out of causal contact, come into causal contact—new regions of the universe appear over the horizon [23]. If our part of the universe were a low entropy fluctuation, then the new parts coming over the horizon would tend to be of higher entropy. This does not seem to be the case. The distant universe seems to be at low gravitational entropy. Our observations that the distant universe is in a state of low entropy is inconsistent with the expected rarity of such low entropy states. This rarity can be quantified by the ratio of the probability of the high entropy state (with W_{hi} microstates) to the probability of the low entropy state (with fewer W_{lo} microstates) [24]:

$$P(S_{hi})/P(S_{lo}) = W_{hi}/W_{lo} = \exp[(S_{hi} - S_{lo})/k] \tag{22.4}$$

Low entropy regions of the universe are not only rare, they are also much more likely to fluctuate to higher entropy than to fluctuate to lower entropy. How much more likely is given by the fluctuation theorem [25]:

$$P(dS_i/dt = \sigma)/P(dS_i/dt = -\sigma) = \exp(\sigma t/k) \tag{22.5}$$

which can be cosmologically interpreted as follows: If some part of the universe (indexed by the subscript i) is not at equilibrium ($S_i < S_{i,max}$), then during a subsequent time t, this part of the universe is much more likely to increase its entropy at a positive rate σ and fluctuate toward equilibrium ($S_{i,max}$) than it is to fluctuate further from equilibrium at a rate $-\sigma$. How much more likely is given by the expression $\exp(\sigma t/k)$.

The Feynman quote ends with an unresolved issue: "For some reason, the universe at one time had a very low entropy for its energy content..." To resolve the issue of the initial entropy of the universe, Carroll [16] has suggested that either we just accept the initial condition without asking why, or that the big bang is not the beginning. The first is the abandonment of scientific cosmology and the second is a very poorly supported speculation. Penrose and Tegmark [14, 20] use anthropic reasoning, but it seems like overkill since it should only apply to the minimal sized local patch needed to create us. However, as mentioned earlier, the inflationary origin of matter from unclumped false vacuum energy may produce a low gravitational entropy universe everywhere it has produced matter. This could be the reason for the initial low entropy of the universe.

22.3 Maximum Entropy Production Principle in Cosmology

22.3.1 Entropy Production Around Supermassive Blackholes

Mass spiralling around a black hole in an accretion disk, can only fall into the black hole if there are mechanisms to remove its angular momentum and load it onto other mass that is then ejected from the system. How efficient those mechanisms are is the main issue. Since the largest component of the current entropy of the universe is the entropy of supermassive black holes, their growth by accretion of mass is the largest source of entropy in the universe. Since the entropy of a black hole is proportional to the square of the mass, $S_{BH} \sim M_{BH}^2$ (Eq. 22.3), the entropy produced during the formation and growth of a black hole is $dS_{BH}/dt \sim M_{BH}\, dM_{BH}/dt$. Thus, dS_{BH}/dt is a maximum when $M_{BH}\, dM_{BH}/dt$ is a maximum. Therefore, to evaluate the Maximum Entropy Production Principle (MaxEP), we need to ask if the structure of accretion disks around black holes of a given mass, maximizes dM_{BH}/dt. Less ambitiously, we can try to use MaxEP predictions to identify new constraints that need to be included in accretion disk models.

How can we determine whether the structure of an accretion disk arranges itself such that $dM_{BH}/dt = (dM_{BH}/dt)_{max}$? We need to understand the details of the angular momentum transfer and to evaluate if, under the constraints given, the material around a black hole arranges itself optimally to transport angular momentum and concentrate it into a relatively small amount of mass that gets ejected from the system.

For mass to accrete onto a black hole, the angular momentum and energy of the mass has to be gotten rid of. Energy from accretion can easily be radiated away through the high luminosity of the inner edge of accretion disks. So the rate limiting step controlling mass infall is the transfer of angular momentum. The angular momentum L, of the mass that is going to fall in, has to be transferred to mass that will be ejected (Figs. 22.1, 22.4). Therefore, to evaluate MaxEP, we need to ask if black hole accretion disks are structured in such a way that they are maximally efficient at exporting angular momentum. The efficiency of an accretion disk can be quantified by how much L it can concentrate in the smallest amount of ejected mass.

Accretion discs are ubiquitous structures in the astrophysics of black holes (i.e. quasars, active galactic nuclei, binary X-ray sources), star formation and even massive planet formation. When an accretion disk around a star runs out of mass to accrete and is no longer able to transport angular momentum, the skeleton it leaves behind is a angular-momentum dominated disk of material, also known as a planetary system. Jupiter and Saturn have been stranded with ~85 % of the angular momentum of our solar system.

Accretion disks are differentially rotating Keplerian disks. That is, the velocity of material at a distance r from the central mass M is $v(r) \sim \sqrt{(GM/r)}$. Since

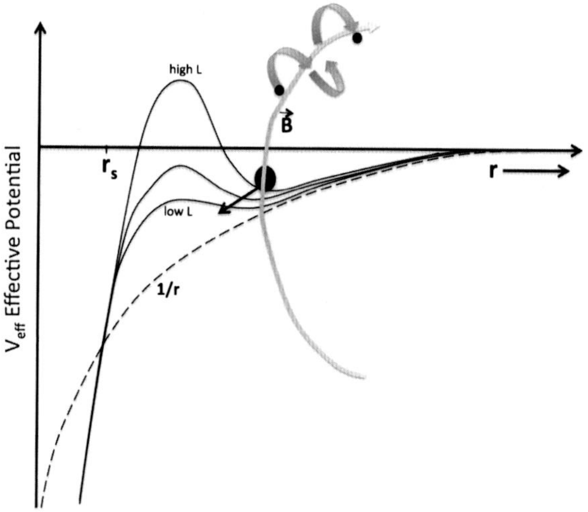

Fig. 22.4 Effective potential (Eq. 22.6) of material in an accretion disk for three values of angular momentum L. The Newtonian 1/r gravitational potential is shown for comparison (*dashed line*). The grey representative magnetic field line ("B") is anchored to the partially ionized material of the accretion disk (large *black circle*, also "m" in Eq. 22.6). Mass m is whipping around the black hole at Keplerian velocities v(r) $\sim$ $r^{-1/2}$ carrying the magnetic field line with it. Partially ionized particles above and below the accretion disk spiral around the magnetic field lines. Since the magnetic field line is rotating, centrifugal forces accelerate and eject these ionized particles like beads on a bullwhip. The acceleration of these particles comes at the expense of the deceleration of the particles anchoring the field lines in the disk. Thus, the transfer of angular momentum from material in the accretion disk to material ejected above and below the disk, occurs through rotating magnetic field lines

velocity is not a constant but depends on radius, we have the frictional sheer of molecular viscosity in the disk. This dissipation has been parametrized in the earliest accretion disk models as the dimensionless parameter alpha [26]. However, ordinary molecular viscosity is not sufficient to explain the amount of angular momentum transport needed to account for the observed accretion rate in accretion disks [27]. Blandford and Payne [28] showed that magnetic stresses are more efficient at transporting angular momentum as they convert centrifugal outflow into the oft-observed collimated jets (see Figs. 22.1, 22.4).

The role of angular momentum in preventing accretion can be seen in the effective potential (Fig. 22.4) of a mass m, with angular momentum L in the accretion disk at a distance r from a black hole of mass M_{BH}, located at r = 0, with an event horizon radius (=Schwarzschild radius) r_s [29]:

$$V_{\mathrm{eff}} = -GM_{\mathrm{BH}}m/r + L^2/\left(2mr^2\right)[1 - r_s/r] \tag{22.6}$$

For the mass m to sink into the potential well of the black hole, we need to reduce the angular momentum L that m has. This reduction lowers the hill of high angular momentum associated with the centrifugal force felt by the orbiting mass.

The "high L" curve drops down to become the "low L" curve. In Fig. 22.4, a representative magnetic field line, is threaded through the large black circle (mass "m"). As m circles around the black hole, it carries the B-field with it. Ionized particles (represented by the small black circles) above (and below) the plane of the accretion disk spiral around the B-field line and get accelerated out and up, like beads on a whip. This transfers some of the angular momentum of m to the bead, lowering the L of m. In this way, magnetic braking of m allows it to accrete onto the central black hole [28].

The efficiency with which partially ionized material can be magnetically whipped to high velocities and thus simultaneously loaded with angular momentum is difficult to quantify because it depends on the complex profiles of ionization, magnetic field strength, density, pressure and temperature above, below and in the disk. It depends on an impedance matching between the magnetic braking of material near the black hole and the magnetic acceleration of material further away. For example, if the ionization fraction is too low, there will not be much material to spiral around the field lines and get ejected. If the density of neutral particles is too high in the region of acceleration, collisions with neutral particles produces an 'atmospheric friction' that will slow down the acceleration (it is difficult to crack a bullwhip underwater in order to accelerate a bead on it). The high density impedes the transport of angular momentum. Magnetohydrodynamics (MHD) is needed to model the system and feedback is important since magnetic fields accelerate the spiraling particles, while at the same time, the spiraling particles maintain the magnetic fields.

Since the amount of matter that could fall into a black hole is limited to how much matter is nearby, the most "efficiently structured" accretion disks (the ones that MaxEP would predict) are the ones that can concentrate angular momentum into the smallest amount of mass and then eject only the smallest fraction of the mass available. This allows a larger fraction of the mass to lose enough angular momentum to fall into the hole and contribute to entropy production (Fig. 22.4). One way to quantify the efficiency of L-transport in an accretion disk is to estimate its ratio of mass accretion to mass ejection. In protostellar accretion disks (e.g. around T-Tauri stars) this ratio is $\sim$5–10 [30]. In the accretion disks of SMBHs, it may be comparable, but high angular resolution observations and modeling of these systems are not good enough to say more. The efficiency cannot be infinite. All the angular momentum cannot be concentrated in one ejected proton. The constraints of the MHD angular momentum transfer, combined with MaxEP would predict that there will be a maximum value to the mass accretion/mass ejection ratio (somewhat analogous to the Carnot efficiency of a reversible heat engine).

Angular momentum is also transported magnetically within the disk. Modeling by Balbus and Hawley [31, 32] showed that the magneto-rotational instability (MRI) produces turbulent viscosity and accounts for additional outward angular momentum transport [33]. The cause of MRI is the tendency of a weak magnetic field to try to enforce corotation on displaced fluid elements. This results in excess centrifugal force at large radii, and a deficiency of centrifugal force at smaller radii. This drives fluid elements away from their equilibrium positions and produces interpenetrating fingers of high and low angular momentum fluid—leading to angular momentum transport [31].

Can we arrange the magnetic field and all the other characteristics of an accretion disk (in the context of the given specific environments around supermassive black holes) to maximize dM_{BH}/dt? Or does Nature do that by herself as MaxEP would predict? As we obtain higher angular resolution images of a significant sample of nearby supermassive blackholes, and as we make more accurate and detailed computer MHD models of their mass accretion, we will get closer to answering this question.

References

1. Lineweaver, C.H., Egan, C.: Life, gravity and the second law of thermodynamics. Phys. Life Rev. **5**, 225–242 (2008)
2. Niven, R.K.: Minimization of a free-energy-like potential for non-equilibrium flow systems at steady state. Phil. Trans. R. Soc. B. **365**, 1323–1331 (2010). (Chap. 7, this volume)
3. Kleidon, A.: Life, hierarchy and the thermodynamics machinery of planet Earth. Phys. Life Rev. (2010). doi:10.1016/j.plrev.2010.10.002
4. Hogg, D.W., et al.: Cosmic homgeneity demonstrated with luminous red galaxies. ApJ **624**, 54–58 (2005)
5. Egan, C., Lineweaver, C.H.: A larger entropy of the universe. Astrophys. J. **710**, 1825–1834 (2010)
6. Kolb, E.W., Turner, M.S.: The early universe. Addison-Wesley, New York (1990)
7. Bekenstein, J.S.: Generalized second law of thermodynamics in black-hole physics. Phys. Re. D **9**, 3292 (1974)
8. Hawking, S.W.: Black holes and thermodynamics. Phys. Rev. D **13**, 191 (1976)
9. Strominger, A., Vafa, C.: Microscopic origin of the Bekenstein-Hawking entropy. Phys. Lett. B. **379**, 99 (1996)
10. Basu, B., Lynden-Bell, D.: A survey of entropy in the universe. QJRAS. **31**, 359 (1990)
11. Smoot, G.F., et al.: Structure in the COBE differential microwave radiometer first-year maps. Astrophys. J. **396**, L1–L5 (1992)
12. Jarosik, N., et al.: Seven-year Wilkinson microwave anisotropy probe (WMAP) observations: sky maps, systematic errors, and basic results. ApJS **192**, 14 (2011)
13. Lineweaver, C.H.: A simple treatment of complexity: cosmological entropic boundary conditions on increasing complexity. In: Edt Lineweaver, C.H., Davies, P.C.W., Ruse, M. (eds.) Complexity and the Arrow of Time, Cambridge University Press, pp. 42–67 (2013)
14. Penrose R.: The big bang and its thermodynamic legacy. In: Road to Reality: A Complete Guide to the Laws of the Universe, pp. 686–734 [Chapter 27]. Vintage Books, London (2004). Plot used in Fig. 1, panel c, from Thomas, A. (2009). http://www.ipod.org.uk/reality/reality_arrow_of_time.asp
15. Guth, A.H.: The Inflationary Universe. Jonathan Cape, London (1997)
16. Carroll, S.M.: From Eternity to Here: The Quest for the Ultimate Theory of Time. Dutton, Penguin, New York (2010)
17. Gron, O., Hervik, S.: Gravitational entropy and quantum cosmology. Class. Quantum Grav. **18**, 601–618 (2001)
18. Gron, O., Hervik, S. The Weyl Curvature Conjecture, arXiv:gr-qc/0205026v1. (2002)
19. Amarzguioui, M., Gron, O.: Entropy of gravitationally collapsing matter in FRW universe models. Phys. Rev. D **71**, 083011 (2005)

20. Tegmark, M.: The Second Law and Cosmology, arXiv 0904.3931v1, see video and slides at http://mitworld.mit.edu/watch/the-second-law-and-cosmology-9279/. (2009)
21. Sagan, C.: Cosmos (1980)
22. Feynman, R.: Feynman Lectures, vol. I (46-8, -9) (1969)
23. Davis, T.M., Lineweaver, C.H.: Expanding Confusion: Common Misconceptions of Cosmological Horizons and the Superluminal Expansion of the Universe. Pub. Astron. Soc. Aust. **21**, 97–109 (2004). See Fig. 1
24. Jaynes, E.T.: Macroscopic prediction in computer systems—operational approaches. In: Haken, H. (ed.) Neurobiology, Physics and Computers, pp. 254–269. Springer, Berlin (1985), Eq. 5
25. Evans, D.J., Searles, D.J.: Equilibrium microstates which generate second law violating steady states. Phys. Rev. E **50**(2), 1645–1648 (1994)
26. Shakura, N.I., Sunyaev, R.A.: Astron. Astrophys. **24**, 337 (1973)
27. Pringle, J.E.: Accretion discs in astrophysics. Ann. Rev. Astron. Astrophys. **19**, 137–162 (1981)
28. Blandford, R.D., Payne, D.G.: Hydrodynamic flows from accretion discs and the production of radio jets. MNRAS **199**, 883–903 (1982)
29. Taylor, E.R., Wheeler, J.A.: Exploring Black Holes: Introduction to General Relativity. Addision Wesley Longman, San Franciso (2000). (Chaps. 4 and 5)
30. Cabrit, S.: The accretion-ejection connexion in T Tauri stars: jets models vs. observations. In: Bouvier, J., Appenzeller, I. (eds.) Star-Disk Interaction in Young Stars, Proceedings of the IAU Symposium No. 243 (2007)
31. Balbus, S.A., Hawley, J.F.: A powerful local shear instability in weakly magnetized disks. I linear analysis. ApJ. **376**, 214–222 (1991)
32. Balbus, S.A., Hawley, J.F.: Instability, turbulence, and enhanced transport in accretion disks. Rev. Mod. Phys. **70**(1), 1–53 (1998)
33. http://en.wikipedia.org/wiki/Magnetorotational_instability

Index

R. C. Dewar et al. (eds.), *Beyond the Second Law*,
Understanding Complex Systems, DOI: 10.1007/978-3-642-40154-1,

N

O

P